ENVIRONMENTAL IMPACTS FROM THE DEVELOPMENT OF UNCONVENTIONAL OIL AND GAS RESERVES

The development of unconventional oil and gas shales using hydraulic fracturing and directional drilling is currently a focal point of energy and climate change discussions. While this technology has provided access to substantial reserves of oil and gas, the need for large quantities of water, emissions, and infrastructure raises concerns over the environmental impacts. Written by an international consortium of experts, this book provides a comprehensive overview of the extraction from unconventional reservoirs, providing clear explanations of the technology and processes involved. Each chapter is devoted to different aspects including global reserves, the status of their development and regulatory framework, water management and contamination, air quality, earthquakes, radioactivity, isotope geochemistry, microbiology, and climate change. Case studies present baseline studies, water monitoring efforts, and habitat destruction. This book is accessible to a wide audience, from academics to industry professionals and policymakers interested in environmental pollution and petroleum exploration.

JOHN F. STOLZ is a professor in the Department of Biological Sciences and Director of the Center for Environmental Research and Education at Duquesne University, Pittsburgh. He is a geomicrobiologist best known for his research on the biogeochemical cycles of arsenic and selenium and the biogenesis of microbiolites. He has published 96 journal articles, 40 book chapters, and edited 2 books. He received the Clean Water Action Dewey Award for environmental stewardship in 2020.

DANIEL J. BAIN is an associate professor of geology and environmental science at the University of Pittsburgh. His main research interests are urban systems, energy production landscapes, and trace metals in environmental systems. He has authored 51 peer-reviewed papers and 2 book chapters.

DR. W. MICHAEL GRIFFIN was Executive Director of the Green Design Institute at CMU. Previously, he was President of GMS Technologies, Vice President for Research and Development at Applied CarboChemicals Inc., Research Director at CHMR and NETAC, and a research scientist at BP America. He has published work on energy and environmental policy, and systems analysis of energy production.

ENVIRONMENTAL IMPACTS FROM THE DEVELOPMENT OF UNCONVENTIONAL OIL AND GAS RESERVES

Edited by

JOHN F. STOLZ
Duquesne University

DANIEL J. BAIN
University of Pittsburgh

W. MICHAEL GRIFFIN
Carnegie Mellon University

CAMBRIDGE
UNIVERSITY PRESS

University Printing House, Cambridge CB2 8BS, United Kingdom

One Liberty Plaza, 20th Floor, New York, NY 10006, USA

477 Williamstown Road, Port Melbourne, VIC 3207, Australia

314–321, 3rd Floor, Plot 3, Splendor Forum, Jasola District Centre, New Delhi – 110025, India

103 Penang Road, #05-06/07, Visioncrest Commercial, Singapore 238467

Cambridge University Press is part of the University of Cambridge.

It furthers the University's mission by disseminating knowledge in the pursuit of education, learning, and research at the highest international levels of excellence.

www.cambridge.org
Information on this title: www.cambridge.org/9781108489195
DOI: 10.1017/9781108774178

© Cambridge University Press 2022

This publication is in copyright. Subject to statutory exception and to the provisions of relevant collective licensing agreements, no reproduction of any part may take place without the written permission of Cambridge University Press.

First published 2022

Printed in the United Kingdom by TJ Books Limited, Padstow Cornwall

A catalogue record for this publication is available from the British Library.

Library of Congress Cataloging-in-Publication Data
Names: Stolz, John F., editor. | Bain, Daniel J., editor. | Griffin, Michael (Michael W.), editor.
Title: Environmental impacts from the development of unconventional oil and gas reserves / edited by John F. Stolz, Duquesne University, Pittsburgh, Daniel J. Bain, University of Pittsburgh, Michael W. Griffin, Carnegie Mellon University, Pennsylvania.
Description: First edition. | Cambridge ; New York : Cambridge University Press, 2022. | Includes bibliographical references and index.
Identifiers: LCCN 2021051814 (print) | LCCN 2021051815 (ebook) | ISBN 9781108489195 (hardback) | ISBN 9781108774178 (epub)
Subjects: LCSH: Oil fields–Production methods–Environmental aspects. | Gas fields–Production methods–Environmental aspects.
Classification: LCC TD195.P4 E5567 2022 (print) | LCC TD195.P4 (ebook) | DDC 363.6–dc23/eng/20211207
LC record available at https://lccn.loc.gov/2021051814
LC ebook record available at https://lccn.loc.gov/2021051815

ISBN 978-1-108-48919-5 Hardback

Cambridge University Press has no responsibility for the persistence or accuracy of URLs for external or third-party internet websites referred to in this publication and does not guarantee that any content on such websites is, or will remain, accurate or appropriate.

Contents

List of Figures	*page* ix
List of Tables	xxi
List of Contributors	xxiii
Preface	xxvii
JOHN F. STOLZ, W. MICHAEL GRIFFIN, DANIEL J. BAIN	

Part I Overview

1 Global Unconventional Oil and Gas Reserves and Their Development — 3
JOHN F. STOLZ, CASSANDRA ZIEGLER, AND W. MICHAEL GRIFFIN

2 Unconventional Shale Gas and Oil Extraction in the Appalachian Basin — 19
JOHN F. STOLZ AND W. MICHAEL GRIFFIN

3 An Overview of Unconventional Gas Extraction in Australia: The First Decade — 44
GERALYN MCCARRON AND SHAY DOUGALL

4 The Governance of Fracking: History, Differences, and Trends — 79
JOHN D. GRAHAM AND JOHN A. RUPP

Part II Environmental Analysis

5 Air Quality — 105
ALBERT A. PRESTO AND XIANG LI

6 Methane and Climate Change — 132
ROBERT W. HOWARTH

7	Water Usage and Management JESSICA M. WILSON AND JEANNE M. VANBRIESEN	150
8	Seismicity Induced by the Development of Unconventional Oil and Gas Resources DAVID W. EATON	173
9	Naturally Occurring Radioactive Material (NORM) NATHANIEL R. WARNER, MOSES A. AJEMIGBITSE, KATHARINA PANKRATZ, AND BONNIE MCDEVITT	214
10	Metal Isotope Signatures as Tracers for Unconventional Oil and Gas Fluids BRIAN W. STEWART AND ROSEMARY C. CAPO	246
11	Isotopes as Tracers of Atmospheric and Groundwater Methane Sources AMY TOWNSEND-SMALL	272
12	The Microbiology of Shale Gas Extraction SOPHIE L. NIXON	292

Part III Case Studies

13	Evaluation of Potential Water Quality Impacts in Unconventional Oil and Gas Extraction: The Application of Elemental Ratio Approaches to Pennsylvania Pre-Drill Data DANIEL J. BAIN, TETIANA CANTLAY, REBECCA TISHERMAN, AND JOHN F. STOLZ	313
14	A Baseline Ecological Study of Tributaries in the Tenmile Creek Watershed, Southwest Pennsylvania BRADY PORTER, ELIZABETH DAKIN, SARAH WOODLEY, AND JOHN F. STOLZ	340
15	The Effects of Shale Gas Development on Forest Landscapes and Ecosystems in the Appalachian Basin MARGARET C. BRITTINGHAM AND PATRICK J. DROHAN	363

| 16 | Managing TDS and Sulfate in the Monongahela River Three Rivers QUEST | 386 |

PAUL ZIEMKIEWICZ, MELISSA O'NEAL, TAMARA VANDIVORT, JOSEPH KINGSBURY, AND RACHEL PELL

Index 411

Colour Plates section to be found between pp. 290 and 291

Figures

1.1 Global distribution of tight oil and gas reserves (US EIA 2014) (A black and white version of this figure will appear in some formats. For the colour version, refer to the plate section.) *page* 11
2.1 Map of United States shale plays in the lower 48 states (US EIA 2016) (A black and white version of this figure will appear in some formats. For the colour version, refer to the plate section.) 22
2.2 Well location plat pages from a conditional use permit for a mega pad with 42 wells in Southwestern Pennsylvania. (A) map of the location of the proposal well pad in relation to other nearby surface structures including information of landowners, (B) map showing lateral for one of the wells (MON404H1) and the number of wells on the pad (upper righthand corner), (C) diagram of the well bore showing specifics such as top hole, kick off point, and bottom hole with depths and lengths 26
2.3 Trucks and equipment associated with unconventional drilling. (A) a caravan of thumper trucks for seismic testing, (B) water tanks at a watering station at the edge of the Monongahela River (note hose leading to the river – arrow), (C) "Sand king" proppant truck, (D) sand (proppant) truck, (E) pressure pump truck, (F) chemical mixing truck, (G) kettle truck for mixing cement, and (H) brine (residual waste) truck. A, C, D, G, and H courtesy of Robert Donnan; E and F courtesy of Bill Hughes; B the author (JFS) 27
2.4 Stages in the development of a well pad. (A) Impoundment for fresh water (note the white pipeline that goes to the adjacent well pad, middle center of the image), (B) Pad preparation with liners in place and cellars for eight wells (w). (C) Drilling has commenced with a "triple" rig. The pad is surrounded by a sound dampening barrier, the drilling rig is in the center, and an impoundment (arrow) for drilling fluids and waste is to the left of the rig. (D) The completed well site with remaining infrastructure; (w) well heads cordoned off by jersey barriers, (s) separators, and (c) condensate tanks. (E) Close up of a well head (note how the cellar can fill with water). B, C, and D, courtesy of Robert Donnan; A and E, the author (JFS) 28
2.5 The well pad during the hydraulic fracturing of the well. (A) well pad in Ohio with some permanent infrastructure already in place, (s) separators,

(c) condensate tanks, (cn) conditioners. (B) closer look at a well pad in Amwell
Pennsylvania (cc) control center, (b) blenders, (wt) water tanks, (p) pressure pumps,
(pp) proppant (sand) containers and trucks. Courtesy of Robert Donnan 31
2.6 Ancillary infrastructure associated with unconventional shale gas
and oil development. (A) impoundment for water, (B) transmission pipeline,
(C) compressor station in West Virginia with 12 compressors under construction,
(D) compressor station across the street from a farmhouse in Ohio,
with six compressors, (E) pigging station in Ohio. Courtesy of Robert Donnan 34
2.7 Infrastructure for processing oil and wet gas. (A) Blue Racer Midstream
in Proctor, West Virginia, (B) Midstream processing facility near Moundsville,
West Virginia, (C) Midstream processing facility Mobley, West Virginia,
(D) propane (left) and butane (right) storage tanks at a facility in Cadiz, Ohio.
Courtesy of Robert Donnan 35
2.8 Processing plants for oil and wet gas. (A) Seneca, Sommerville Ohio,
(B) Fort Beeler, West Virginia, (C) Mobley, West Virginia, and
(D) Sherwood, West Virginia. Courtesy of Robert Donnan 36
2.9 Shell ethane cracker facility under construction, Beaver, PA. Courtesy
of Robert Donnan 37
3.1 Estimates of conventional and unconventional gas reserves, and
production in 2014 (tcf) (Geoscience Australia, 2019) (A black and
white version of this figure will appear in some formats. For the colour version,
refer to the plate section.) 45
3.2 Prospective conventional and unconventional gas resources (tcf) (Geoscience
Australia, 2018) (A black and white version of this figure will appear in
some formats. For the colour version, refer to the plate section.) 46
3.3 Oil and gas tenements (areas with a permit or authority from
the government and under the relevant legislation to allow exploration
or production of oil and gas) (Energy Resource Insights, 2020) (A black and
white version of this figure will appear in some formats. For the colour version,
refer to the plate section.) 47
3.4 Workforce by subdivision in Queensland (Australian Bureau of Statistics, 2018) 54
3.5 Spillover job impacts per coal seam gas (CSG) job (data sourced from Ogge, 2015) 54
3.6 Revenues created across industries (data sourced from Chang, 2019) 55
3.7 QGC Kenya Water Treatment Facility Storage Ponds (Wieambilla, Queensland),
(Dean Draper, used with permission) 61
5.1 Map of unconventional oil and gas wells in Pennsylvania
(map generated May 14, 2020). As in other states, unconventional oil
and gas infrastructure is located away from major population centers
in Allegheny County (Pittsburgh) and Philadelphia County.
Source: PA DEP Oil and Gas Mapping tool (www.depgis.state.pa.us/
PaOilAndGasMapping/, 2020) 108
5.2 Daily maximum NO_2 concentrations measured in (A) Allegheny County (Pittsburgh,
Latitude: 40.617488, Longitude: −79.727664) and (B) Beaver County (Latitude:
40.747796, Longitude: −80.316442) from January 1, 2005 to
September 30, 2019. Annual averages are shown as black dots centered
on June 1 of each year. The black vertical line denotes the end date
of the data shown in Carlton et al. (2014) 109

List of Figures xi

5.3 The extent of the study area (within 2 km of a sampled point) is shown as shaded. In the left panel, areas within 2 km of a conventional or unconventional well are shown in blue or yellow, respectively. Areas within 2 km of both types of wells are shown in green. Areas within 2 km of any well are classified as "near-well." In the right panel, average traffic density increases as the color gets darker. In general, more heavily traveled roads occur near Pittsburgh (in the center of Allegheny County) and on interstate highways. AADT = Annual Average Daily Traffic reported by Pennsylvania Spatial Data Access (www.pasda.psu.edu/, 2015) (A black and white version of this figure will appear in some formats. For the colour version, refer to the plate section.). Map generated in ESRI ArcGIS Pro 111

5.4 Background and near-well measurements for low and high traffic volumes. The amount of traffic on the roads significantly affects the underlying sampled distribution for both black carbon (left) and PAHs (right) (A black and white version of this figure will appear in some formats. For the colour version, refer to the plate section.) 112

5.5 VOC (in ppb) and methane concentrations (in ppm) measured in (a) the Denver-Julesburg Basin, (b) the Uintah Basin, and (c) Northeastern PA Marcellus Shale (NEPA). Box-and-whisker plots represent distribution of measurements at gas production facilities. The red line in the middle of the box represents the median concentration; the top and the bottom of the box represent 75th and 25th percentile, respectively. The background measurements of each basin are shown as green crosses (A black and white version of this figure will appear in some formats. For the colour version, refer to the plate section.) 118

5.6 Comparison of the VOC concentration measured in this study with (a) VOC measured in 28 US cities (Baker et al. 2008), (b) VOC measured at Boulder Atmospheric Observatory (BAO) (Gilman et al. 2013), and (c) VOC measured in Horse Pool in winter 2012 and winter 2013 (Helmig et al. 2014). Data from this study are presented using box-whisker plots, and data from previous studies are presented as symbols with standard deviation indicated by the error bars (A black and white version of this figure will appear in some formats. For the colour version, refer to the plate section.) 120

5.7 Facility-level VOC emission rates measured at (a) gas wells in the Denver-Julesburg Basin, (b) gas wells in the Uintah Basin, (c) compressor stations in the Uintah Basin, and (d) gas wells in NEPA. The emission rates of each VOC species are presented with box-whisker plots. The red line in the middle of the box represents the median emission rate; the top and the bottom of the box represent 75th and 25th percentile, respectively 121

6.1 Greenhouse gas footprint of natural gas (including shale gas), diesel oil, and coal per unit of heat energy released as the fuels are burned. Direct emissions of carbon dioxide are shown in light gray. Methane emissions expressed as carbon dioxide equivalents are shown in dark gray. As discussed in the text, the methane emission rate used here for natural gas, 3.2% of production, is conservative. Emission estimates are from Howarth (2020) 141

6.2	Stylized comparison of the global temperature response over time from carbon dioxide and methane emitted for a one-year pulse of each gas at time zero. Top shows area under curve integrated for the time zero to year 20 time period, with methane shown solid gray and carbon dioxide as striped. Bottom panel is the same except through the 100-year time period. Note that the integrated area for carbon dioxide in both panels also underlies the area for methane, except for the extreme left-handed side of the curves. Adapted from IPCC (2013) and based on the absolute global temperature change potential. Reprinted from Howarth (2020)	143
7.1	Water use and wastewater generation in the development of oil and gas resources. Specific activities in the "Wastewater disposal and reuse" inset are: (a) disposal via injection well, (b) wastewater treatment with reuse or discharge, and (c) evaporation or percolation pit disposal (US EPA 2015) (A black and white version of this figure will appear in some formats. For the colour version, refer to the plate section.)	151
7.2	Increasing publications related to "oil and gas" and "water" over the past 50 years. Figure courtesy Xiaoju (Julie) Chen (A black and white version of this figure will appear in some formats. For the colour version, refer to the plate section.)	151
7.3	Hydraulic fracturing water use per well. Figure shows average of the data (ISH Energy, 2014) within 1st–99th percentile of water volumes used per well from Jan. 2011 to Aug. 2014 in watershed (8 HUCs) with at least two wells reporting water use. (Reproduced with permission from Gallegos et al. 2015)	153
7.4	Volume of produced water by state for 2007, 2012, and 2017. Boxed inset focuses on those states not in the top 10 produced water generators (note x-axis scale). Data from Clark and Veil (2009); Veil (2015); Veil (2020) (A black and white version of this figure will appear in some formats. For the colour version, refer to the plate section.)	155
7.5	Produced water per well in various US shale plays. (Reproduced with permission from Kondash et al. 2017)	156
7.6	Volume of produced water from conventional (light gray) and unconventional (dark gray) natural gas development in Pennsylvania	157
7.7	Salinity of produced waters in the United States (Allison and Mandler 2018) (A black and white version of this figure will appear in some formats. For the colour version, refer to the plate section.)	158
7.8	Produced water quality (TDS, chloride, and bromide) in Pennsylvania from conventional (blue) and unconventional (orange) wells. Data from Hayes (2009) (A black and white version of this figure will appear in some formats. For the colour version, refer to the plate section.)	159
7.9	Onsite and offsite management of produced water from shale gas development. Modified VanBriesen and Hammer (2012)	161
7.10	Produced water management practices for 2007, 2012, and 2017. (A black and white version of this figure will appear in some formats. For the colour version, refer to the plate section.) Data from Clark and Veil (2009); Veil (2015); Veil (2020)	162
7.11	Produced water management options used in Pennsylvania 2010–2012. Data from PA DEP (2018)	163

List of Figures xiii

7.12 Observed changes in chloride (a) and bromide (b) loads in the
Monongahela River during 2009–2012. Data from Wilson and VanBriesen (2013)   165

8.1 Locations of seismicity induced by saltwater disposal and hydraulic fracturing
(A black and white version of this figure will appear in some formats.
For the colour version, refer to the plate section.)   175

8.2 Schematic diagram illustrating basic earthquake terminology (A black and
white version of this figure will appear in some formats. For the colour version,
refer to the plate section.)   176

8.3 Magnitude-frequency diagrams for Oklahoma, Alberta and northeastern
British Columbia (NE BC)   179

8.4 World stress map (Heidbach 2016) showing direction of SHmax, stress
regime and method of stress determination, with overlay showing selected focal
mechanisms for injection-induced events (A black and white version of this figure
will appear in some formats. For the colour version, refer to the plate section.)   181

8.5 Graphs illustrating the calculation of a 1-D effective vertical stress (S'_V) profile
for a hypothetical sedimentary basin. Symbols are defined in the text. (A black
and white version of this figure will appear in some formats. For the colour
version, refer to the plate section.)   182

8.6 Mohr diagrams for depths of 1.0 km, 2.5 km, and 5.0 km   184

8.7 Schematic illustration (not to scale) of three proposed mechanisms of
fault activation by fluid injection   187

8.8 $M \geq 2.5$ earthquakes in Oklahoma (OK) and adjacent states including
Texas (TX) and Kansas (KS) (A black and white version of this figure will
appear in some formats. For the colour version, refer to the plate section.)   190

8.9 $M \geq 2.5$ earthquakes (red circles) in central Alberta, Canada (A black and white
version of this figure will appear in some formats. For the colour version,
refer to the plate section.)   193

8.10 $M \geq 2.5$ earthquakes (red circles) in northeastern British Columbia (BC), Canada
(A black and white version of this figure will appear in some formats.
For the colour version, refer to the plate section.)   195

8.11 Variability in input parameters (left) and tornado plot (right) showing sensitivity
of fault B (Figure 9.6) to input parameters for Monte Carlo fault slip
potential analysis   199

8.12 Probability of nominal pore-pressure increase required to exceed Mohr-Coulomb
criterion on faults A and B (Figure 9.6), at 5.0 km depth   200

8.13 Maximum seismic moment (M_0) versus injected volume (V) for some reported
cases of large-volume waste brine disposal and hydraulic-fracturing induced
seismicity   202

9.1 Three types of radioactivity that originate in the nucleus of an atom; alpha,
a particle with two protons and two neutrons, beta, a negatively charged particle
that changes a neutron to a proton, and gamma an electromagnetic energy that
originates from the nucleus (A black and white version of this figure will
appear in some formats. For the colour version, refer to the plate section.)   215

9.2 ^{226}Ra and ^{228}Ra decay chains showing the primordial parent nuclides,
^{238}U and ^{232}Th. Adapted from Nelson et al. (2015) (A black and white version
of this figure will appear in some formats. For the colour version, refer to the
plate section.)   217

9.3a Map of the continental United States with published NORM data for produced water. Note the wide range in values reported in (pCi/L) but also the lack of publicly available data for many oil and gas basins. Most data is sourced from the USGS PW database and Fisher 1998 (A black and white version of this figure will appear in some formats. For the colour version, refer to the plate section.) 223

9.3b There are over 100,000 oil and gas wells with published Chloride (Cl) data available in the USGS PW database. Using the relationship between Cl and radium-226 described by Fisher (1998) for PW from Texas, an estimate of the radium 226 in other basins can be calculated (A black and white version of this figure will appear in some formats. For the colour version, refer to the plate section.) 223

9.3c There are over 100,000 oil and gas wells with published TDS data available in the USGS PW database. Using the relationship between TDS and radium-226 described by Fisher (1998) for PW from Texas, an estimate of the radium 226 in other basins can be calculated (A black and white version of this figure will appear in some formats. For the colour version, refer to the plate section.) 224

9.4 Top 10 states of oil and gas produced water volume generation (in barrels/year for 2012) (A black and white version of this figure will appear in some formats. For the colour version, refer to the plate section.) 225

9.5 Produced water management for 2007 (based on Clark and Veil 2015) (A black and white version of this figure will appear in some formats. For the colour version, refer to the plate section.) 226

9.6 Depiction of NORM management and disposal. Reprinted with permission from Center for Coal Field Justice (A black and white version of this figure will appear in some formats. For the colour version, refer to the plate section.) 227

10.1 Simple major element mixing curves for the cases of mixing produced water from (a) hydraulically fractured Marcellus Shale gas wells and (b) Upper Devonian conventional wells with adjacent shallow ground water. The enrichment of each element relative to the uncontaminated ground water is plotted as a function of the mixing ratio. All data are from a site in Greene County, Pennsylvania, reported by Kolesar Kohl et al. (2014) 250

10.2 Ca/Sr vs. Ca/Mg of groundwater and conventional (Upper Devonian) and unconventional (Marcellus) produced water from a single site in Greene County, Pennsylvania. (a) The three data clusters define a linear trend. (b) Mixing curves between the groundwater and each of the produced water types. Labeled percentages on the curve are the % produced water needed to move the mixture to the indicated point on the curve. The limit of resolution is the approximate zone above which the two produced waters cannot be differentiated from each other in the mixture. Data are from Kolesar Kohl et al. (2014) 252

10.3 Strontium isotope mixing diagram for groundwater contamination with either Marcellus or Upper Devonian produced water for the Greene County, Pennsylvania site. The labeled percentages on the mixing lines indicate the % of each produced water needed to shift the mixture to the indicated point. Because the isotopic compositions of the two produced water sources fall on opposite sides of that of the groundwater, the direction of mixing (black arrows) allows this method to effectively discriminate between sources. Data are from Kolesar Kohl et al. (2014)   253

10.4 Comparison of the sensitivity of individual element concentrations and Sr isotope ratios to mixing of groundwater with Upper Devonian and Marcellus produced water at the Greene County, Pennsylvania site. Major element concentrations in the groundwater increase significantly with contamination from either produced water source, and could be detectable with, for example, a TDS meter. The Sr isotope data provide information about which produced water source is the contaminant, even at very low levels of contamination (relative to measurement uncertainty). Data are from Kolesar Kohl et al. (2014)   254

10.5 Mixing diagram for hypothetical Appalachian Basin stream water contaminated by either Marcellus Shale produced water (values shown here for four different counties in Pennsylvania), Upper Devonian conventional well produced water (endmember values far off scale), or acid mine drainage from SW Pennsylvania. The numbers indicated on the mixing curves are the percent of contaminant added to the stream water. Note that even in this worst-case scenario where the stream isotope composition falls wholly within the Marcellus field, it is still possible to differentiate among contamination sources. Data are from Chapman et al. (2012) (A black and white version of this figure will appear in some formats. For the colour version, refer to the plate section.)   256

10.6 Time series data for Sr concentrations and isotope compositions from hydraulically fractured Marcellus Shale gas wells from three locations in Greene County, southwestern Pennsylvania. Data are from Chapman et al. (2012) and Capo et al. (2014)   257

10.7 Range of δ^7Li values measured in oil and gas produced waters from throughout the world (modified from Pfister et al. 2017). Data are from Chan et al. (2002), Millot et al. (2011), Macpherson et al. (2014), Warner et al. (2014), Phan et al. (2016), and Pfister et al. (2017)   259

10.8 Li-Sr isotope mixing diagram for a freshwater source mixing with unconventional Marcellus Shale and conventional Upper Devonian produced water in the Appalachian Basin. The percentage values labeled on the curves indicate the amount of produced water contaminant added to the fresh water. The freshwater Sr endmember is the spring sample from Kolesar Kohl et al. (2014) and the Li endmember is the global average value for rivers. Produced water Sr data are from Chapman et al. (2012), Kolesar Kohl et al. (2014), and Capo et al. (2014), with corresponding Li data from Phan et al. (2016)   260

10.9 Barium isotope mixing diagram for river water (Ohio River, SW Pennsylvania) contaminated with produced water, using average values for Marcellus Shale and Upper Devonian produced waters. Individual data points represent samples collected throughout the Appalachian Basin. The values on the curves indicate the percentage of produced water mixed in to achieve the observed offset. Data are from Tieman et al. (2020) (A black and white version of this figure will appear in some formats. For the colour version, refer to the plate section.)   263

List of Figures

11.1 Monthly average methane concentrations as measured at the Mauna Loa Observatory in Hawaii since 1983.
Source: www.esrl.noaa.gov/gmd/ccgg/trends_ch4/   274

11.2 Carbon and hydrogen stable isotopic composition of methane collected from various sources in southern California. Solid symbols represent fossil-derived CH_4, and open symbols represent biological sources. Isotope ratios represent source signatures corrected for the presence of background air. (Townsend-Small et al. 2012)   275

11.3 Methane concentrations in air observed at Mount Wilson Observatory (a) as measured continuously with TLS (black dotted line) and in discrete samples with GC-FID (red points). Also shown is the approximate concentration of CH_4 in unpolluted air (1.82 ppm) as measured at Trinidad Head, California, by the NOAA flask sampling network. Shown are (b) $\delta^{13}C$, (c) δD, and (d) $\Delta^{14}C$ in discrete samples, shown with CH_4 concentrations measured by TLS. (Townsend-Small et al. 2012)   276

11.4 Keeling plots of $1/CH_4$ concentration versus (a) $\delta^{13}C$-CH_4 and (b) δD-CH_4. (Townsend-Small et al. 2012)   277

11.5 Keeling plots of $\delta^{13}C$ and δ^2H composition of CH_4 versus $1/CH_4$ (ppm^{-1}) collected downwind of (a and b) oil and gas, (c and d) landfill, and (e and f) cattle sources in the Denver-Julesburg basin. Each point represents an individual sample. (Townsend-Small et al. 2016)   279

11.6 Keeling plots of δ^2H composition of CH_4 at three ground sites (a-c) in the Front Range. (d) δ^2H-CH_4 for samples taken during aircraft flights within the Front Range. All flights shown are significant at $p < 0.05$. Flight tracks can be viewed using the following link: www-air.larc.nasa.gov/missions/discover-aq/kmz/FRAPPE_C130_2014_ALL_July26-August18.kmz. (Townsend-Small et al. 2016)   280

11.7 Carbon and hydrogen stable isotopic composition of methane from potential groundwater contamination sources, including agricultural and landfill sources as well as oil and gas sources. The isotopic composition of atmospheric methane is also shown. Theoretical range of values of methane sources shown in black and dotted lines from Whiticar (1999). Data are from Townsend-Small et al. (2016); Sherwood et al. (2017); Botner et al. (2018) (A black and white version of this figure will appear in some formats. For the colour version, refer to the plate section.)   282

11.8 Time-series map of Ohio counties where sampling occurred. Red circles are active natural gas wells. Blue diamonds are sites where time series groundwater samples were taken, and light blue circles represent sites where single sample groundwater measurements were made. Groundwater sample locations are noted when samples were taken between the years noted in each map. There was a large increase in active natural gas wells from 2013 to 2014. (Botner et al. 2018) (A black and white version of this figure will appear in some formats. For the colour version, refer to the plate section.)   283

11.9 Carbon stable isotopic composition of CH_4 versus CH_4 concentration in samples collected along the time series. Each symbol represents a different groundwater well. Also included are measurements of the $\delta^{13}C$ of CH_4 from

	a shale gas well ($-47.3‰$) and a conventional gas well ($-41.3‰$) in the study area for comparison. The radiocarbon content of CH_4 ($\Delta^{14}C\text{-}CH_4$) and radiocarbon age (in years bp) of four water samples is also shown. (Botner et al. 2018)	284
11.10	(a) Relationship of dissolved CH_4 concentration with distance from the nearest active shale gas well in groundwater wells sampled only once. (b) Carbon stable isotopic composition of CH_4 in the same samples. (Botner et al. 2018)	285
11.11	Carbon and hydrogen isotopic compositions of groundwater samples taken during this study, along with isotopic composition of several sources (value for background air is from Townsend-Small et al. (2016)). Also shown are approximate literature values for different endmembers of CH_4 sources from Whiticar (1999). Groundwater samples are shown in blue circles, for time series samples, and purple diamonds, for single samples, with shading corresponding to sample CH_4 concentration, with increasing shading corresponding to increasing concentration. (Botner et al. 2018) (A black and white version of this figure will appear in some formats. For the colour version, refer to the plate section.)	286
12.1	(A) Global location of shale formations with assessed resource estimations (dark red) and those with resources yet to be estimated (light brown). Shale formations where microbiology investigations have been assessed are marked with yellow circles (adapted from Mouser et al. 2016), and the numbers refer to the following references (full details in reference list at the end of the chapter): [1] Zhong et al. 2019; [2] Strong et al. 2013; [3] Lipus et al. 2018; [4] Wang et al. 2017 [5] Struchtemeyer et al. 2011; [6] Davis et al. 2012; [7] Struchtemeyer and Elshahed 2012; [8] Santillan et al. 2015; [9] Fichter et al. 2012; [10] Schlegel et al. 2011; [11] Daly et al. 2016; [12] Booker et al. 2017; [13] Murali Mohan et al. 2013a; [14] Murali Mohan et al. 2013b [15] Cluff et al. 2014; [16] Akob et al. 2015; [17] Tucker et al. 2015; [18] Vikram et al. 2016; [19] Lipus et al. 2017b; [20] Nixon et al. 2019; [21] Wuchter et al. 2013; [22] Kirk et al. 2012; [23] Zhang et al. 2017. Base map modified from the US Energy Information Services Assessed Resources Basin Map, 2013 (public domain). (B) Number of microbiology studies conducted for each formation, where a survey of the population size, composition, or both, has been carried out (A black and white version of this figure will appear in some formats. For the colour version, refer to the plate section.)	295
13.1	Chemistries of "Pre-drill" data reported to the Pennsylvania Department of Environmental Protection by oil and gas companies prior to hydraulic fracturing (likely all wells within 762 meters (2,500 feet) and others at the operator's discretion (58 Pa. C. S. §§ 3218(c)) plotted in Br/SO_4 vs Ba/Cl ratio space (Brantley et al. 2014). Pre-drill groundwater water samples are shown as grey symbols and surface water samples as black symbols. Lines show areas where various end members are expected to plot (see Brantley et al. (2014)). Ellipsis surrounding the clusters (areas) of different sources of contamination and ground and surface water samples were done using the 2D Confidence Ellipse application in OriginPro software (OriginLab, Northampton, Massachusetts)	320

List of Figures

13.2 Chemistries of "Pre-drill" data reported to the Pennsylvania Department of Environmental Protection by oil and gas companies prior to hydraulic fracturing (likely all wells within 762 meters (2,500 feet) and others at the operator's discretion (58 Pa. C. S. §§ 3218(c)) plotted in Ca/Mg vs Ca/Sr ratio space (Tisherman and Bain 2019). Pre-drill groundwater water samples are shown as grey stars and surface water samples as black stars. Ellipsis surrounding the clusters (areas) of different sources of contamination and ground and surface water samples were done using the 2D Confidence Ellipse application in OriginPro software (OriginLab, Northampton, Massachusetts)   321

13.3 Chemistries of "Pre-drill" data reported to the Pennsylvania Department of Environmental Protection by oil and gas companies prior to hydraulic fracturing (likely all wells within 762 meters (2,500 feet) and others at the operator's discretion (58 Pa. C. S. §§ 3218(c)) plotted in List of Br/SO_4 vs Mg/Li ratio space (Cantlay et al. 2020b). Pre-drill groundwater water samples are shown as grey stars and surface water samples as black stars. Ellipses show areas where various end members plot as summarized and reported in Cantlay et al. (2020b). Only Bradford County data could be plotted here owing to limited Br and Li data   322

13.4 Chemistries of "Pre-drill" data reported to the Pennsylvania Department of Environmental Protection by oil and gas companies prior to hydraulic fracturing (likely all wells within 762 meters (2,500 feet) and others at the operator's discretion (58 Pa. C. S. §§ 3218(c)) plotted in SO_4/Cl vs Mg/Li ratio space (Cantlay et al. 2020b). Pre-drill groundwater water samples are shown as grey stars and surface water samples as black stars. Ellipses show areas where various end members plot as summarized and reported in Cantlay et al. (2020a, 2020b, 2020c). Only Bradford County data could be plotted here owing to limited data for Li   323

13.5 Chemistries of "Pre-drill" data reported to the Pennsylvania Department of Environmental Protection by oil and gas companies prior to hydraulic fracturing (likely all wells within 762 meters (2,500 feet) and others at the operator's discretion (58 Pa. C. S. §§ 3218(c)) plotted in Mg/Na vs SO_4/Cl ratio space (Cantlay et al. 2020b). Pre-drill groundwater water samples are shown as grey stars and surface water samples as black stars. Ellipses show areas where various end members plot as summarized and reported in Cantlay et al. (2020a, 2020b, 2020c)   324

13.6 PADEP pre-drill data plotted in the Ca/Mg vs Ca/Sr ratio space (Tisherman and Bain 2019) for the three counties (Bradford, Lycoming, and Mercer) with available data. Important differences to note: Bradford County groundwaters plot densely even in ratio space areas characteristic of unconventional brines. Some Lycoming County groundwaters plot in those mixing regions, but the majority of the data are >100 Ca:Sr, or typical of waters not impacted by O&G activities. In Mercer County, almost all these data plot in the "unimpacted portion of the ratio space"   325

13.7 PADEP pre-drill data plotted in the Mg/Na vs SO_4/Cl ratio space (Cantlay et al. 2020b) for the two counties (Bradford and Lycoming) with available data. Note the strong excursion toward lower Mg/Na ratios in the Bradford County data relative to Lycoming County   326

13.8	Map of Pennsylvania Counties discussed in this chapter	328
13.9	Groundwater chemistry data from the Bradford (Williams et al. 1998), Lycoming (Lloyd, Jr. and Carswell 1981), and Greene (Stoner 1987) county groundwater reports plotted in the a) Ca/Mg vs Ca/Sr and b) Mg/Na vs SO_4/Cl ratio spaces. All these chemistries were measured long before unconventional extraction activities began. There were insufficient Br and Li data in these reports to evaluate the other ratio spaces	329
13.10	Data collected from Tenmile Creek (Greene County, PA) between 2010 and 2012. Individual samples collected through time are shown with start and end dates and arrows indicating direction for the (a) Ca/Mg vs Ca/Sr, (b) Mg/Na vs SO_4/Cl, and (c) SO_4/Cl vs Mg/Li ratio spaces. Note, these plots show a small portion of the spaces shown in the rest of the chapter, blown up to allow visualization of the water chemistry evolution	331
13.11	Surface water chemistry evolution in New York/Pennsylvania in streams influenced by road salting and salt springs. Stream water chemistry as reported in Johnson et al. (2015) for Silver Creek (reference conditions), Apalachin-D (road salt), Fall Brook (salt spring), and Wilkes (salt spring). Begin and end dates are shown, and arrows indicate direction of movement	334
14.1	Tenmile Creek basin with candidate small watershed-pairs indicated with hatching. Shading indicates the density of Marcellus wells (SPUDs/km^2) in the basin and circles indicate existing permits granted as of 2009. Triangles and squares indicate sites previously sampled by Kimmel and Argent (2010, 2012) for fish and macroinvertebrates. The selected Bates Fork and Fonner Run stream-pair is delineated inside the box. Figure by Dan Bain with permission (A black and white version of this figure will appear in some formats. For the colour version, refer to the plate section.)	343
14.2	Genetic diversity (average observed and expected heterozygosity, effective number of alleles, and allelic richness) of fantail darter (*Etheostoma flabellare*) and Johnny darter (*E. nigrum*) populations in Fonner Run and Bates Fork	353
15.1	Forest cover (green) overlaying Marcellus (blue) and Utica (red) shale plays with wells drilled in PA as of 2019 (A black and white version of this figure will appear in some formats. For the colour version, refer to the plate section.)	364
15.2	Combined abundance of 13 forest interior specialists based on abundance data from the second breeding bird atlas (Shen 2015, top figure). Abundance is greatest in dark red blocks. Bottom shows the blocks which support the greatest abundance of forest interior specialists above the Marcellus shale layer (A black and white version of this figure will appear in some formats. For the colour version, refer to the plate section.)	372
16.1	Timeline of significant events relating to total dissolved solids in the Upper Ohio River Basin from 2008 to present	388
16.2	TDS concentrations for all Monongahela River mainstem sampling locations (river miles 11, 23, 61, 82, 89, and 102) and discharge (cfs) at USGS gage 03072655 at river mile 82 near Masontown, PA, from 2009 through 2021.	

	The vertical dotted lines represent: long dash = initiation of the discharge management plan in January 2010; medium dash = Pennsylvania restricted flowback water at Public Owned Treatment Works (POTWs) in May 2011; short dash = CONSOL Energy's reverse osmosis plant went online in May 2013	390
16.3	Sulfate concentrations for all Monongahela River mainstem sampling locations (river miles 11, 23, 61, 82, 89, and 102) and discharge (cfs) at USGS gage 03072655 at river mile 82 near Masontown, PA, from 2009 through 2021. The vertical dotted lines represent: long dash = initiation of the discharge management plan in January 2010; medium dash = Pennsylvania restricted flowback water at Public Owned Treatment Works (POTWs) in May 2011; short dash = CONSOL Energy's reverse osmosis plant went online in May 2013	391
16.4	The Three Rivers QUEST coverage of the Upper Ohio River Basin covering parts of Maryland, Ohio, New York, Pennsylvania, and West Virginia	392
16.5	A complete diagram of all sites and the relative inputs related to each site. The Allegheny and the Monongahela converge within the city of Pittsburgh, PA, to form the Ohio River	393
16.6	3RQ sampling sites within the Monongahela River Basin	396
16.7	3RQ sampling sites within the Allegheny River Basin	397
16.8	3RQ sampling sites within the Ohio River Basin	398
16.9	The average chloride (Cl) to sulfate (SO_4) ratios in mmol/L between July 2009 and March 2021 in the Monongahela River Basin	399
16.10	Total dissolved solids and flow for the mainstem sampling stations; dots represent TDS (mg/L), solid line represents flow (cfs). (A) Monongahela River from 2009 to 2021 (M11, M23, M61, M82, M89, M102), (B) Allegheny River (A6, A30, A45, A83) from 2013 to 2021, (C) Ohio River from 2013 to 2020 (O12 and O85)	400
16.11	Compilation of TDSsdc (mg/L) and flow (cfs) correlation for (A) Monongahela River mainstem sampling locations (M11, M23, M61, M82, M89, M102) from 2009 to 2021, (B) Allegheny River mainstem sampling locations (A6, A30, A45, A83) from 2013 to 2021, (C) Ohio River mainstem sampling locations (O12 and O85) from 2013 to 2020	401
16.12	Compilation of sulfate (mg/L) and flow (cfs) correlation for (A) Monongahela River mainstem sampling locations (M11, M23, M61, M82, M89, M102) from 2009 to 2021, (B) Allegheny River from mainstem sampling locations (A6, A30, A45, A83) from 2013 to 2021, (C) Ohio River mainstem sampling locations (O12 and O85) from 2013 to 2020	403

Tables

1.1	Global distribution of tight oil and gas reserves and the current status of development (US EIA 2015a)	page 4
2.1	Shale plays in the United States	21
2.2	Upstream, midstream, and downstream processes and products	24
3.1	Percentage splits of gas consumption by sector (not including UAFG) in 2018 (AEMO 2019)	46
3.2	Inquiries and status of moratorium on UG industry in Australia	56
3.3	Data gaps in PM2.5 monitoring program	66
5.1	Hazard Ratio (HR) of VOC measured downwind of two high emitters in Denver-Julesburg basin. Bold values indicate HR $>$ 1	122
6.1	Top-down estimates for upstream emissions of methane from natural gas systems, including studies based on aircraft flyovers and satellite remote-sensing data. Estimates are the percentage of the methane in natural gas that is produced	135
6.2	Shale gas production and upstream methane emissions from various major shale gas producing fields in 2015	136
11.1	Properties of C and H isotopes and IAEA standards	273
12.1	A summary of key microbiology studies carried out on shale gas wells in formations in the USA, Canada, and China, including information on formation age, well depths, sample types and age since hydraulic fracturing, maximum salinity, and dominant microorganisms. Locations of formations are given in Figure 12.1	296
13.1	Pennsylvania Department of Environmental Protection Pre-Drill prefixes inferred from Shale Network HIS source description and used in this analysis	318
13.2	Counts of all data points (i.e., individual measurement of an individual constituent) per county or region (SWPA, NE-non-brad, NWPA are regions designated in the Shale Network HIS)	324
13.3	Bromide detection limits reported in shale network PADEP pre-drill data	334
14.1	Summary of fish survey data from Bates Fork and Fonner Run across five sample dates with IBI scores and designations for both the Ohio Headwater IBI (Ohio EPA 1987) and Monongahela IBI (Kimmel and Argent 2007) and Tolerance Indicator Values for chloride (Meador and Carlisle 2007)	348

14.2 Microsatellite markers tested in fantail and Johnny darters for inclusion in the study. Markers used in this study for a given species are marked "Y," while those that failed to amplify in more than 33% of individuals were not included in the analysis 351

14.3 Locus-specific and overall genetic characteristics of fantail darter populations in Fonner Run and Bates Fork. Observed (H_o) and expected (H_e) heterozygosity, alleles per locus (A), allelic richness (A_r), effective number of alleles per locus (A_e), and genetic differentiation between populations (F_{st}) 352

14.4 Locus-specific and overall genetic characteristics of Johnny darter populations in Fonner Run and Bates Fork. Observed (H_o) and expected (H_e) heterozygosity, alleles per locus (A), allelic richness (A_r), effective number of alleles per locus (A_e), and genetic differentiation between populations (F_{st}) 352

14.5 Number of Two-lined Salamanders (*Eurycea bislineata*) captured at Bates Fork and Fonner Run on seven surveys of three quadrats at each stream. In each cell, the first number indicates Bates Fork and the second number indicates values at Fonner Run. See Pascuzzi 2012 for quadrat GPS coordinates 354

16.1 Example of the Monongahela River Basin TDS model output for dry (left) and wet (right) periods 387

16.2 Site code and site descriptions by river basin 394

Contributors

Moses A. Ajemigbitse, AquaBlok Ltd, Swanton, Ohio

Daniel J. Bain, Geology and Environmental Science, University of Pittsburgh, Pennsylvania

Margaret C. Brittingham, Department of Ecosystem Science and Management, Penn State University

Tetiana Cantlay, Department of Biological Sciences, and the Center for Environmental Research and Education, Duquesne University, Pittsburgh, Pennsylvania

Rosemary C. Capo, Geology and Environmental Science, University of Pittsburgh, Pennsylvania

Elizabeth Dakin, Department of Biological Sciences, and the Center for Environmental Research and Education, Duquesne University, Pittsburgh, Pennsylvania

Shay Dougall, Molliwell, Brisbane, Queensland

Patrick J. Drohan, Department of Ecosystem Science and Management, Penn State University, Pennsylvania

David W. Eaton, Department of Geoscience, University of Calgary, Alberta

John D. Graham, Paul H. O'Neill School of Public and Environmental Affairs, Indiana University

W. Michael Griffin, Engineering and Public Policy, Carnegie Mellon University, Pennsylvania

Robert W. Howarth, Ecology and Environmental Biology, Corson Hall, Cornell University, Ithaca, NY

Joseph Kingsbury, Water Resources Institute, West Virginia University, Morgantown, WV

Xiang Li, South Coast Air Quality Management District, Diamond Bar, CA

Geralyn McCarron, Fellow of the Australian College of General Practitioners, Castle Hill Medical Centre, Murrumba Downs, Queensland

Bonnie McDevitt, U.S. Geological Survey, Reston, VA

Sophie L. Nixon, Department of Earth and Environmental Sciences, Manchester Institute of Biotechnology, The University of Manchester, Manchester, UK

Melissa O'Neal, Water Resources Institute, West Virginia University, Morgantown, WV

Katharina Pankratz, Civil and Environmental Engineering, Penn State University, University Park, PA

Rachel Pell, Water Resources Institute, West Virginia University, Morgantown, WV

Brady Porter, Department of Biological Sciences, and the Center for Environmental Research and Education, Duquesne University, Pittsburgh, PA

Albert A. Presto, Department of Mechanical Engineering, Carnegie Mellon University, Pittsburgh, PA

John A. Rupp, Indiana University, Paul H. O'Neill School of Public and Environmental Affairs, Bloomington, IN

Brian W. Stewart, Geology and Environmental Science, University of Pittsburgh, Pittsburgh, PA

John F. Stolz, Department of Biological Sciences, and the Center for Environmental Research and Education, Duquesne University, Pittsburgh, PA

Rebecca Tisherman, Geology and Environmental Science, University of Pittsburgh, Pittsburgh, PA

Amy Townsend-Small, Department of Geology and Geography, University of Cincinnati, Cincinnati, OH

Jeanne M. VanBriesen, Department of Civil & Environmental Engineering and Department of Engineering and Public Policy, Carnegie Mellon University, Pittsburgh, PA

Tamara Vandivort, Water Resources Institute, West Virginia University, Morgantown, WV

Nathaniel R. Warner, Civil and Environmental Engineering, Penn State University, University Park, PA

Jessica M. Wilson, Civil and Environmental Engineering, Manhattan College, Riverdale, NY

Sarah Woodley, Department of Biological Sciences, Duquesne University, Pittsburgh, PA

Cassandra Ziegler, Center for Environmental Research and Education, Duquesne University, Pittsburgh, PA

Paul Ziemkiewicz, Water Resources Institute, West Virginia University, Morgantown, WV

Preface

The combination of directional (e.g., horizontal) drilling and hydraulic fracturing ("fracking") has revolutionized oil and gas exploration, especially in the last two decades. They have been applied to conventional reserves, allowing for greater recovery, and for the development of "tight" deposits, primarily shales, releasing previously untapped reserves. The Energy Information Administration (EIA) has estimated that globally, shale gas reserves may contain 7,577 Tcf (trillion cubic feet) and shale oil reserves may contain 419 billion bbl (barrels). Unconventional shale extraction has reached commercial-level production in the United States and Canada, while increasing development is happening in China and Argentina. Australia has been developing its coal bed methane deposits, especially in Queensland, and there has been exploratory drilling in England (Bowland Basin), Germany (Niedersachsen), and Poland. At the same time Scotland, France, and parts of Australia have a moratorium or outright ban on the process. While many celebrate the potential economic benefits, concerns about environmental impacts that include water contamination, air quality degradation, habitat fragmentation, and the continued contribution to climate change have been raised. The "slick water" stimulation and "fracking" rely on a complicated mix of chemicals and "proppant" (fine grained silica sand), while the shales themselves contain salt brines (e.g., sodium, chloride, bromide), distinct trace element content (e.g., barium, strontium), and heavy metals including naturally occurring radioactive materials (NORMs). The extraction and distribution operations depend on complex infrastructure such as compressor stations, cryogenic processing plants, and an extensive network of pipelines for water and gas (e.g., gathering lines, transmission lines). Solid and liquid waste disposal has also presented challenges, with some solutions resulting in unexpected adverse consequences such as the generation of trihalomethanes in municipal water and radium (^{226}Ra) contamination as a result of road brining. Climate change has also put the focus on the global impacts of continued extraction and use of fossil fuels, with many nations signing the Paris Climate Agreement of 2016, promising significant carbon dioxide emissions reductions.

Despite the global expansion of unconventional oil and gas exploration, the literature has been scrambling to keep pace. There are numerous popular books providing some of the history behind it, personal stories, and even fiction. There are also a number of industry-published books that delve into the technical aspects such as proppant and fluid characterization. We realized there was a need and interest in a volume that addressed the environmental impacts. The impetus for this book initially came out of a two-day conference held at Duquesne University in November 2013 ("Facing the Challenges: research on shale gas extraction symposium"). There were twenty-two scholarly presentations covering a broad range of topics, the majority of which dealt with the environmental impacts. While a few of these presentations were published as a special issue of the journal *Environmental Science and Health, Part A* (2015, volume 50, issue 5), the symposium provided a framework for a comprehensive compendium and a potential list of contributors. We also reached out to other colleagues working in the field, especially several outside the United States to include contributions from Europe, Canada, and Australia. The Marcellus and Utica Shales of the Appalachian Basin have been, in many respects, the testing grounds for unconventional shale development in other parts of the country and the world. The authors we have solicited chapters from are known for their pioneering work. In the end we compiled 16 contributions. The book is divided into three sections: Overview, Environmental Analysis, and Case Studies. The Overview comprises four chapters. Chapter 1 provides an overview of the global unconventional oil and gas reserves and the status of their development at the time of this publication. Chapter 2 is an introduction to the development of unconventional oil and gas reserves based primarily on experience with the Appalachian Basin. Chapter 3 covers developments in Australia, where both gas shales and coal bed methane deposits are being tapped. Chapter 4 looks at the trends and challenges in governance, addressing issues related to mineral rights ownership, royalties, and regulations. The Environmental Analysis section, the bulk of the book, comprises nine chapters. Chapter 5 covers air quality issues. Chapter 6 tackles fugitive methane and its impact on climate change and lifecycle assessment. Chapter 7 provides a comprehensive look at water usage and management. Induced seismicity, as a result of hydraulic stimulation and waste injection facilities, is addressed in Chapter 8. Both drill cuttings and produced water from shales are known to contain naturally occurring radioactive materials (NORMs), a subject covered in Chapter 9. The next two chapters focus on the use of isotopes as tracers to identify sources, namely metal isotopes (Chapter 10) and methane isotopes (Chapter 11). The last chapter in this section discusses the microbiology (Chapter 12). The last section, Case Studies, provides assessments from a more holistic approach. The first chapter in this section evaluates water chemistry using mass ratio analyses to identify potential sources of contamination (Chapter 13). The second is a baseline study of a paired stream system in southwestern Pennsylvania, which was completed early on in the development of the Marcellus shale (Chapter 14). The next chapter (Chapter 15) addresses the effects of shale gas development on forest landscapes and ecosystems. The final chapter reports on the activities of the Three Rivers Quest Project, a consortium of several regional universities that have been monitoring the water quality of the three rivers of the Ohio River Basin (Allegheny, Monongahela, Ohio) for the last decade (Chapter 16).

In putting together this volume our goal was to present a broad picture of the development of unconventional oil and gas shales. Without this background it will be difficult to address the challenges, especially considering the legacy of environmental impacts from conventional oil and gas extraction. It is our hope that this book will be accessible to a wide audience of readers, from industry to academics, as well as laypersons interested in the subject matter. The authors would like to thank all the contributors, as well as the Colcom Foundation and Heinz Endowments for support over the years. A special thanks to Robert Donnan for his amazing photographs, and to Dr. David Kahler whose assistance with the formatting of the equation-heavy chapters was greatly appreciated.

Part I
Overview

1

Global Unconventional Oil and Gas Reserves and Their Development

JOHN F. STOLZ, CASSANDRA ZIEGLER, AND W. MICHAEL GRIFFIN

1.1 Introduction

The United States Energy Information Administration (EIA) comprehensively assessed global shale reserves in 2011 and followed up in 2013 (U.S. Energy Information Administration 2015). The 2011 assessment focused on 14 regions outside the United States, and the 2013 assessment expanded to 137 formations in 95 basins from 41 countries (Table 1.1). The EIA estimates the global shale gas reserves at 7,577 Tcf and the global shale oil reserves at 419 billion bbl. The 10 largest reserves for shale gas are China (1,115 Tcf), Argentina (802 Tcf), Algeria (707 Tcf), United States (623 Tcf), Canada (573 Tcf), Mexico (545 Tcf), Australia (437 Tcf), South Africa (370 Tcf), Russia (285 Tcf), and Brazil (245 Tcf). The European Union combined has significant reserves (435 Tcf), with Poland (146 Tcf) and France (137 Tcf) having the largest reserves. The 10 largest reserves for shale oil are United States (78 Bbbl), Russia (75 Bbbl), China (32 Bbbl), Argentina (27 Bbbl), Libya (26 Bbbl), United Arab Emirates (23 Bbbl), Chad (16 Bbbl), Australia (17.5 Bbbl), Venezuela (13 Bbbl), and Mexico (13 Bbbl). To date, only the United States and Canada have reached commercial level production for shale gas and oil, followed to a lesser extent by China and Argentina. While the estimated ultimate recovery (EUR) of these resources may be significant, the current status of their development relies on several factors. These factors include whether the reserves are technically and economically recoverable (i.e., readily accessible geographically and geologically), the status of existing infrastructure, such as refining capacity and pipelines, and a favorability of regulatory frameworks and political climate. The Society of Petroleum Engineers (SPE) defines proved reserves as those that are commercially recoverable "under current economic conditions, operating methods, and government regulations" (SPE 1997). The EIA did not consider other low permeability tight formations, such as sandstones and carbonates, and only assessed those with sufficiently studied geology (EIA 2013). It also considered the current technology available for the development of these formations, including the advances in drilling and hydraulic fracturing technology (Stolz and Griffin, see Chapter 2). The EIA calculated three values: OIP/GIP concentration, Risked OIP/GIP, and Risked Recoverable (EIA 2013). Oil-in-place (OIP) and gas-in-place (GIP) estimates were based primarily on the thickness of the organic-rich shale, porosity, pressure, and temperature. Risked OIP/GIP estimates were based on two additional factors: the play

Table 1.1. *Global distribution of tight oil and gas reserves and the current status of development (US EIA 2015a)*

Area	Country	Basin	Formation	Status of Development
North America	Canada	Horn River	Muskwa/Otter Park Evie/Klua	Provence dependent mix of active development or ban. 1,2,3
		Cordova	Muskwa/Otter Park	
		Liard	Lower Besa River	
		Deep Basin	Doig Phosphate	
		Alberta Basin	Banff/Exshaw	
		East and West Shale Basin	Duvernay	
		Deep Basin	North Nordegg	
		NW Alberta Area	Muskwa	
		Southern Alberta Basin	Colorado Group	
		Williston Basin	Bakken	
		Appalachian Fold Belt	Utica	
		Windsor Basin	Horton Bluff	
	Mexico	Burgos	Eagle Ford Shale Tithonian Shales	Active development. 1,4,5
		Sabinas	Eagle Ford Shale Tithonian La Casita	
		Tampico	Pimienta	
		Tuxpan	Tamaulipas Pimienta	
		Veracruz	Maltrata	
South America	Colombia	Middle Magdalena Valley	La Luna/Tablazo	Moratorium. 1
		Llanos	Gacheta	
	Colombia/ Venezuela	Maracaibo Basin	La Luna/Capacho	
	Argentina	Neuquen	Los Molles Vaca Muerta	Active development.1,4
		San Jorge Basin	Aguada Bandera Pozo D-129	
		Austral-Magallanes Basin	L. Inoceramus-Magnas Verdes	
		Parana Basin	Ponta Grossa	
	Brazil	Parana Basin	Ponta Grossa	State dependent with active development or ban. 1,6
		Solimoes Basin	Jandiatuba	
		Amazonas Basin	Barreirinha	
	Paraguay Uruguay	Parana Basin	Ponta Grossa Cordobes	Banned. 1,2,3
	Paraguay/ Bolivia	Chaco Basin	Los Monos	
	Chile	Austral-Magallanes Basin	Estratos con Favrella	

Table 1.1. (cont.)

Area	Country	Basin	Formation	Status of Development
Eastern Europe	Poland	Baltic Basin/ Warsaw Trough	Llandovery	Allowed but currently inactive. 2,3,4,7
		Lublin	Llandovery	
		Podlasie	Llandovery	
		Fore Sudetic	Carboniferous	
	Lithuania/ Kaliningrad	Baltic Basin	Llandovery	Allowed but currently inactive. 7,8
	Russia	West Siberian Central	Bazhenov Central	Allowed but currently inactive. 9,10,11
		West Siberian North	Bazhenov North	
	Ukraine/ Romania	Carpathian Foreland Basin	L. Silurian	Allowed but currently inactive. 2,9
	Ukraine	Dniepr-Donets	L. Carboniferous	Allowed but currently inactive. 11
	Ukraine/ Romania	Moesian Platform	L. Silurian	Allowed but currently inactive. 2,3,9
	Romania/ Bulgaria		Etropole	Banned in Bulgaria. 1,2,3,12
Western Europe	UK	N. UK Carboniferous Shale Region	Carboniferous Shale	Moratorium. 1,13,14
		S. UK Jurassic Shale Region	Lias Shale	
	Spain	Cantabrian	Jurassic	Allowed but currently inactive. 1,2,3
	France	Paris Basin	Lias Shale	Banned. 1,2,3
			Permian-Carboniferous	
		Southeast Basin	Lias Shale	
	Germany	Lower Saxony	Posidonia	Banned. 2,3,15
			Wealden	
	Netherlands	West Netherlands Basin	Epen	Moratorium 1,2,3,4
			Geverik Member	
			Posidonia	
	Sweden	Scandinavia Region	Alum Shale - Sweden	Allowed but currently inactive. 2,3,4
	Denmark		Alum Shale - Denmark	Moratorium. 1,2,3,4
Africa	Morocco	Tindouf	L. Silurian	
		Tadla	L. Silurian	
	Algeria	Ghadames/ Berkine	Frasnian	Allowed but currently inactive. 16,17
			Tannezuft	
		Illizi	Tannezuft	
		Mouydir	Tannezuft	
		Ahnet	Frasnian	
			Tannezuft	

Table 1.1. (*cont.*)

Area	Country	Basin	Formation	Status of Development
		Timimoun	Frasnian	
			Tannezuft	
		Reggane	Frasnian	
			Tannezuft	
		Tindouf	Tannezuft	
	Tunisia	Ghadames	Tannezuft	
			Frasnian	
	Libya	Ghadames	Tannezuft	
			Frasnian	
		Sirte	Sirte/Rachmat Fms	
			Etel Fm	
		Murzuq	Tannezuft	
	Egypt	Shoushan/Matruh	Khatatba	Allowed but currently
		Abu Gharadig	Khatatba	inactive. 18,19
		Alamein	Khatatba	
		Natrun	Khatatba	
	Chad	Termit	L. Cretaceous	
			U. Cretaceous	
		Bongor	L. Cretaceous	
		Doba	L. Cretaceous	
		Doseo	L. Cretaceous	
	South Africa	Karoo Basin	Prince Albert	Allowed but currently
			Whitehill	inactive.1,4
			Collingham	
Asia	China	Sichuan Basin	Qiongzhusi	Active
			Longmaxi	development. 4,20
			Permian	
		Yangtze Platform	L. Cambrian	
			L. Silurian	
		Jianghan Basin	Niutitang/Shuijintuo	
			Longmaxi	
			Qixia/Maokou	
		Greater Subei	Mufushan	
			Wufeng/Gaobiajian	
			U. Permian	
		Tarim Basin	L. Cambrian	
			L. Ordovician	
			M.-U. Ordovician	
			Ketuer	
		Junggar Basin	Pingdiquan/Lucaogou	
			Triassic	
		Songliao Basin	Qingshankou	
	Mongolia	East Gobi	Tsagaantsav	
		Tamtsag	Tsagaantsav	

Table 1.1. (cont.)

Area	Country	Basin	Formation	Status of Development
	Thailand	Khorat Basin	Nam Duk Fm	
	Indonesia	C. Sumatra	Brown Shale	Active development. 21
		S. Sumatra	Talang Akar	
		Tarakan	Naintupo	
			Meliat	
			Tabul	
		Kutei	Balikpapan	
		Bintuni	Aifam Group	
	India	Cambay Basin	Cambay Shale	Active development. 4,22
		Krishna-Godavari	Permian-Triassic	
		Cauvery Basin	Sattapadi-Andimadam	
		Damodar Valley	Barren Measure	
	Pakistan	Lower Indus	Sembar	
			Ranikot	
	Turkey	SE Anatolian	Dadas	
		Thrace	Hamitabat	
	Kazakhstan	North Caspian (North Margin)	Tournaisian, Radaevskiy-Kosvinskiy L. Serpukhovian, Vereiskiy, Gzelian-Kasimovian	
		North Caspian (SE Margin)	Visean	
		Mangyshlak	Karadzhatyk	
		South Turgay	Karagansay	
			Abaleen	
Middle East	Jordan	Hamad	Batra	
		Wadi Sirhan	Batra	
	Oman	S. Oman Salt	Thuleilat Shale	
			Athel	
			U Shale	
		N. Oman Foreland	Natih	
		Rub' Al-Khali	Sahmah Shale	
	United Arab Emirates	Rub' Al-Khali	Qusaiba	
			Diyab	
			Shilaif	
Australia	Australia	Cooper	Roseneath-Epsilon-Murteree (Nappamerri)	Provence dependent mix of active development, partial ban, or moratorium. 1,2,3
			Roseneath-Epsilon-Murteree (Patchawarra)	
			Roseneath-Epsilon-Murteree (Tenappera)	
		Maryborough	Goodwood/Cherwell Mudstone	

Table 1.1. (*cont.*)

Area	Country	Basin	Formation	Status of Development
		Perth	Carynginia	
			Kockatea	
		Canning	Goldwyer	
		Georgina	L. Arthur Shale (Dulcie Trough)	
			L. Arthur Shale (Toko Trough)	
		Beetaloo	M. Velkerri Shale	
			L. Kyalla Shale	

1) Herrera (2020), 2) Mead and Maloney (2018a), 3) Mead and Maloney (2018b), 4)Vinson and Elkins (2020), 5) Bertram (2019), 6) Dias (2019), 7) Kuznestsov (2013), 8) Reed (2015), 9) Stefan (2015), 10) Rapoza (2019), 11) Thomas (2014), 12) BBC News (2012), 13) Ambrose (2019), 14) Cairney et al. (2018), 15) Gesley (2017), 16) Aczel (2020), 17) Chikhi et al. (2019), 18) Reuters (2014), 19) Gasser (2017), 20) Myers (2019), 21) Campbell (2013), 22) Yadav (2020).

success probability (how likely is the play to produce) and the prospective area success factor (which includes additional risk factors that may affect production). Risk Recoverable, or the estimated technically recoverable oil and gas, was calculated by multiplying the OIP or GIP value by a recovery efficiency factor (EIA 2013). The latter factor was based on the mineralogy of the shale and how efficiently the formation could be hydraulically fractured. Only the data for Risked Recoverable estimates are included here, as they provide a baseline for assessing how successful a play might be (EIA 2013). This chapter provides a brief review of the major shale gas and oil plays and the status of their development as of 2020, focusing on China, Argentina, Algeria, the United States, Canada, Mexico, Australia, South Africa, Russia, and Brazil. Additional discussions can be found in the chapters covering the United States (Stolz and Griffin, see Chapter 2; Graham and Rupp, see Chapter 4), Australia (McCarron and Doughal, see Chapter 3), and France and the United Kingdom (Graham and Rupp, see Chapter 4).

1.2 China

China has seven areas that have been assessed for technically recoverable oil and gas, namely the Sichuan, Tarim, Junggar, Songliao, Jianghan, and Subei basins, and the Yangtze Platform (US EIA 2015e). The Sichuan (626 Tcf), Tarim (216 Tcf), Junggar (36 Tcf), and Songliao (16 Tcf) basins have the greatest potential for shale gas. The major shale oil reserves are found in the Junggar, Tarim, and Songliao basins with an estimated 8 Bbbl, 12 Bbbl, and 11.5 Bbbl, respectively. South China has marine black shales, with major development in the Sichuan Basin and Yangtze Platform. Commercial exploration and production has been underway, reportedly reaching nearly 600 wells and 9 bcm of production in 2017 (Vinson and Elkins 2020). The expansion has been driven by the 13th Five-Year-Plan for Energy Development announced by the National Development and Reform Commission and National Energy Administration in early 2017 with the goal of increasing proven reserves to 1 tcm by 2020 (Vinson and Elkins 2020). The geological complexity of the basins, with their faults and seismic activity, has proven to be a challenge. Development in Sichuan has been hampered by earthquakes (Myers 2019). Other challenges include the extreme depth of the deposits (3,200 m on average), limited accessibility, the lack of water resources, limited geologic data, clay-rich deposits that are more difficult to hydraulically fracture, lack of pipelines in some areas, and the high cost of development (US EIA 2015e). China has made investments in other global shale plays such as the United States in an effort to gain greater expertise. Sinopec is the major oil and gas company involved in the development, as many of the foreign companies such as Shell, Chevron, and BP have dropped out. China has also created its own oilfield services industry, manufacturing the equipment required for shale gas and oil extraction (Vinson and Elkins 2020).

1.3 Argentina

Argentina, traditionally known for its oil production, has four basins, the Neuquén, Golfo San Jorge, Austral, and Paraná (US EIA 2015c). Together, they are estimated to contain 802 Tcf and 27 Bbbl of risked recoverable gas and oil. The Neuquén is a marine shale, Jurassic to Cretaceous in age. The Golfo San Jorge is a lacustrine shale, Jurassic to Cretaceous in age. The Austral basin is a marine black shale, Cretaceous in age, while the Paraná is a black shale of Devonian age (US EIA 2015c). The Neuquén is the most promising for development with commercial production by the Argentine national company YPF SA, as well as foreign companies Apache, EOG Resources, ExxonMobil, and TOTAL. Development for shale oil began in the Vaca Muerta formation in 2010 (US EIA 2015c). The industry, as a whole, has expanded since then with the passage of key legislation. The Hydrocarbon Sovereignty Law, passed in 2012 and enacted by an Executive Branch decree, reclaimed the country's hydrocarbon deposits by public domain (Beller and Schiariti 2012). The law also declared a 51% share of YPF SA and Respol YPF Gas SA, essentially repatriating the companies after several years of private ownership. Further stimulation was provided by the recently decreed "Argentine Natural Gas Production and Demand Scheme Promotion Plan," also known as the 2020–2024 Gas Plan. The plan set production goals for the Austral (20 MMm^3/d) and Neuquén (47.2 MMm^3/d) basins. It has reinvigorated the operations of the regional private companies Tecpetrol SA and Pluspetrol SA. The main challenge limiting development has been financing, as the country continues to suffer from high inflation and interest rates (Newberry 2019). The high cost of drilling and completion also contributes as it continues to be greater than the price break point (Newbery 2019). An emerging issue has been social unrest in response to the rapid inflation and the country's response to COVID-19, with oilfield workers striking and healthcare providers (i.e., nurses, doctors, orderlies) blockading roads in protesting for better pay and health benefits (Otaola 2021).

1.4 Algeria

The geologic history of Algeria has created a patchwork of seven different basins scattered across the country. From west to east, they are the Tindouf, Reggane, Timimoun, Ahnet, Mouydir, Illizi, and Ghadames/Berkine basins. Together, they are estimated to contain 707 Tcf and 5.7 Bbbl of risked recoverable gas and oil, respectively. There are two major shale gas and oil formations in Algeria, the Tannezuft shale (Silurian) and the Frasnian shale (Upper Devonian) within these basins (US EIA 2015b). Exploratory drilling occurred in the Ahnet basin near In Salah in late 2014 (Belakhdar 2020). Despite Algeria's reliance on petroleum exports for their economy, the development of Algerian shales has been met with public protests (Chikhi et al. 2019; Belakhdar 2020) and political unrest (Aczel et al. 2018; Aczel 2020). The National Assembly passed legislation in 2013 to promote shale gas and later actively sought international partners for their development. This resulted, however, in large protests held in the cities of Adrar and Ouargla in 2014 and 2015. There was again political unrest after the 2019 Hydrocarbon Law was passed to encourage development and international support. Further development is on hold (Aczel 2020).

1.5 United States

The United States has several large shale deposits, with more than 24 basins and 38 formations (US EAI 2020; Stolz and Griffin 2021, see Table 2.1). Alabama, Alaska, Arkansas, California, Colorado, Indiana, Kansas, Louisiana, Michigan, Mississippi, Montana, Nebraska, Nevada, New Mexico, North Dakota, Ohio, Oklahoma, Pennsylvania, Tennessee, Texas, Utah, Virginia, West Virginia, and Wyoming all have deposits that are being or could be developed (see map in Chapter 2, Figure 2.1). Most of the shale oil is being produced from seven shale basins, namely, the Permian Basin (Texas), Eagle Ford (Upper Cretaceous) in Texas, Bakken (Upper Devonian) in North Dakota and Montana, Niobrara (Upper Cretaceous) in Colorado and Wyoming, Haynesville (Jurassic) in Louisiana and Texas, and the Utica (Ordovician) and Marcellus (Devonian) both in Pennsylvania, Ohio, and West Virginia (US EIA 2020). While fracking is permitted and there are sizable shale deposits in Alaska (Gryc 1985), the focus has been exclusively on conventional reserves. Greater discussion of unconventional oil and gas extraction in the United States can be found in Chapter 2.

Three states currently ban hydraulic fracturing, Maryland, New York, and Vermont. Other bans or moratoria exist at the city or county level. Attempts to ban fracking in Denton and Dallas, Texas, were overturned by the state legislature (Chapter 4). In Pennsylvania, Pittsburgh City Council passed an ordinance prohibiting the commercial extraction of natural gas in 2010 (Baca 2010). Subsequently, Bucks and Monroe County councils passed similar ordinances. None have been challenged to date. Most recently, the Delaware River

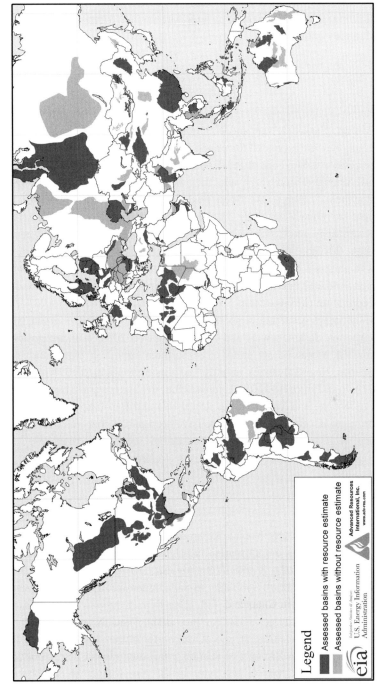

Figure 1.1 Global distribution of tight oil and gas reserves (US EIA 2014) (A black and white version of this figure will appear in some formats. For the colour version, refer to the plate section.)

Basin Commission made their moratorium permanent. In additional to Pennsylvania, the Delaware River Basin covers parts of New York, New Jersey, and Delaware (E360 Digest 2021). In California, fracking is banned in Los Angeles and Santa Cruz, as well as San Benito and Monterey counties.

1.6 Canada

Canada has numerous basins and formations across British Columbia and the Northwest Territories, Alberta, Saskatchewan and Manitoba, Quebec, and Nova Scotia (US EIA 2015g). Together, they are estimated to contain 573 Tcf and 8.8 Bbbl of risked recoverable gas and oil, respectively. The bulk of the deposits are in Western Canada, primarily British Columbia and Alberta (Figure 1.1), and include the Liard Basin, Horn River Basin, Cordova Embayment, Alberta, and Deep Basin (Table 1.1). Major formations are the Muskwa (Upper Devonian) and Otter Park (Upper Devonian), both of which are found in the Horn River Basin and Cordova, and the Lower Besa River (Devonian) in the Liard Basin (US EIA 2015g). The Williston Basin, with the Bakken formation (Upper Devonian), is in southern Saskatchewan and Manitoba. Moving east, Quebec is home to the Appalachian Fold Belt and the Utica shale (Ordovician). The Windsor Basin, with the Horton Bluff formation (Mississippian), is in Nova Scotia (Table 1.1).

Shale development is occurring in six provinces, Alberta, British Columbia, Manitoba, Northwest Territories, Saskatchewan, and Yukon, with over 200,000 unconventional wells having been drilled. Moratoria are under effect in three provinces, Newfoundland, Nova Scotia, and Quebec. New Brunswick initially had a moratorium in 2015, which converted to a ban in 2018. Both Prince Edward Island and Ontario have not considered a moratorium as neither have reserves that could be developed (Natural Resources Canada 2020). The two major hurdles to development in Canada have been public concerns over the environmental impacts (e.g., drinking water contamination, induced earthquakes, landscape impacts) and lack of economically viable deposits (Minkow 2017).

1.7 Mexico

The major basins and deposits of Mexico lie on its eastern side along the Gulf of Mexico. The five major basins, which are all marine in origin, are the Burgos, Sabinas, Tampico, Tuxpan, and Veracruz (US EIA 2015h). They are estimated to contain 545 Tcf of risked recoverable gas and 13.1 Bbbl of combined oil and condensate. The major formations are the Eagle Ford (Upper Cretaceous) and Tithonian (Upper Jurassic) of the Burgos and Sabinas basins, Pimienta (Jurassic) in the Tampico basin, Tamaulipas (Middle Cretaceous) in the Tuxpan basin, and Maltrata (Upper Cretaceous) in the Veracruz basin (US EIA 2015h). Petroleos Mexicano SA de CV (Pemex), the state-owned petroleum company, began exploration in 2011, with a test well in the Eagle Ford Shale. The geologic complexity (i.e., deformities) and extreme depths (>5 km) of some of the shale deposits as well as lack of geophysical data have hampered further development. Economic pressures have also contributed, with the high cost of oil field services in Mexico and the lack of foreign investment (Vinson and Elkins 2020). Concerns over infrastructure security (i.e., safety of the pipelines) and water availability are also in play. While initially promoting the development of their oil and gas reserves with a goal of energy independence, President Andrés Manuel López Obrador is considering a ban on hydraulic fracturing, a move supported by Environment Minister Victor Toledo (NGI 2019).

1.8 Australia

There have been six assessed shale gas basins in Australia, the Cooper, Canning, Georgina, Beetaloo, Perth, and Maryborough (US EIA 2015d). Together, they are estimated to contain 437 Tcf and 17.5 Bbbl of risked recoverable gas and oil, respectively. The Cooper Basin contains lacustrine deposits including the Roseneath, Epsilon, and Murteree shales, which are all Permian in age (Guo and Mccabe 2017). The hydrocarbons have been concentrated in three troughs, the Nappamerri, Patchawarra, and Tenappera (US EIA 2015d). The Canning basin in northwestern Australia, is the largest in area (469,000 km^2) and contains the Goldwyn shale (Middle Ordovician). The Georgina basin, with the Dulcie and Toko Troughs of the L. Arthur shale (Middle Cambrian), is the next in areal extent (324,000 km^2). The Perth, Beetaloo, and Maryborough basins are significantly smaller (52,000, 36,000, and 11,000 km^2, respectively). The M. Velkerri shale and L. Kyalia shale of the Beetaloo basin are the oldest, and are both Precambrian in age. The Perth basin has the Kockatea shale (Late Triassic), while the small Maryborough has the Goodwood/Cherwell mudstone (Cretaceous) (Table 1.1).

Development in Australia varies by province, with active development, moratoria, or bans (Chapter 3). Queensland has been the most active, with major infrastructure already existing. New South Wales allows it, and the Northern Territory is mixed, with 51% open to development. Western Australia does not permit unconventional development, with the exception of existing leases, and the landowners maintain the right of refusal. Victoria has completely banned all onshore unconventional gas development, South Australia has a 10-year ban, and the Tasmanian Government has a moratorium until 2025. Further discussion about Australia can be found in Chapter 3.

1.9 South Africa

South Africa has one major basin, the Karoo, but it covers two-thirds of the country (612,000 km^2). The Karoo basin has three marine shale deposits, the Prince Albert, Whitehill, and Collingham, all Late Permian in age (US EIA 2015j). Together, they are estimated to contain 370 Tcf of risked recoverable gas. The southern portion of the basin has promise for dry gas production, especially the Whitehill formation with its high organic content (6%). The formations have been impacted by past geologic activity in the form of igneous intrusions, which may have affected the quantity and quality, as well as the ability to extract the gas (US EIA 2015j). Economic and energy needs have influenced the country's views on fracking. Most of South Africa's energy is generated from coal. Initially, there was a moratorium, enacted in 2011, on oil and gas exploration in the Karoo Basin, primarily as a result of environmental concerns and water scarcity (Agbroko 2011). The moratorium was lifted in 2012 after the release of the findings of an inter-agency task force, and in 2017 the Mineral Resources Minister issued guidelines regulating permits (Vinson and Elkins 2020). The Supreme Court of Appeal of South Africa subsequently decided in 2019 that the permits issued by Ministry of Mines were not legal. More significant barriers to further development include the lack of technical expertise, limited infrastructure, continued environmental and water scarcity concerns, as well as growing public opposition (Vinson and Elkins 2020).

1.10 Russia

Although Russia has several basins with potential, namely the Timan Pechora, Volga-Urals in the west, and East Siberia in the east, only the centrally located West Siberia Basin has been assessed (US EIA 2015i). In an area that has been a major source of conventional oil and gas since the 1960s, the Central and North Bazhenov shales (Upper Jurassic) of the West Siberia Basin (3.5 million km^2) could potentially hold 75 bbl of oil and 285 tcf of gas (US EIA 2015i). Gazprom Neft successfully drilled a horizontal well in the Bazhenov, demonstrating that Russia was fully capable of unconventional extraction; however, further development has been hampered by political pressures, both foreign and domestic (Rapoza 2019). A major deal between Exxon and the Russian oil producer Rosnef was tabled owing to sanctions imposed by the United States after the Ukraine incursion. More significant is President Vladimir Putin's insistence that Russia does not need to tap unconventional resources as the country has enough conventional resources (Rapoza 2019).

1.11 Brazil

Brazil boasts 18 onshore basins, with 3, the Paraná (Devonian), Solimões (Devonian), and Amazonas (Devonian), providing the bulk of conventional oil and gas (US EIA 2015f). In combination these marine black shales may hold 245 Tcf and 5.4 bbl of risked recoverable gas and oil (US EIA 2015f). Other basins that have not been fully characterized are the Parnaiba, Parecis, Sao Francisco, and Chaco-Parana, which have significant areal extent, as well as the smaller coastal locations of Potiguar, Sergipe- Alagoas, Taubate, and Reconcavo. Unconventional development is allowed in Brazil; however, it has been banned in the states of Paraná and Santa Catarina. When the government auctioned off land in the Jurua Valley for development, the court voided the sales in defense of the indigenous inhabitants, protecting their ancestral territory (Rogato 2016). In February 2020, Brazil's energy ministry formed an agreement with the US Department of Energy to increase investment and provide technical support for the development of Brazilian unconventional reserves (Bnamericas 2020).

1.12 Conclusions

The reserves for tight oil and gas shales are indeed global, with formations known in six continents (Antarctica the sole exception). Based on US EIA assessments, many of these reserves are significant. As has been discussed here, their successful development, however, is dependent on many factors, including the ease of accessibility and feasibility of extraction (i.e., technically recoverable), technical expertise, infrastructure, supporting service industry, financing, government regulations, environmental concerns, and public sentiment.

References

Aczel MR. (2020). Public opposition to shale gas extraction in Algeria: Potential application of France's "Duty of Care Act." *The Extractive Industries and Society.* 7(4): 1360–1368. doi:10.1016/j.exis.2020.09.003.

Aczel MR, Makuch KE, and Chibane M. (2018). How much is enough? Approaches to public participation in shale gas regulation across England, France, and Algeria. *The Extractive Industries and Society.* 5: 427–440.

Agbroko R. (2011). S. Africa imposes "fracking" moratorium in Karoo. Reuters, April 21, 2011. www.reuters.com/article/us-safrica-fracking-idUSTRE73K45620110421. Accessed September, 2020.

Ambrose J. "Fracking halted in England in major government U-turn." *The Guardian*, November 1, 2019. www.theguardian.com/environment/2019/nov/02/fracking-banned-in-uk-as-government-makes-major-u-turn. Accessed June, 2020.

Baca MC. (2010). Pittsburgh bans natural gas drilling. ProPublica, November 16, 2010. www.propublica.org/article/pittsburgh-bans-natural-gas-drilling. Accessed June 2020.

BBC News. (2012). "Bulgaria bans shale gas drilling with 'fracking' method." BBC News, January 19, 2012. www.bbc.com, www.bbc.com/news/world-europe-16626580. Accessed June, 2020.

Belakhdar N. (2020). "Algeria is not for Sale!" mobilizing against fracking in the Sahara. *Middle East Report.* 296 (Fall 2020).

Beller RW and Schiariti M. (2012). Argentina: regulations of the new Hydrocarbons Sovereignty Regime. Mondaq, August 7, 2012. www.mondaq.com/argentina/energy-law/190692/regulation-of-the-new-hydrocarbons-sovereignty-regime. Accessed June, 2020.

Bertram R. (2019). "Will fracking be banned in Mexico?" Energy Transition. April 17, 2019. https://energytransition.org/2019/04/will-fracking-be-banned-in-mexico/ Accessed June, 2020.

Bnamericas. (2020). Brazil-US to cooperate on regulations for unconventional oil and gas. Bnamericas, February 4, 2020. www.bnamericas.com/en/news/brazil-us-to-cooperate-on-regulations-for-unconventional-oil–gas. Accessed June 2020.

Cairney P, Fischer M, and Ingold K. (2018). Fracking in the UK and Switzerland: Why differences in policymaking systems don't always produce different outputs and outcomes. *Policy and Politics.* 46(1): 125–147. 10.1332/030557316X14793989976783.

Campbell C. "Indonesia embraces shale fracking: but at what cost?" *Time*, June, 2013. https://world.time.com/2013/06/25/indonesia-embraces-shale-fracking-but-at-what-cost/ Accessed June, 2020.

Chikhi L, Zhdannikov D, and Bousso R. (2019). "Exxon's talks to tap Algeria shale gas falter due to unrest -sources." Reuters, March 20, 2019. www.reuters.com/article/us-algeria-protests-exxon-mobil-idUSKCN1R11G8. Accessed June, 2020.

Dias P. (2019). "Brazil bans fracking in two states: National Congress hears arguments for a national ban." Fossil Free, August 15, 2019. https://gofossilfree.org/press-release/brazil-bans-fracking-in-two-states-national-congress-hears-arguments-for-a-national-ban/. Accessed June, 2020.

E360 Digest. (2021). Regulators ban fracking permanently in the four-state Delaware River watershed. Yale Environment 360. March 1, 2021.

Esterhuyse S, Avenant M, Redelinghuys N, Kijko A, Glazewski J, Plit L, Kemp M, Smit A, Vos AT, and Williamson R. (2016). A review of the biophysical and socio-economic effects of unconventional oil and gas extraction: Implications for South Africa. *Journal of Environmental Management.* 184: 419–430.

Gasser H. (2017). "Shell positioning itself as Egypt's preferred future partner." Energy Egypt, September 19, 2017. https://energyegypt.net/gasser-hanter-shell-positioning-itself-as-egypts-preferred-future-partner/. Accessed June, 2020.

Gesley J. (2017). "Germany: Unconventional Fracking Prohibited. Global Legal Monitor." March 8, 2017. www.loc.gov/law/foreign-news/article/germany-unconventional-fracking-prohibited/ Accessed June, 2020.

Graham J and Rupp J. (2022). Governance of fracking: Trends and challenges. In Stolz JF, Griffin WM, and Bain DJ (eds.) *Environmental Impacts from the Development of Unconventional Oil and Gas Reserves.* Cambridge University Press.

Gryc G. (1985). *The National Petroleum Reserve in Alaska: Earth-science Considerations. U.S. Geological Survey Professional Paper 1240-C.* U.S. Government Printing Office, Washington, DC.

Guo F and Mccabe P. (2017). Lithofacies analysis and sequence stratigraphy of the Roseneath-Epsilon-Mureree gas plays in the Cooper Basin, South Australia. *Journal of the Australian Petroleum Production and Exploration Association.* 57: 749. Doi:10.1071/AJ16202.

Herrera H. (2020). *Legal Status of Fracking Worldwide: An Environmental Law and Human Rights Perspective.* The Global Network for Human Rights and the Environment. https://gnhre.org/2020/01/06/the-legal-status-of-fracking-worldwide-an-environmental-law-and-human-rights-perspective/. Accessed June 29, 2020.

Kuznestsov S. (2013). "Lithuania: fracking gets trickier." *Financial Times*, August 2, 2013. www.ft.com/content/371065ad-0a57-3c8f-a78e-dd9200fba9fe. Accessed June, 2020.

McCarron G and Dougall S. (2022). An overview of unconventional gas extraction in Australia. In Stolz JF, Griffin WM, and Bain DJ (eds.) *Environmental Impacts from the Development of Unconventional Oil and Gas Reserves.* Cambridge University Press.

Mead LJ and Maloney M. (2018a). Violations of nature's rights. In *The Permanent Peoples' Tribunal Session on Human Rights, Fracking and Climate Change.* Earthlawyers.org. https://earthlawyers.org/wp-content/uploads/2018/04/PPT-Natures-Rights-Final-Submission-31.03.18.pdf. Accessed June, 2020.

Mead LJ and Maloney M. (2018b). Appendix 2. In "The Permanent Peoples' Tribunal Session on Human Rights, Fracking and Climate Change." https://earthlawyers.org/wp-content/uploads/2018/04/APPENDIX-2-Legal-Status-of-UOGE-across-the-world-31.03.18.pdf. Accessed June, 2020.

Minkow D. (2017). What you need to know about fracking in Canada. The Narwhal. April 6, 2017. https://thenarwhal.ca/what-is-fracking-in-canada/. Access September, 2020.

Myers SL. (2019). "China experiences a fracking Boom, and all the problems that go with it." *The New York Times*, March 8, 2019. NYTimes.com www.nytimes.com/2019/03/08/world/asia/china-shale-gas-fracking.html. Accessed June, 2020.

Natural Resources Canada. (2020). Shale and tight resources in Canada. www.nrcan.gc.ca/our-natural-resources/energy-sources-distribution/clean-fossil-fuels/natural-gas/shale-tight-resources-canada/17669. Accessed November, 2020.

Newbery C. (2019). *Argentina Faces Big Challenges to Develop Vaca Muerta for Export Growth*. S&P Global.

NGI. (2019). Mexico environment minister wants to prohibit fracking. Natural Gas Intelligence. December 10, 2019. www.naturalgasintel.com/mexico-environment-minister-wants-to-prohibit-fracking/. Accessed September, 2020.

Orthofer CL, Huppmann D, and Krey V. (2019). South Africa after Paris: Fracking its way to the NDCs? Frontiers in Energy Research. doi.org/10.3389/fenrg.2019.00020.

Otaola J. (2021). Argentina health works reject wage offer, keep road blockades at Vaca Muerta shale deposit. Reuters. April 26, 2021.

Rapoza K. (2019). "Putin: 'We'll never frack.'" *Forbes*, November 20, 2019. www.forbes.com/sites/kenrapoza/2019/11/20/putin-well-never-frack/. Accessed June, 2020.

Reed S. (2015). "Chevron to abandon shale natural gas venture in Poland." *The New York Times*, January 30, 2015. www.nytimes.com/2015/01/31/business/international/chevron-to-abandon-shale-venture-in-poland-a-setback-to-fracking-europe.html. Accessed June, 2020.

Reuters. (2014). "Egypt Signs First Gas Fracking Contract with Apache, Shell." *Reuters*, December 17, 2014. www.reuters.com/article/egpyt-shale-deals-idUSL6N0U11NR20141217. Accessed June, 2020.

Rogato M. (2016). Brazil does a U-turn on fracking, indigenous lands protected from oil and gas exploration. *Lifegate*. www.lifegate.com/brazil-fracking-stopped-amazon-jurua-valley. Accessed June, 2020.

SPE (Society of Petroleum Engineers). (1997). Petroleum Reserves Definitions [Archived 1997]. www.spe.org/en/industry/petroleum-reserves-definitions/. Accessed June, 2020.

Stefan M. (2015). "Fracking in Romania: Gone, but hardly forgotten." Food & Water Action Europe, September 28, 2015. www.foodandwatereurope.org/blogs/fracking-in-romania-gone-but-hardly-forgotten/. Accessed June, 2020.

Stolz JF and Griffin WM. (2022). Unconventional shale and gas oil extraction in the Appalachian Basin. In Stolz JF, Griffin WM, and Bain DJ (eds.) *Environmental Impacts from the Development of Unconventional Oil and Gas Reserves*. Cambridge University Press.

Thomas AR. (2014). "Fracking, Ukraine, and Russia." *Industry Week*, March 2, 2014. www.industryweek.com/the-economy/article/21962437/fracking-ukraine-and-russia. Accessed June, 2020.

United Nations. (2018). Shale Gas. Commodities at a glance: Special issue on shale gas, No. 9. https://unctad.org/en/PublicationsLibrary/suc2017d10_en.pdf. Accessed July 17, 2020.

U.S. Energy Information Administration. (2013). Shale gas and shale oil resource assessment methodology. www.eia.gov/analysis/studies/worldshalegas/pdf/methodology_2013.pdf. Accessed June 29, 2020.

U.S. Energy Information Administration. (2014). Shale oil and gas resources are globally abundant. www.eia.gov/todayinenergy/detail.php?id=14431. Accessed June 29, 2020.

U.S. Energy Information Administration. (2015a). World Shale Resource Assessments www.eia.gov/analysis/studies/worldshalegas/. Accessed June 29, 2020.

U.S. Energy Information Administration. (2015b). *Technically Recoverable Shale Oil and Shale Gas Resources: Algeria*. U.S. Department of Energy.

U.S. Energy Information Administration. (2015c). *Technically Recoverable Shale Oil and Shale Gas Resources: Argentina*. U.S. Department of Energy.

U.S. Energy Information Administration. (2015d). *Technically Recoverable Shale Oil and Shale Gas Resources: Australia*. U.S. Department of Energy.

U.S. Energy Information Administration. (2015e). *Technically Recoverable Shale Oil and Shale Gas Resources: Brazil*. U.S. Department of Energy.

U.S. Energy Information Administration. (2015f). *Technically Recoverable Shale Oil and Shale Gas Resources: Canada*. U.S. Department of Energy.

U.S. Energy Information Administration. (2015g). *Technically Recoverable Shale Oil and Shale Gas Resources: China*. U.S. Department of Energy.

U.S. Energy Information Administration. (2015h). *Technically Recoverable Shale Oil and Shale Gas Resources: Mexico*. U.S. Department of Energy.

U.S. Energy Information Administration. (2015i). *Technically Recoverable Shale Oil and Shale Gas Resources: Russia*. U.S. Department of Energy.

U.S. Energy Information Administration. (2015j). *Technically Recoverable Shale Oil and Shale Gas Resources: South Africa*. U.S. Department of Energy.

U.S. Energy Information Administration. (2020). *Assumptions to the Annual Energy Outlook 2020: Oil and Gas Supply Module*. U.S. Department of Energy.

Vinson and Elkins. (2020). Global Fracking Resources www.velaw.com/shale-fracking-tracker/resources/. Accessed June, 2020.

Yadav, S. (2020). "Fracking in India." *The Ecologist*, August14, 2020. https://theecologist.org/2020/aug/14/fracking-india Accessed August, 2020.

2

Unconventional Shale Gas and Oil Extraction in the Appalachian Basin

JOHN F. STOLZ AND W. MICHAEL GRIFFIN

2.1 Introduction

The development of natural gas and oil from tight shale reservoirs in the United States has increased exponentially over the last two decades. This growth was facilitated by advances in directional drilling technology coupled with the optimization of a slick water stimulation technique known as hydraulic fracturing as well as by market forces and a favorable regulatory environment. Energy markets were anticipating steep increases in commodity prices at the beginning of the new millennium as a result of depleting conventional reserves (Auzanneau 2018). The price of crude at the end of the year 2000 was $50 a barrel with the specter of rising prices on the horizon; by June of 2008 it had risen to almost $165 a barrel. Similarly, natural gas prices were in the $2.20 range in January 2000, only to rise to over $13 bcf (thousand cubic feet) in October 2005. In response, a special task force, headed by then Vice President Dick Cheney, was convened in 2001, and it drafted the Energy Policy Act of 2005, an act "to ensure jobs for our future with secure, affordable, and reliable energy" (Public Law 109-58 – August 8, 2005). The Act contained two major provisions to help the industry. The first, known as the "Halliburton Loophole," owing to Vice President Cheney's leadership role at Halliburton before being elected vice president, exempted hydraulic fracturing from the Safe Drinking Water Act (Section 322). The same section exempted the underground injection of natural gas for "the purposes of storage." This latter provision was also important as it provided a lower cost means (relative to liquified natural gas or LNG) for storage. The second provision, under Subtitle J, was to provide funding, through revenues generated by federal government oil and gas leases, to improve technologies for ultra-deepwater, unconventional natural gas, and other petroleum resources. The funds would go primarily to the National Energy Technology Laboratory (NETL). The results were impressive, with natural gas production going from about 3 Tcf to almost 15 Tcf per year by 2010, and 30 Tcf by 2015 (US EIA 2020). An additional incentive was provided when the Department of Energy (DOE) authorized export of natural gas and other products such as ethane. US exports of ethane went from almost nothing in 2013 to over 260,000 bbls per day in 2018, mostly to Canada, India, and the United Kingdom but also to Norway, Brazil, Sweden, and Mexico (US EIA 2019). Further, developments in drilling technology allowed for more wells per pad and laterals in excess of 5 km (discussed in Section 2.4.1.2). This review presents an overview of the processes involved in unconventional gas extraction as currently practiced, with particular focus on the Appalachian Basin.

2.2 Unconventional Oil and Gas Plays in the United States

The United States has an abundance of shale deposits with over 25 basins and 30 formations (Table 2.1) scattered across the country (Figure 2.1). Together, they may hold upward of 78 Bbbl of oil and 623 Tcf of natural gas (US EIA 2020). Unconventional gas and oil drilling is now occurring in over 20 states; however, most activity is occurring in North Dakota (Bakken), Texas (Permian, Barnett, Eagle Ford), Louisiana (Haynesville), Arkansas (Fayetteville), Michigan (Antrim), Oklahoma (Woodford, Mississippian), Wyoming (Green River), Colorado (Niobrara-Codell), Pennsylvania, Ohio, and West Virginia (Upper Devonian, Marcellus, Utica). The exponential increase in dry shale gas production that began shortly after 2005 has leveled off in the last several years to an average of around 70 bcf/d (US EIA 2020). The Marcellus shale has been the most productive, producing 22.84 bcf/d of gas over the last two years (2019–2020). This is followed by the Permian (11.74 bcf/d), Haynesville (9.32 bcf/d) Utica (6.92 bcf/d), Eagle Ford (3.81 bcf/d) Woodford (2.57 bcf/d), and Barnett (2.11 bcf/d) (US EIA 2020). While the COVID-19 pandemic caused a brief downturn in production, drilling was again on the rise in the latter part of 2020.

2.3 Horizontal Drilling and Hydraulic Fracturing: A Potent Combination

Shale reserves are known as "unconventional" or "tight" because the gas and oil is trapped in a rock with low porosity and permeability and thus cannot be extracted using conventional methods originating with Drake (Soeder 1988). The term "unconventional" is sometimes legally defined, such as in Pennsylvania, which uses both rock type (i.e., shale) and depth (i.e., a geologic formation below the Elk Sandstone) to identify unconventional activities (Oil and Gas Act of 2012, Act 13). While there are many tight gas and oil reservoirs globally (See Chapter 1) and throughout the United States (Figure 2.1) (National Petroleum Council 1980), their development had been constrained by their limited thickness and depth. For example, the Marcellus Shale, with an aerial extent of over 95,000 square km, ranges in thickness from 15 to 100 m and in depth from only 600 m (outcropping in upstate New York) to over 3 km below the surface (Milici and Swezey 2006). These two hurdles were overcome by combining directional (i.e., horizontal) drilling with hydraulic fracturing. The idea of directional drilling is not new, going as far back as the early 1930s (Gleason 1934). While the methods were crude, using a whipstock (i.e., deflecting tool) downhole, and cumbersome, with limited control, they allowed multiple wells to be drilled from the same pad. Today, a computer guided hydraulic or mud motor with a surface-adjustable bent housing is used to direct the drill bit, allowing for greater dexterity and command. Further, state of the art computer programs can provide real-time adjustments to improve drilling efficiency and performance, and quick responses to downhole issues such as vibration and drill string buckling (Jeffery and Creegan 2020).

The second innovation was hydraulic fracturing, or "fracking." Again, the idea of breaking up a formation to release gas or oil even from a conventional play is not new. The credit is given to E. A. L. Roberts, who in 1865 received a patent to use explosives with

Table 2.1 *Shale plays in the United States*[1]

Basin	Formation	Age
Appalachian	Devonian (Ohio)	Upper Devonian
	Marcellus	Devonian
	Utica	Middle Ordovician
Black Warrior	Chattanooga	Devonian
	Floyd-Chattanooga	Mississippian
	Floyd-Neal	Mississippian
	Conesauga	Cambrian
Michigan	Antrim	Upper Devonian
Illinois	New Albany	Devonian-Mississippian
Forest City	Excello	Pennsylvanian
	Mulky	Pennsylvanian
Arkoma	Fayetteville	Mississippian
	Woodford	Devonian
	Caney	Mississippian
East Texas	Haynesville-Bossier	Jurassic
Anadarko	Woodford	Devonian
Fort Worth	Barnett	Mississippian
Palo Duro	Bend	Permian
Permian Basin	Barnett	Mississippian
Morfa	Woodford	Devonian
Maverick	Eagle Ford	Upper Cretaceous
	Pearsall	Lower Cretaceous
Willston	Gammon	Upper Cretaceous
	Bakken	Upper Devonian
Denver-Julesburg	Niobrara	Upper Cretaceous
Raton	Pierre	Upper Cretaceous
Montana Thrust Belt	Cody	Upper Cretaceous
Greater Green River	Hilliar	Eocene
	Baxter	Upper Cretaceous
	Mancos	Late Cretaceous
Big Horn	Mowry	Upper Cretaceous
Paradox	Hermosa	Pennsylvanian
Uinta	Mancos	Upper Cretaceous
	Manning Canyon	Upper Mississippian
Piceance	Niobrara	Upper Cretaceous
San Juan	Lewis	Upper Cretaceous
San Joaquin	Monterey-Tremblor	Middle Tertiary
Santa Maria	Monterey	Miocene

[1] US **EIA (2020)**

his "exploding torpedo." Initially, gunpowder was used, but was replaced by nitroglycerine despite its relative instability, to fracture the formation at depth and improve well production (Wells 2020). For hydraulic fracturing, the first field testing was in 1947, a gas field in a limestone formation at 2,500' (762 m) in Hugoton KS (Veatch et al. 2017). According to the

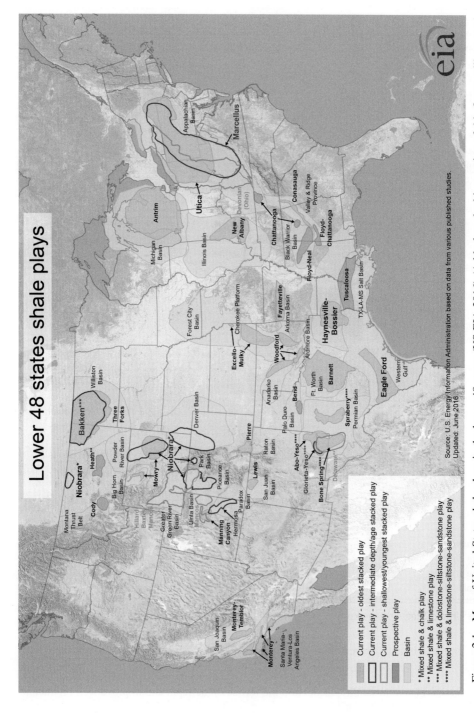

Figure 2.1 Map of United States shale plays in the lower 48 states (US EIA 2016) (A black and white version of this figure will appear in some formats. For the colour version, refer to the plate section.)

American Oil and Gas Historical Society, the first commercial hydraulic fracturing of an oil well dates to March 17, 1949, in Duncan, OK and Holliday, TX (Wells 2020). The process was patented by the Stanolind company and licensed to Halliburton. Its application to tight gas shale, involving the combination of directional drilling and hydraulic fracturing, was pioneered in the Barnett Shale (Texas) by George Mitchell in the early 1980s. While a variety of fluids have been used, including diesel, currently, it typically involves a mixture of water, sand (proppant), and a proprietary mixture of chemicals, injected into the ground at very high pressures (5–10,000 psi) (NYSDEC 2011). Alternatives have included liquid nitrogen (Yan et al. 2020) and a "green" fracking fluid developed by the Pacific Northwest National Laboratory named "StimuFrac^{TM}" that uses the reaction of carbon dioxide with poly(allylamine) (Shao et al. 2015). Successful fracture propagation is crucial for well production. Therefore, modern hydraulic fracturing is often monitored by microseismic measurements and computer modeling, and areas with poor fracture propagation are refracked (Mayerhofer et al. 2006; Veatch et al. 2017). An array of surface tiltmeters can be deployed to measure fracture orientation, while downhole tiltmeters can provide greater resolution of the fracture height and length (Evans et al. 1982; Halliburton 2011). A number of fracture propagation models are available to create both 2-D and 3-D reconstructions (Veatch et al. 2017).

2.4 Upstream, Midstream, and Downstream

The industry may be divided up into three sectors: upstream, midstream, and downstream (Table 2.2). Upstream is the exploration and production of oil and gas, including initial surveying and obtaining of mineral rights all the way through to production. Midstream provides the link between upstream and downstream processes and includes the processing of the gas or oil, gathering and transmission pipelines, transport, and storage. Downstream includes refining, production of products (e.g., gasoline, plastics), marketing, and retail sales. Each will be discussed in this section, with particular focus on upstream processes.

2.4.1 Upstream

2.4.1.1 Predrilling Activities

The first stage of upstream activities is well site planning and preparation. Before any pad preparation or drilling can begin, a company must acquire the mineral rights to the gas or oil. This is usually facilitated by a landman, who negotiates the lease, the signing bonus, and the royalty rate. It is not a given that a landowner will also own their mineral rights. In the United States this is known as "split estate," where the mineral rights are separate from the surface rights. Lease agreements can vary greatly from state to state, and within a given state. Pennsylvania has no set minimum and some lease agreements have gone for as little as $7 an acre, but in some cases over $5,000 an acre (Lampe and Stolz 2015). Pennsylvania does mandate a minimum royalty rate of 12.5% (Oil and Gas Lease Act, 1979); however, companies may deduct production and development costs from the royalties (Lampe and Stolz 2015). Further, the royalties are solely for methane produced from the well but not

Table 2.2 *Upstream, midstream, and downstream processes and products*

Upstream	Midstream	Downstream
Seismic surveys	Transport	Refineries
Geophysical surveys	Rail, Truck, Barge, Tanker	Liquid Natural Gas facilities
Test drilling	Storage	
Surface and mineral rights	Pipelines: gathering, feeder, transmission	Distributors, distribution pipelines
Leases	Compressor stations	
	Pigging operations	
Pad Preparation	Cryogenic Gas plants	Natural gas
Drilling	Methane	Petroleum products
Casing	Ethane	Gasoline
Fracking	Propane	Diesel
Completion	Butane	Kerosene
Production	Drip gas	Jet fuel
Plugging & abandonment	Odorant	Heating oil
		Lubricants
	Hazardous waste landfill	Ethane cracker plants
	Class II Injection wells	Post-processing (products)
	Brine treatment facilities	Marketing
		Retail

other products. Marcellus wells in Southwester PA have "wet" gas that includes ethane, propane, butane, and drip gas. Another aspect of leasing is unit consolidation or compulsory integration, otherwise known as "forced pooling." Thirty-eight states have some form of compulsory integration that allows a company holding the majority of the mineral rights of a given unit or pool to drill under the unleased properties. Forced pooling was originally developed for conventional oil and gas plays as protection for mineral owners against "rule of capture," where it was legal for a well on an adjacent property to extract the gas or oil from another without compensation ("I drink your milkshake"). In a recent court case, the Pennsylvania Supreme Court ruled that rule of capture does apply to tight oil and gas formations; however, a driller might be open to litigation if their well crossed property lines without a lease (*Briggs* vs *Southwestern Energy Production Co*, J-48-2019). Nevertheless, the ruling did give credence to what landowners are often told as an encouragement to sign a lease, that the companies "will get your gas anyway." Forced pooling does not currently apply to the Marcellus Shale in Pennsylvania, but the Pennsylvania Oil and Gas Conservation Law of 1961 allows for unit consolidation for the Utica Shale and formations below it (Lampe and Stolz 2015). Surface leases are also acquired for infrastructure such as impoundments for water storage and pipelines. Permits may also be required for water withdrawals in areas that have river basin commissions or are under state regulatory control. Water withdrawals in eastern Pennsylvania are managed by the Delaware River Basin Commission and the Susquehanna River Basin Commission, while the rest of the state is managed by the PA DEP. Further information about water management can be found in Chapter 7.

In addition to the leases, the company has to obtain a drilling permit from the appropriate regulatory agency. In Texas, it is the Rail Road Commission; in Colorado, the Oil and Gas Conservation Commission (COGCC) that is part of the Department of Natural Resources; in North Dakota, the Industrial Commission (NDIC); in Ohio, the Department of Natural Resources (ODNR); in West Virginia, the Department of Environmental Protection (WV EPA); and in Pennsylvania, the Department of Environmental Protection (PA DEP). In Pennsylvania, there are two basic types of permits: permitted use and conditional use (Chapter 32 Title 58 Oil and Gas Act). In the former, the area in question is either not subject to zoning restrictions (i.e., the municipality does not have zoning) or is zoned for industry. A conditional use permit is usually needed when the property is zoned for residential, agriculture, or both. In that case, the permit requires conditions and variances to zoning and local ordinances (Governor's Center for Local Government Services, 2001). The document, which may be quite lengthy (hundreds of pages), must be prepared by a licensed engineer (usually in the employ of an environmental consulting firm) and contain the following elements: a letter of transmittal, the application form, project narrative (basically a short overview of the project), conditional use plan, primary land development plan, permit application for the well pad, permit application for the pipeline, and permit application for each well that includes the well location plat (Figure 2.2), in addition to the preparedness, prevention, and contingency plan, master emergency response plan, and site-specific emergency response (i.e., evacuation) plan. Both the well pad and pipeline permits include erosion and sediment control plans, wetland delineation report, and site restoration plan. Additional information may be requested and provided such as easements and bonding requirements. The company must also notify and provide a copy of the plat to municipalities, landowners, storage operator, and coal operators within 3,000' (914.4 m) from the well location (58 Pa CS 32). In Pennsylvania, water testing of private wells is required (i.e., "predrill test") for homes within 2,500' (762 m) of the well pad.

Once the leases have been obtained, a seismic survey is conducted to create 2D or 3D maps of the rock strata and identify the optimal location to drill. This can involve thumper trucks (Figure 2.3) or small explosive charges, usually 2 lb (0.9 kg) of dynamite buried at 20 ft (6 m) depth, to generate seismic waves that can be measured by multiple sensors distributed across the land surface. These wave measurements are used to create 2D and 3D maps, respectively (Lampe and Stolz 2015). After the drilling location is determined, permits and any special variances to local municipal ordinances are obtained (i.e., the zoning board has approved the permit). The site is cleared of trees, the surface is leveled, and the pad liner laid out and covered with a uniform substrate (Figure 2.4). Two small impoundments, one for the drilling fluids and one for the drilling waste fluid and rock cuttings, and the cellar for each of the wells, are prepared Figure 2.4 The dimensions of the pad ultimately depend on the number of wells to be drilled. Initially, one to a few wells were drilled on each pad, but today, mega pads may have over 40 wells (Figure 2.2).

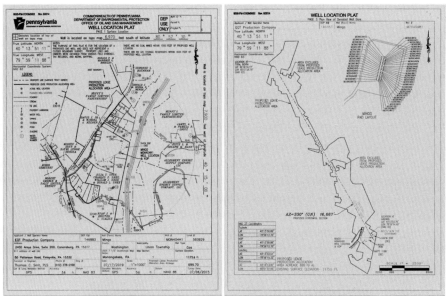
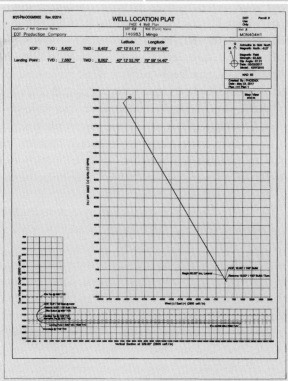

Figure 2.2 Well location plat pages from a conditional use permit for a mega pad with 42 wells in Southwestern Pennsylvania. (A) map of the location of the proposal well pad in relation to other nearby surface structures including information of landowners, (B) map showing lateral for one of the wells (MON404H1) and the number of wells on the pad (upper righthand corner), (C) diagram of the well bore showing specifics such as top hole, kick off point, and bottom hole with depths and lengths

Figure 2.3 Trucks and equipment associated with unconventional drilling. (A) a caravan of thumper trucks for seismic testing, (B) water tanks at a watering station at the edge of the Monongahela River (note hose leading to the river – arrow), (C) "Sand king" proppant truck, (D) sand (proppant) truck, (E) pressure pump truck, (F) chemical mixing truck, (G) kettle truck for mixing cement, and (H) brine (residual waste) truck. A, C, D, G, and H courtesy of Robert Donnan; E and F courtesy of Bill Hughes; B the author (JFS)

Figure 2.4 Stages in the development of a well pad. (A) Impoundment for fresh water (note the white pipeline that goes to the adjacent well pad, middle center of the image), (B) Pad preparation with liners in place and cellars for eight wells (w). (C) Drilling has commenced with

2.4.1.2 The Well: Drilling and Hydraulic Fracturing (Fracking)

The next stage is the drilling, which is done in several phases. It involves considerable infrastructure (e.g., the rig or derrick) and specialized equipment (e.g., blow out preventer, choke manifold). Between 100–150 truckloads of heavy equipment and another 100–1000 truckloads of fracturing chemicals, water, and proppants are used to drill and frack the well (Groat and Grimshaw 2012). The rig can come in several sizes and with one rig type (i.e., single) being used to drill the vertical sections of the well and a larger (i.e., triple) used for the horizontal section (Figure 2.4). Alternatively, a single multipurpose rig is used. Another innovation has been to increase the number of wells drilled on a single pad. Initially, an area below the rig, known as the cellar, is excavated that provides a space for the casing and the blow out protector to sit. The platform, which is suspended above the cellar, has a variety of features, including the iron roughneck, which is used to add or remove pipe and provides torque; the mouse hole, which holds the next sections of drill pipe; and the rat hole. The rat hole holds the Kelly, a polygonal steel tubular apparatus suspended through the rotary table that provides fluid and rotation to the drill stem. The mast is the tower that supports the hoisting system, and holds the traveling block that lifts the drill string pipe into place, the derrick board further up the mast, and the crown block, the set of pulleys at the very top (Sharpe 2014).

The well itself comprises a series of nested pipes and cement fillings known as casings. Minimally, there are four: the conductor, the surface, the intermediate, and the production casing (Belyadi et al. 2019). The conductor hole is drilled first and cased with steel piping secured into place by a layer of concrete, to protect groundwater (i.e., aquifer). The hole may be 18 to 36" (45 to 90 cm) in diameter and from 20 to 50' (6 to 15m) in length (Belyadi et al. 2019). The surface casing is a 26" (66 cm) hole that may run up to 2,000' (50 m) in length. In Pennsylvania, multiple casings are required, with an additional intermediate string needed if coal seams are encountered (Oil and Gas Act of 2012, Act 13). The intermediate casing, usually 9 5/8" (24 cm) in width, provides protection against abnormal pressures that might be encountered downhole (Belyadi et al., 2019). Lastly, the production hole is drilled and cased. The horizontal portion of the well may require a larger drilling rig, named the "triple," as it can hold three lengths of piping (Figure 2.4), and a directional drill (as discussed in Section 2.3). The "kick off" point is the depth at which the drill is turned, with the "landing point" at the beginning of the horizontal section, ideally well within the desired rock formation (i.e., shale; Belyadi et al. 2019). The horizontal may extend outward several thousand feet. Eclipse Resources reported a lateral length of 19,500' (6 km) in the Utica formation in Ohio, for a total length of 27,750' (8.5 km), and more recently, Olympus Energy reported a Marcellus well with a lateral length of

---

Figure 2.4 (*cont.*) a "triple" rig. The pad is surrounded by a sound dampening barrier, the drilling rig is in the center, and an impoundment (arrow) for drilling fluids and waste is to the left of the rig. (D) The completed well site with remaining infrastructure; (w) well heads cordoned off by jersey barriers, (s) separators, and (c) condensate tanks. (E) Close up of a well head (note how the cellar can fill with water). B, C, and D, courtesy of Robert Donnan; A and E, the author (JFS)

20,060' (6.1 km). Each well can produce 1,000s of tons of drilling waste, depending on the depth of the formation and the length of the horizontal. The drill cuttings and formation water released during the drilling may contain a variety of salts, heavy metals, and naturally occurring radioactive materials (NORM) such as uranium, radium, and radon (Brown 2014; Warner et al., Chapter 9). The drilling waste may be buried onsite or, more typically, transported to landfills. Pennsylvania allows sanitary landfills to take up to 80% of their total daily volume, while some has been exported to West Virginia and New York. New York, however, passed legislation in the summer of 2020 aimed at prohibiting this practice (S3392/A2655). Once the horizontal leg of the well is drilled, a production casing, or inner string, usually with a 5 ½" (14 cm) width, is laid from the top hole at the surface to the bottom hole at the end of the well bore (NYSDEC 2011). A drilling log is kept and provided as part of the well record, indicating the different formations (with top and bottom depth), as well as any gas, oil, or subsurface water encountered.

The well is then ready for completion and stimulation, which are done in stages. The well pad becomes crowded with a variety of trucks, containers, and equipment (Figures 2.3 and 2.5). These include the control center, water tanks, pressure pumps, proppant containers, and the blenders, which mix the water, chemicals, and proppant (Figure 2.5). Once the production casing is in place, the well bore needs to be exposed to the formation. This is accomplished by "perfing" with a perforating gun, a device that sets off explosive charges in all directions, at depth in the casing. The perforating gun string, which may be as short as 200' (60 m) to as long as 1000' (300 m), is lowered into the hole by a wireline. Each charge creates a hole through the pipe, the cement, and the rock. The perforating gun is then withdrawn from the hole and the well is stimulated. This involves injection of the hydraulic fracturing mixture at high pressure, 5,000 to 10,000 psi (35,000 to 70,000 kPa), along with proppant (fine-grained sand). The fluids may be pumped in at rates upward of 3,000 gallons (11,000 L) per minute (NYSDEC 2011). This process enlarges the natural fractures and induces new fractures in the rock, releasing the trapped gas. The first stage is at the farthest end of the well bore, and once completed it is temporarily plugged and the next stage is prepared. This is known as "conventional plug and perf" and involves the use of a composite bridge plug (Belyadi et al. 2019). Early on, both the length of the laterals and the stages were shorter, requiring ten stages to hydraulically fracture a 3,000' (100 m) well. Today, a well may require more than 40 stages, based on completion reports filed with the PA DEP, which were reviewed for this chapter. Two innovations have improved the efficiency and lowered the costs. The first was increasing the number of wells per pad (Figure 2.3). Well spacing is carefully engineered to maximize formation fracturing but minimize cross communication between parallel horizontal bores. The second involved hydraulic fracturing multiple wells at the same time. In zipper fracking, one well is being stimulated while a second well is perfed and has its plug set, alternating back and forth between the wells (Belyadi et al. 2019). In cube fracking, multiple wells, as many as 20, are drilled at different depths. While it may save time, it increases the amount of water, chemicals, and proppant. During completion, the well may be flared to control the back pressure. Once the well is in production, it is connected to the gathering pipeline, the trucks and containers are removed, and the site is reclaimed, leaving behind the well heads,

Figure 2.5 The well pad during the hydraulic fracturing of the well. (A) well pad in Ohio with some permanent infrastructure already in place, (s) separators, (c) condensate tanks, (cn) conditioners. (B) closer look at a well pad in Amwell Pennsylvania (cc) control center, (b) blenders, (wt) water tanks, (p) pressure pumps, (pp) proppant (sand) containers and trucks.
Courtesy of Robert Donnan

separators, and condensate tanks (for the produced water), as well as any additional infrastructure needed for onsite treatment (Figure 2.4). The condensate tanks must be regularly serviced by brine trucks (Figure 2.3), as produced water will be generated throughout the lifetime of the well. This produced water can be "recycled" (i.e., used to frack other wells), transported to a brine treatment facility, or disposed of at a Class II injection well. In Pennsylvania, once a well is in production, a completion report is submitted. The report includes information on the total amount of fluids used, the perforation record (number and length of the stages), composition of the stimulation fluid, and information on the specifics of each stage, including pump rates, pressures, and proppant sizes. The well record has information on the types of casings (e.g., conductor, surface, coal protective, intermediate, production) and includes the drilling log.

Thousands of chemicals have been reported to have been used in the fluid formulations for hydraulic stimulation (Colborn et al. 2011, GWPC 2009 US EPA 2016); however, there are only about a dozen critical components (FracFocus 2020). The basic constituents include surfactant, friction reducer, scale inhibitor, corrosion inhibitor, crosslinker and gelling agents, acid, breaker, iron control, and biocide (e.g., glutaraldehyde, to control bacterial growth). Surfactants are nonionic detergents that are used to reduce surface tension and increase fluid recovery. Friction reducers, such as granulated anionic polyacrylamide and petroleum distillates, reduce the friction in the pipe and provide the "slick" in slick water stimulation. Scale inhibitors, such as ethylene glycol, prevent formation of calcium carbonates and calcium, barium, and strontium sulfates. Crosslinkers and gelling agents, usually guar gum, are used to control the viscosity of the fluid. Acid, such as hydrochloric, is used to aid in fracturing and dissolving the rock by enhancing fissure formation. It can also aid as a corrosion inhibitor; however, additional chemicals (e.g., N, Dimethyl formamide) may be used. Biocides are needed to control bacterial growth (Chapter 12). Glutaraldehyde is commonly used; however, many microbes associated with shale gas and oil development, especially those adapted to the high salt environment, are resistant and may actually use it for food. In addition, proprietary tracers, such as metalloids (e.g., antimony) and rare earth elements (e.g., iridium, gadolinium), may be employed to aid in tracking the flowback fluids from each stage. The exact formulation depends on the company and the characteristics of the rock formation. Most of the additives are known and can be found in the national database maintained by the FracFocus Chemical Disclosure Registry, which currently lists 175,353 disclosures (FracFocus 2020). Nevertheless, there are several components that are proprietary in nature and have designations on the Material Safety Data Sheet (MSDS) that are undefined and often without a Chemical Abstract Service (CAS) number. The US EPA also knows of 41 chemicals whose identities were obtained through a nondisclosure agreement but have thus far not been publicly revealed (Horwitt 2016).

Proppant is composed of fine-grained particles that are pumped downhole along with the slick water formulation to prop open the fractures and, ideally, improve fracture conductivity and production (Veatch et al. 2017). Three main characteristics define the different types of proppant and determine their utility for a given formation, namely, composition, strength, and size. Proppant may be made of sand, resin-coated sand (precured or curable), or ceramic. Sand is the lowest strength, and is capable of sustaining pressures of up to 6,000 psi (Belyadi et al. 2019). Resin-coated sand can sustain a bit higher pressure, up to 8,000 psi. Ceramics come in three varieties, intermediate-strength, lightweight, and high-strength. A high-strength proppant such as sintered bauxite may handle pressures up to 20,000 psi (Belyadi et al. 2019). Size is measured as mesh with 100 mesh being the smallest. Larger sizes, in increasing order, are 40/70, 30/50, 20/40, 12/20, and 6/12. The 100 mesh is used first, as it is the finest grain size and seals off microfractures. Subsequent stages will use larger and larger mesh, as the larger particles allow for greater flow. Two other important characteristics are uniformity and shape. Ideally, rounded particles have greater interstitial spaces, allowing for greater conductivity. The last consideration is the amount of proppant to be used in each stage, and again, this depends on the formation and the complexity of the design. For example, a "400k" frac type will use 400,000 lbs (180 metric tons) of proppant, comprising 50,000 lbs of 100 mesh, 200,000 lbs of 30/50 mesh, and 150,000 lbs of 20/40 mesh (Belyadi et al. 2019).

2.4.1.3 Estimated Ultimate Recovery and Decline Rates

Oil and gas wells developed by hydraulic fracturing have the interesting characteristic that they are very productive initially but have rapid decline rates (Hughes 2013; Zammerilli et al. 2014). The estimated ultimate recovery per well, or EUR, may be in the billions of cubic feet but 50–75% of that is usually recovered in the first year of production (Baihly et al. 2010). Wells in the Haynesville (Louisiana) have a EUR of 6.5 Bcfe with an initial decline rate of 85% (Chesapeake Energy 2009). Wells from the Barnett (Texas) and Fayetteville (Arkansas) are comparable, with EURs of 2.65 and 2.4 Bcfe and decline rates of 70% and 68%, respectively (Chesapeake Energy 2009). Production rates from the Marcellus (Pennsylvania) vary by the geological location, as wells in the eastern part of the state produce dry gas (i.e., methane), while those in the western part of the state produce "wet" gas, with other products including ethane, propane, butane, and drip gas. Swindell reported a state average of 6.2 Bcf, with some in the northeastern part of the state as high as 20 Bcf, based on his evaluation of 5,000 wells (Swindell 2018). Cabot Oil and Gas, with the major of their operations in the northern part of the state, have reported a EUR of between 35 to 40 Bcf, or 4.4 Bcf per 1,000' (300 m). This is based on their typical 8,000' horizontal well, with 50 frac stages, and 2,000 lbs of proppant per foot (907 kg per 0.3 m) (Cabot Oil and Gas, Q2 2017). However, their decline rates are hyperbolic, similarly to other Marcellus wells, with a first year decline rate of around 75%. Thus, the wells are most productive in the first three to five years. An assessment of PA DEP data for gas production in Washington County PA over the period 2005–2018 showed that 133,000 Mm^3 was produced from the 1811 Marcellus Shale wells drilled. However, by the end of that period, 74 had already been plugged, 52 were "regulatory inactive," and two were listed as abandoned (Pratt 2019). The data also revealed that the tipping point (for when a well would be plugged) was when the production level went below a threshold of 0.66 Mm^3/d or had only 11 production days per month (Pratt 2019). The amount of gas or oil produced and the lifetime of well have been increased by increasing the length of the horizontal and improvements in fracking design as described in this chapter.

2.4.2 Midstream and Downstream

As outlined in Table 2.2, a significant amount of infrastructure is required for both midstream and downstream processing (Figures 2.6–2.9). Midstream activities include

Figure 2.6 Ancillary infrastructure associated with unconventional shale gas and oil development. (A) impoundment for water, (B) transmission pipeline, (C) compressor station in West Virginia with 12 compressors under construction, (D) compressor station across the street from a farmhouse in Ohio, with six compressors, (E) pigging station in Ohio.
Courtesy of Robert Donnan

Figure 2.7 Infrastructure for processing oil and wet gas. (A) Blue Racer Midstream in Proctor, West Virginia, (B) Midstream processing facility near Moundsville, West Virginia, (C) Midstream processing facility Mobley, West Virginia, (D) propane (left) and butane (right) storage tanks at a facility in Cadiz, Ohio.
Courtesy of Robert Donnan

Figure 2.8 Processing plants for oil and wet gas. (A) Seneca, Sommerville Ohio, (B) Fort Beeler, West Virginia, (C) Mobley, West Virginia, and (D) Sherwood, West Virginia. Courtesy of Robert Donnan

Figure 2.9 Shell ethane cracker facility under construction, Beaver, PA.
Courtesy of Robert Donnan

pipeline construction and their maintenance, and processing that separates the products from the natural gas mixture. Processing facilities are quite large, especially in areas with "wet gas" such as Southwestern Pennsylvania, Eastern Ohio, and West Virginia (Figures 2.6–2.8). Downstream processes include oil refineries and ethane cracker facilities for plastics production (Figure 2.9). Gas is transported primarily by pipelines. Gathering pipelines bring gas or oil from the well pad to a processing plant, refinery, or storage facility. They are usually small in diameter, 10–15 cm, and low pressure (Afework et al. 2020). Feeder lines, which are 15–30 cm in diameter, connect the processing plant or storage facility to transmission lines. Transmission lines (Figure 2.6) are larger, 30 cm to 1 m in diameter, and under higher pressures (e.g., 200–1,500 psi) requiring compressor stations (Figure 2.6) to move the product. The larger pipelines are maintained by pigging stations (Figure 2.6), facilities that deploy devices that clean the insides of the pipeline (e.g., removing scale buildup). Pennsylvania already has about 8,600 miles of intra- and interstate transmission pipelines but estimates suggest an additional 10,000 to 25,000 miles will be needed, depending on the number of wells eventually drilled (Johnson et al. 2011). Interstate pipelines are regulated by the Federal Energy Regulatory Commission (FERC), the federal agency that regulates the interstate transmission of electricity, natural gas, and oil (FERC 2020). In 2004, the Pipeline and Hazardous Materials Safety Administration (PHMSA) was created by the Department of Transportation (DOT) to develop and enforce regulations for the 4.2 million km of pipeline in the United States.

Processing prior to distribution is often required, as the gas may contain contaminants or additional, and desirable, hydrocarbons (NaturalGas.org 2013). Water is a common contaminant, especially in mature plays (Abbad et al. 2015). Three phase separators are used to separate the water/gas/oil mixture. If necessary, further drying of the gas can be performed using glycol or solid desiccant treatment. Diethylene glycol or triethylene glycol are used in a glycol contactor, which are often found as part of the infrastructure on the well pad (Figure 2.5). The gas may be contaminated with hydrogen sulfide and/or carbon dioxide, which need to be removed. Sulfur, in the form of hydrogen sulfide, is particularly noxious in that it is both toxic and corrosive. Its removal is known as "sweetening" and involves passing the gas through a solution of monoethanolamine or diethanolamine. The commercial product Selexol contains dimethyl ether of polyethylene glycols that can remove hydrogen sulfide and carbon dioxide, which takes place in a stepwise process (Manning and Thompson 1991).

Portions of both the Marcellus and Utica shales produce wet gas, a mixture of hydrocarbon products, or natural gas liquids (NGL) that need to be separated out. This takes place in midstream processing plants, known as cryogenic plants owing to the use of refrigerants to cool the different liquids from the gas. The process involves multiple fractionators, namely, the deethanizer, depropanizer, debutanizer, and deisobutanizer. After the initial cooling to around -85°C, where the methane is separated out (it remains in the gas stream), the mixture of natural gas liquids is warmed in stages to their respective boiling points (ethane -89°C, propane -42°C, butane -1°C). The remaining natural gas condensate, also known as drip gas, contains a mixture of butane, pentane, hexane, and heavier hydrocarbons (i.e., C5 or higher). Butane is most commonly used in camping stoves and lamps. An alternative method, known as the absorption method, uses an oil to remove the natural gas liquids from the methane. In this case, the oil/NGL mixture is either heated or cooled for maximum separation of the products.

A major driver of the development in the Appalachian Basin has been an expanded export market for natural gas, usually as liquified natural gas (LNG), and other products such as ethane. The Natural Gas Act of 1938 made it illegal to export natural gas or its products without authorization from, at that time, the Federal Power Commission, and now the Department of Energy. It had always been a goal of the unconventional gas industry to expand global markets, and the United States became a net exporter of natural gas and natural gas products in early 2011 (EIA 2019). In 2018, FERC and PHMSA signed a memorandum of understanding to expedite the application process and permitting of new LNG facilities (DOT 2020). Several LNG processing facilities have been approved (i.e., Gibbstown Logistics Center in Gloucester County, NJ) or are in the planning stages. As mentioned earlier in this chapter, ethane export has also seen a dramatic increase. In addition, interest in ethylene and polyethylene production facilities has also grown, with several proposed in Pennsylvania, Ohio, and West Virginia by Shell Chemical, PTT Global Petrochemical, and ExxonMobil, respectively. Shell's ethane cracker facility is under construction in Beaver, PA (Figure 2.9) and is scheduled to open in 2022. According to the permit, once it is fully operational, the cracker plant will process 4.41 million gallons

(16.7 million liters) of ethane and produce nurdles of polyethylene. The facility will be serviced by the Falcon Pipelines, delivering ethane from processing facilities (similar to those shown in Figures 2.7 and 2.8) in Houston, PA (the southern branch of the pipeline), and Cadiz, OH (the western branch of the pipeline). Based on eight years of production records for wet gas wells in Washington County, PA (2009–2016) with an average content of 15 to 18% ethane (Mosquera Netzkarsch 2016), an estimated 1,000 fully producing wet gas wells will be needed to provide the daily requirement. Plans are currently underway for an underground ethane storage facility that would provide an uninterrupted source of ethane.

2.5 Conclusions

Unconventional oil and gas drilling now occurs in over 30 states. This expansion has been the result of both technological advances, initially favorable market conditions, and governmental encouragement. The Energy Policy Act of 2005 was especially important as it exempted the process of hydraulic fracturing and subsurface natural gas storage from the Safe Drinking Water Act. Further, it provided financial incentive for the DOE and NETL to improve technology, efficiency, and ultimately production. The technological advances included the combination of horizontal drilling and hydraulic fracturing, new slick water formulations, and improvements in drilling technologies and fracking design such as "zipper fracking." The United States is once again a global leader in oil and natural gas production, an expansion that has involved all three business sectors, upstream, midstream, and downstream, with major infrastructure development. Upstream processes include pre-drilling activities (geophysical surveys, leasing, title searches), pad preparation, drilling, hydraulic fracturing, and completion. Midstream processes include gathering and transmission pipeline construction, pigging operations, and processing facilities. Downstream processes include oil refineries and ethane cracker facilities for plastics production.

However, this is not 1859 (when Colonel Drake drilled the first commercial oil well in Pennsylvania) or even the 1970s, when conventional oil production reached its peak in the United States (Auzanneau 2018). There is now greater social awareness about the environmental impacts of continued reliance on fossil fuels (McCarron and Dougall 2022; Graham and Rupp 2022). Each sector of the industry comes with its own set of issues. From the increased truck traffic, emissions, noise and light pollution at the well pad, the water, chemicals, and sand needed for the well stimulation, to the toxicity and radioactivity of the solid and liquid wastes, and the eventual fate of the well itself. The environmental impact of this development is the focus of this volume, with chapters devoted to specific topics, including air quality (Presto and Li 2022), fugitive methane (Howarth 2021), water usage and management (Wilson and VanBriesen 2022), seismic activity (Eaton 2022), naturally occurring radioactive materials (Warner et al. 2022), isotopes (Stewart and Capo 2022; Townsend-Small 2022), and microbiology (Nixon 2022), and water contamination (Bain et al. 2022).

References

Abbad M, Dyer S, Ligneul P, and Badri M. (2015). In-line water separation: a new promising concept for water debottlenecking close to the wellhead. SPE-172687-MS.

Afework B, Hanania J, Stenhouse K, and Donev J. (2020). Pipeline. https://energyeducation.ca/encyclopedia/Pipeline. Retrieved August 2, 2020.

Auzanneau M. (2018). *Oil, Power, and War, a Dark History*. Chelsea Green Publishing.

Baihly J, Altman R, Malpani R, and Luo F. (2010). Shale gas production decline trend comparison over time and basin. *Society of Petroleum Engineers SPE*. 135555.

Bain DJ, Cantlay T, Tisherman R, and Stolz JF. (2022). Deciphering water contamination in areas with legacy conventional development. In Stolz JF, Griffin WM, and Bain DJ (eds.) *Environmental Impacts from the Development of Unconventional Oil and Gas Reserves*. Cambridge University Press.

Belyadi H, Fathi E, and Belyadi F. (2019). *Hydraulic Fracturing in Unconventional Reservoirs: Theory, Operations, and Economic Analysis*, 2nd ed. Elsevier.

Brown VJ (2014) Radionuclides in fracking wastewater: managing a toxic blend. *Environmental Health Perspectives* 122:A50-A55.

Cabot Oil and Gas. (2017). Q2 earnings call transcript.

Carter KM, Harper JA, Schmid KW, and Kostelnik J. (2011). Unconventional natural gas resources in Pennsylvania: The backstory of the modern Marcellus Shale play. *Environmental Geosciences*. 18: 217–257.

Chapter 32 title 58 Oil and Gas. (2020). www.legis.state.pa.us/WU01/LI/LI/CT/HTM/58/00.032.HTM. Retrieved August 6, 2020.

Chesapeake Energy. (2009). Operational release 7-30-09.

Colborn T, Kwiatkowski C, Schultz K, and Bachran M. (2011). Natural gas operations from a public health perspective. *Human Ecological Risk Assessment Journal*. 17: 1039–1056.

Eaton D. (2022). Seismicity induced by the development of unconventional oil and gas resources. In Stolz JF, Griffin WM, and Bain DJ (eds.) *Environmental Impacts from the Development of Unconventional Oil and Gas Reserves*. Cambridge University Press.

Engelder T. (2009). Marcellus 2008: Report card on the breakout year for gas production in the Appalachian Basin: Fort Worth Basin Oil and Gas Magazine, Issue 20, pp. 18–22.

Evans K, Holzhausen G, and Wood GM. (1982). The geometry of large-scale nitrogen gas hydraulic fracture formed in Devonian Shale: An example of fracture mapping with tiltmeters. *Society of Petroleum Engineers Journal*. 22: 755–763.

Federal Energy Regulatory Commission (FERC). (2020). Landowner topics of interest. www.ferc.gov/industries-data/natural-gas/landowner-topics-interest. Retrieved August 3, 2020.

FracFocus Chemical Disclosure Registry. (2020). www.fracfocus.org/index.php?p=learn/what-is-fracturing-fluid-made-of. Retrieved July 31, 2020.

George E. (2016). *Fracking 101: A Beginners Guide to Hydraulic Fracturing*. Q Press Publishing.

Gleason S. (1934). Slanted oil wells. *Popular Science Monthly*. 124(5): 40–41. www.drillingformulas.com/what-are-positive-displacement-mud-motors-in-drilling-for-oil-and-gas/ Retrieved July 31, 2020.

Governor's Center for Local Government Services. (2001). Special exceptions, conditional uses and variances. Planning Series #7, Eight edition, Pennsylvania Department of Community and Economic Development.

Graham J and Rupp J. (2022). Governance of fracking: Trends and challenges. In Stolz JF, Griffin WM, and Bain DJ (eds.) *Environmental Impacts from the Development of Unconventional Oil and Gas Reserves*. Cambridge University Press.

Groat CG and Grimshaw TW. (2012). *Fact-Based Regulation for Environmental Protection in Shale Gas Development*. The Energy Institute of The University of Texas at Austin.

Groundwater Protection Council (GWPC). (2009). *Modern Shale Gas Development in the United States: A Primer, prepared for the U.S. Department of Energy*. National Energy Technology Laboratory (NETL).

Halliburton. (2011). Tilt fracture mapping, measuring fracture-induced rock deformation to provide downhole clarity. Pinnacle Technologies Bulletin.

Horwitt D. (2016). Toxic secrets: Companies exploit weak US chemical rules to hide fracking risks. Partnership for Policy Integrity..

Hughes JD. (2013). *Drill, Baby, Drill: Can Unconventional Fuels Usher in a New Era of Energy Abundance*. Post Carbon Institute.

Hughes JD. (2019). *Shale Reality Check 2019: Drilling into the U.S. Government Optimistic Forecasts for Shale and Tight Oil Production through 2050*. Post Carbon Institute.

Jeffrey C and Creegan A. (2020). Adaptive drilling application uses AI to enhance on-bottom drilling performance. *Journal of Petroleum Technology*. 72(8) https://pubs.spe.org/en/jpt/jpt-article-detail/?art=7392

Johnson N, Gagnolet T, Ralls R, and Stevens J. (2011). Pennsylvania energy impacts assessment report 2: natural gas pipelines. The Nature Conservancy.

Lampe DJ and Stolz JF. (2015). Current perspectives on unconventional shale gas extraction in the Appalachian Basin.

Manning FS and Thompson RE. (1991). *Oilfield Processing of Petroleum Volume 1: Natural Gas*. Pennwell Publishing Company.

Mayerhofer MJ and Lolon EP. (2006). *Integration of Microseismic Fracture Mapping Results with Numerical Fracture Network Production Modeling in the Barnett Shale*. Society of Petroleum Engineers.

McCarron G and Dougall S. (2022). An overview of unconventional gas extraction in Australia. In Stolz JF, Griffin WM, and Bain DJ (eds.) *Environmental Impacts from the Development of Unconventional Oil and Gas Reserves*. Cambridge University Press.

Milici RC and Swezey CS. (2006). Assessment of Appalachian Basin oil and gas resources: Devonian shale-Middle and Upper Paleozoic Total Petroleum System. Open-File Report Series 2006-1237. United States Geological Survey.

Mosquera Netzkarsch. (2016). *Impact on the Air Quality of the new Shell Ethane Cracker Plant in Beaver County*. Capstone Report, Center for Environmental Research and Education, Duquesne University.

National Academies of Sciences, Engineering, and Medicine. (2016). *Flowback and Produced Waters: Opportunities and Challenges for Innovation: Proceedings of a Workshop*. The National Academies Press. doi:org/10.17226/24620

National Academies of Sciences, Engineering, and Medicine. (2018a). *Onshore Unconventional Hydrocarbon Development: Legacy Issues and Innovations in Managing Risk-Day 1: Proceedings of a Workshop*. The National Academies Press. doi:org/10.17226/25067.

National Academies of Sciences, Engineering, and Medicine. (2018b). *Onshore Unconventional Hydrocarbon Development: Induced Seismicity and Innovations in Managing Risk-Day 2: Proceedings of a Workshop*. The National Academies Press. doi:org/10.17226/25083

National Petroleum Council. (1980). *Unconventional Gas Sources, Volume III: Devonian Shale: Report Prepared for the U.S. Secretary of Energy*, June, 1980, National Petroleum Council.

NaturalGas.org. (2013). Processing Natural Gas. http://naturalgas.org/naturalgas/processing-ng/ Retrieved August 3, 2020.

New York State Department of Environmental Conservation (NYSDEC). (2011). Revised Draft General Environmental Impact Statement (SGEIS) on the Oil, Gas, and Solution Mining Regulatory Program: Well Permit Issuance for Horizontal Drilling and High-Volume Hydraulic Fracturing to Develop the Marcellus Shale and Other Low-Permeability Gas Reservoirs.

Nixon S. (2022). The microbiology of shale gas extraction. In Stolz JF, Griffin WM, and Bain DJ (eds.) *Environmental Impacts from the Development of Unconventional Oil and Gas Reserves*. Cambridge University Press.

Pennsylvania Oil and Gas Lease Act. (1979). www.legis.state.pa.us/WU01/LI/LI/US/HTM/1979/0/0060..HTM Retrieved June 2, 2020.

Pennsylvania Oil and Gas Act. (2012) (Act 13). www.legis.state.pa.us/cfdocs/legis/li/uconsCheck.cfm?yr=2012&sessInd=0&act=13 Retrieved June 2, 2020.

Pratt D. (2019). *Unconventional Gas Extraction in Southwestern Pennsylvania: Production, Water Quality, and Complaints*. Master's thesis, Duquesne University.

Presto AA and Li X. (2022). Air quality. In Stolz JF, Griffin WM, and Bain DJ (eds.) *Environmental Impacts from the Development of Unconventional Oil and Gas Reserves*. Cambridge University Press.

Shao H, Kabilan S, Stephens S, Suresh N, Beck AN, Varga T, Martin PF, Kuprat A, Jung HB, Um W, Bonneville A, Heldebrant DJ, Carroll KC, Moore J, and Fernandez CA. (2015). Environmentally friendly, rheoreversible, hydraulic-fracturing fluids for enhanced geothermal systems. *Geothermics*. 58: 22–31.

Sharpe E. (2014). How oil drilling works. *Energy Ink Magazine*. Spring 2014. https://energyink.us/4-Issue-Spring-14xC.html. Retrieved August 18, 2020.

Soeder DJ. (1988). Porosity and permeability of eastern Devonian gas shale, Society of Petroleum Engineers Formation Evaluation. *Society of Petroleum Engineers*. 3: 116–124.

Stewart BW and Capo RC. (2022). Metal isotope signatures for unconventional oil and gas fluids. In Stolz JF, Griffin WM, and Bain DJ (eds.) *Environmental Impacts from the Development of Unconventional Oil and Gas Reserves*. Cambridge University Press.

Swindell GS. (2018). Estimated ultimate recovery (EUR) study of 5,000 Marcellus Shale wells in Pennsylvania. www.gswindell.com/marcellus_eur.htm. Retrieved August 3, 2020.

Townsend-Small A. (2022). Isotopes as tracers of atmospheric and groundwater methane sources. In Stolz JF, Griffin WM, and Bain DJ (eds.) *Environmental Impacts from the Development of Unconventional Oil and Gas Reserves*. Cambridge University Press.

U.S. Energy Information Administration (EIA). (2013). U.S. shale gas production. www.eia.gov/naturalgas/. Retrieved August 3, 2020.

U.S. EIA. (2016). U.S. Map, lower 48 states shale plays. www.eia.gov/maps/images/shale_gas_lower48.jpg. Retrieved August 3, 2020.

US EIA. (2019). The United States expands its role as world's leading ethane exporter. www.eia.gov/todayinenergy/detail.php?id=38232 Accessed June 2, 2020.

US EIA. (2020). Natural Gas Annual, September 2020.

U.S. Department of Transportation. (2020). PHMSA, FERC sign memorandum of understanding to strengthen safety review and permitting process for LNG facility proposals. www.phmsa.dot.gov/news/phmsa-ferc-sign-memorandum-understanding-strengthen-safety-review-and-permitting-process-lng-0. Retrieved August 3, 2020.

U.S. Environmental Protection Agency. (2016). Hydraulic fracturing for oil and gas: Impacts from the hydraulic fracturing water cycle on drinking water resources in the United States (Final Report). Report EPA/600/R-16/236F.

U.S. Public Law 109-58. Energy Policy Act of 2005, 109th Congress www.govinfo.gov/content/pkg/PLAW-109publ58/pdf/PLAW-109publ58.pdf Retrieved June 2, 2020.

Veatch RW, King GE, and Holditch SA. (2017). *Essentials of Hydraulic Fracturing*. PennWell Books.

Warner N, Ajemigbitse MA, Pankratz K, and McDevitt B. (2022). NORMS. In Stolz JF, Griffin WM, and Bain DJ (eds.) *Environmental Impacts from the Development of Unconventional Oil and Gas Reserves*. Cambridge University Press.

Wells BA. (2020). Shooters – a "fracking" history. American Oil & Gas Historical Society. URL: https://aoghs.org/technology/hydraulic-fracturing. Retrieved August 3, 2020.

Wilson JM and VanBriesen JM. (2022). Waste usage and management. In Stolz JF, Griffin WM, and Bain DJ (eds.) *Environmental Impacts from the Development of Unconventional Oil and Gas Reserves*. Cambridge University Press.

Yan H, Tian L, Feng R, Mitri H, Chen J, He K, Zhang Y, Yang S, and Xu Z. (2020). Liquid nitrogen waterless fracking for the environmental protection of arid areas during unconventional resource extraction. *Science of the Total Environment*. 721: 137719, https://doi.org/10.1016/j.scitotenv.2020.137719

Zammerilli A, Murray RC, Davis T, and Littlefield J. (2014). Environmental impacts of unconventional natural gas development and production. U.S. Department of Energy, National Energy Technology Laboratory. DOE/NETL-2014/1651.

3

An Overview of Unconventional Gas Extraction in Australia

The First Decade

GERALYN MCCARRON AND SHAY DOUGALL

3.1 Introduction

A conundrum exists regarding the unconventional gas (UG) industry in Australia that results in a series of paradoxes. Australia, now vying with Qatar to be one of the biggest producers and exporters of liquefied natural gas (LNG) in the world, has a perceived problem with domestic gas supply and pricing to the extent that it faces the prospect of building five terminals to reimport its own gas. The lauded "once in a generation" opportunity to extract Australia's resource wealth (QLD Gov 2010b) has resulted in miniscule royalties for the Australian people. UG, promoted as an active strategy to combat climate change, has resulted in rising annual Australian emissions for the past five years. (Climate Action Tracker, Hannan 2019). In Australia, UG is vigorously promoted in some states and banned in others. In this chapter, we chronicle and rationalize the development of the UG industry in Australia, focusing on the specifically Australian experience with unconventional gas (parallel to the USA experience) and the multidimensional challenges across the spectrum for the government, domestic end users, and those required to host the burgeoning export industry.

3.1.1 Scale of the Unconventional Gas Industry in Australia

Australia is an ancient and geologically complex continent with coal seam gas (CSG) reserves (coal bed methane) dominating the geology of the eastern states, notably the Bowen and Surat basins in Queensland (QLD) and the Gunnedah and Sydney Basins in New South Wales (NSW) (Figure 3.1). Conversely, shale gas reserves are predominant in Western and Central Australia with prospective shale reserves in the Canning and Perth Basins of WA, the Amadeus, Georgina, and Beetaloo sub-basins of the Northern Territory (NT), and the Cooper-Eromanga Basin, which spans South Australia (SA), NT, and Qld (Hunter 2016) (Figure 3.1). In contrast to the United States of America, where shale gas has been the dominant form of unconventional gas exploration and production, the focus in Australia to date has been liquefied natural gas from coal seams (Figure 3.1). Currently, most of the unconventional gas produced in Australia is from Queensland, with remarkably little consumed domestically. Most of the gas is exported. (Table 3.1)

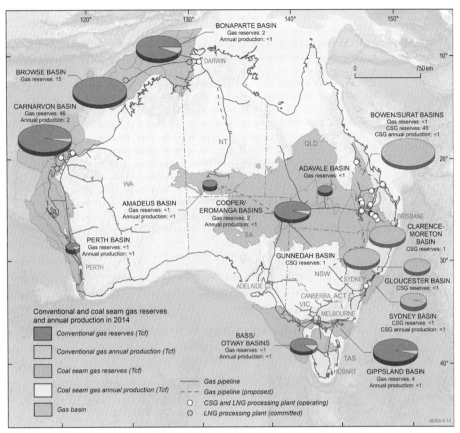

Figure 3.1 Estimates of conventional and unconventional gas reserves, and production in 2014 (tcf) (A black and white version of this figure will appear in some formats. For the colour version, refer to the plate section.)
(Geoscience Australia, 2019)

Australia became the largest exporter of LNG in the world for the first time in November 2018 and will likely continue to vie for a position in the top three with Qatar and the United States into the near future (Jaganathan 2018; AERA 2019). Producers advise that rates of drilling in CSG fields continue to increase, while production rates hold steady. Increasing exploration and production are required to maintain export targets with intense speculation on the prospect of shale and tight gas development across Australia. Figure 3.2 details prospective unconventional gas resources, including shale, tight gas, and CSG, along with prospective conventional resources, while Figure 3.3 demonstrates the extensive industry footprint in terms of prospective petroleum production and exploration tenements.

3.2 Politics, Policy, and Promises of UG in Australia

The gas industry in Queensland has enjoyed long-term bipartisan political support. In parallel with the nascent shale gas industry in the United States, from the early 2000s there

Table 3.1. *Percentage splits of gas consumption by sector (not including UAFG[1]) in 2018 (AEMO 2019)*

Regional	Residential/ commercial (%)	Industrial (%)	GPG (%)	LNG (%)	Regional gas consumption[2]
Queensland	<1	7	2	90	1,380PJ
New South Wales	42	48	10	0	116PJ
South Australia	12	27	62	0	93PJ
Tasmania	8	51	41	0	10PJ
Victoria	58	31	11	0	212PJ
Total	10	14	7	68	1,811PJ

[1] UAFG, or Unaccounted for Gas, refers to metered gas that enters the gas network but does not reach consumers.

[2] Consumption totals are based on metered load for the full 2018 calendar year, except for transmission-connected industrial consumers, based on survey (estimated for October–December 2018) and GPG, based on estimated gas usage for each plant's metered electricity generation into the NEM.

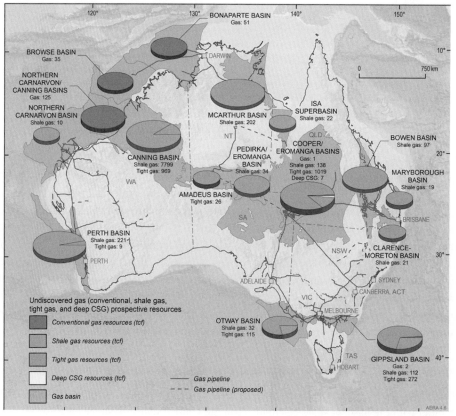

Figure 3.2 Prospective conventional and unconventional gas resources (tcf) (A black and white version of this figure will appear in some formats. For the colour version, refer to the plate section.)

(Geoscience Australia, 2018)

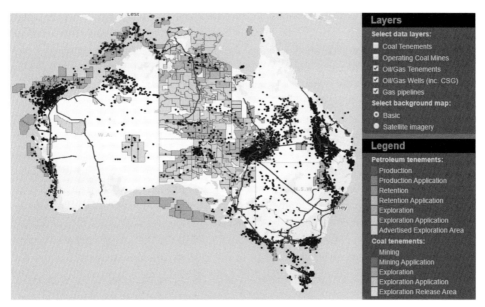

Figure 3.3 Oil and gas tenements (areas with a permit or authority from the government and under the relevant legislation to allow exploration or production of oil and gas) (A black and white version of this figure will appear in some formats. For the colour version, refer to the plate section.)
(Energy Resource Insights, 2020)

was significant political interest in the small-scale commercial exploitation of methane produced from coal seams in Queensland's Surat Basin (Hunter 2016). In 2000, as part of their "Smart State Policy," and as a proposed means of actively reducing emissions in order to combat anthropogenic climate change, the Queensland government established energy policy requiring electrical retailers to source 13% of their electricity from gas-fired generators (Queensland Government 2007). This created a fixed gas demand supporting the fledgling industry (Drinkwater 2015).

By 2010 four consortia were actively investigating the export potential of Queensland CSG as LNG.

- Queensland Curtis LNG Consortium (QCLNG), British Gas/Queensland Gas Company (QGC)
- Gladstone LNG (GLNG) (Santos/Petronas/Total/Kogas)
- Asia Pacific LNG (APLNG) (Origin/ConocoPhillips/Sinopec)
- Arrow LNG (Arrow Energy/PetroChina/Shell)

In a critical point in the genesis of the LNG export industry in Australia, on March 24, 2010 in Beijing, British Gas entered into a contract with the China National Offshore Oil Corporation to export 72 million tonnes of coal seam liquefied natural gas from the Surat Basin over the next 20 years (Hansard a, Hansard b). This contract preceded any rights by British Gas to the actual resource, as they did not have a production license, and environmental assessment of the risks of production had not taken place.

Since the early 1990s the Queensland Government had a firm strategy for ecologically sustainable development. These principles included integrated and long-term decision making, inter- and intra-generational equity, the precautionary principle, conserving biological diversity and ecological integrity, and internalizing environmental costs (QLD Gov Regional Policies). However, the development of the UG industry challenged fundamental elements of these principles. Overt political pressure was exerted by British Gas to ensure that decisions were made before their June 2010 board meeting, otherwise "customers will begin to go away," as stated by QLD Director General Ian Fletcher, recounting a conversation with David Maxwell, Senior Vice President of QGC (Marsh 2018 re: Fletcher e-mail 2010). The precautionary approach was short circuited, and in order to ensure that commercial contracts were realized, the Queensland government acceded to an accelerated approvals process. This process significantly breached the requirements of *Environmental Protection Act 1994*, which mandated baselines and rigorous assessment of the cumulative impacts of development (Marsh 2016). On May 28, 2010 the first major coal seam gas to LNG project (GLNG) received approval from the Queensland Government (Santos) and within months two additional projects (QCLNG and APLNG) were approved (Queensland Government 2010a).

Subsequently, in lieu of mandated pre-development assessment and planning, the government embarked on a policy referred to as "adaptive management" (Queensland Government 2013a), though it will be demonstrated (in this chapter) that the initiative is adaptive in name only, and does not conform to international best practice. An expanding and cascading range of legislative and regulatory changes were enacted to enable the production required. This included the Mines Legislation (Streamlining) Amendment Bill 2012, which involved amendments to 16 pieces of legislation. Written public submission on the amendments were limited to a six-day period, notably held over a long weekend and during the 2012 Olympic Games. The Petroleum and Gas (Production and Safety) Act (2004) (PGPSA) has been subject to over 1,000 major amendments (Hunter 2016). Such repeated modification of the relevant legislation, coupled with actions to avoid public scrutiny, are a demonstration of the Queensland government's role in enabling rather than regulating the industries' development.

Concurrently with changes in legislation, the Queensland government created the Queensland Gas Scheme with the objective of creating a mature gas industry and encouraging new gas sources and infrastructure in the state. The Land Access Framework and Multiple Land Use Framework were established that "codified coexistence" between agricultural land uses and UG mining (Council of Australian Governance, COAG). The Gasfields Commission came into existence as a response to problems identified at the interface between rural landholders and the newly emerging coal seam gas industry. It was intended that the Commission would "*manage and improve the sustainable coexistence of landholders, regional communities and the onshore gas industry in Queensland*" (Scott 2016). Just three years later, in a report on the Management of the UG industry (Managing coal seam gas activities Report 12: 2019–2020), the Queensland Audit Office reported that the need for the GasFields Commission Queensland should be reconsidered, since many of its functions are duplicated elsewhere, and its one specific duty, which was to provide transparency and certainty that regulators of the UG industry are performing their roles effectively, was not being fulfilled and, further, the public doubt its independence and transparency, and the commission's reports provide no in-depth analysis, insights, advice, or recommendations.

In September 2012 the Queensland Government formed a special Resources Cabinet Committee under Deputy Premier Jeff Seeney (QLD Gov 2012). This Committee was tasked with identifying ways to fast-track investment in the resources sector, review existing policies to cut red tape, remove impediments to project assessment and approvals, mobilize resources to accommodate timely approvals, and ensure an integrated Queensland Government approval regime. Within six months the Committee had cut the terms of reference for Environmental Impact Statements from 100 to 25 pages, and introduced, with industry assistance, a standardized approvals process for issuing environmental authorities for petroleum exploration. The new approvals process reduced the number of conditions from 300 to 65 with the timeframe for issuing approvals cut from 18 months to 30 days. They lifted the moratorium on oil shale development and finalized amendments to the Lake Eyre Basin Wild Rivers Declaration to improve the efficiency of oil and gas operations. The Wild Rivers Declaration was a 2011 initiative by Queensland's government of the time intended to permanently protect the unique and spectacular environment of the Lake Eyre Basin (QLD Gov 2011).

Changes to the Petroleum and Gas (Production and Safety) Act 2004 gave the industry the right to unlimited take of ground water as a necessary activity in the process of extracting petroleum and gas. This water (altered through the extraction process) became an above-ground waste product of the process with no direct use. In order to facilitate the disposal of the vast amount of waste water, in 2013, environmental regulations were altered so that raw, treated, or amended water from CSG operations was no longer treated as a regulated waste and could be used under the CSG beneficial use policy[1] (QLD Gov 2013b; Qld Gov 2014a). Additionally, the definition of Category C Environmentally Sensitive areas and essential petroleum activities was "clarified," allowing essential petroleum activities to take place in 98% of the affected areas. Buffer zones were reduced from 300 m to 200 m. The Strategic Cropping Land Act 2011 was revised. The Nature Conservation Act 1992 (NCA) was amended to remove "red tape," which included removing the conservation of nature as the primary purpose of the NCA but added protecting prescribed persons from civil liability (QLD Gov 2013c).

In 2012–2013, the Australian Government Productivity Commission undertook a 12-month inquiry into the nonfinancial barriers to mineral and energy resource exploration in order to remove "red and green tape" that was discouraging interest in "the dash for gas" (AUS Gov 2013). Despite the Productivity Commission's legislative mandate to ensure the industry developed in a way that was ecologically sustainable (AUS Gov 1998), it ignored warnings from medical professionals, Doctors for the Environment (2013), that "current regulations may be insufficient … " In their July 2013 submission to the Productivity Commission, DEA stated, "The social impact assessment guidelines dated July 2013 are grossly inadequate. In effect, the Newman Government (the QLD state government at the time) is making their already inadequate health impact assessment process even more inadequate under the guise of cutting red tape, which does not exist." Furthermore, they stated "We are astonished and are concerned that the panel failed to understand the proper functioning of the environmental impact assessment (EIA) process and failed to recognize that the environment is inseparable from human health." Alternatively, the Industry lobby group the Australian Petroleum Production and Exploration Association (APPEA) noted it was pleased with the Australian Federal government's pledge to cut red tape, its dedicated sitting day to cut more than 8,000 regulations, and its rapid move to secure the agreement of all states and territories for one-stop shops for industry environmental approvals (APPEA 2014). This raft of changes removing environmental and health protections were accompanied by 14,000 public service jobs cuts (Hawthorne 2012), defunding of the Environmental Defenders Office (EDO 2013),[2] and abolition of the Climate Commission (Arup 2013).

3.2.1 Adaptive Environmental Management Regime

A major element of the regulatory changes meant that the Queensland Government now regulates unconventional gas (predominantly CSG) under its Adaptive Environmental Management Regime (AEMR). In its ideal implementation, "adaptive management" is the practice of incorporating recent information into an ongoing decision-making process, especially in environments of high uncertainty. As noted by senior lecturer in law Dr. Nicola Swayne, "a truly adaptive environmental management approach must be able to embrace the hard decisions that go with 'learning by doing' including the ultimate decision of ceasing CSG activities in Queensland in the face of significant information gaps and/or an unacceptably high risk of cumulative adverse impacts" (Swayne 2012).

According to the Environmental Defender's Office lawyers, the government's version of AEMR is a policy framework that is extremely unclear, and is not set out in legislation regulating unconventional gas (EDO 2016). As identified in analysis by Marsh, and documents presented by Marsh (Hansard C 2014; Marsh 2016; Marsh 2018), the industry and government were constrained by the original policy of environmental protection and Environmental Impact Statements. These policies required detailed modeling and cumulative impact assessment processes that the industry either could not (or would not) provide. As a concession, the industry would give "in principle" information as a basis for approval, with detailed data promised later (often well after the physical completion of the works). Some environmental experts, such as the Environmental Defenders Office and other academics, have identified this process as problematic given that it hampers risk assessment during the EIS process (particularly in relation to groundwater) with decisions deferred to post approval (EDO 2016), and prevents genuine community engagement in EIAs. Accompanying the AEMR, in 2013 the government department responsible for environmental protection partnered with the APPEA to develop streamlined model conditions (streamlined conditions) for the UG industry. This process was overtly aimed at red tape reduction benefits for the UG industry as well as reducing departmental decision-making timeframes (EDO 2016). The following schedules of conditions have been streamlined: General environmental protection; Waste management; Protecting acoustic values; Protecting air values; Protecting land values; Protecting biodiversity values; Protecting water values; and Rehabilitation. Deferring of the front-end risk assessment and community engagement processes (on what is seen as acceptable risk) also impacts landholders who are required to host the industry. From the beginning, the individual families who live with the immediate, short- and long-term impacts to their homes, businesses, and natural resources have not been included or consulted as stakeholders, making it very difficult for them to access meaningful recognition of issues or remedies.

Without a thorough EIA process, baseline studies, and the appropriate legal decision-making framework against which AEMR could be tested and amended (Swayne 2012), the AEMR policy has left regulators "virtually incapable of ceasing CSG activities in the face of significant uncertainty and high risks of cumulative environmental impacts"(Drinkwater 2015). This has resulted in a new industry with little to no conventional environmental management processes and very few checks and balances to address legacy issues that will continue well beyond the relatively short term of the industry's existence.

3.2.2 Coexistence and Conflict of Duties

One of the most significant impacts on landholders affected by CSG activities, and one of the issues addressed in the cascade of regulatory changes, is that of land access and compensation. Land access in Queensland in relation to resources and rural landholders has been described by K. Witt as a series of epochs, namely, pre-1957, "socialist" Queensland; 1957–1989, Queensland "countrymindedness"; 1989–2000, Queensland "reformed." and that a fourth epoch is underway as a result of "new values in coal seam gas in rural Queensland" (Witt 2012).

Under Australian law, the Crown retains rights to Petroleum and grants licenses for exploration and production over Crown Land, land held under native title, and privately held freehold land (Hunter 2016). This means the legal right of the petroleum titleholder creates conflict with the legal rights of the private landholder titleholder. These issues are exacerbated by the long duration of the impacts from the petroleum titleholder's activities. It is this fundamental issue that caused the government to develop The Land Access Code or to "codify *coexistence*" (Turton 2014). However, integral to the codification is the establishment and maintenance of good relationships, adequate consultation, and negotiation. This means that voluntary goodwill on the farmers' behalf is still a crucial part of managing that relationship. The Land Access Code only facilitates *the requirement* for landholders to allow UG operators access to their properties, and thereby fails to address the legal rights of the landholder, or an awareness of the limits of the petroleum title holders' rights (or, indeed, methods to enforce them). Because of this, a conflict is created with no remedy.

A consequence of active governmental support of the nascent commercial industry exploiting a crown-owned resource is the development of conflicts of duties. This refers to instances where one agency "promotes" the mining and resource industry, as well as permitting, licensing, regulating, and enforcing breaches by the industry. Conflicts of duties in the situation of required "coexistence" are highlighted by Huth et al. (2018) and Dougall (2019). The unconventional gas footprint is co-located on private property that creates a "shared space that is a farm business, a home and a resource extraction network."

The regulatory management of the industry is challenged in its ability to address the Workplace Health and Safety implications for the host farmer and the potential for UG companies to induce hazards, or compound existing ones, by operating amid the host farmers' workplace.

An example of the consequence of such a conflict of duties is clear when an analysis of the responsible department's prosecutions is undertaken. The main regulation under which the industry operates in Queensland is the Environmental Protection Act. An indicator of levels of enforcement can be gauged by assessment of the prosecutions publicly listed on the department's web site (QLD Gov prosecution bulletins). This shows a total of 79 Prosecutions between 2012–2019. Only five of the prosecutions relate to the UG industry, with the majority of those relating to the handling of waste, a point discussed further in Section 3.6.

3.3 Promised Economic Benefits

In tandem with the political drive behind the gas industry, and in order to gain public acceptance, government and industry stakeholders engaged in a sustained campaign to highlight the purported benefits of the industry to the wider community. Development of the resource was presented as a once in a generation opportunity with the catch phrase "Jobs and growth" (QLD Gov 2010b) and promised,

1. The gas industry would increase gas supplies, thereby making gas cheaper.
2. The increased gas production would act as a bridging fuel in the transition to renewables.
3. The industry would create more jobs.
4. The industry would boost government revenue.

To date the Australian experience has differed significantly, as reported in the Australia Institute's report Fracking the future (Grudnoff 2014) and the IEEFA update: the staggering cost of gas in Australia (Robertson 2019).

3.3.1 Make Gas Cheaper[3]

Despite now being one of the world's biggest producers, gas prices in Australia have tripled in recent years (Robertson 2019). There is no benchmark price, with the price on the East Coast of Australia being controlled by a small number of producers. Australian Energy Market Operator's media releases (AEMO 2017) warn Australia is facing a gas shortage with dire implications for general industry, with reports of manufacturers being offered short-term contracts for gas at a wholesale price of $20 per gigajoule, up from a historical stable average of $3–$4 (Chang 2017). Since peaking in 2017, prices have fallen to $8–$12/GJ but remain well above global parity. In Japan, consumers are paying less for Australian LNG than domestic Australian consumers (Chang 2017). The economic incentives are perverse to the extent that it is now economically feasible to reimport LNG from Japan that was previously shipped from Australia in the first place, with proposals for the imminent construction of no less than five import terminals (MacDonald-Smith 2019, Robertson 2019). Despite this, AEMO has predicted that such measures will not lower domestic gas prices (AEMO 2017).

3.3.2 Bridging Fuel

Australia has promoted the development of "natural gas" as a strategy to combat climate change. However, the stated intent of the industry as a "transition fuel" must be assessed in conjunction with consideration to the fugitive emissions for the industry, which is currently unmeasured in Australia (discussed in Section 3.7). The claimed advantage of burning gas over coal is that it is a cleaner burning fuel. However, the global warming potential of methane is 86 times that of CO_2 over a 20-year period, or 34 times that of CO_2 over 100 years. The climate impacts of gas depend not only on the amount of methane burned (and therefore converted into carbon dioxide) but also the amount of methane released directly into the atmosphere. Research by the University of Melbourne indicates that if just 3% of produced gas escapes into the atmosphere, the adverse impact on climate change is at least as bad as the climate impact of burning (and converting to CO_2) the remaining 97% (Lafleur 2016). At or above 3% of emissions, burning gas loses any climate advantage over burning coal. What is more, liquification of natural gas (i.e., LNG), which involves cooling the gas to -160 °C, is an extremely energy intensive process requiring about 8% of gas produced to be burned. With the positive promotion of fossil fuel production, and without an active plan to phase it out, consideration must be given to the compounding effect the gas industry has as a disruptive technology in the uptake of renewables.

3.3.3 More Jobs

With an Australian population of around 25 million, the Australian oil and gas industry employs 20,700 people. In Queensland, compared to other sectors of employment (and contrary to popular belief), the contribution of fossil fuel-related industries is seen to be relatively minor (see Figure 3.4). Australian Bureau of Statistics (ABS) data indicate that in Queensland in 2017/18 Oil and gas extraction (~10,000) and coal mining (~20,000) employed significantly less than Accommodation and Food Services (~200,000), Manufacturing (~170,000), or even Rental, Hiring, and Real Estate Services (~40,000). Additionally (Figure 3.5), the "spillover" effects to the job market (i.e., other jobs created to support the additional oil and gas jobs) are largely offset by the reduced agricultural employment (minus 1.8 jobs) per CSG job, with retail and manufacturing jobs remaining flat (Fleming and Measham 2015).

3.3.4 Boost to Government Revenue

With regard to increasing government revenue, in Queensland, the epicenter of Australia's LNG industry to date, the State Government received a *total* of $418 million in royalties between 2010 and 2017 (The Gasfields Commission 2018). In contrast, State revenue from

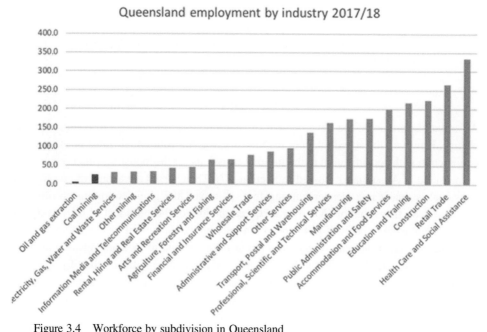

Figure 3.4 Workforce by subdivision in Queensland
(Australian Bureau of Statistics, 2018)

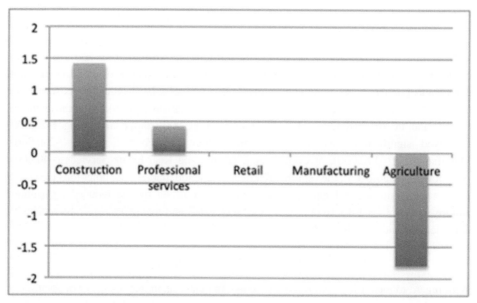

Figure 3.5 Spillover job impacts per coal seam gas (CSG) job
(data sourced from Ogge, 2015)

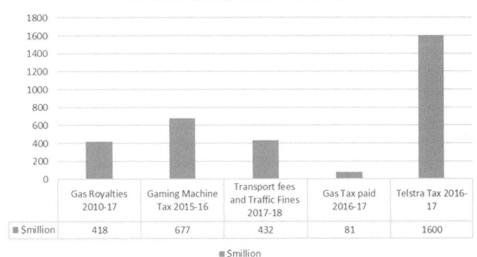

Figure 3.6 Revenues created across industries
(data sourced from Chang, 2019)

what might be considered a relatively trivial source, gaming machine tax, was $677 million in the *single* fiscal year 2015–2016 (QLD Gov Budget a). State Budget papers reveal that in 2017–2018 royalties from the CSG industry were a mere 29% of predicted returns, coming in at $187 million instead of the predicted $636 million projected in 2014/15 (QLD Gov Budget b). Again, in contrast, transport fees and traffic fines added $432 million to State revenue in 2017–2018 (Figure 3.6). Attempts to lift the state's royalty on gas from 10% to 12.5% of production measured at market value predictably resulted in the gas giants claiming "the shock move puts in doubt future investment in the state and threatens the competitiveness of the industry," while they also mounted legal challenges (Ludlow 2019; Toscano 2019). According to data from the federal Australian Taxation Office, Australian telecommunications company Telstra, with revenue of $26.9 billion, paid $1.6 billion in tax in 2016/17. With similar revenue, the oil and gas industry paid just $81 million in tax.

3.4 Inquiries and Research

Public concern about the health and environmental consequences of the "dash for gas" has resulted in numerous public inquiries and reviews both federally and at state level with in some cases moratoria and bans (Table 3.2).

The Federal Senate Select Committee on Unconventional Gas Mining (2016), which took detailed testimony from the Queensland gas fields, lapsed following the double dissolution of federal parliament, and was not reinstated. The wide-ranging recommendations made in the Committee's interim report were not enacted and the unmitigated expansion of the gas industry in Queensland has continued. Internationally, research into

Table 3.2 *Inquiries and status of moratorium on UG industry in Australia*

Commonwealth	State	Moratorium
Senate Rural Affairs and Transport References Committee, Management of the Murray Darling Basin Interim report: The impact of mining coal seam gas on the management of the Murray Darling Basin (2011).	**New South Wales**[1] 2012 inquiry into coal seam gas. 2014, the NSW Chief Scientist and Engineer, Professor Mary O'Kane, conducted an independent review of CSG activities.	
Standing Council on Energy and Resources (now COAG Energy Council), National Harmonised Regulatory Framework for Natural Gas from Coal Seams (2013).	2014, Mr. Bret Walker SC completed an independent review of the process for arbitrating land access arrangements for mining and petroleum exploration.	
Productivity Commission, Mineral and Energy Resource Exploration (2014).	**Victoria**[2] 2015, inquiry into unconventional gas.	Permanently ban all onshore unconventional gas exploration and development, including fracking and coal seam gas (CSG).
Senate Select Committee into Certain Aspects of Queensland Government Administration related to Commonwealth Government Affairs (2015).	2013, the Hon Peter Reith AM chaired a Victorian Gas Market Taskforce inquiry that considered gas supply issues.	Conventional gas exploration will be permitted in Victoria from July 2021. But fracking and coal seam gas drilling will be permanently banned.
EPBC Water Trigger Review.	2012, an inquiry into greenfields mineral exploration and project development in Victoria.	
	Queensland 2014, the Queensland Competition Authority has reviewed the regulation of the CSG industry.	
	Western Australia[3] 2013 the implications for Western Australia of hydraulic fracturing for unconventional gas.	Fracking will not be permitted over 98 percent of Western Australia.
	South Australia 2015, an inquiry into the potential risks and impacts in the use of fracking to produce gas.	Ban on fracking in the South-West, Peel and Perth metro area remains. Broome and the Dampier Peninsula will be protected.

Table 3.2 (*cont.*)

Commonwealth	State	Moratorium
	2015, a review of hydraulic fracturing.	Moratorium on hydraulic fracturing lifted on existing petroleum titles. traditional owners and farmers will have the right to say no to oil and gas production from fracking on their land.
	Northern Territory[4] 2014, inquiry into hydraulic fracturing. 2016 the independent Scientific Inquiry into Hydraulic Fracturing of Onshore Unconventional Reservoirs.	Following a separate scientific inquiry, the government lifted its own moratorium on 51% of the territory, with the remaining 49% still covered by the ban.
	Tasmania[5]	Moratorium on fracking until 2025.
	South Australia[6]	10-year fracking ban across the agricultural-rich Limestone Coast region in the south-east of the state.

[1] The NSW Chief Scientist and Engineer's 2014 Inquiry, tasked with identifying and managing the risks of UG development in the State, concluded that the technical challenges and risks could be managed through the implementation of 16 recommendations. The NSW Government welcomed the Chief Scientist Report, supporting all recommendation. However, a follow up Inquiry five years later found that only two of the 16 recommendations had been implemented. In the interim, popular resistance across NSW has resulted in early decommissioning of planned projects and buyback of petroleum licenses. By 2020 the area of the state covered by petroleum licenses had reduced from 60% to 7% (Hansard NSW). Significantly for gas major Santos, who keenly continue to pursue their quest for a production license in the NSW Pilliga State forest, the 7% of exploration licenses still in place would not be subjected retrospectively to the consideration of economic, environmental, and social factors that could rule them off limits for CSG (Hansard NSW).

[2] With ongoing resistance to unconventional gas in Victoria, a clear lack of a social license (Luke 2018), and a wish to protect Victoria's "clean, green" agricultural reputation (VIC Gov), moratoria on fracking in new CSG exploration wells have been in place since 2012, with a legislated permanent ban on all onshore unconventional gas exploration and development in 2017 (the Resources Legislation Amendment (Fracking Ban) Act 2017 (Vic)).

[3] The Western Australia Standing Committee of Environment and Public Health Affairs (2015), while finding community concerns justified, considered the risks manageable. In November 2018, the moratorium in Western Australia was lifted, permitting fracking on existing petroleum leases covering 50,000 square kilometers. The seven LNG facilities currently operating in WA include three LNG facilities onshore on the Pilbara coast, the massive Gorgon plant on Barrow Island, and a floating LNG facility (which is the largest in the world) (*The Straits Time* 2018).

Table 3.2 (cont.)

[4] Despite widespread community opposition and a clear lack of social license (Luke 2018), at the end of the Scientific Inquiry into Hydraulic Fracturing in the Northern Territory (2018), Justice Pepper concluded that the risks could be managed if all 135 recommendations were adopted and implemented. The Moratorium in place was lifted with a commitment by the State Government to adopt all recommendations. During Senate Estimate hearings in 2019, it was disclosed that the State government had already rescinded on their commitment (McLennan 2019).

[5] In Tasmania the Government has extended a moratorium on the use of fracking for the purposes of hydrocarbon resource extraction, e.g., shale gas and petroleum, for a further five years, until March 2025.

[6] The 2015 Inquiry by the South Australian government concluded that the UG industry could be developed in the southeast of South Australia "without significant or unacceptable controlled risks" and would deliver "significant benefits" to the wider community (Luke 2018). The "fierce resistance" (Aston 2017) to the UG industry in SA (where in 2017–2018 the gross food and wine revenue totaled $20.3 billion) (Primary Industries 2019) resulted in the sitting state government offering royalty sharing with farmers. However, the government lost the 2018 State Election to an opposition promising a 10-year ban on UG: a ban subsequently limited to seven of SA's 68 council areas. (South Australia Petroleum and Geothermal Energy (Ban on Hydraulic Fracturing) Amendment Bill 2018).

shale gas and its various environmental and health impacts has been prolific in recent years. In Australia, despite the documented and profound public concerns, academic research into the impacts of CSG, particularly the health impacts, has been remarkably sparse.

It is generally accepted that there are some differences between coal seam gas (CSG) and shale gas extraction. CSG is extracted at relatively shallow depths compared to shale gas. With CSG the initial means of stimulation after drilling to the source rock is dewatering the coal seam aquifer, whereas with shale gas, hydraulic fracturing is always required. However, when the flow of CSG inevitably reduces, the secondary method of stimulation is hydraulic fracturing and/or infilling with more wells (Arrow Bowen Gas Project EIS). Similarities between shale and CSG extraction include the massive environmental footprint and industrialization of the landscape, with the extensive range of supporting infrastructure (e.g., pipelines, compressor stations, power stations, water treatment plants, liquification plants, export ports). Similarities also include chemicals used (surfactants, foaming agents, biocides) and exhumed (petroleum products, radioactive material etc.), water used, the waste stream created, fugitive emissions, and reported health impacts.

Proponents of the gas industry in Australia historically have used the differences in geology and method to dismiss the outcomes of international gas research as irrelevant to the Australian CSG experience. However, this has not led to equivalent research being undertaken in Australia. Funded by industry and government, the Commonwealth Scientific and Industrial Research Organisation's (CSIRO) subsidiary group, the Gas Industry Social and Environmental Research Alliance (GISERA) along with the University of Queensland's Centre for Coal Seam Gas, are the main groups producing research output regarding the Australian gas industry. Despite early and ongoing reports of health impacts (McCarron 2013), which parallel the US experience, major data gaps in health information remain. Werner et al. (2015) highlighted the absence of methodologically rigorous studies and stressed "absence of evidence does not mean evidence of absence." Claudio et al. (2018) noted that the Queensland Health Report (QLD Gov 2013d) was a missed opportunity resulting in an ongoing lack of adequate data as it failed to meet Health Impact Assessment international best practice because seven of nine key steps were omitted. As confirmed by Keywood et al. (2018), "An in-depth health impact study is yet to be conducted in an Australian CSG region."

A limited number of studies have been published on physical health impacts. Werner et al. (2016) found that certain hospital admission rates (neoplasms and blood/immune diseases) increased more quickly in the CSG area than in the other study area after adjusting for key sociodemographic factors. Further research by Werner et al. (2018) into hospitalization of children suggests potential age-specific health impacts, with the strongest associations found for respiratory disease and blood/immune disorders. The researchers noted strong associations with respiratory disease up to the age of 14, with a 7–11% increase in hospitalization for respiratory disease in very young children (0–4 years). Disturbingly, they documented a 467% increase in the prevalence of blood/immune diseases in the 5–9 year-old age group, compared to children in areas without CSG activity.

Research into the mental health impacts of coexisting with the gas industry has identified impacts on mental health and wellbeing. Walton (2014) documented that 48.5% of gas field residents surveyed reported that their community was "only just coping," "not coping," or resisting the industry. Morgan et al. (2016) found "clinically significant levels of psychological morbidity" in some farmers.

In contrast to the dearth of research into physical and mental health impacts and the environmental determinants of health, the body of literature on social impacts and "social license to operate" has been growing steadily in Australia and has been recently reviewed by Luke et al. (Luke 2018). Funding for research into unconventional gas has been focused on industry engagement approaches, with a number of academics involved in analyzing the application of the concept of "social license to operate" to gas and energy developments. (Luke 2018). In the conclusion to their overview of the UG industry through the lens of the "social license to operate," Luke et al. note "many unanswered questions" as to "which socioeconomic, political, and cultural factors" result in swift enforcement of bans in some states and not in others. Luke queries the geospatial patterns of UG development and to what extent "do they follow the contours of prevailing power dynamics in Australia's political economy?"

As discussed later, the Office of Groundwater Impact Assessment (OGIA), an independent entity established under the Water Act 2000, undertakes evidence-based independent scientific assessments of groundwater impacts from resource operations in the Surat Basin, and produced Underground Water Impact Reports (UWIR) in 2012, 2016, and 2019. The work of OGIA is the exception rather than the rule in Queensland. Whereas monitoring and reporting of real-time outcomes of the industry impacts in NSW have resulted in the activities being terminated, for example, AGL's irrigation trial with produced water (Casson 2015), in Queensland the same type of monitoring has not occurred, and the activities have continued unabated. Notable by its absence in Queensland is any program of seismic monitoring similar to that applied to Cuadrilla's site in Lancashire, England, which resulted in termination of the project (Gosden 2019).

3.5 Water Usage

Australia is the driest inhabited continent on earth (AUS Gov environment). The impact of the gas industry on water resources has been and remains a major community concern, and has been the focus of multiple inquiries, including the Scientific Inquiry into Hydraulic Fracturing in the Northern Territory (2018), Western Australian Standing Committee of Environment and Public Health Affairs (2015), NSW Chief Scientist and Engineer (2014), and the Senate Select Committee on Unconventional Gas Mining (Bender Inquiry) (2016).

CSG production requires depressurizing the coal seams, which necessitates dewatering the aquifer that the coal seam lies within. The amount of water removed from the aquifer by the CSG companies in order to do this is related to the permeability of the coal seam as well as the connectivity of the target aquifer with other aquifers above and below the targeted area. The Walloons aquifer, the main target aquifer for the current Queensland CSG activity, lies within the Great Artesian Basin. The Walloons aquifer and other aquifers within the GAB provide domestic and stock water supplies for many towns and farming and grazing operations across the state that are otherwise low rainfall and drought prone areas (Towler et al. 2016).

Under the Queensland Water Act 2000, all rights to the use, flow, and control of all water in Queensland are vested in the State (Qld Gov 2000). The State may authorize persons to take water through legislation and statutory instruments, water allocations, licenses, and permits. However, as previously discussed, the extractive gas industry has a statutory right to unlimited amounts of "associated water" from coal seams during the course of its normal operations. This statutory right to take water is outside the State's water licensing requirements, which apply to other users.

In Queensland's Surat and Southern Bowen Basin's Cumulative Management Area (CMA) the cumulative groundwater impacts from petroleum and gas (P&G) development have been monitored and modeled by OGIA since 2011. In their UWIR, OGIA report that extraction of groundwater by the CSG industry has averaged 60,000 ML annually, while extraction by all other users (stock and domestic, agriculture, irrigation, town water supply, industrial) is estimated at 164,000 ML/year. Currently, OGIA predicts 571 privately operated and licensed stock and domestic water bores[4] to be no longer available owing to CSG activity in the long term, an increase of 10% on 2016 predictions (UWIR 2018). OGIA reports that "in some areas water levels have declined up to 250 m in the immediate vicinity of CSG production." OGIA predicts that eight groups of springs will be impacted by a drop of pressure of more than 0.2 m in their source aquifers. OGIA's long-term monitoring of springs notes changes at many locations, including "dead trees, salt scalded soil and spring vents that have stopped flowing." OGIA states that "groundwater levels within CSG production areas may take more than 1,000 years to fully recover." Interestingly, extraction of gas and water has been linked to subsidence; however, currently

Figure 3.7 QGC Kenya Water Treatment Facility Storage Ponds (Wieambilla, Queensland), (Dean Draper, used with permission)

there are very limited publicly available subsidence data for Australian coal seam gas developments, though subsidence monitoring is widely proposed for Australian coal seam gas developments (IESC 2014).

In addition to the removal of the existing water resource, the industry poses some risks to the contamination of water resources. Contamination can occur through surface spills and discharge of operationally used and produced chemicals, fluids, and produced water. Gas migration from the depressurized coal seam into aquifers may unintentionally stimulate gas flow in other domestic water reservoirs as a result of the extraction activities (Towler et al. 2016).

While large-scale evaporation ponds (Figure 3.7) were nominally banned in Queensland with directions to be remediated by 2011, in practice they still exist (QLD Gov 2008). However, the Queensland State government requires the CSG companies to, where possible, put the vast amounts of "produced water" from the dewatering activities to some "beneficial" use. "Beneficial" use of this water is constrained by the salt content and endogenous toxins, as well as introduced contaminants, which require the water to be treated before use. As a result of further treatment or evaporation, large volumes of salt and other concentrated secondary waste products are produced. While details on the quantities of these waste products are not available in any form other than widely varying estimates, there is a local development approval in place for the town of Chinchilla that details a waste facility with permission to receive 4.5 million tonnes of Product Salt from the reverse osmosis water treatment plant at 900 000 tonnes per annum for 5 years, and an additional 900,000 tonnes of waste landfill salt at 45,000 tonnes per annum for 20 years (Discussed further in Industry Waste and Conflict of Duties).

3.6 Industry Waste

The report prepared for the Department of Environment and Energy, Hazardous Waste in Australia (2017), describes Queensland as a *"unique hazardous waste jurisdiction."* The coal seam gas industry *"provides enormous waste management challenges not present in other states and territories"* with the CSG industry making up close to 20% of *"apparent waste generation"* in Queensland. These figures, however, only include waste that has been subject to hazardous waste tracking. CSG waste waters contain a combination of geogenic and introduced chemicals including hydrocarbons, heavy metals, radioactive elements, drilling/"fracking" chemicals, and salts present in the formation water and drilling muds. There are significant data gaps and inadequate knowledge about the toxicity of mixtures of CSG waste, their breakdown products, and metabolites.

In Australia, there has never been a comprehensive study on exposures to the unconventional gas industry's chemical waste releases. Vast volumes of CSG waste are not regulated as hazardous waste and are not required to be tracked (Ascend Waste & Environment 2017). While 231,054 tonnes of CSG waste was declared in 2014–2015 (principally drilling muds), it appears from the 2017 Hazardous waste report that an estimated 25 million tonnes per annum of strongly saline CSG waters, contained in onsite storage or managed offsite through the "Beneficial Use Approval" program (QLD Gov 2014b) (see this chapter), was neither registered as hazardous waste nor tracked as such. The hazardous waste 2017 report notes that "tracking data suggests treatment of salty waters by desalination to enable reuse occurs only in a minority of cases," and even then "leaves a salt brine or solid waste as a byproduct." Of note, critically missing from the hazardous waste assessment are naturally occurring radioactive materials (NORM). While NORM are a recognized problem in the United States, and despite early research (Tait 2013) showing elevated radon levels in Queensland's gas fields, they would appear to be not monitored or tracked in Australia.

The subsequent 2019 Hazardous Waste in Australia report (Ascend Waste & Environment 2019) indicates a further weakening in the definition of, management, and tracking of hazardous waste from the gas fields. Queensland's Beneficial Use Approval system for regulating CSG drilling muds and extraction waters (sometimes known as "associated water") is being progressively replaced by a regime of "End of Waste Codes." One such code exists for drilling mud that facilitates processing through composting. Drilling muds managed according to the End of Use Code appear to fall out of the regulated waste framework. Whether this will remove the tracking requirement for drilling muds is unclear, but if it is no longer classified as regulated waste, it would appear so.

Large-scale reverse osmosis desalination plants, which have been online since 2014, are an attempted solution to evaporation ponds. These reverse osmosis plants can lower salts in the associated water to a level that enables a range of uses such as irrigation, stock drinking water, and dust suppression. However, this comes at the cost of concentrating the original problem – salt, or brine, particularly significant in view of the vast volumes of CSG waters treated by desalination. Suffice to say, the volumes of brines currently stored in regulated dams by the major operators in Queensland is extremely large, particularly since inputs to these facilities have only started to occur at scale from late 2014. Since inception of the industry in Queensland it has been, so far unsuccessfully, working on permanent solutions to brine management. Dam aggregation is not a sustainable solution, while plans for purification and resale of the product salt were also unrealized. The current focus is on possible solidification and crystallization of salts for long-term storage in dedicated (onsite) facilities, and geological repositories, a form of permanent isolation of waste that is soon to be operational in Queensland.

With the CSG industry responsible for around 20% of apparent waste generation "in this unique hazardous waste jurisdiction," before the reclassification outlined earlier, the level of investigation and enforcement by the regulators as demonstrated by the paucity of prosecutions (five relate to the CSG industry out of a total of 79 between 2012 and 2019) is notable.

3.7 Emissions

As noted by Luke (2018), the unconventional gas industry is a significant player at the water, energy, and climate nexus. CSG extraction, compression, pumping, liquefication, and shipping, along with the ancillary processes including reverse osmosis of wastewater, are extreme energy requiring activities with significant GHG emissions. Despite being signatories to the Paris Agreement, Australia's acknowledged greenhouse gas emissions have been rising year by year. Annual emissions to March 2019 were 561 million tonnes of carbon dioxide equivalent, up from 554.5 million tonnes in 2018, and 551.2 million tonnes in 2017. The rise in emissions is directly attributable to the LNG industry (Hannan 2019).

In 2016, Chevron, ExxonMobil, and Shell were permitted to operate the Gorgon LNG plant on Barrow Island, 60 kms off the Western Australia coast. This was under the key condition that 80% of the CO_2 extracted from the project's gas reservoirs would, over a five-year rolling average, be captured and sequestered. However, two years into the project all the emissions have been vented to the atmosphere. Chevron have been removing BTEX and mercury from the gas produced for export, but along with the greenhouse gases they have been venting vapors containing 300 ppm BTEX and 13,000 µg/m^3 mercury into the local atmosphere (Gov Western Australia 2019). Analysis by energy consulting firm Energetics indicates that the annual 6.2 million tonnes in emissions saved by all the solar panels in Australia have been wiped out by this single oil and gas project (Diss 2018).

Analysis by the Climate Council of WA (CCWA) of the climate impacts of WA's LNG industry indicates that emissions from current LNG facilities in WA make up 36% of WA's total emissions (Clean State 2019). WA's emissions have risen 23% above the 2005 baseline as a result of the LNG industry. Massive LNG projects proposed in WA, in addition to the projects in that state that have come online since 2005, would see a predicted 8% increase in total Australian emissions above the 2005 baseline. Analysis by Western Australia's peak environment group (Clean State 2019) indicates that the proposed Western Australian Browse Basin and Burrup Hub liquid natural gas (LNG) development would make it one of the largest sources of carbon pollution in the world with 20 million tonnes local and 80 million tonnes Scope 3 emissions.[5]

In 2018, the Northern Territory's gas inquiry concluded that proposed extraction of gas from the Mc Arthur basin could contribute more than 6% of Australia's emissions (Pepper 2018). However, on re-analysis of the Northern Territory data, Professor Ian Lowe (2019) concluded that extracting and burning gas on the scale of the proposed shale gas development (even without accounting for fugitive emissions) would result in 600 million tonnes of CO_2 equivalent per year, more than Australia's most recent and highest ever recorded total national emissions at 560 million tonnes.

The extent of greenhouse gas emissions emanating from the massive LNG projects in Queensland, including the processing plants and export terminal on Curtis Island, remains unclear. Calculation of emissions by industry-funded GISERA (Schandl 2019), using commercial in confidence information from a single unnamed project, estimates the direct and indirect emissions (scope 1 and 2) from this project within Australia at 4.38 Mt CO_2-e/year (MRIO) and 5.94 Mt CO_2-e/year (LCA). Scope 3 emissions within Australia were calculated by GISERA at 0.16 Mt CO_2-e/year with a further 38.76 Mt CO_2-e/year (LCA) emissions "generated by shipping LNG, re-gasification and combustion of gas in Asia."

With the officially acknowledged greenhouse gases rising, the magnitude of the actual problem is unclear owing to the extent of unacknowledged emissions. Estimated official leakage of methane reported in Australia is 0.5% of total oil and gas production. This contrasts significantly with measured leakage in US unconventional gas fields of 2–30% of production. The Melbourne Energy Institute's 2016 review of current and future methane emissions from Australian unconventional oil and gas production highlighted the serious deficits in data and under reporting of Australian gas field emissions (Lafleur et al. 2016). They note that "government regulators do not require oil and gas companies to actually measure emissions as a basis for accounting and reporting. Rather, assumptions and simple factors may be used" (Forcey 2018a).

Critically, methane emissions from the extensive network of gas field pipelines and compressor stations are entirely ignored or "set at zero" in calculations of emissions by the gas industry and regulators (Forcey 2018a). This zero-emission accounting includes gas field infrastructure specifically engineered to emit methane and associated gases (and demonstrably doing so), including gas and water gathering pipelines, water gathering vents, produced water storage, and gas and water treatment plants (Forcey 2018b). Methane emissions are also ignored (set at zero) for the thousands (exact number unknown) of decommissioned and abandoned wells. Likewise, migratory emissions are set at zero despite the visual evidence of gas bubbling from water sources proximal to sites of unconventional gas extraction (for instance, the Condamine River in the Western Darling Downs) (Forcey 2018a). Despite these migratory emissions being set at zero, peak flow rates at the main Condamine River seep have been measured at 2,000 L/minute (equivalent to approximately 3 million liters of methane per day) (APLNG). The bubbling was "mitigated" by collecting and flaring the gas onshore, and eventually by drilling 12 gas wells directly south of the river. As confirmed by CSIRO (Day 2014), despite the rapid expansion and massive scale of the Australian gas industry, and the critical importance of greenhouse gas emissions, "reliable measurements on Australian oil and gas production facilities are yet to be made."

As noted previously, Australia has promoted the development of "natural gas" as a strategy to combat climate change. However, the stated intent of the industry as a "transition fuel" is suspect given that the chronic under reporting of fugitive emissions in Australia makes a reliable assessment of this claim impossible. Doctors for the Environment Australia warn, "It is not possible to overemphasise the enormity of health, economic, security and environmental costs of an inadequate response to global warming" (Haswell 2018). The Climate Council of Australia highlighted the vulnerability of Australia to the impacts of climate change, a vulnerability tragically demonstrated in the 2019–2020 bushfire season. With Eastern Australia in drought with no significant rain forecast, the Bureau of Meteorology forecast higher than average temperatures. The NSW and Queensland 2019–2020 bushfire season (usually occurring in the heat of summer) commenced in winter with multiple lives lost, thousands of homes destroyed, and millions of acres of land and more than half a billion native animals incinerated (Flanagan 2020). Hazardous levels of toxic smog blanketed South Eastern Australia, with a 30% increase in hospital presentations for cardiac and respiratory conditions in Sydney (Aubusson 2019). With Australia already in climate crisis, the fact that reliable measurements on Australian oil and gas production facilities have not yet been made has significant implications for adequate response.

3.8 Air Quality

There has been an ambient air monitoring network in South East Queensland since the 1980s (National Environment Protection Measures, NEPM). However, it was not until late in the gas field boom–bust cycle, in February 2015 (when over 6,000 wells had already been drilled, infrastructure construction was completed, and thousands of FIFO workers had already left) that the first ambient air monitor was situated in the gas fields of South West Queensland. The data collected in a collaboration between CSIRO and GISERA is owned by industry and vetted by them before publication on the Queensland government website. In their analysis of Ambient air quality in the Surat Basin 2014–2018, CSIRO/GISERA claim that "air quality in the region is well within relevant air quality objectives for the majority of the time for a wide range of gaseous pollutants that are potentially emitted by CSG" (Lawson 2018). An analysis of the data reported by CSIRO/GISERA as acceptable shows striking information gaps that make the validity of their conclusion questionable. Table 3.3 shows the number of exceedances for $PM_{2.5}$ as reported by CSIRO/GISERA, with the data gaps highlighted in red. Similar data gaps exist for PM_{10}, TSP, NOx, CO, and O_3.

As reported by one of the authors of this chapter (McCarron), between 2007 and 2014 there was a striking increase in hospitalization rates of Queensland's Darling Downs residents for acute respiratory and circulatory conditions. This was coincident with the escalation of gross calculated air pollutants that was acknowledged by the gas industry and reported by them to the National Pollutant Inventory (NPI).

There remains a yawning gap among a) estimated gross emissions reported to the NPI, b) ambient air monitoring, and c) the unmeasured, unmonitored, and unmitigated emissions from the thousands of pieces of gas field infrastructure (high point vents, hydraulic power

Table 3.3. *Data gaps in $PM_{2.5}$ monitoring program*

PM$_{2.5}$ exceedances Air EPP (Environmental Protection (Air) Policy) objectives and NEPM objectives with data gaps in monitoring program. (Averaging period 24hrs)

	Hopeland	Miles Airport	Condamine	Tara	Burncluith
2014	No data	No data	No data	No data	No data
2015	1 No data for January through to August. Inadequate data for September and October	1 No data January through to August. Inadequate data in December	Not Measured No data	No data	No data
2016	2 Inadequate data January	1 Inadequate data January and March	1 No data January and February. Inadequate data October and November	No data	No data
2017	0 Inadequate data for January March, April, June, and July	1 Inadequate data for August and September	0 No data for July, August, September, October, November, and December	No data	No data
2018 January and February only	0 Inadequate data February	0	Not Measured No data	No data	No data
total	3	3	1	–	–

units, etc.) to which the resident population is exposed. People are exposed to spikes of multiple chemicals, and mixtures of chemicals, from potentially hundreds of sources, for varying timeframes, and under varying weather conditions including temperature inversions. Despite ongoing pleas from gas field residents and the specific recommendation by Queensland Health in 2013 that exposure monitoring should occur, it has still not been done.

3.9 The People

As Dr Wayne Somerville, a specialist clinical psychologist with deep personal experience engaging with the UG Industry impacts on people in Northern New South Wales, stated in his 2016 submission to the Senate Select Committee on Unconventional Gas Mining,

The transformation of rural landscapes into industrialized gas fields profoundly changes the lives of the people who live there. The people threatened by, or who suffer losses or injuries from gas field development are those who suffer the symptoms of emotional, economic distress and physical ill-health. Those people are not the multinational gas companies, their well funded lobbyists, or the politicians or government department[s] responsible for the industry. (*Dr Wayne Somerville, 2016*)

The unconventional gas industry footprint has tended to be superimposed over areas with a long agricultural history. The normal expectation of the families and small agribusinesses required to host the industry has been that government would uphold their presumed rights, act as an independent and impartial arbiter, and ultimately ensure remediation in instances of damage. On the contrary, residents have found themselves subject to industry imposition with nearly no protections in place. The re-engineering of the regulatory environment around the development of the industry (previously described in Politics, Policy and Promises section) worked in tandem with the failure of the Australian Government to have any regulatory requirement to consider the human health and wellbeing impacts of the industry via the absence of strong human rights jurisprudence.

A prescriptive and contemporary legislative representation of this was identified by Dougall (2019), in that

application of the constructs of risk identification and management in relation to the WHS [Work Health and Safety] of host farmers from the hazards generated by the UG industry is required under the WHS Act, but is not in practice routinely considered by the industry or the administering agencies. It also showed that such constructs and their application are obscured jurisdictionally, which is borne out in the lack of subsequent prosecutions and court actions or the range of other compliance and enforcement activities investigated. The research showed that the host farmer is left in a position where she is potentially exposed to WHS risk via the interface with the UG industry, but without a clearly defined path for remedy via the P&G Act, nor clearly defined path of jurisdictional support from the WHS Act.

People with lived experience have struggled to be heard over the louder regulatory supported proponents of the industry. In unconventional gas this is particularly present because of the "societal power and efforts to influence decision making processes in the circumstances of unsettled science and risk debates" that the industry represents. Therefore, there exists a particularly strong imaginary line between the legitimacy of experts and the lived experience of those individuals hosting the industry (Espig et al. 2016).

In May 2018, the International Permanent Peoples' Tribunal held a special session on Human Rights, Fracking (Unconventional Gas) and Climate Change (Permanent Peoples Tribunal, 2018). Individual Australians contributed to the evidence heard by the Tribunal, which included testimony from Queensland, New South Wales, Western Australia, and the Northern Territory. Ultimately, the findings were that the unconventional gas industry violates the full spectrum of human, social, economic, civil, political, and cultural rights that are endowed on the people that host the industry. The serious rights violations suffered by people and nature are accompanied by little, if any, economic and social benefits to those communities. Further, the investigation found that because Government policies and industry operations are procedurally and operationally organized, integrated, and implemented around the violation of these rights, it creates a situation whereby both Government and industry are fundamentally incapable of enacting and upholding these human rights.

Demonstrably, through the lens of "the people" vs the industry, there remains an underacknowledged and underpursued field of conflict.

3.10 Sustainability and Legacy

At this relatively early milestone in the development of the industry (10 years), it is interesting to note that the Petroleum and Gas Inspectorate has undertaken a review of the legacy issues that the industry is now of an age to be experiencing. The Inspectorate recently conducted inspections of aging oil and gas production facilities in Western Queensland, and started a review of the abandonment practices of petroleum wells while concurrently issuing a revised version of the Code of practice for the construction and abandonment of petroleum wells and associated bores in Queensland (Petroleum and Gas Inspectorate 2019). The results of these were reported to indicate multiple areas for improvement, including reports not being submitted by the due date; lack of clear evidence being provided within the abandonment reports to show that legislative requirements have been met; wells requiring better signage post-abandonment; failure to properly decommission equipment and monitor/maintain the facilities during the shutdown period, including factors such as corrosion, erosion, fatigue, equipment obsolescence, normalization of deviance (accepting degraded conditions as being normal), changes in codes and standards, and lack of data to forecast future risks.

Considering this data together with producers advising that rates of drilling in CSG fields continue to increase while production rates hold steady, such issues directly relate to the government and the industry's ability to ensure an acceptable standard in long-term well integrity, containment of petroleum, and protection of groundwater resources. The Government's ability to have a meaningful impact on such compliance and resultant legacy issues is demonstrably the case in point provided by the Linc Energy Contamination Incident in the heartland of the area that is currently the focus of the unconventional gas industry (Western Downs, Queensland). Systemic failures in the government's approvals, monitoring, and enforcement of conditions resulted in one of the largest environmental contamination incidents in the country's history (EDO 2018). The sentencing judge found that Linc Energy, who operated an underground coal gasification plant (UCG) at Chinchilla Queensland, had willfully and unlawfully caused serious environmental harm over a period of seven years. Judge Shanahan noted that "Each gasifier was operated in a manner that resulted in explosive and toxic gases, tars and oils escaping into the landform." The resulting contamination of the groundwater system will require monitoring and remediation for many years, and at the time of sentencing in 2018 there was an ongoing risk "of explosion and toxicity at the site and in the landforms" (Smith 2018). Similarly, in 2013 Cougar Energy had been found guilty of contamination of groundwater with benzene and toluene from its UCG plant at Kingaroy, Queensland (Queensland Government 2013e). Limited redress was pursued in both these cases. These incidents highlight the failure in continuity of accountability regulations that allow the contaminator to simply close the doors and walk away with impunity.

Given the intensity and scale of the footprint of the industry, and the failure by government planners to address cumulative impacts, the implications for long-term legacy and sustainability issues post-production are significant. Legacy issues relate to the change in land use and extensive industrialization. While the industry is short term, the residual degrading infrastructure (much of it buried) poses intergenerational problems for people living and working on the land. The question of intergenerational equity is profound. Already there have been extensive changes to the landscape, and state forest use, including limitation of public access. Along with the contaminated waste produced by the industry, the already documented (and ever increasing) loss of water bores imperils the future productivity of the land, and sustainable food production. In Queensland, expansion of the gas industry has seen the major transfer of land equity from multi-generational families to corporations. In the Northern Territory, Australian First Peoples with a history dating back more than 65,000 years have maintained a strong and distinct cultural identity and spiritual life. Their culture and wellbeing are inextricably linked to the land and their custodial roles and responsibility for country. The expansion of the industry into the Northern Territory not only threatens intergenerational equity but also foreshadows the disintegration of the cultural, spiritual, and physical wellbeing of indigenous Australians.

3.11 Conclusion and Recommendations

Responses to the identified issues with the unconventional gas industry tend to focus on gaps in regulation and recommendations toward a rational regulatory framework. However, it is the opinion of the authors of this chapter that the regulation specifically required is legislation to urgently phase out and, in a timely and controlled manner, ultimately ban unconventional gas along with other fossil fuels. No new wells should be permitted. Climate disruption is recognized as an existential threat. It is the single biggest long-term threat to global health this century. The critical need to limit further temperature rise is established. In this changing climate, the fundamental importance of water to our lives and wellbeing, and the need to protect it, cannot be overemphasized. Fossil fuel extraction and usage is a major driver of climate change. The global unconventional gas industry shows no signs of voluntarily slowing down, with a focus on expanding the use of gas in the production of plastics (resulting in global pollution and ocean acidification) (Lloyd-Smith 2018), as well as promoting the use of fracked gas to produce hydrogen. It is urgent that emissions are reduced. We support the recommendation of DEA that in the short term in Australia, limited, existing, and already developed reserves should be used judiciously to assist in the rapid transition of Australia and its workforce toward clean energy resources (wind, solar, hydroelectric) and that Australia should urgently assist developing countries to transition away from gas power. However, the transition should be entered into "with eyes wide open" to the true environmental and human cost of alternative energy sources with inbuilt mechanisms to minimize, reclaim, recycle, reuse, and protect earths' precious resources. The United Nations' 16 Framework Principles on Human Rights and the Environment are apt in this scenario (United Nations 2018). These principles have the capacity to underwrite a rational regulatory framework across all developments, and properly applied would give humankind its best chance to thrive in this turbulent millennium.

3.12 Postscript

While this chapter was in the late stages of preparation, the COVID-19 pandemic seriously impacted the world. As with all sectors of the economy and society, its effect on the LNG industry in Australia has been significant. In the context of the risk to the health of the remote indigenous communities of the Northern Territory from FIFO workers, the gas exploration industry was deemed a nonessential service and gas giant Santos's program in the NT has been put on hold (McLennan 2020). The CEO of Santos, Kevin Gallagher, was noted to have sold $2.2 million of his shares (Simply Wall St. 2020), while the Australian Financial review reported that the LNG exports are to take a 20 billion dollar hit as prices nosedive (MacDonald-Smith 2020). These developments would suggest that the UG industry may not be financially viable under the new conditions.

However, COVID provided an agile demonstration of opportunism in disaster management. On March 20, 2020, in the midst of the COVID crisis the Australian Parliament was indefinitely adjourned (*until a date and hour to be fixed by the Speaker*), with the Senate suspended until August 11, 2020 (Horne 2020). An (unelected) National COVID-19 Coordination Commission, announced by Prime Minister Scott Morrison on March 25, 2020, included significant personnel from the resources sector such as Chairman Neville Power (Deputy Chairman of Strike Energy Ltd and ex-CEO of Fortescue Metals Group) and Catherine Tanna (Managing Director of Energy Australia and ex-Managing Director of BG Group's Australian business, QGC Pty Limited) (National COVID-19 Coordination Commission 2020). This was rapidly followed by proposals of a "gas-fired COVID recovery"(Foley 2020), accompanied by "aggressive deregulation" (Curtis 2020; Gocher 2020).

The Queensland Government used the cover of a COVID media update to laud a planned $10bn, 5,000 PJ Gas Project (PetroChina, Shell, and Arrow) in the Surat Basin Queensland, despite the fact that it was not yet operational and had been announced 14 months previously. ("It's supplying gas to manufacturers to make PPE." Qld Minister for Natural Resources Mines Anthony Lynham) (News.com 2020).

Notes

1 "Better quality water" for irrigation under general beneficial use approval conformed to limited standard conditions. Electrical conductivity (EC) of <950 μs/cm^3 as a 95 percentile over a one-year period. Sodium adsorption ratio (SAR) 6 or less for heavy soils as a 95 percentile over a one-year period, 12 or less for light soils as a 95 percentile over a one-year period. pH within the 6.0–8.5^4 accounting for atmospheric equilibration as a 95 percentile over a one-year period. Heavy metals do not exceed the values prescribed. Variations to the standard water quality conditions were permitted when a water quality parameter could not be met and a report was provided to the administering authority.
2 Environmental Defenders Office (EDO) is the largest environmental legal center in the Australia-Pacific, providing access to justice, running groundbreaking litigation, and leading law reform advocacy. They are an accredited community legal service and a nongovernment, not-for-profit organization that uses the law to protect and defend Australia's wildlife, people, and places.
3 All dollar values are in $AUD.
4 In Australia, a private water well used by a farmer (via a registration process for government authorisation for use) for household and farm animal water is referred to as a "stock and domestic water bore." Water bores, in contrast to wells, are drilled to extract water from underground sources.

5 Scope 1 emissions are direct GHG emissions from all sources owned or controlled by a company, Scope 2 are indirect GHG emissions from the generation of electricity purchased and used by a company, and Scope 3 emissions are indirect greenhouse gas emissions occurring as a consequence of the activities of a facility from sources not owned or controlled by it

References

Arrow Energy. (2013). Arrow Bowen Gas Project EIS Project 4. www.arrowenergy.com.au/__data/assets/pdf_file/0009/28944/Section_04-Project-Description.pdf (accessed April 23, 2020).

Arup T. (2013). Abbott Shuts Down Climate Commission. *The Sydney Morning Herald*. www.smh.com.au/politics/federal/abbott-shuts-down-climate-commission-20130919-2u185.html (accessed April 23, 2020).

Ascend Waste & Environment. (2017). Hazardous Waste in Australia. Department of the Environment and Energy. www.environment.gov.au/system/files/resources/291b8289-29d8-4fc1-90ce-1f44e09913f7/files/hazardous-waste-australia-2017.pdf (accessed April 23, 2020).

Ascend Waste & Environment. (2019). Hazardous Waste in Australia. Department of Environment & Energy. www.environment.gov.au/protection/publications/hazardous-waste-australia-2019 (accessed April 23, 2020).

Aston H. (2017). Barnaby Joyce's Forecast of CSG Royalty Riches Compared to '$100 Lamb Roast' Claim. *The Sydney Morning Herald*.

Aubusson K. (2019). NSW bushfires: 'Apocalyptic' Health Effects of Sydney's Toxic Air. *The Sydney Morning Herald*. www.smh.com.au/national/nsw/nsw-bushfires-apocalyptic-health-effects-of-sydney-s-toxic-air-20191211-p53ixc.html (accessed April 23, 2020).

Australia Pacific LNG. (2017). Condamine River Seeps. www.aplng.com.au/topics/coal-seam-gas/condamine-river-seeps.html?fbclid=IwAR2Lb86uv6yJSYg22rR4k_oibSFKJdUo4AGv3-Sw3XBcGDwmJ2xYISkRpLE (accessed April 23, 2020).

Australian Bureau of Statistics (ABS). (2018). Labour Force, Australia, Detailed, Quarterly, Nov 2018 www.abs.gov.au/AUSSTATS/abs@.nsf/DetailsPage/6291.0.55.003Nov%202018?OpenDocument (accessed April 23, 2020).

Australian Energy Market Opperator (AEMO). (2017). Gas development required to meet future demand. Aemo. https://aemo.com.au/-/media/media-hub/documents/2017/more-gas-development-required-to-meet-future-demand/media-release-gas-development-required-to-meet-future-energy-demand.pdf (accessed April 23, 2020).

AEMO. (2019). Gas Statement of Opportunities for Eastern and South Eastern Australia. Tech. rep. Melbourne, Australia: Australian Energy Market Operator. www.aemo.com.au/-/media/Files/Gas/National_Planning_and_Forecasting/GSOO/2019/2019-GSOO-report.pdf (accessed April 23, 2020).

Australian Energy Resources Assessment (AERA). (2019). Website. https://aera.ga.gov.au/#!/energy-resources-and-market (accessed 23 April 2020)

Australian Government. (1998). Productivity Commission Act 1998 (Cth), subsection 10 (1)(i).

Australian Government. (2005). www.environment.gov.au/land/rangelands (accessed 23 April 2020).

Australian Government. (2013). Mineral and Energy Resource Exploration Productivity Commission Inquiry Report No. 65. www.pc.gov.au/inquiries/completed/resource-exploration/report/resource-exploration.pdf (accessed April 23, 2020).

Australian Government. (2016). Senate Select Committee on Unconventional Gas Mining (Bender Inquiry). www.aph.gov.au/Parliamentary_Business/Committees/Senate/Gasmining (accessed April 23, 2020).

Australian Petroleum Production and Exploration Association (APPEA). (2014). Red tape shackles economic growth. www.appea.com.au/tags/green-tape/ (accessed April 23, 2020).

Casson R. (2015). AGL's irrigation trial ends in doubt. *Manning River Times*. www.manningrivertimes.com.au/story/3044536/agls-irrigation-trial-ends-in-doubt/ (accessed April 23, 2020).

Chang C. (2017). How Australia is being screwed over it's gas. News.com.au. www.news.com.au/finance/economy/australian-economy/how-australia-is-being-screwed-over-its-gas/news-story/4187e60617aec18e87d57453cfca0167 (accessed April 23, 2020).

Chang C. (2019). Tax and royalty systems for Australia's gas and oil industries need reform, experts argue. news.com.au. www.news.com.au/finance/business/mining/tax-and-royalty-systems-for-australias-gas-and-oil-industries-need-reform-experts-argue/news-story/a900c328f1a01bf4e3aee8b867138262 (accessed April 23, 2020)

Claudio F, de Rijke K, and Page A. (2018). *The CSG Arena: A Critical Review of Unconventional Gas Developments and Best-Practice Health Impact Assessment in Queensland*. Impact Assess. Project Appraisal 36, 105–114.

Clean State. (2019). Runaway Train: The impact of WA's LNG industry on meeting our Paris targets and national efforts to tackle climate change. CCWA & Clean State Report: October 2019 https://apo.org.au/sites/default/files/resource-files/2019-11/apo-nid266691.pdf (accessed April 23, 2020).

Climate Action Tracker. (2020). https://climateactiontracker.org/countries/australia/ (accessed April 18, 2020).

COAG. (2013). Multiple Land Use Framework. Online publication. www.coagenergycouncil.gov.au/sites/prod.energycouncil/files/publications/documents/Multiple%20Land%20Use%20Framework%20-%20Dec%202013.pdf (accessed April 23, 2020).

Curtis K. (2020). Virus shadow looms over economy for years. Australian Associated Press. www.aap.com.au/tax-cuts-deregulation-on-road-to-recovery/ (accessed April 26, 2020).

Day S, Dell'Amico M, Fry RA, and Javanmard Tousi H. (2014). *Field Measurements of Fugitive Emissions from Equipment and Well Casings in Australian Coal Seam Gas Production Facilities*. GISERA.

Diss K. (2018). How the Gorgon gas plant could wipe out a year's worth of Australia's solar emissions savings. ABC www.abc.net.au/news/2018-06-21/gorgon-gas-plant-wiping-out-a-year-of-solar-emission-savings/9890386 (accessed April 20, 2020).

Doctors for the Environment Australia (DEA). (2013). Submission 70. Australian Government Productivity Commission. Doctors for the Environment Australia Inc. Hon Secretary David Shearman. www.pc.gov.au/inquiries/completed/resource-exploration/submissions/submissions-test2/submission-counter/subdr070-resource-exploration.pdf (accessed April 23, 2020).

Doctors for the Environment Australia (DEA). (2019). Submission 0014. Inquiry into the implementation of the recommendations contained in the NSW Chief Scientist's independent review of CSG activities in NSW. www.parliament.nsw.gov.au/lcdocs/submissions/66403/0014%20Doctors%20for%20the%20Environment%20Australia.pdf (accessed April 23, 2020).

Dougall S. (2019). Workplace health and safety (WHS) implications for farmers hosting unconventional gas (UG) exploration & production. *Policy and Practice in Health and Safety*. **17**(2): 156–172. DOI: 10.1080/14773996.2019.1649903

Drinkwater R. (2017). Unconventional Gas. Australian Parliament Library. www.aph.gov.au/About_Parliament/Parliamentary_Departments/Parliamentary_Library/pubs/BriefingBook45p/UnconventionalGas (accessed April 23, 2020).

Drinkwater R. (2015). Understanding Environmental Risks Associated with Unconventional Gas in Australia. Master of Science (Engineering) Thesis, Department of Civil Engineering, Monash University, Clayton, VIC, Australia, 115 pages.

Environmental Defenders Officer (EDO) (2013). EDO offices face closure after federal funding cuts. EDO article online. www.edonsw.org.au/edo_offices_face_closure_after_government_funding_cuts (accessed April 23, 2020).

Environmental Defenders Officer (EDO). (2016). Submission 56. Select Committee on Unconventional Gas Mining. www.aph.gov.au/Parliamentary_Business/Committees/Senate/Gasmining/Gasmining/Submissions (accessed April 23, 2020).

Environmental Defenders Officer (EDO). (2018). Spotlight on prosecutions of Linc Energy. EDO Article. www.edoqld.org.au/spotlight_on_prosecutions_of_linc_energy (accessed April 23, 2020).

Energy Resource Insights (ERI). (2020). http://data.erinsights.com/maps/fossilfuels-au.html (accessed April 14, 2020).

Espig M and de Rijke K. (2016). Unconventional gas development and the politics of risk and knowledge in Australia. *Energy Research & Social Science*. **20**: 82–90. doi:10.1016/j.erss.2016.06.001

Flanagan R. (2020). Australia is Committing Climate Suicide. *The New York Times*. www.nytimes.com/2020/01/03/opinion/australia-fires-climate-change.html (accessed April 23, 2020).

Fleming DA and Measham TG. (2015). Local economic impacts of an unconventional energy boom: The coal seam gas industry in Australia. *Australian Journal of Agricultural and Resource Economics*. **59**(1): 78–94.

Foley N. (2020). Gas to fire economic recovery and capitalise on cheap oil prices. *The Sydney Morning Herald*. www.smh.com.au/politics/federal/gas-to-fire-economic-recovery-and-capitalise-on-cheap-oil-prices-20200421-p54lw8.html. (accessed April 26, 2020).

Forcey T. (2018a). Greenhouse Gas Footprint: What we do and don't know about gas. ReNew magazine. http://renew.org.au/articles/greenhouse-gas-footprint-what-we-do-and-dont-know-about-gas/ (accessed April 23, 2020).

Forcey T. (2018b). Submission 548. Northern Territory Fracking Inquiry. https://frackinginquiry.nt.gov.au/?a=479474 (accessed April 23, 2020).

The Gasfields Commission. (2018). Queensland's Petroleum & Gas Industry Snapshot, Petroleum and Gas Royalties Paid to the State 2010–2017. https://gasfieldscommissionqld.org.au/gas-industry/gasindustryfacts (accessed April 23, 2020).

Geoscience Australia. (2019). www.ga.gov.au/scientific-topics/energy/resources/petroleum-resources/coal-seam-gas (accessed April 14, 2020).

Gocher D. (2020). A gas-fired recovery? Seriously? Lobby Watch. https://medium.com/lobbywatch/a-gas-fired-recovery-seriously-27cee36e8712 (accessed April 26, 2020).

Gosden E. (2019). Regulator digs into Cuadrilla's fracking earthquake. The Times. www.thetimes.co.uk/article/regulator-digs-into-cuadrillas-fracking-earthquake-xtvnkr72p (accessed April 23, 2020).

Government of West Australia. (2019). Decision Report, Gorgon LNG Project, Works Approval Number W6199/2018/1.

Grudnoff M. (2014). Fracking the future: Busting industry myths about coal seam gas. The Australia Institute Online Institute Paper No. 16. ISSN 1836-8948. www.tai.org.au/sites/default/files/IP%2016%20Fracking%20the%20future%20-%20amended_0.pdf (accessed April 15, 2020).

Hannan P. (2019). Australia's greenhouse emissions set new seven-year highs on gas boom. *The Sydney Morning Herald*. www.smh.com.au/environment/climate-change/australia-s-greenhouse-emissions-set-new-seven-year-highs-on-gas-boom-20190830-p52mbe.html (accessed April 23, 2020).

Hansard. (2010a). Queensland Parliament. page 1140, www.parliament.qld.gov.au/documents/hansard/2010/2010_03_25_WEEKLY.pdf (accessed April 23, 2020).

Hansard. (2010b). Queensland Parliament. Letter from BG Group to Premier Anna Bligh. www.parliament.qld.gov.au/Documents/TableOffice/TabledPapers/2010/5310T1958.pdf (accessed April 23, 2020).

Hansard. (2014). Queensland Parliament. Page MARSH, Ms Simone, Private capacity, https://parlinfo.aph.gov.au/parlInfo/search/display/display.w3p;db=COMMITTEES;id=committees%2Fcommsen%2F9e7a99b6-f87b-4ac0-a713-d192a8889516%2F0001;query=Id%3A%22committees%2Fcommsen%2F9e7a99b6-f87b-4ac0-a713-d192a8889516%2F0000%22 (Accessed 15/04/2020) (accessed April 23, 2020).

Hansard NSW Parliament. (2019). REPORT ON PROCEEDINGS BEFORE PORTFOLIO COMMITTEE NO. 4 – INDUSTRY, 3rd December 2019.

Haswell M. and Shearman D. (2018). *The Implications for Human Health and Wellbeing of Expanding Gas Mining in Australia: Onshore Oil and Gas Policy Background Paper*. Doctors for the Environment Australia. www.dea.org.au/wp-content/uploads/2018/12/DEA-Oil-and-Gas-final-28-11-18.pdf (accessed April 23, 2020).

Hawthorne M. (2012). 14,000 jobs to go but no sackings. ABC news. www.abc.net.au/news/2012-09-14/no-qld-public-servants-sacked-newman-says/4261346 (accessed April 23, 2020).

Horne N. (2020). Parliamentary Library. www.aph.gov.au/About_Parliament/Parliamentary_Departments/Parliamentary_Library/FlagPost/2020/April/COVID-19_and_parliamentary_sittings (accessed April 26, 2020).

Huth N, Cocks B, Dalgliesh N, Poulton P, Marinoni O, and Garcia J. (2018). Farmers' perceptions of coexistence between agriculture and a large scale coal seam gas development. *Agriculture and Human Values*. **35**(1): 99–115.

Hunter T. (2016). The development of shale gas and coal bed methane in Australia: Best practice for international jurisdiction. *Houston Journal of International Law*. **38**(2): 367–424.

Independent Expert Scientific Committee (IESC). (2014). Knowledge report. Monitoring and management of subsidence induced by coal seam gas extraction. www.environment.gov.au/system/files/resources/632cefef-0e25-4020-b337-80a9932d1c67/files/knowledge-report-csg-extraction_0.pdf (accessed April 22, 2020).

Jaganathan J. (2018). Reuters. www.reuters.com/article/us-australia-qatar-lng/australia-grabs-worlds-biggest-lng-exporter-crown-from-qatar-in-nov-idUSKBN1O907N (accessed April 23, 2020).

Keywood M, Grant S, Walton A, Aylward L, Rifkin W, Witt K, Kumar A, and Williams M. (2018). *Human Health Effects of Coal Seam Gas Activities: A Study Design Framework*. CSIRO. Task 4 Report for Health Project (H.1).

Lafleur D, Forcey T, Saddler H, and Sandiford M. (2016). *A Review of Current and Future Methane Emissions from Australian Unconventional Oil and Gas Production*. Melbourne Energy Institute. http://climatecollege.unimelb.edu.au/review-current-and-future-methane-emissions-australian-unconventional-oil-and-gas-production (accessed April 23, 2020).

Lawson SJ, Powell JC, Noonan J, Dunne J, and Etheridge D. (2018). Ambient air quality in the Surat Basin, Queensland, Overall assessment of air quality in region from

2014–2018. Report for the Gas Industry Social and Environmental Research Alliance (GISERA), Project No G.3.

Lloyd-Smith M and Immig J. (2018). Ocean Pollutants Guide, Toxic Threats to Human Health and Marine Life. IPEN. Online. https://ntn.org.au/wp-content/uploads/2018/10/ipen-ocean-pollutants-v2_1-en-web.pdf (accessed April 23, 2020).

Lowe I. (2019). Climate Change Impacts of Proposed Shale Gas Development in the Northern Territory. Lock the Gate Online. https://d3n8a8pro7vhmx.cloudfront.net/lockthegate/pages/6323/attachments/original/1571177037/LTG_NT_ShaleGas_2019_A4_SML.pdf?1571177037&fbclid=IwAR3HH7k6EUoXQ5wIe7RgqWCqnh-txV5ZMVh49PW7WLPMG24z2BniA0_AjCY (accessed April 23, 2020).

Ludlow M. (2019). Gas consortium wins royalty decision against Queensland government. Financial Review. www.afr.com/business/energy/gas-consortium-wins-royalty-decision-against-queensland-government-20190524-p51qru?fbclid=IwAR2f47OKQbN-l53FI3qSguIg4FYVgxjzLyE1YhPF4s2jtXkhPFVjg5JcdF0 (accessed April 23, 2020).

Luke H. (2018). Unconventional gas development in Australia: A critical review of its social license. *The Extractive Industries and Society*. **5**(4): 648–662. https://doi.org/10.1016/j.exis.2018.10.006.

MacDonald-Smith A. (2019). Forrest backed Port Kembla LNG import project gets green light. Financial Review. www.afr.com/business/energy/forrest-backed-port-kembla-lng-import-project-gets-green-light-20190429-p51i38 (accessed April 23, 2020).

MacDonald-Smith A. (2020). LNG exports to take $20b hit as prices nosedive. Financial Review article. www.afr.com/companies/energy/lng-exports-to-take-20b-hit-as-prices-to-nosedive-20200416-p54kcn (accessed April 18, 2020).

Marsh S. (2016). Submission 316. Senate Select Committee on Unconventional Gas Mining. www.aph.gov.au/Parliamentary_Business/Committees/Senate/Gasmining/Gasmining/Submissions (accessed April 23, 2020).

Marsh S. (2018). The Second British Invasion: how royal cronies and the gas debacle took Australia for billions (including the Fletcher email). MichaelWest.com.au www.michaelwest.com.au/the-second-british-invasion-how-royal-cronies-and-the-gas-debacle-took-australia-for-billions/ (accessed April 23, 2020).

McCarron G. (2013). Symptomatology of a gasfield. Self-published article. http://d3n8a8pro7vhmx.cloudfront.net/lockthegate/pages/49/attachments/original/1367333672/2013-04-symptomatology_of_a_gas_field_Geralyn_McCarron.pdf (accessed April 23, 2020).

McCarron G. (2018). Air Pollution and human health hazards: a compilation of air toxins acknowledged by the gas industry in Queensland's Darling Downs. *International Journal of Environmental Studies*. **75**(1): 171–185. DOI: 10.1080/00207233.2017.1413221

McLennan C. (2019). Another backflip on promises to stick to fracking rules. *The Katherine Times*. https://bit.ly/2yICORK. (accessed June 24, 2019).

McLennan C. (2020). Fracking comes to an abrupt halt in the NT. *The Katherine Times*. www.katherinetimes.com.au/story/6710441/fracking-comes-to-an-abrupt-halt-in-the-nt/ (accessed April 18, 2020).

Morgan MI, Hine DW, Bhullar N, Dunstan DA, and Bartik W. (2016). Fracked: Coal seam gas extraction and farmers' mental health. *Journal of Environmental Psychology*. **47**: 22–32.

National COVID-19 Coordination Commission. 2020. Australian Government Website. www.pmc.gov.au/nccc/who-we-are (accessed April 26, 2020).

National Pollutant Inventory. (2015). Emissions by individual facility. www.npi.gov.au/npidata/action/load/emission-by-individual-facility-result/criteria/state/QLD/year/2015/jurisdiction-facility/Q012QGC015 (accessed August 6, 2019).

NEPM. National Environment Protection (Ambient Air Quality) Measure. *Report of the Risk Assessment Taskforce*. Appendix 5. Ambient Air Monitoring Programs around Australia. Online report. www.nepc.gov.au/system/files/resources/9947318f-af8c-0b24-d928-04e4d3a4b25c/files/aaq-ratf-rpt-appendix5-ambient-air-monitoring-programs.pdf (accessed April 23, 2020).

News.com video. (2020). Arrow Energy announces $10bn Surat Gas project in Queensland. Online video clip. https://bit.ly/2KwN4iK (accessed April 18, 2020).

NSW Chief Scientist and Engineer. (2014). *Final Report: Office of the Chief Scientist and Engineer*. New South Wales Government.

Ogge M. (2015). Be careful what you wish for. The economic impacts of unconventional gas in Queensland and implications for Northern Territory policy makers. The Australia Institute Online Discussion paper. www.tai.org.au/sites/default/files/Be%20careful%20what%20you%20wish%20for%20FINAL_0.pdf (accessed April 15, 2020).

Pepper R. (2018). *Final Report of the Scientific Inquiry into Hydraulic Fracturing in the Northern Territory*. https://frackinginquiry.nt.gov.au/inquiry-reports/final-report

Permanent Peoples Tribunal. (2018). 47th Session on human rights, fracking, and climate change. http://permanentpeoplestribunal.org/47-session-on-human-rights-fracking-and-climate-change-14-18-may-2018/?lang=en *(accessed 23 April 2020)*

Petroleum and Gas Inspectorate. (2019). December 2019 Newsletter. www.vision6.com.au/v/23788/1774178580/email.html?k=rQmALtW4AuN5i35dwhsMTkTnaPb4f8OOw1Hdw8m-kEQ (accessed April 23, 2020).

Primary Industries, South Australia. (2019). Fast Facts Overview.https://pir.sa.gov.au/__data/assets/pdf_file/0011/339842/PIRSA_Primary_Industries_in_SA_Fast_Facts_Overview_-_1_April_2019.pdf (accessed December 27, 2019).

Queensland Government. (2000). Queensland Water Act. www.legislation.qld.gov.au/view/pdf/inforce/current/act-2000-034 (accessed April 22, 2020).

Queensland Government. (2007). Queensland gas scheme proves a winner. http://statements.qld.gov.au/Statement/2007/4/3/queensland-gas-scheme-proves-a-winner (accessed April 23, 2020).

Queensland Government. (2008). CSG water to be put to good use. http://statements.qld.gov.au/Statement/Id/61113 (accessed April 23, 2020).

Queensland Government. (2010a). Premier announces major milestone for fledgling LNG industry. http://statements.qld.gov.au/Statement/2010/6/25/premier-announces-major-milestone-for-fledgling-lng-industry (accessed April 23, 2020).

Queensland Government. (2010b). Queensland's LNG industry, A once in a generation opportunity for a generation of employment. www.cabinet.qld.gov.au/documents/2010/nov/lng%20blueprint/Attachments/LNG%20Blueprint.pdf (accessed April 23, 2020).

Queensland Government. (2011). Wild Rivers Declaration. Queensland Government, 2011, Georgina and Diamantina Basins Wild River Declaration. www.diamantina.qld.gov.au/documents/800087/4666855/087%20-%20Declaration.pdf (accessed April 23, 2020).

Queensland Government. (2012). Queensland Resources Cabinet Committee to examine regulation impacts.http://state.governmentcareer.com.au/archived-news/queensland-resources-cabinet-comittee-to-examine-regulation-impacts (accessed April 23, 2020).

Queensland Government. (2013a). Adaptive Management. www.ehp.qld.gov.au/management/non-mining/adaptive-management.html (accessed April 23, 2020).

Queensland Government. (2013b). Resources Cabinet Committee, 6 month report card: July to December 2013. www.dlgrma.qld.gov.au/resources/report/6-month-report-card-jul-dec-2013.pdf (accessed April 23, 2020).

Queensland Government. (2013c). Nature Conservation and Other Legislation Amendment Act (No. 2) 2013. www.legislation.qld.gov.au/view/pdf/asmade/act-2013-055 (accessed April 23, 2020).

Queensland Government. (2013d). *Coal Seam Gas in the Tara Region: Summary Risk Assessment of Health Complaints and Environmental Monitoring Data.* Queensland Government Department of Health.

Queensland Government. (2013e). Cougar Energy fined $75,000 for breaching Environmental Protection Act. http://statements.qld.gov.au/Statement/2013/9/24/cougar-energy-fined-75000-for-breaching-environmental-protection-act (accessed April 15, 2020).

Queensland Government. (2014a). General Beneficial Use Approval: Irrigation of Associated Water (including coal seam gas water). https://environment.des.qld.gov.au/__data/assets/pdf_file/0037/89389/wr-ga-irrigation-associated-water.pdf (accessed April 20, 2020).

Queensland Government. (2014b). General beneficial use approval Associated water (including coal seam gas water). www.ehp.qld.gov.au/assets/documents/regulation/wr-ga-associated-water.pdf (accessed April 23, 2020).

Queensland Government. (2016). Underground coal gasification banned in Queensland. http://statements.qld.gov.au/Statement/2016/4/18/underground-coal-gasification-banned-in-queensland (accessed April 15, 2020).

Queensland Government (2009) Central West Regional Plan, Part E- Regional Policies and Strategies www.dlgrma.qld.gov.au/resources/plan/central-west/cw-plan-part-e-regional-policies-and-strategies.pdf (accessed April 23, 2020).

Queensland Government Budget Strategy and Outlook a. (2017–2018).

Queensland Government Budget Strategy and Outlook b. (2019–2020). https://budget.qld.gov.au/files/BP2.pdf (accessed April 23, 2020).

Queensland Government prosecution bulletins (2012–2019). https://environment.des.qld.gov.au/management/compliance-enforcement/prosecution-bulletins (accessed 23 April 2020)

Robertson B. (2019). IEEFA update: The staggering cost of gas in Australia. Institute for Energy Economics and Financial Analysis, http://ieefa.org/the-staggering-cost-of-gas-in-australia/ (accessed April 23, 2020).

Santos. (2010). GLNG Wins Queensland Environmental Approval. www.santos.com/media/1802/280510_glng_wins_queensland_environmental_approval.pdf (accessed April 23, 2020).

Schandl H, Baynes T, Haque N, Barrett D, and Geschke A. (2019). Whole of Life Greenhouse Gas Emissions Assessment of a Coal Seam Gas to Liquefied Natural Gas Project in the Surat Basin, Queensland, Australia. Final Report. GISERA. Project G2. https://gisera.csiro.au/wp-content/uploads/2019/07/GISERA_G2_Final_Report-whole-of-life-GHG-assessment.pdf (accessed April 23, 2020).

Scott R. (2016). Independent Review of the Gasfields Commission Queensland and associated matters.https://cabinet.qld.gov.au/documents/2016/Oct/RevGasComm/Attachments/Report.PDF (accessed April 23, 2020).

Simply Wall St. (2020). MD, CEO & Director Kevin Gallagher Just Sold A Bunch Of Shares In Santos Limited (ASX:STO). https://simplywall.st/news/md-ceo-director-kevin-gallagher-just-sold-a-bunch-of-shares-in-santos-limited-asxsto/ (accessed April 27, 2020).

Smith P and Cunningham-Foran K. (2018). Court Processes for Environmental and Resource Issues in Queensland, Australia. www.courts.qld.gov.au/__data/assets/

pdf_file/0011/587207/lc-sp-pas-court-processes-for-environmental-and-resource-issues-in-qld.pdf (accessed April 20, 2020).

Somerville W. (2016). Submission 16, Senate Select Committee on Unconventional Gas Mining. www.aph.gov.au/Parliamentary_Business/Committees/Senate/Gasmining/Gasmining/Submissions (accessed April 23, 2020).

Swayne N. (2012). Regulating coal seam gas in Queensland: Lessons in an adaptive environmental management approach. *Environmental and Planning Law Journal*. **29**(2): 163–185.

Tait D, Santos I, Maher D, Cyronak T, and Davis, R. (2013). Enrichment of radon and carbon dioxide in the open atmosphere of an Australian coal seam gas field. *Environmental Science & Technology*. **47**(7): 3099–3104. http://dx.doi.org/10.1021/es304538g (accessed 23 April 2020)

The Straits Times. (2018). World's largest floating LNG plant starts production. *The Straits Times*. www.straitstimes.com/business/economy/worlds-largest-floating-lng-plant-starts-production (accessed April 23, 2020).

Toscano N. (2019). Gas giants sound warnings on Queensland royalty hike, *The Sydney Morning Herald*. www.smh.com.au/business/companies/gas-giants-sound-warnings-on-queensland-royalty-hike-20190612-p51wyu.html (accessed April 23, 2020).

Towler B, Firouzi M, Underschultz J, Rifkin W, Garnett A, Schultz H, Esterle J, Tyson S, and Witt K. (2016). An overview of the coal seam gas developments in Queensland. *Journal of Natural Gas Science and Engineering*. 31: 249–271.

Turton D. (2014). Codifying coexistence: Land access frameworks for Queensland mining and agriculture in 1982 and 2010. *Journal of Australasian Mining History*. 12.

United Nations, Framework Principles on Human Rights and the Environment. (2018). www.ohchr.org/EN/Issues/Environment/SREnvironment/Pages/FrameworkPrinciplesReport.aspx (accessed April 23, 2020).

Underground Water Impact Report (UWIR), 2012, 2016, (2018). www.business.qld.gov.au/industries/mining-energy-water/resources/environment-water/ogia (accessed April 23, 2020).

Victorian Government. (2017). Media Release. Fracking banned in Victoria, giving certainty to farmers. www.premier.vic.gov.au/fracking-banned-in-victoria-giving-certainty-to-farmers/ (accessed April 23, 2020).

Walton A, McRae R, and Leonard R. (2014). *Survey of Community Wellbeing and Responding to Change: Western Downs Region in Queensland*. CSIRO. https://gisera.csiro.au/wp-content/uploads/2018/03/Social-2-Final-Report.pdf (accessed April 23, 2020).

Werner AK, Vink S, Watt K, and Jagals P. (2015). Environmental health impacts of unconventional natural gas development: A review of the current strength of evidence. *Science Total Environment*. 505: 1127–1141.

Werner AK, Watt K, Cameron CM, Vink S, Page A, and Jagals P. (2016). All-age hospitalization rates in coal seam gas areas in Queensland, Australia, 1995–2011. *BMC Public Health*. 16: 125. https://doi.org/10.1186/s12889-016-2787-5

Werner A, Watt K, Cameron K, Vink S, Page A, and Jagels P. (2018). Examination of Child and Adolescent Hospital Admission Rates in Queensland, Australia, 1995–2011: A Comparison of Coal Seam Gas, Coal Mining, and Rural Areas. *Maternal and Child Health Journal*. 22: 1306–1318 https://doi.org/10.1007/s10995-018-2511-4 (accessed April 23, 2020).

Western Australian Standing Committee of Environment and Public Health Affairs. (2015). *Implications for Western Australia of Hydraulic Fracturing for Natural Gas*, Report 42. Western Australian Standing Committee of Environment and Public Health Affairs.

Witt K. (2012). Understanding responsibility in land ownership and natural resource management. PhD thesis, University of Queensland, pp. 42–49.

4

The Governance of Fracking

History, Differences, and Trends

JOHN D. GRAHAM AND JOHN A. RUPP

4.1 Introduction

In the last fifteen years, advances in drilling and completion technologies, colloquially termed "fracking," have been responsible for a rapid increase in US production of oil and natural gas. In 2011 the United States surpassed Russia as the world's leading producer of natural gas; in 2018 the United States surpassed Saudi Arabia as the world's leading producer of oil (EIA 2019). The result has been a sustained plunge in oil and natural gas prices, which have boosted the welfare of oil- and gas-consuming economies around the world. In contrast, the widespread commercial usage of these technologies has been unsuccessful in Europe for political, geological, and economic reasons. Even in the United States, fracking has been widely embraced and applied in some states (e.g., Pennsylvania) yet prohibited in others (e.g., New York).

In this chapter, we seek to explain why fracking is flourishing in some jurisdictions but not in others. Drawing from the tenets of the risk-perception paradigm in cognitive psychology, we show that some degree of public stigmatization of fracking was predictable. We also point to several legal, economic, and political factors that may help explain why oil and gas development has thrived in the United States, factors that operate outside of the risk-perception paradigm. We conclude with some observations about the future of governance of oil and gas development.

4.2 Why Some Public Opposition Was Predictable

Starting in the 1970s, a cadre of social scientists in North America and Europe set out to explain why some technologies are considered by society to be too risky to allow while others retain a social license to operate with less consternation. The specific paradox that those social scientists sought to resolve was that formal risk analysis showed that nuclear power and genetically modified organisms (GMOs) are far less dangerous to society in general, measured by actuarial risks, than coal-fired power and tobacco products, but the perceptions of the risks by the public and politicians of nuclear power and GMOs are much greater than those of coal power and tobacco products (Fischhoff et al. 1978). This led to the restriction and – in some cases – prohibition of nuclear power and GMO products.

As an explanation, the risk-perception paradigm posits that the extent of public outrage about a given technology is less a function of attributable mortality and morbidity and more related to several qualitative attributes of hazards that trigger concern (Slovic 1987). Among others, the following attributes of hazards are each predicted to trigger public concern:

- whether a technology's hazards are imposed on people involuntarily;
- whether the hazards are seen as uncontrollable by the individual citizen;
- whether the hazardous technology is unfamiliar to laypeople;
- whether the hazardous technology could cause multiple people to be harmed at the same time and place; and
- whether the hazards threaten future generations.

A technology that exhibits several of these attributes is expected to stimulate greater opposition than a technology that exhibits only one of them. The manner in which these risk-perception attributes unfold in a public process can lead to further "social amplification" of perceived risk (Kasperson and Kasperson 2005).

A single adverse incident at one time and in one location can be amplified by the mass media and communicated throughout a community, a nation, or the world, thereby raising public perception of risk beyond what experts might see as proportionate to the frequency or severity of the adverse incident. What matters are feelings and emotions, not actuarial risk (Slovic, Finucane, Peters, and MacGregor 2004). The amplification typically unfolds through a process whereby the media portrays villains and victims in ways that stimulate public interest in the drama.

Regulators may be inhibited in their ability to protect the public as they must also accommodate the ability of developers to commercially implement their trades and therefore are often portrayed as villains, captured by commercial interests that they are supposed to be regulating. Advocates of an incriminated technology may inadvertently inflame the drama by making exaggerated claims of safety or by concealing the truth about a hazard or by conveying the notion that they care more about profit or power than protecting public health, safety, and the environment. Environmental and consumer groups typically play a role in the amplification, gathering background information for reporters or informing legislators of regulatory conflicts of interest or organizing grassroots protests in order to draw media attention to a technological hazard (Baker 2005).

As public concerns build, policymakers may decide it is necessary to invoke the "precautionary principle" (implicitly or explicitly) and restrict or remove the social license to operate, at least temporarily (Bourguignon 2015). The public may be reassured by the regulatory determination that the technology will not be used unless and until it is proven safe. If a given technology is perceived to be unsafe, exclusive of objective reality, a phenomenon known as "technological stigma" sets in (Garrick 1998). Once a technology is stigmatized in the eyes of the public and politicians, it is extremely difficult to attenuate or remove that stigma. The stigmatization process is asymmetric, since it is much easier to stigmatize a technology thought to be safe than it is to persuade people that a technology was wrongly incriminated (Slovic et al. 2001).

Elsewhere, we have explained why the risk-perception paradigm, coupled with social-amplification processes and technological stigma, predicts a high level of public concern about unconventional oil and gas development (Graham, Rupp, and Schenk 2015). Because the actual and perceived risks exist at a variety of scales, there are numerous avenues by which people can perceive risks that could cause the harm. These include the fact that the technology is unfamiliar to laypeople; drilling can be employed near someone's home without the resident's consent; concerns about pollution of surface water, groundwater, and/or drinking water are plausible to laypeople; it is difficult for people to control their level of risk without moving out of their residence or converting completely to bottled water; multiple people could be harmed by polluted water at the same time; and future generations could be put at risk by pollution and ecosystem damage. Thus, concerns about localized water contamination alone can be sufficient to trigger social amplification of risk perception.

Other public concerns may arise from much broader negative implications and therefore stimulate a more widespread stigmatization. For starters, the boomtown development, including heavy streams of truck traffic, induced by fracking operations may disrupt the character of small rural communities. In addition, the handling of fracking waste may trigger offsite concerns such as spills from trucks and earthquakes induced by deep-well injection of wastes. Moreover, construction of the infrastructure necessary for successful fracking operations (e.g., pipeline construction) may trigger additional concerns beyond the location of fracking operations. Insofar as oil and gas usage culminate in global climate change, local fracking operations may be perceived negatively in that they contribute to the global scale and long-term impacts of climate change.

Thus, the perceived risks of fracking that contribute to stigmatization vary enormously in breadth, scale, and duration. They range from the very localized (e.g., the quality of drinking water in a neighborhood near an oil and gas development site) to the long-term, global scale (e.g., impact of combustion emissions of greenhouse gases on climate change).

To illustrate the explanatory power of the risk-perception paradigm, we explore several jurisdictions where the social licenses to use advanced fracking technologies have been halted owing to public perceptions of risk, amplification, and stigmatization. Then we take on the harder task of explaining how and why other political jurisdictions, located predominantly in North America, managed to build and sustain sufficient public support for oil and gas development.

4.3 The French Ban First

In 2010, as US and Canadian companies began to employ fracking techniques in North America, several large international energy companies persuaded the Conservative government of President Nicolas Sarkozy to allow some exploratory wells to be "fracked" at several locations in France. Domestic sources of oil and gas were certainly needed in France, as the country was meeting 99% of its needs with imports. It was believed that reserves in France might be as large as 5 trillion cubic meters of technically recoverable shale gas, one of the largest reserve totals among countries in Europe. Two French territories, the Paris Basin and the Southeast, were considered the most promising for exploration.

In March 2010 the French Minister for Ecology, Energy and Sustainability (Jean-Louis Borloo, a centrist French politician) granted France's initial three exploratory licenses for shale gas. Each license was for three to five years in duration and the companies involved were Total (a French multinational), Schuepbach Energy (an American multinational), and Devon Energy (also an American multinational). Meanwhile, in Brussels, the European Commission – as reflected in the public remarks of Energy Commissioner Gunther Oettinger (a well-known German politician) – was encouraging European countries to consider shale gas.

A catalyzing event for opposition to shale gas was the June 2010 release in the US of an HBO documentary called *Gasland*. The movie features interviews of suffering families living near drilling sites in Pennsylvania, Colorado, Wyoming, and Texas; one scene features a family that can actually set their tap water on fire with a lit match, allegedly owing to drinking water contamination from "fracking" (here we use the word fracking in its broader sense to refer to the entire development process). With only limited persuasive effect, the US industry tried to explain that methane contamination of drinking water can occur in any area where natural gas is deposited near drinking water supplies, regardless of whether drilling is occurring.

The release of *Gasland* in June 2010 and the movie's subsequent nomination for an Oscar in February 2011 caused two massive social media spikes around the world (Vasi et al. 2015; Reuters 2015). The amplification occurred not just in the United States but also in France.

Local activists and government officials in France were already disturbed that they had not been consulted as to whether their communities would be sites for fracking experiments. Greenpeace and Friends of the Earth launched their French anti-fracking campaigns by going into town meetings throughout the country, showing excerpts from *Gasland*. The implication was that drinking water contamination by multinational companies could occur throughout France.

Anti-fracking activists also targeted the French mining code, arguing that it was archaic for not facilitating public involvement before exploratory licenses are granted. Under the terms of the code, the "research permits" were granted after a round of competition among companies, without local consultation. The Mayor of a town is informed only at the stage of a declaration of research tasks planned for the municipality. Local citizens must review notices at the town hall if they are interested in tracking developments.

There was much debate as to whether the French mining code was consistent with European law. The Aarhaus Convention calls for public consultation on oil and gas projects but not at the stage of research, development, and testing (which can last for up to two years) unless exploratory work is likely to cause a significant adverse effect on human health and the environment. Borloo's decision to award the three permits seems to be legally defensible but the political prudence of the decision seems questionable given how little French local officials and citizens knew about shale gas development. Since the companies were at the early exploratory phase, without any assurance of commercial success, they were not prepared to invest millions of euros in public relations at such an early stage.

The French public's mobilization against shale gas was rapid, intense, and highly creative. It involved grassroots activists, national NGOs, ordinary citizens, local elected officials, and members of the French Parliament from multiple parties. Local governments enacted decrees prohibiting shale gas exploration. The mayors of Ardeeche and Gard adopted such decrees even though the Texas-based Shuepbach Energy had been awarded a permit to explore in their communities. The larger activist agenda was reform of the mining code, more emphasis on alternative energy sources, and a "frack ban."

From Sarkozy's perspective, a public rebellion against his administration's shale gas policies would certainly not help his forthcoming re-election campaign. While France was suffering from roughly 10% unemployment, any economic stimulus from shale gas was many years away, long after French citizens would go to the polls in April of 2012. Sarkozy's margin of victory in 2007 was not huge (53.1% to 46.9%), and his troubled term was dominated by financial crises and economic turmoil.

The Sarkozy administration quickly reversed its stance on shale gas. Energy Minister Borloo stepped down in November 2010 and expressed regret that he had issued the three shale gas permits (Erlanger 2010). Sarkozy announced in February 2011 a temporary freeze on shale gas exploration. Sarkozy's environment minister Nathalie Kosciusko-Morizet (a former mayor) became the administration's spokesperson on the issue (she later ran Sarkozy's re-election campaign). Her message was hardly nuanced: "I'm against hydraulic fracturing. We have seen the results in the US. There are risks to the water tables and these risks we don't want to take" (Badker 2011).

The Sarkozy administration proceeded to work with the Parliament on a frack ban. It passed the lower house 287 to 146 and later the Senate 176 to 151 (Badker, 2011). The "opponents," largely from the Socialist Party, argued that the ban should have also covered oil and gas development that occurs without hydraulic fracturing (Castelvecchi 2011). One of the American energy companies challenged the ban on constitutional grounds but lost in 2013 at the French constitutional court (Patel and Viscusi 2013).

Sarkozy was defeated in his 2012 re-election effort (51.6% to 48.4%) by Francois Hollande of the Socialist Party. In 2017 the Socialist majority, led by President Emmanuel Macron, legislated a ban on all new oil and gas permits and a termination of all existing permits by 2040 (France-Presse 2017). The measure was largely symbolic as there were only 63 ongoing oil and gas projects in the entire country (Corbet 2017).

In short, the stigmatization of oil and gas development in France, facilitated by the anti-fracking campaign, was complete. The history proceeded in a way that is entirely consistent with what would be predicted by the risk-perception paradigm.

As we lack space, we shall not present the history of the German ban on fracking. It follows a similar script except that it took longer to be enacted (June 2016), Russian interests may have helped stoke opposition to shale gas in Germany (Harvey 2014; Johnson 2014), and the Merkel government secured an exemption for some small-scale test probes, to be done for scientific purposes, and the ban must be reviewed in 2021 (France-Presse 2016).

4.4 The United Kingdom Abandons Fracking

Nowhere in Europe was the interest in fracking stronger than it was in the United Kingdom. Offshore production of oil and gas in the North Sea experienced more than a decade of decline and British energy policymakers, based on the favorable North American experience, looked to revitalize the country's onshore program. Hydraulic fracturing in the UK began as part of technologies used to develop conventional oil and gas fields in the North Sea in the 1970s. It was later used in numerous conventional onshore vertical oil and gas wells.

The scale of potential shale gas reserves in the UK is substantial. The Bowland shale discovery in the north of England by Cuadrilla Resources Ltd was seen as potentially matching the natural gas reserves in the North Sea.

Under UK law, mineral resources are owned by the Crown Estate. Developers must win competitive bids for exclusive drilling rights and acquire landowner and local planning authority permission. Unfavorable local decisions may be appealed to the Secretary of State (Whitton and Colton 2017).

The first license for onshore shale gas development in the UK was issued in January 2008 by the government of Prime Minister Gordon Brown of the Labour Party. The license was granted to Cuadrilla Resources for a test drill in Lancashire in northwest England.

The proposed project triggered significant local opposition owing to five concerns: heavy truck traffic in and out of the community; the industrialization of landscape; risk of water pollution; fear of localized air pollution and site noise; and possible earthquakes (Short et al. 2019). Activists expressed concern that the exploration was not accompanied by an environmental impact assessment; such a document is not required under UK law because the drilling is small in scale and exploratory in nature. Despite the concerns, the exploratory activity was permitted, and the company began drilling at Preese Hall (Lancashire) in August 2010.

In April and May of 2011 several earthquakes (one magnitude 2.3; the other magnitude 1.5) occurred near the Preese Hall drilling site (Marshall 2011). The tremors were not harmful, and tremors of this magnitude are not uncommon in the UK (about 15 per year nationwide). However, these induced earthquakes were a signal that neither the company nor the UK government fully understood what was happening as the technology was being employed. The UK government promptly issued a temporary moratorium on shale gas development and an investigation was undertaken. The government determined that the exploratory drilling and hydraulic fracturing (especially the injection of large volumes of water into the shale) caused the tremors.

In 2013 the coalition government of Prime Minister David Cameron of the Conservative Party lifted the moratorium on shale gas development. New controls were required to minimize seismic risk, including the completion of a pre-project seismic survey and the monitoring of seismic activity both during and post-exploration (Smith-Spark and Boulden 2013). The Cameron-led Coalition government released an ambitious plan to stimulate corporate and local community support for fracking (Scott 2014). It included new economic incentives for companies as well as sharing of financial benefits with councils and affected site communities. A Shale Wealth Fund would distribute 10% of all shale gas tax revenues to local communities. Payments could go directly to households or to local authorities (Pickard 2013). A new Office for Unconventional Gas and Oil (OUGO) was established in the government to oversee and promote shale gas, as Prime Minister Cameron was "going all out for shale." Conservative Party politician George Osborne, Cameron's Chancellor of the Exchequer, took a leadership role on the initiative.

With the new policy, the French energy company Total became the first multinational to commit to shale gas development in the UK. The European petrochemical giant Ineos also expressed interest in UK shale (Jahncke 2014). By mid-2015 there were 75 applications for the next round of onshore leasing bids in the UK (Williams 2015).

More pro-development news occurred in June 2013 when the British Geological Survey upped its estimates of shale gas resources in the UK to 1.3 quadrillion cubic feet, although the amount of recoverable reserve was unknown. Since the UK consumed 3 trillion cubic feet of gas in 2013, even 10% of the 1.3 figure would be equivalent to 40 years of current yearly consumption (Werber and Kent 2013). More pro-development news came from a study by the UK public health department, which concluded that groundwater was not at risk of contamination as long as best practices were followed by regulators and operators. The study emphasized that adverse water impacts in the United States were attributable to failure of operators and regulators to adhere to best practices (Harvey 2013). It was inferred that the UK could learn from the mistakes in the United States and institute a potentially safer program.

The economic promise of significant amounts of natural gas being produced domestically was appealing to UK politicians. Ernst and Young estimated that 64,500 direct and indirect jobs could be created in the UK if 4,000 new onshore wells, developed using advanced technologies, were brought into production (Williams 2015).

Cuadrilla Resources followed with a formal request for exploratory drilling in Lancashire. Under its new policy with the UK government, Cuadrilla offered $151,000 for each well site explored and 10% of any resulting revenues to the local community (Reed 2013).

Cuadrilla's request triggered well-organized protests by anti-fracking groups elsewhere, some of which led to arrests. Protestors met at the village of Balcombe in West Sussex and superglued themselves to buildings, intentionally getting themselves arrested. The protestors included Caroline Lucas, the UK's only Green Party MP (Darby 2014). The Diocese of Blackburn (Church of England) issued a leaflet highlighting the perceived downsides of fracking but did not explicitly endorse or oppose fracking. The leaflet emphasizes the Christian duty to be "stewards of the earth." Activists repeatedly expressed concerns that local stakeholder views were given little weight because decisions were ultimately determined by George Osborne in London (Whitton and Colton 2017).

The anti-fracking campaign gave priority to the Lancashire area where the county council planning committee was divided. A first round of delays occurred as Cuadrilla worked hard to obtain the necessary environmental permits (Graeber 2015). The company was seeking permission to explore at two Lancashire sites between Preston and Blackpool near Little Plumpton and Roseacre. Anti-fracking campaigners urged the county council to refuse the permission. The campaign organization Avaaz gathered more than 43,000 signatures from across the country in less than 24 hours. Senior campaign organizer Bert Wander charged that "George Osborne wants to transform Lancashire into Texas, covering Britain's green and pleasant with gas wells across our hills and valleys" (Staff and Agencies 2015).

In June 2015 the county council planning committee voted nine to three (with two abstentions) against the Cuadrilla applications, citing "unacceptable impact" on landscape and noise levels (Williams 2015). The vote overruled the recommendation of the council's planning staff; it occurred at a meeting that drew a large number of protesters and spectators. For Cuadrilla the setback was frustrating, as it had convened public consultations and submitted thousands of pages of documents over a two-year period. Cuadrilla's only hope was to appeal the unfavorable decision to an independent planning inspectorate and the UK Secretary of State for communities and land (Williams 2015).

Cuadrilla decided to appeal but that process took more than another year. In the interim, a different company, the privately held Third Energy of the UK, was granted permission by North Yorkshire County Council to use hydraulic fracturing at an existing natural gas well in Kirby Misperton (northern England). The council overcame opposition on a seven to four vote (Williams 2016). Ultimately, Lancashire County Council was overruled by the UK Minister of Energy and Growth, Claire Perry; Cuadrilla was also granted permission to proceed (Chestney 2018). Perry's decision was contrary to the recommendations of the National Infrastructure Commission, a group of independent advisors. Jonathan Bartley, co-leader of the Green Party, complained that Perry's decision "was snuck out on the last day of Parliament" (Cockburn 2018).

Meanwhile, opposition to fracking was building in the UK Parliament. The Environmental Audit Committee of the Parliament issued a report calling for a moratorium on fracking. Ineos responded that the report was partial and partisan, as it left out the benefits of natural gas development using fracking (Scott 2015). Three months later, a report organized by the nonprofit group Medact called for a moratorium until public health concerns were addressed. Two dozen health and medical experts signed a letter supporting the moratorium call (Mathiesen 2015). In the Parliament, the Cameron-led Coalition granted modest concessions to anti-fracking sentiments in order to pass a pro-fracking infrastructure bill. The concessions included a ban on fracking in national parks and other protected areas (Staff and Agencies 2015). Scotland, which imposed a fracking moratorium in 2015, went further than the UK Parliament and banned fracking in 2017.

The UK shale gas program came to a surprising and abrupt halt in the summer of 2019 under the Conservative government of Prime Minister Boris Johnson. A large tremor (2.9 magnitude) that caused houses to shake was recorded near Cuadrilla's site near Blackpool, the largest fracking-related tremor experienced in Britain. Operations at the Preston New Road site had already been suspended a few days earlier after a series of "microseismic events." A spokesperson for Frackfree Lancashire said, "Enough is enough ... We are sick of being treated as human guinea pigs" (Halliday 2019).

The UK government reacted strongly, again imposing a moratorium on shale gas development. The UK's Oil and Gas Regulatory Authority followed with a report indicating that they could no longer ensure local communities that there would not be "unacceptable" impacts. Nor could the Authority accurately predict the magnitudes of future earthquakes that might be triggered. Moreover, the government disclosed that they had already expended $32 million to date on the shale gas program, with no production of energy accomplished.

Considering the circumstances, the Johnson government made the determination that the UK's shale program was simply not worth continuing. Further fracking was suspended indefinitely "until compelling new evidence" is provided about safety. The decision was a major U-turn for the Conservative Party and Boris Johnson (Ambrose 2019).

In summary, the UK ultimately landed in the same place as France and Germany, though after much more effort was taken to launch a shale gas development program. The UK experience brought a new risk to the forefront, induced seismicity, which triggers several of the attributes in the risk-perception paradigm. This risk, in combination with a series of unknown risks and the perceptions that these could result in significant harm to health and human welfare in the development regions, led to the decision by the national government to halt onshore oil and gas development using fracking. The UK experience underscores the critical role of national political leadership coupled with the sentiments of local constituents.

4.5 The United States Experience

We have argued that the risk-perception paradigm provides an important part of the explanation of why France, Germany, and the UK opted to prohibit shale gas development using fracking, at least for the foreseeable future. Now we turn to the more difficult question of why shale-gas development flourished in parts of the United States.

Before considering what has transpired in individual states, a few overarching points are important to consider about development of oil and gas in the United States. One point concerns ownership of mineral rights; the other concerns national political leadership.

4.5.1 Mineral Rights and Royalty Income

The United States is one of only a few countries in the world that associates the rights to minerals underneath a landowner's real estate with their land ownership. This policy dates to the initial formation of the United States when the new nation sought to elevate the ability of individuals to exploit their private commercial interests over those of the British Crown. As a result, oil and gas production agreements in the United States are private agreements (contracts) between a commercial developer and the mineral estate owner and are governed by civil law. Only about 15% of oil and gas reserves are located on federally owned lands, and thus most royalties from the production of oil and gas in the United States are paid by developers to private parties rather than to the federal or state governments. These royalties are usually paid to the landowners who reside above the oil and gas deposits.

If a developer wants to drill on a private parcel of land, they must purchase or lease the rights to the minerals from the landowner (King 2020). Usually, the rights are leased. The developer negotiates with the landowner to lease the right to access and produce the oil and gas from under their property. Typically, the landowner and the mineral rights owner are the same person, but sometimes they are different people, a circumstance termed "severed" mineral rights. In addition to royalties on produced oil and gas, there is often a one-time cash incentive to lease, termed a "signing bonus," which is paid at the signing of the lease, typically $200–$500 per acre. If the drilling is unsuccessful, the bonus may be the only money that is paid by the developer to the resident.

Oil and gas royalties are typically a percentage of the value of the oil or gas produced and are usually paid monthly. The amounts can be quite significant, depending on the productivity of the site. The amount is typically a negotiated percentage (a minimum of 12.5% in Pennsylvania, but more commonly in the range of 18 25%) of the revenue the developer makes from sale of the gas. It is not uncommon for royalty checks to be $5,000 to $10,000 per well per month (2017). And such revenue streams last over the life of a well, which may be many years. Approximately 12 million landowners in the United States receive royalties for the exploitation of oil, gas, or other minerals under their property (Cusick 2018).

The financial benefits to surface rights and mineral rights owners in the United States have important ramifications for the sustainability of the social license to operate, ramifications that are not considered fully by the risk-perception paradigm. Recipients of signing bonuses and royalties often become ardent advocates of shale gas development. In addition, as beneficiaries are typically residents and citizens in the local jurisdiction where the development occurs, a local political constituency in support of shale gas development is often created.

The risk-perception literature does include some studies suggesting that, if citizens perceive benefits from taking technological risks, they are sometimes inclined to take those risks. Some of this literature concerns financial compensation to nearby residents for the siting of a noxious facility but other literature considers more generally the perceived benefits of hazardous technology (Kunreuther 1987; Himmelberger 1991; Kunreuther et al. 1993). This risk benefit literature has not yet been formally incorporated into the risk-perception paradigm, but it needs to be considered to understand the US acceptance of shale gas.

One might be tempted to conclude that the United States' exceptional royalty scheme is the dominant factor in explaining the difference between the politics of shale gas development in the United States versus that in Europe. It is undoubtedly an important factor, but the issue is more complicated. The Canadian government has also developed a robust oil and gas program using advanced technologies, but in Canada 89% of mineral rights are owned by the government (the Crown) and only 11% by private individuals and corporations (Wyatt 2015). On the other hand, the State of New York issued a ban on fracking in late 2014 even though there were numerous private landowners and local officials in the depressed communities near the Pennsylvania border – the "Southern Tier" counties – that advocated for shale gas development to be permitted, at least in their communities (Kaplan 2014). Thus, private ownership of mineral rights is only one of several factors that are different in the United States.

4.5.2 National Political Leadership

The Arab Oil Embargo of 1973–1974 stimulated strong public demand for alternatives to US reliance on imported oil from the Middle East and Venezuela. A series of US presidents supported unconventional oil and gas development, even though it has become politically riskier to do so in recent years.

The Jimmy Carter, Ronald Reagan, and George Herbert Walker Bush administrations, with crucial support from the US Congress, allocated $137 million in federal monies from 1978 to 1992 to the Department of Energy's Eastern Shale Gas program. The knowledge generated helped develop and commercialize many of the advanced technologies that are in widespread use in the United States today. The applied R&D showed how to optimize directional drilling, undertake micro-seismic monitoring of multi-stage hydraulic fracturing, and model what is happening 10,000 feet below the earth's surface as a shale gas development unfolds.

Under the Bill Clinton, George W. Bush, and Barack Obama administrations the Department of Energy's Office of Fossil Energy was funded to engage in a wide range of research relevant to developing a viable shale gas/oil industry: water quality and availability; management of wastewater; induced seismicity, methane emissions, and subsurface science; minimizing ecological disruption at the surface; and transportation and storage. Shale gas development received bipartisan support in the US political system.

The "fracking boom" did not start until the very end of President George W. Bush's second term but Bush set the stage for the boom in several ways (Graham 2010; Haskins 2014; Rapier 2016). Bush invested significant political capital in passage of the Energy Policy Act of 2005, which deferred regulation of fracking to the states and exempted developers from selected EPA regulations under the Clean Water Act, the Clean Air Act, the Safe Drinking Water Act, the Resource Conservation and Recovery Act, and the Superfund Law. The 2005 law had the practical effect of allowing state regulators to treat fracking as an extension of existing oil and gas development methods, making them seem more familiar. Bush's view was that state regulators were better positioned to address these issues than the EPA but he took enormous criticism from environmentalists for these exemptions. The Act also gave the Federal Energy Regulatory Commission new authority to supersede state and local decision-making on new pipelines and infrastructure for natural gas. The tax cuts passed by Congress at Bush's request also provided tax advantages to oil and gas developers that invested in innovative technologies. And Bush expanded leasing of federal land for oil and gas development, a step that was important to getting North Dakota's surge of production off to a strong start.

Much to the surprise of environmental activists, President Barack Obama became a determined advocate of fracking, and was proud of the boom in US oil and gas production that occurred during his presidency. The importance of Obama's stance was not related to specific fracking-related legislation that he advocated or to specific regulations that he adopted that facilitated fracking. What was important is that the leader of the national Democratic party, through his public speeches, was defining the mainstream position of the Democratic party, thereby isolating the positions of anti-fracking groups as extremist

(Cockerham 2013). After Obama's re-election in 2012, he was confronted by anti-fracking protesters outside a town hall event at Binghamton in upper-state New York. Protestors greeted Obama with signs saying "No Fracking Way." The protests had no effect, as Obama remained pro-fracking throughout his presidency; indeed, Obama and his secretary of state, Hillary Clinton, sought to expand fracking throughout the world, especially in China.

GOP President Donald Trump was cautious as a candidate, taking the stance in Colorado that local communities should be able to decide whether they wish to allow fracking. Hillary Clinton took a similar position. As president, Trump became more strongly pro-fracking as he tried to repeal or loosen several of the Obama-era regulations regarding methane emissions from oil and gas sites and fracking on federal public lands (Milman 2017; Reuter 2018).

The stances of national US politicians are not the whole story. There are also state-specific reasons why oil and gas production surged in the United States from 2008 to 2020. In 2018 the five leading oil-producing states, measured by millions of barrels of crude oil, were Texas (1.61), North Dakota (0.46), New Mexico (0.25), Oklahoma (0.20), and Colorado (0.18). The leading natural gas producers, measured by trillion cubic feet of dry natural gas, were Texas (6.84), Pennsylvania (6.12), Louisiana (2.78), Oklahoma (2.70), and Ohio (2.35). We consider here some of the key public acceptance factors in Texas, North Dakota, and Pennsylvania.

4.5.3 Texas

The risk-perception paradigm posits that familiarity with a technology tends to foster public acceptance. In no state is oil and gas development more familiar than the state of Texas. Development is not concentrated in any single part of the state: the Barnet Shale (from Dallas west and south), Eagle Ford Shale (from the Mexican border to East Texas), Permian Basin (west Texas), Granite Wash (Texas Panhandle), and Haynesville Shale (East Texas).

As fracking came into widespread use after 2010, the State's regulatory body, the Texas Railroad Commission (TRR), did not treat it as a new technology that required an entirely distinct regulatory system. Instead, TRR used basically the same standards and permitting procedures, and considered this application of advanced drilling and completion technologies to be only an extension of conventional technologies.

The mission, leadership, and structure of the TRR is also unusual. The multi-faceted mission of the agency is to prevent waste of natural resources, protect the rights of property owners, prevent pollution, and provide safety. The agency does not report to the Governor. It is an independent agency overseen by three commissioners who are selected by voters to six-year, overlapping terms in statewide elections. The Republican Party in Texas is historically much stronger than the Democratic Party in statewide politics, though the Democrats periodically capture the Governorship (e.g., Ann Richards from 1991 to 1995) or a Senate seat (e.g., Lloyd Bentsen from 1971 to 1993). The elected commissioners tend to be Republicans with experience in the oil and gas sector. A full-time executive director runs the agency, supervises a highly experienced career staff, and reports to the Commissioners. Daily activities relate primarily to permitting, field inspections, testing, monitoring, and remediation of abandoned wells, but a shortage of inspectors and an antiquated computer system have hampered the agency's effectiveness (Handy 2017). Until 2017 the budget of the TRR was generated predominantly from industry fees; when revenues slumped during a temporary downturn in the oil business, the state legislature responded with a 46% budget increase through a special appropriation. In fiscal year 2020 the budget of the TRR was $162 million, including about 700 full-time employees, by far the largest oil and gas regulatory agency in the United States.

The most important threat to the oil and gas boom in Texas occurred in 2014 in Denton, Texas, a city (population 140,000) in the Dallas-Fort Worth metropolitan area. Residents of Denton, where 280 oil wells operated within the city limits, became annoyed about oil and gas drilling (Eaton 2014a, b). It had nothing to do with water pollution, earthquakes, or climate change; the issues were nuisances, noise, late-night work, and proximity of drilling to churchyards, school properties, college campuses, and suburban developments. One particular developer, Range Resources, annoyed local residents in multiple incidents (Goldenberg 2014).

City officials tried to address the concerns by adopting a 1,200-feet setback requirement. Many citizens were dissatisfied, in part because the new setback rule could not solve a situation in South Denton where drilling was occurring 200 feet from homes. Years before the houses were built, vertical oil wells were drilled in the area and then abandoned. The homes were built later. In 2013 Eagle Ridge Energy re-drilled the same wells deeper and horizontally through the Barnet Shale. Since the well site existed before the homes, the city's new setback requirement did not apply (Sakelaris 2014). Local activists took matters into their own hands and forced an anti-fracking proposition on to the ballot, over objections from the city.

In a stunning result, the proposed ban on fracking passed on November 4, 2014 by a comfortable margin of 59% to 41% (Eaton 2014b). The same city had voted in 2012 for Mitt Romney over Barack Obama 65% to 35%. If a fracking ban could pass in this community, could it pass in numerous communities in Texas and throughout the country?

The powerful oil-and-gas lobby in Texas, with support from GOP Governor Greg Abbott, decided that an aggressive response was necessary. There were numerous owners of surface and mineral rights who supported aggressive action (Sakelaris 2015). Within six months, the Texas legislature passed, and Governor Abbott signed, a new law that prohibited local bans of hydraulic fracturing; the law does allow local governments to enact "reasonable" restrictions (Gold 2015). Some experts say that the courts may have overruled the Denton ban even if the legislature had not acted. The ban might be construed as a legitimate use of local government police or zoning power, but it is also possible that the local ban would have been ruled a "taking" under the Fifth Amendment of the US Constitution, since owners of surface and mineral rights were harmed without compensation (Matthews 2014).

In summary, Texas was successful in commercializing fracking because the public is familiar with oil and gas development, the use of these new technologies was framed as an extension of the application of existing conventional technologies, and the TRR was viewed as capable of regulating the industry. The attempted frack ban in Denton was quashed by the GOP-controlled Texas legislature.

4.5.4 North Dakota

North Dakota's surge to #2 among US states in oil production is explained by the rich oil resources contained within the Bakken Formation, a gas-rich unit that underlies a 200,000 square mile region underlying parts of Montana, North Dakota, Saskatchewan, and Manitoba. It resides entirely in the subsurface and was not even discovered until 1953. The application of advanced technologies was necessary to make the Bakken resources into an economically recoverable reserve. The boom began in 2007–2008; the state first reached 1 million barrels per day (bpd) in 2014 and topped 1.5 million before tapering temporarily when global oil prices collapsed. Some forecasters project the state will surpass 2 mbpd prior to 2030 (Springer 2019).

For the historically depressed economy of North Dakota, development of Bakken oil led to a boom in economic performance. Even in the depths of the Great Recession (2007–2009) and its aftermath, which pushed US unemployment over 10%, the rate of unemployment in North Dakota remained below 4%. Since 2010, the state has led all 50 states with an annual rate of economic growth of 5.8%. The population of the state also surged 13% from 2010 to 2018 alone, creating what is now called the "Third Dakota Boom" (the first two were railroad booms) (Springer 2019).

The Third Dakota Boom has not been without its challenges and drawbacks. The pace of economic and population growth has outstripped the state's meager infrastructure and created tensions in small-town life. The North Dakota Industrial Commission, though designed in a similar way to the Texas Railroad Commission, has been poorly resourced and thus unable to mount a serious enforcement effort against unscrupulous developers or protect the state from growth in the frequency of oil spills and oil-waste overflows during extreme weather events (Sontag 2014). There have also been ecosystem damages at historic sites and adverse impacts on Native American tribes (Dawson 2014; Rusya 2015).

From a climate-policy perspective, a significant drawback of the North Dakota oil boom has been the widespread practice of flaring natural gas (methane) into the atmosphere, as methane is a highly potent greenhouse gas. Developers in the state are focused on oil production, since that is what generates profits, but the same well that produces oil often produces natural gas as well. There is a shortage of pipeline and processing capacity to handle the gas, as low natural gas prices have undercut the profitability of gas production. Well operators are no longer permitted to flare in North Dakota but state policy allows variances when unavoidable circumstances arise and for other reasons (DOE 2019).

The state government's target is to minimize flaring to less than 12% of the total gas produced, but this target is difficult to meet. In mid-2019 flaring spiked to 24%, double the state goal, in part owing to the shutdown of several natural gas processing facilities and pipelines (Sisk 2019). If the United States were to adopt a national carbon tax or stringent regulation of greenhouse gases, and if the practice of flaring were covered by such programs, the oil boom in North Dakota could experience a significant setback.

Overall, the successful commercialization of fracking in North Dakota is not readily explained by the risk-perception paradigm. The state did not have a recent history of conventional oil and gas production, so familiarity is hard to argue. However, the state does have a strong history of lignite coal mining as well as a culture of natural resource use for agriculture. The philosophy is that it is permissible for human beings to access natural resources to enhance the human condition. Moreover, the economic benefits of the oil boom to a state with below-average income are a contributing factor to the acceptance of fracking.

4.5.5 Pennsylvania

The emergence of Pennsylvania as a leading state in production of natural gas is a remarkable development. The state is reputed to have drilled the world's first oil well near Titusville in 1859 and the state was the nation's leading producer of oil in the 1890s. But, by the end of World War II, oil and gas development in Pennsylvania had receded to the point of national insignificance.

The turnaround is related to the discovery of the Marcellus Shale, which extends throughout much of the Appalachian Basin beneath parts of Pennsylvania, West Virginia, New York, and Ohio. The U.S. Geological Service estimates that Marcellus Shale contains 84 trillion cubic feet of technically recoverable reserves as well as 3.4 billion barrels of natural gas liquids (USGS 2020). Without fracking, this shale resource is not recoverable. The first well was drilled in 2004 followed by four more in 2005. The surge in drilling began in 2008 under the administration of two-term Democratic Governor Ed Rendell (2003–2011), with 3,000 wells drilled in 2010 alone (Woodall 2016).

The industry grew so rapidly that the state's Department of Environmental Protection did not have sufficient staffing, resources, or expertise to evaluate the surge in permit applications that occurred. (Wypijewski 2018) The result was serious mismanagement of wastewater that put the entire Marcellus Shale opportunity at risk of public rejection.

Specifically, when a hydraulic fracturing operation on a well is performed, it generates "flowback water." Approximately 10% of the water injected in the stimulation – often 300,000 to 800,000 gallons – may return to the surface. A typical operation in Pennsylvania has more flowback water than a typical operation in Texas, which partly explains why Pennsylvania was not well prepared. Since that water contains the small concentrations of the chemical fracking fluids used during the stimulation, care is necessary as those chemicals can be toxic and carcinogenic. In addition, during the productive life of the well, the flow of natural gas may be accompanied by "produced water" that contains natural contaminants such as radiative elements and heavy metals. Most of the wastewater generated from shale gas development is produced water. These wastewaters must be managed carefully and responsibly to avoid significant environmental and public health risks.

For several years, the state of Pennsylvania allowed the inappropriate practice of disposing of untreated wastewater in sewage treatment facilities. The practice is inappropriate because those facilities are not effective at handling many of the contaminants in the wastewater. The practice ended in 2011 pursuant to a voluntary agreement (EPA banned the practice nationwide in 2016), but not before radioactive materials, toxic metals and other contaminants were detected in surface waters at several locations in the state, including at drinking water intake points along the Monongahela River (Johnson 2013; Wilson and Vanbriesen 2013). The discovery of this problem caused considerable public backlash against both industry and government, forcing new solutions for the wastewater problem.

The preferred solution for managing these produced waters is deep-well injection but Pennsylvania has few sites that are considered geologically suitable for deep-well injection. In the 2010–2014 period, Pennsylvania had only seven active sites compared to 74 in West Virginia, 159 in Ohio, and over 7,000 in Texas (Cheremisinoff and Devletshin 2015). Promising sites are often wells that were previously drilled for oil and gas recovery using conventional methods.

In the near term, the solution for Pennsylvania's developers was a blend of water recycling and trucking of wastewater to deep-well injection sites in Ohio. In 2012 Portage County, Ohio (part of the Akron area), with its 15 active waste-injection wells, accounted for 16.6% of the state's injected wastewater, about two-thirds from out of state (primarily Pennsylvania and West Virginia). Portage has a disproportionate share of injection wells because the wells sit at the northern edge of the Clinton sandstone that was drilled extensively in the 1970s and 1980s. The volume of wastewater injected in the County rose 18% in 2012 compared to 2011, and two additional wells were permitted but not yet active (Downing 2013).

Grassroots opposition to deep-well injection of wastes in Ohio grew out of a fear that the practice was a threat to aquifers that supply drinking water. An unexpected, frightening experience in Youngstown, Ohio triggered new lines of public concern.

From March to November 2011, nine small earthquakes – none severe enough to cause loss of life or property damage – were recorded in and around Youngstown. No earthquakes had previously been detected in the area. The Ohio Department of Natural Resources (ODNR) investigated the situation and concluded that a deep-well injection site, Northstar 1, the only such facility in the Youngstown area, was the likely cause of the tremors. In January 2012 ODNR suspended disposal of wastewater (which was largely from oil and gas operations) within a five-mile radius of the Northstar 1 well (Narcisco 2014). Further investigations uncovered something that was previously unknown: a fault line ran beneath the site of Northstar 1.

An in-depth study of the Youngstown area reported 109 small earthquakes from January 2011 to February 2012 (Kim 2013). Six of the tremors were large enough to be felt. The study correlated the timing of the tremors to the operations at the Northstar 1 site. The authors concluded that the deep-well injection of wastes increased pore pressure along the preexisting subsurface faults located close to the wellbore. The issue was complicated further by evidence that some fracking wastes were dumped illegally into the Mahoning River. A national group, Food and Water Watch, led 35 other groups (including Buckeye Forest Council, Frack-Free Ohio, and Frack Action Columbus) in an effort to persuade the Governor of Ohio, Republican John Kasich, to ban underground injection of fracking wastes in the state of Ohio. Food and Water Watch's efforts were part of "Global Frackdown," an international anti-fracking event (Bell 2013).

The Kasich administration declined the request for a ban, arguing that deep-well injection is the "safest method" known for dealing with fracking wastes. At the time of Kasich's decision, Ohio was beginning to develop its own shale gas development activity along the Utica Shale in eastern Ohio.

Instead of relying on trucking wastes to Ohio, the administration of Republican Governor Tom Corbett worked with developers to find more suitable sites for deep-well injection in Pennsylvania. But activist groups were spreading the word that deep-well injection is unsafe. As fast as promising new sites could be identified, local communities would act to oppose or ban them. Such bans were enacted in Grant Township PA, Indiana County PA, and Highland Township, Elk County PA. The Pennsylvania DEP pushed back on the grounds that the agency had the authority to permit the conversion of former producing wells into drilling-waste disposal sites. Litigation ensued that took several years to resolve (Phillips 2015). The state did not begin to permit new sites for deep-well injection of wastes until 2018 (Legere 2018).

The environmental, health, and safety concerns about fracking in Pennsylvania created strong fissures in the state Democratic Party. In the summer of 2013, the Democratic state committee voted 115–81 for a position statement that included a moratorium on fracking until the health and environmental concerns were resolved. Several Democrats seeking the nomination to challenge GOP Governor Corbett criticized the moratorium stance, and former Democratic Governor Ed Rendell called the position statement "very ill advised."

The risk-perception paradigm, which would certainly predict a moratorium in Pennsylvania, given the events that transpired, does not account for several factors that operate in favor of a continuation of fracking. The factors are more political and economic in nature and relate less to the psychology of risk perception.

First, the President of the United States, Democrat Barack Obama, was traveling the country urging support of his pro-fracking policies, "We've got to tap into this natural gas revolution that's bringing energy costs down in this country" (Jacksonville, Florida, July 25, 2013). Obama made a similar statement in his February 2013 State of the Union address (Vickers 2013).

Many Democratic leaders were inclined to support Obama's position. Marilyn Levin voted against the moratorium in the state committee in her capacity as party Chairwoman of Dauphin County, located in the Harrisburg area. Levin argued, "It would be wrong for us to be in opposition to our president who is trying very hard to lead this country out of the biggest economic crises since the Depression" (Vickers 2013).

Second, the Marcellus Shale gas boom was creating new jobs. From 2007 to 2012 alone, the number of jobs in Pennsylvania's oil and gas industry grew from 5,829 to 15,114, a 259.3% increase. The percentage gain was the largest in any oil and gas producing state except North Dakota, which was in the midst of the Third Dakota Boom (Cruz, Smith, and Stanley 2014). Each job created in the oil and gas sector creates related jobs in natural gas processing, pipelines, power plants, and chemical factories that use natural gas as a feedstock.

In Pennsylvania politics, the "Steamfitters Union," typically a core constituency of the Democratic Party, is strongly pro-fracking. Shell's large ethane cracker in Beaver County, Pennsylvania was employing 1,500 steamfitters during the peak point of its construction. Since industrial labor unions typically join with oil and gas companies to advocate fracking, some political experts describe it as "suicidal" for an elected Democratic official in Pennsylvania to oppose fracking (Frazier 2018).

Finally, in 2011 the Corbett administration and the GOP legislature collaborated on creation of a new Impact Fee that helped defuse local opposition to shale gas development. The Impact Fee is assessed on each well in the state. Unlike revenues from severance taxes, which are typically allocated to statewide needs, about 60% of the revenues from the Impact Fee in Pennsylvania are allocated directly to county and municipal governments; 40% goes to statewide projects. The counties where shale gas development occurs receive a disproportionate share of the county revenues (Foran 2014). In 2018, the Impact Fee generated a record total of $243 million; about $130 million of the collected fees in 2018 were allocated to county and municipal governments that host shale wells (Legere 2019). Without fracking, such revenues would disappear, which helps explain local public acceptance of fracking (Paydar et al. 2016). Note, however, that the Impact Fee's allocations do not necessarily ensure that state and local regulators of oil and gas development have sufficient resources to do their job.

Given these considerations, it should not be surprising that Democrat Tom Wolf (2015–2018; re-elected 2019–2022) did not advocate or seek a moratorium on fracking in his first term. His administration tightened regulations on fracking to reduce methane emissions and facilitated construction of new pipelines to bring natural gas to out-of-state markets. Nor has Wolf advocated a moratorium in his second term even though the Republican-controlled legislature blocked his proposal to supplement the Impact Fee with a new severance tax. Much to the consternation of environmental groups, Wolf played an important role in the construction of the 300-mile Mariner East pipelines that are intended to transport natural gas liquids from western Pennsylvania to a huge refinery in the Philadelphia area for potential export to European and other global markets (Associated Press 2019).

4.6 Looking Forward

The bans on fracking in France, Germany, and the United Kingdom were predictable given the previous literature on public risk-perception of new technologies. Several legal, political, and economic (royalty-allocation) factors in the United States acted to sustain the application of advanced technologies for oil and gas development, factors that are not necessarily considered in the risk-perception literature. Some of those factors are operating nationwide while others are relevant only in selected states.

The private ownership of land and mineral rights has created a potent constituency that supports continuation of oil and gas development in the United States. Citizens who receive royalty payments (or own land and mineral rights that could generate royalty payments in the future) will not allow fracking to be prohibited without a major political fight. Oil and gas companies also will contribute to the pro-fracking side of the fight, but the role of private citizens interested in their royalties may be at least as important as the power of the big energy companies. More research is needed on the political activism of royalty recipients and whether their activism attenuates during periods of depressed oil and gas prices.

The pro-fracking stances of both Democratic and Republican presidents have played an influential role in marginalizing the stances of anti-fracking activists. Those activists have been forced to play a state-local level strategy rather than a national strategy of legislation to ban fracking, as was adopted in the European countries. It is much more expensive and time consuming for activists to fight fracking in each of the 50 states and thousands of localities than it is to persuade a sympathetic US president to champion their cause.

Since oil- and gas-producing states are crucial to any GOP candidate's path to the presidency, one should not expect any serious GOP candidate for president to be anti-fracking. However, future Democratic presidents may not be as enthusiastic about fracking as Obama was. The major candidates for the Democratic nomination in 2020 were split between those advocating a ban on fracking (e.g., Sanders, Warren, Steyer, and Buttigieg) and those unwilling to advocate a ban (e.g., Biden, Klobuchar, Bennett) but supporting stronger regulation. Notably, in the 2020 general election campaign against Trump, Biden advocated science-based regulation of fracking on private lands rather than a ban; he did support a ban of fracking on public lands, which accounts for a limited share of fracking in the United States.

Could surprising new scientific evidence (e.g., pollution-related diseases) or an unexpected calamity (e.g., an earthquake with substantial loss of life and property damage) turn the tide against the use of fracking in the United States? Possibly, but the more likely scenario is that such new circumstances would lead to a targeted tightening of the permitting regime for fracking in order to address the new concerns. This is what happened in both Pennsylvania and Ohio when adverse events occurred.

In the major oil-producing and gas-producing states, significant factors omitted from the risk-perception paradigm suggest that continuation of fracking is likely in the United States. Those factors include the perception that these advanced technologies are not completely new technologies but an extension or refinement of conventional drilling techniques; a perception that expanded oil and gas production provided a much needed boost to US economic performance, both nationally and locally; significant support for fracking from industrial labor unions, a core constituency of the Democratic party; and a normative view in resource-extracting states that it is appropriate for mankind to use the land for the material improvement of the human condition.

As mentioned earlier, the perceived risks associated with oil and gas development can exist at multiple temporal and spatial scales. The much localized, short-term implications of methane contamination of drinking water are quite different from the linkage to long-term, global scale impacts of climate change. Since perceptions of the different kinds of risks vary widely, the opponents of fracking operations will vary the anti-fracking message by emphasizing those risks that are most likely to be persuasive to the audience.

If there is a single concern that could derail or curtail shale gas/oil development in the United States, it is climate-change policy. Producing oil and gas using hydraulic stimulation techniques is highly energy-intensive and a significant source of the methane emissions linked to climate change. The use of oil and gas is also a major source of carbon dioxide emissions. Any national climate policy that puts a stiff price on greenhouse gas emissions would raise the cost of oil and gas compared to renewable sources of energy, making fracking less economical and reducing the demand for oil and gas. The risk-perception literature suggests that major changes in public knowledge and climate literacy are required to stimulate grassroots public engagement on climate change (Lee et al. 2015). Predicting when the United States will adopt a serious national climate policy is fraught with unknowns and beyond the scope of this chapter.

References

Ambrose J. (2019). *Fracking Halted in England in Major Government U-Turn*. The Guardian. November 1, 2019.
Associated Press. (2019). *FBI Investigating Gov. Tom Wolf's Administration over Mariner East Pipeline Permits*. Triblive.com. November 13.
Badker M. *France to Ban Fracking*. (2011). Businessinsider. May 12, 2011.
Baker V. (2005). Greenpeace v Shell: Media exploitation and the Social Amplification of Risk Framework (SARF). *Journal of Risk Research*. 8(7–8): 679–691.
Bell J. (2013). Anti-Fracking Movement Asks Kasich to Ban Injection of Waste in Underground Wells. October 18, www.bizjournas.com/columbus/news/2013/10/18
Bourguignon D. (2015). The Precautionary Principle: Definitions, Applications and Governance. Think Tank. European Parliament. Europarl.europa.ed. September 12.
Castelvecchi D. (2011). *France Becomes First Country to Ban Extraction of Natural Gas by Fracking*. Scientific American. June 30, 2011.
Cheremisinoff NP and Davletshin A. (2015). *Hydraulic Fracturing Operations: Handbook of Environmental Management Practices*. Scrivener Publishing.
Chestney N. (2018). *First Fracked Shale Gas Site Gets Go-Ahead in UK*. Energy and Environment News. July 25, 2018.
Cockburn H. (2018). *Government Gives Last-Minute Go-Ahead to UK Fracking Site*. The Independent. July 24, 2018.
Cockerham S. (2013). Obama Position on Fracking Leaves both Sides Grumbling. *McClatchy DC Bureau*. August 23, 2013.
Corbet S. (2017). *France to Ban Oil, Gas Output on Home Soil in Symbolic Step*. Apnews. September 6, 2017.
Cournil C. (2013). Adoption of Legislation on Shale Gas in France: Hesitation and/or Progress? European Energy and Environmental Law Review. *Kluwer Law International*. 22(14): 141–151.
Cruz J, Smith PW, Stanley S. (2014). *Shale Gas Boom in Pennsylvania: Employment and Wage Trends. US Bureau of Labor Statistics*. Bls.gov. February.
Cusick M. (2018). *Millions Own Gas and Oil under Their Land. Here's Why Only Some Get Rich*. NPR.com. March 15.
Darby M. (2014). *Fracking Hell: What's the Future?* Climatechangenews.com. July 28, 2014.
Dawson C. (2014). North Dakota Reacts to Drilling Critics. Wall Street Journal. January 29, A3.
Department of Energy. (2019). *North Dakota Natural Gas Flaring and Venting Regulations: Office of Oil and Natural Gas*. Energy.gov. Retrieved December 15, 2019.

Downing B. (2013). Portage County is No. 1 in Ohio for Injecting Drilling Wastes. *Akron Beacon Journal*. July 14.
Eaton L. (2014a). Fracking's Unlikely Battle Ground. *Wall Street Journal*. November 4, B1.
Eaton L and Molinski D. (2014b). Texas Fracking Ban Faces Industry Challenge. *Wall Street Journal*. November 6, B2.
ECONorthwest. (2019). *The Economic Costs of Fracking in Pennsylvania*. Report for Delaware Riverkeeper Network. May 14.
EIA. (2019). The US Leads Global Petroleum and Natural Gas Production with Record Growth in 2018. Eia.gov. August 20, 2019.
Erlanger S. (2010). *Sarkozy Appoints a More Rightist Cabinet in Reshuffle*. New York Times. November 14, 2010.
Fischhoff B, Slovic P, Lichtenstein S, Read S, and Combs B. (1978). How safe is safe enough? A psychometric study of attitudes towards technological risks and benefits. *Policy Sciences*. 8: 127–152.
Foran C. (2014). Is PA Wasting Its Fracking Wealth? *National Journal*. April 15.
France-Presse Agence. (2016). *German Government Agrees to Ban Fracking After Years of Dispute*. The Guardian.com. June 24, 2016.
France-Presse Agence. (2017). *France Bans Fracking and Oil Extraction in All of Its Territories*. The Guardian.com. December 20, 2017.
Frazier R. (2018). *In Pennsylvania, Unions Throw Political Weight behind Natural Gas*. Stateimpact.npr.org. March 27.
Garrick J. (1998). Technological stigmatism, risk perception and truth. *Reliability Engineering and System Safety*. 59(1): January, 41–45.
Gold R. (2015). Texas Prohibits Local Fracking Bans. *Wall Street Journal*. May 19, A3.
Goldenberg S. (2014). *How a Ruby-Red Texas Town Turned against Fracking*. The Guardian.com. December 2.
Graeber DJ. (2015). *British Fracking Takes Early Steps*. UPI.com. January 16, 2015.
Graham JD. (2010). *Bush on the Home Front: Domestic Policy Triumphs and Setbacks*. Indiana University Press.
Graham JD, Rupp JA, and Schenk O. (2015). Unconventional gas development in the USA: Exploring the risk perception issues. *Risk Analysis*. 35(10).
Halliday J. (2019). *Latest Fracking Tremor Believed to Be UK's Biggest Yet. The Guardian*. August 26.
Handy RM. (2017). *Shortage of Inspectors Leaves Thousands of Oil, Gas Wells Unmonitored*. Houston Chronicle. February 8.
Harvey F. (2013). *Dangers of Shale Gas Extraction Mainly Come from Operational Failure, According to Public Health England. The Guardian*. October 31, 2013.
Harvey F. (2014). *Russia 'Secretly Working with Environmentalists to Oppose Fracking'. The Guardian*. June 19, 2014.
Haskins J. (2014). *Like Low Gas Prices? Blame Bush*. USNEWS.com. December 5.
Herron J. (2012). Jitters Threaten Fracking in UK. *Wall Street Journal*. April 4, 2012, B9.
Himmelberger JJ, Ratick SJ, White AL. (1991). Compensation for risks: Host community benefits in siting locally unwanted facilities. *Environmental Management*. 15(5): 647–658.
Hopey D and Templeton D. (2019). *Governor Wolf Wants More Study about How Gas Drilling Impacts Citizens' Health*. Post-gazette.com. June 19.
Jahncke R. (2014). Plotting an American-Style Fracking Revolution. *Wall Street Journal*. October 13, 2014, A17.
Johnson K. (2014). *Russia's Quiet War Against European Fracking*. Foreignpolicy.com. June 20, 2014.

Johnson J. (2013). Study warns of fracking water contaminants. *Chemical and Engineering News*. 91(40): 41.

Kahya D. (2011). *Can Europe Benefit from Shale Gas?* BBC News. January 20, 2011.

Kaplan T. (2014). *Citing Health Risks, Cuomo Bans Fracking in New York State*. New York Times. December 17, 2014.

Kasperson J and Kasperson RE. (2005). *The Social Contours of Risk: Publics, Risk Communication, and the Social Amplification of Risk*. Volume 1. Earthscan.

Kim W-Y. (2013). Induced seismicity associated with fluid injection into a deep well in Youngstown, Ohio. *Journal of Geophysical Research: Solid Earth*. 118: 1–13.

King HM. (2020). *Mineral Rights*. Geology.com. retrieved, January 22, 2020.

Kunreuther H. (1987). A compensation mechanism for siting noxious facilities: Theory and experimental design. *Journal of Environmental Economics and Management*. 14(4): 371–383.

Kunreuther H, Fitzgerald K, and Aarts TD. (1993). Siting noxious facilities: A test of the facility siting Creda. *Risk Analysis*. 13(3): 301–318.

Lee ML et al. (2015). Predictors of climate change awareness and risk perception around the world. *Nature Climate Change*. 5: 1014–1020.

Legere L. (2018). *Despite Opposition, Feds OK Marcellus Wastewater Disposal Well in Plum*. Pittsburgh Post Gazette. March 8.

Legere L. (2019). *Shale Gas Impact Fees Raise $252 Million, Breaking Record*. Post-gazatte.com. June 27, 2019.

Lerner JS and Kiltner D. (2001). Fear, anger, and risk. *Journal of Personality and Social Psychology*. 81(1): 146–159.

Marshall M. (2011). *How Fracking Caused Earthquakes in the UK*. New Scientist.com. November 2, 2011.

Mathiesen K. (2015). Doctors and Academics Call for Ban on "Inherently Risky" Fracking. *The Guardian*. May 30, 2015.

Matthews M. (2014). Anti-Fracking Laws Versus Property Rights. *Wall Street Journal*. August 1, A9.

Milman O. (2017). Trump Proposes Scrapping Obama-Era Fracking Rule on Water Pollution. *The Guardian*. July 25.

Narcisco D. (2014). Fracking Halted Near Small Quakes. *The Columbus Dispatch*. March 11.

Nieves LA, Himmelberger JJ, Ratick SJ, and White A. (1992). Negotiated compensation for solid-waste disposal facility siting: An analysis of the Wisconsin experience. *Risk Analysis*. 12(4): 505–511.

Patel T and Viscusi G. (2013). *France's Fracking Ban "Absolute" After Court Upholds Law*. Bloomberg.com. October 11, 2013.

Paydar NH, Clark A, Rupp JA, and Graham JD. (2016). Fee disbursements and the local acceptance of unconventional gas development: Insights from Pennsylvania. *Energy Research and Social Science*. 20: 31–44.

Phillips S. (2015). *DEP Delays Deep Injection Well Decisions*. Stateimpact/npr.org. September 8.

Pickard J. (2013). *Osborne Looks to Create World's Most Generous Shale Tax Regime*. Financial Times. July 19, 2013.

Rapier R. (2016). The Irony of President Obama's Oil Legacy. January 15.

Reed S. (2013). *British Company Applies for Shale Gas Fracking Permit*. New York Times. July 5, 2013.

Reuters. (2015). *Gasland: HBO Documentary Key Driver of Opposition to Fracking, Study Finds*. The Guardian.com. September 2, 2015.

Reuters. (2018). *Trump's EPA Proposes Weaker Methane Rules for Oil and Gas Wells*. Reuters.com. September 11.
Rusya E. (2015). *A Look at the Complexity of Fracking in North Dakota*. Envirocenter.yale.edu. October 14.
Sakelaris N. (2014). Railroad Commission Head Talks Denton Frack Ban, What Agency Did Wrong. *San Antonio Business Journal*. November 10.
Sakelaris N. (2015). Denton Frack Ban Spawns Another Bill that Limits City Petitions. *Dallas Business Journal*. May 11.
Scott A. (2014). Fracking: Seeking to Exploit Shale Gas, the UK Brings Big Oil Company on Board. *Chemical and Engineering News*. January 20, 2014, 9.
Scott A. (2015). Fracking under pressure in the UK. *Chemical and Engineering News*. 93 (5): 18.
Short D and Szolucha A. (2019). *Fracking Lancashire: The Planning Process, Social Harm and Collective Trauma*. Geoforum. 98. January 2019, 264–276.
Sisk AR. (2019). *Flaring Reaches Record High amid Pipeline, Gas Plant Shutdowns*. Bismarcktribune.com. August 15.
Slovic P. (1987). Perception of Risk. *Science* 236 (4799): 280–285.
Slovic P, Flynn J, and Kunreuther H. (2001) (eds.) *Risk, Media and Stigma: Understanding Public Challenges to Modern Science and Technology*. Routledge.
Slovic P, Finucane ML, Peters, E, and MacGregor DG. (2004). Risk as analysis and risk as feelings: Some thoughts about affect, reason, risk and rationality. *Risk Analysis*. 24: 311.
Smith-Spark L and Boulden J. (2013). *UK Lifts Ban on Fracking to Exploit Shale Gas Reserves*. CNN.com. May 3, 2013.
Sontag D and Gebeloff R. (2014). *The Downside of the Boom. New York Times*. November 22, 2014.
Springer P. (2019). *Leading the Nation in Economic Growth for the Decade, North Dakota Saw its Third Historic Boom. Grand Forks Herald*, December 23.
Staff and Agencies. (2015). *Fracking Decision Due as Lancashire Mulls Major Expansion in North-West*. The Guardian.com. January 28, 2015.
US Geological Service. (2020). How Much Gas is in the Marcellus Shale? Usgs.gov. retrieved January 27, 2020.
Vasi IB, Walker ET, Johnson, JS et al. (2015). "No Fracking Way!" Documentary Film, Discursive Opportunity, and Local Opposition Against Hydraulic Fracturing in the United States, 2010 to 2013. *American Sociological Review*. 80(5), September 1, 2015.
Vickers RJ. (2013). *Democrats Square Off Over Call for Fracking Moratorium in Pennsylvania*. Pennlive.com. August 17, 2013.
Werber C and Kent S. (2013). UK Raises Estimate of Shale-Gas Reserves. *Wall Street Journal*. June 28, 2013, B7.
Whitton J and Colton M. (2017). Shale gas governance in the United Kingdom and the United States: Opportunities for public participation and the implications for social justice. *Energy Research and Social Science*. 26: 11–22.
Williams S. (2015a). UK Panel to Decide on Fracking. *Wall Street Journal*. June 24, 2015, B1.
Williams S. (2015b). UK County Council Rejects Cuadrilla's Fracking Applications. *Wall Street Journal*. June 30, 2015, B3.
Williams S. (2016). UK Region Approves First Fracking Permit Since 2011. *Wall Street Journal*. May 24, 2016, B3.
Wilson J and VanBriesen J. (2013). Source water changes and energy extraction activities in the Monongahela River, 2009–2012. *Environmental Science and Technology*. 47(21), October.

Woodall C. (2016). *The Rise and Fall – and Rise? – of Pennsylvania's Oil and Gas Industry*. Pennlive.com. March 17.
Wyatt CC. (2015). Mineral Rights in Canada. Pipelinenews. November 10, 2015.
Wypijewski J. (2018). *What Happened When Fracking Came to Town. New York Times.* July 31.
Zubrin R. (2014). *Putin's Anti-Fracking Campaign*. Nationalreview.com. May 5, 2014.

Part II
Environmental Analysis

5
Air Quality

ALBERT A. PRESTO AND XIANG LI

5.1 Introduction: Air Quality and Air Pollution Control

This chapter addresses impacts of unconventional oil and gas development and operations on ground-level air pollution and air quality. Exposure to air pollution can negatively impact human and ecosystem health and the built environment (Pope III and Dockery 2006; Pope III, Ezzati, and Dockery 2009; Di et al. 2017; Lefler et al. 2019; Pope et al. 2019). While air pollution has generally decreased in the United States over the past few decades, there are concerns that emissions from unconventional oil and gas may slow or even reverse these downward trends (Moore et al. 2014), thereby increasing risks to human health (Adgate et al. 2014).

In the United States, air quality is regulated by the Environmental Protection Agency (EPA) under the Clean Air Act (CAA) and its amendments. The EPA defines two broad classes of regulated air pollutants: criteria pollutants and hazardous air pollutants. Criteria pollutants are widely distributed, harmful to human health, and regulated by ambient outdoor concentration standards called the National Ambient Air Quality Standards (NAAQS). The criteria pollutants are CO, NO_2, SO_2, O_3, $PM_{2.5}$ (particulate matter smaller than 2.5 μm diameter), PM_{10}, and lead (Pb). Of these, $PM_{2.5}$ dominates human health burden from environmental exposures both in the United States and globally (Brauer et al. 2012, 2016; Burnett et al. 2014). $PM_{2.5}$ exposure is associated with cardiovascular disease (Pope III et al. 2004; Grahame and Schlesinger 2010), lung cancer (Pope III et al. 2002; Raaschou-Nielsen et al. 2013), asthma (Schildcrout et al. 2006; Clark et al. 2010; Anenberg et al. 2018), and decreased life expectancy (Pope III et al. 2009; Correia et al. 2013). O_3, NO_2, and SO_2 are all respiratory irritants that can exacerbate asthma symptoms (Anenberg et al. 2018). Of these, O_3 has the largest public health burden in the United States, second only to $PM_{2.5}$.

Outdoor concentrations of criteria pollutants are regulated by the NAAQS. Each criteria pollutant has its own NAAQS limit designed to protect human health. The NAAQS include both a concentration and a time horizon. For example, the $PM_{2.5}$ NAAQS has both an annual standard of 12 μg m^{-3} and a daily (24-hr) standard of 35 μg m^{-3}; the former protects against health effects resulting from chronic exposure (e.g., cardiovascular disease) while the latter protects against more acute effects such as asthma attacks. The NAAQS are reviewed on a regular basis and updated to reflect the evolving understanding of the relationship between air pollution and health. For example, the annual $PM_{2.5}$ NAAQS was reduced from 15 to 12 μg m^{-3} in 2012, and 1-hour standards for SO_2 and NO_2 were introduced in 2010.

Areas that do not meet the NAAQS are deemed "nonattainment." There are nine nonattainment areas for the annual $PM_{2.5}$ standard. Approximately 20 million people live in these nonattainment areas (www.epa.gov/green-book, 2020). Most $PM_{2.5}$ nonattainment areas are in California. Allegheny County, PA, which is in the Marcellus Shale play, is also nonattainment for the 2012 annual $PM_{2.5}$ NAAQS. O_3 nonattainment is a larger problem; there are 52 nonattainment areas for the 2015 standards, spread across 22 states and the District of Columbia. O_3 nonattainment areas include parts of California; major southern cities such as Houston, Dallas, and Atlanta; and the east coast megalopolis. Approximately 120 million people live in O_3 nonattainment areas. Nonattainment is less of an issue for other criteria pollutants. For example, there are no nonattainment areas for CO or NO_2.

A critical distinction for criteria pollutants is whether they are primary or secondary (Seinfeld and Pandis 2006). Primary pollutants are directly emitted from sources. CO is directly emitted by most combustion systems, and SO_2 is emitted from sources that burn sulfur-containing fuels such as coal. Secondary pollutants are formed in the atmosphere via chemical conversion of primary emissions. O_3 is a secondary pollutant formed by photochemical oxidation of volatile organic compounds (VOCs) in the presence of nitrogen oxides (NOx).

Some pollutants can be both primary and secondary; NO_2 and $PM_{2.5}$ are examples. Most NO_2 in the atmosphere is secondary. It is formed from the rapid reaction of NO, which is emitted directly from combustion systems, with O_3. Combustion systems such as diesel engines also have small direct emissions of NO_2 (Dallmann et al. 2012). $PM_{2.5}$ has significant contributions from both primary emissions and secondary processes, though secondary $PM_{2.5}$ dominates (Fine et al. 2008; Jimenez et al. 2009; Kanakidou et al. 2005). Primary $PM_{2.5}$ is emitted from combustion and wind-blown dust. The secondary portion is formed from the oxidation of both inorganic and organic precursors, including VOCs, in the atmosphere.

The divide between primary and secondary pollutants has obvious impacts for air pollution control. Reducing concentrations of primary pollutants requires reducing the direct emissions, either by reducing activity (e.g., fewer miles driven by cars and trucks) or by reducing the emissions intensity (e.g., grams of pollutant emitted per mile driven). Reducing concentrations of secondary pollutants requires reductions in the precursors to the pollutant. A challenge with secondary pollutants, especially ozone, is that the chemistry is highly nonlinear. Thus, a clear understanding of the relevant precursors and their atmospheric chemical mechanisms is required.

In addition to criteria pollutants, the EPA defines a second class of air pollutants known as hazardous air pollutants (HAPs). HAPs are also known as air toxics. The EPA defines 187 pollutants as HAPs because of known or suspected carcinogenicity, mutagenicity, or teratogenicity. While criteria pollutants are ubiquitous and have "high" concentrations (parts per billion [ppb] for gases or μg m^{-3} for PM), HAPs often have much lower concentrations (e.g., ng m^{-3}). Additionally, some HAPs are specific to certain sources or industries, so these species are less ubiquitous. HAPs are commonly regulated by emissions standards, rather than ambient concentration standards (www.epa.gov/haps, 2020).

Many HAPs are VOCs that can contribute to the formation of O_3 and $PM_{2.5}$. One important group of HAPs is known as BTEX: benzene, toluene, ethyl benzene, and the isomers of xylene. These species are all toxic to humans, with benzene in particular being a carcinogen (Hanninen et al. 2014; Bolden et al. 2015). As VOCs, these species can also participate in the chemical reactions that form atmospheric O_3. Some products of the atmospheric oxidation of BTEX are of sufficiently low volatility that they condense onto existing $PM_{2.5}$, thereby generating secondary organic $PM_{2.5}$ (Ng et al. 2007; Hildebrandt et al. 2009).

Air pollution monitoring and control in the United States has traditionally focused on urban areas more than rural areas. One potential concern for unconventional oil and gas development is that this is largely a rural industry (Figure 5.1). Thus, much of the development, and therefore emissions, from this industry occurs in areas with minimal air pollution monitoring. The existing monitoring network may therefore be insufficient to capture the magnitude and geographical extent of increased concentrations owing to emissions from unconventional oil and gas sources (Carlton et al. 2014). An extreme example of the impacts of oil and gas emissions on air quality occurred in Wyoming's Upper Green River Basin, where oil and gas emissions contributed to ozone concentrations exceeding 150 ppb (Schnell et al. 2009), more than double the current NAAQS standard of 70 ppb.

The remainder of this chapter is divided based on potential impacts of unconventional oil and gas emissions on criteria pollutant concentrations, with specific focus on NO_2, $PM_{2.5}$, and O_3. Unconventional oil and gas emissions are expected to have minimal impacts on CO, SO_2, and Pb, so this chapter does not address these species directly.

5.2 Nitrogen Dioxide (NO_2)

NO_2 and NO are the two components of atmospheric nitrogen oxides ($NOx = NO + NO_2$). As noted in Section 5.1, NOx emissions are dominated by NO (Dallmann et al. 2012; May et al. 2014). In the atmosphere, NO rapidly reacts with ozone to form NO_2.

NO_2 is a criteria pollutant. It can also influence concentrations of O_3 and $PM_{2.5}$ (Seinfeld and Pandis 2006). NO_2 photolysis generates O_3 via,

$$NO_2 + h\nu \rightarrow NO + O$$

$$O + O_2 + M \rightarrow O_3 + M$$

VOCs help recycle NO back to NO_2, leading to daytime increases in O_3 that are driven in part by NOx concentrations. NO_2 can also react to form inorganic nitrate, which can contribute to $PM_{2.5}$ (Seinfeld and Pandis 2006).

Nationwide, major sources of NOx include vehicles and electric power plants. Owing to the large number of on-road vehicles and the rapid conversion of vehicle-emitted NO to NO_2, NO_2 has long been used as a proxy for exposures to traffic related air pollution in both

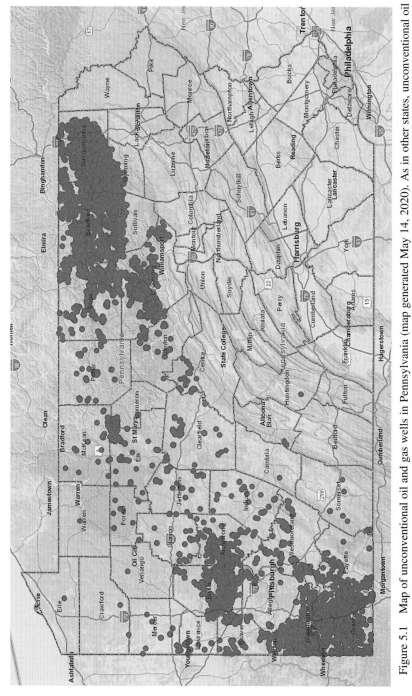

Figure 5.1 Map of unconventional oil and gas wells in Pennsylvania (map generated May 14, 2020). As in other states, unconventional oil and gas infrastructure is located away from major population centers in Allegheny County (Pittsburgh) and Philadelphia County.
Source: PA DEP Oil and Gas Mapping tool (www.depgis.state.pa.us/PaOilAndGasMapping/, 2020).

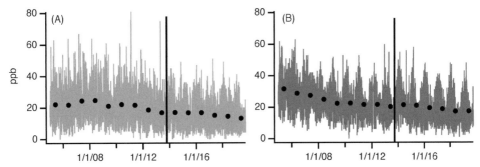

Figure 5.2 Daily maximum NO_2 concentrations measured in (A) Allegheny County (Pittsburgh, Latitude: 40.617488, Longitude: −79.727664) and (B) Beaver County (Latitude: 40.747796, Longitude: −80.316442) from January 1, 2005 to September 30, 2019. Annual averages are shown as black dots centered on June 1 of each year. The black vertical line denotes the end date of the data shown in Carlton et al. (2014)

US and European urban areas (Gehring et al. 2013; Jerrett et al. 2013; Penard-Morand et al. 2006; Su et al. 2009; Shekarrizfard et al. 2018).

In discussing the potential "data gap" in air pollutant monitoring that can arise from the growth of unconventional gas and oil as a largely rural industry, Carlton et al. (2014) pointed to increases in NO_2 in southwestern Pennsylvania as potential evidence of impacts from this industry on downwind air quality. They showed that NO_2 concentrations increased slightly between 2010 and 2012, coincident with a period of growth in the natural gas industry in the Marcellus Shale, at a monitoring location in Beaver County, PA. Figure 5.2 shows similar data through September 2019 for the same site used by Carlton et al., as well as a nearby site in Pittsburgh.

Figure 5.2 shows that over the period from 2005–2019, there were significant reductions in NO_2 concentrations. In both Pittsburgh and Beaver, the daily average maximum NO_2 has fallen by roughly a factor of two, from 22 to 13 ppb in Pittsburgh and 32 to 17 ppb in Beaver. As noted by Carlton et al., the rate of decrease slowed in Beaver around 2010, coincident with an increase in shale gas activities. There was also a slight increase in NO_2 between 2013 and 2014–2015. Concentrations have steadily declined since 2015. The same trend was not observed in Pittsburgh, where NO_2 decreased more rapidly since 2010 than before. This is likely a result of reductions in diesel vehicle emissions following strict NOx emissions limits put in place for the 2010 model year (Bishop et al. 2013).

There are multiple possible reasons for the NO_2 trend in Beaver from 2010–2015, including increased diesel truck activity associated with unconventional natural gas production in the region. However, without additional data about traffic volumes or the age of trucks, it is difficult to clearly attribute the change to unconventional natural gas. Nonetheless, the trend in Beaver was clearly different than the trend in Pittsburgh, which is less impacted by unconventional natural gas activities.

Since NOx is formed during combustion, NOx sources associated with unconventional oil and gas development are related to parts of the industry that rely on combustion engines. This includes well drilling and hydraulic fracturing, trucks that service wells, and compressors. Therefore, it is reasonable to expect that unconventional oil and gas development might lead to local NO_2 enhancements in areas with more intense development (e.g., more wells). As noted previously and by Carlton et al. (2014), many of these areas lack regulatory air quality monitoring stations.

Roy et al. (2014) constructed an emissions inventory for the Marcellus Shale Basin for 2009 and 2020. They estimated basinwide NOx emissions of 58 (95% confidence interval: 23–123) tons/day in 2009 and 129 (56–211) tons/day in 2020. Their 2009 estimate was dominated by emissions from trucks and well drilling, whereas the 2020 estimate also included significant emissions from compressor stations.

While these emissions represent a NOx source that was new to the basin (e.g., NOx emissions from unconventional oil and gas in the Marcellus were effectively zero in 2000), they represent a small portion of overall NOx emissions. Roy et al. estimated total NOx emissions of ~1400 tons/day in 2009 and ~1100 tons/day in 2020. However, while emissions from electric power plants and traffic were expected to decrease over the 2009–2020 period, emissions from unconventional oil and gas were predicted to increase. These new emissions eroded some of the emissions reductions achieved by controls on vehicles and power plants. The predicted reduction in NOx emissions from 2009–2020 was ~300 tons/day when Marcellus shale gas sources were included but would have been ~400 tons/day without those sources.

While oil and gas emissions of NOx are small, they are not inconsequential. Later, we describe the impacts of oil and gas emissions of VOCs on regional ozone formation in Colorado, Utah, and Wyoming. Since VOCs and NOx are co-emitted (or emitted from the same sites), both contribute to ozone episodes in those areas (Rappenglück et al. 2014).

5.3 $PM_{2.5}$

As noted previously, ambient $PM_{2.5}$ has contributions from both primary (direct) emissions and secondary formation in the atmosphere. This section treats primary and secondary PM separately.

5.3.1 Primary $PM_{2.5}$

Overall, unconventional oil and gas operations appear to be a minor source of primary $PM_{2.5}$. As with NOx, primary $PM_{2.5}$ associated with unconventional oil and gas operations is emitted from combustion sources. The inventory of Roy et al. (2014) estimated primary $PM_{2.5}$ emissions for 2009 (2 tons/day) and 2020 (3.5 tons/day) in the Marcellus basin. These emissions were predicted to mostly be associated with well drilling and hydraulic fracturing: Both of these are processes that use diesel engines that do not necessarily have to meet the strict $PM_{2.5}$ emissions standards that are in place for on-road diesel vehicles. The shale-related emissions of $PM_{2.5}$ are even a smaller fraction of overall emissions than NO_2; Roy et al. estimated that primary $PM_{2.5}$ emitted by all sources in the Marcellus basin exceeded 150 tons/day in both 2009 and 2020.

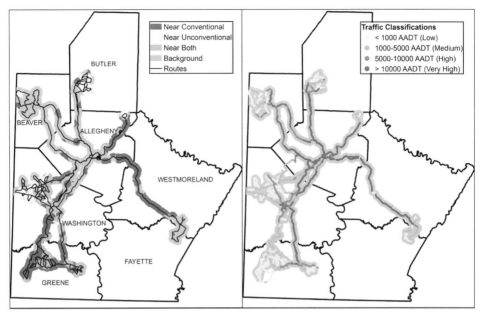

Figure 5.3 The extent of the study area (within 2 km of a sampled point) is shown as shaded. In the left panel, areas within 2 km of a conventional or unconventional well are shown in blue or yellow, respectively. Areas within 2 km of both types of wells are shown in green. Areas within 2 km of any well are classified as "near-well." In the right panel, average traffic density increases as the color gets darker. In general, more heavily traveled roads occur near Pittsburgh (in the center of Allegheny County) and on interstate highways. AADT = Annual Average Daily Traffic reported by Pennsylvania Spatial Data Access (www.pasda.psu.edu/, 2015) (A black and white version of this figure will appear in some formats. For the colour version, refer to the plate section.)
Map generated in ESRI ArcGIS Pro

Our group conducted mobile sampling of $PM_{2.5}$ components near unconventional and conventional gas wells in the Marcellus shale basin. A sampling campaign was conducted during 15 separate days from May 9–August 20, 2012. Sampling was conducted in seven counties in Southwestern Pennsylvania: Allegheny, Beaver, Butler, Fayette, Greene, Washington, and Westmoreland (Figure 5.3). Sampling equipment was outfitted on a gasoline-powered mobile laboratory with an onboard power system for in-motion sampling. Full details of the mobile laboratory are available in Tan et al. (2014).

We measured concentrations of two components of $PM_{2.5}$ that are associated with diesel combustion: black carbon (BC) and particle-bound polycyclic aromatic hydrocarbons (PAHs). Sampling areas were classified both by proximity to gas wells (near or far/background) and traffic volume (low, medium, high, or very high; Figure 5.3). There was not a trend of elevated BC concentrations near gas wells. BC would be expected from heavy diesel traffic, or possibly from heavy diesel-powered equipment onsite in early production such as generators and drill rigs. Figure 5.4 shows that well locations are not the determinant for BC in the study area; instead, a strong positive relationship is seen with traffic levels on the sampled roads. Traffic volume, not well proximity, is a primary driver of exposure to BC.

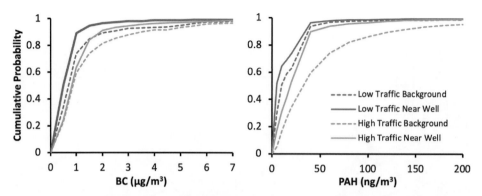

Figure 5.4 Background and near-well measurements for low and high traffic volumes. The amount of traffic on the roads significantly affects the underlying sampled distribution for both black carbon (left) and PAHs (right) (A black and white version of this figure will appear in some formats. For the colour version, refer to the plate section.)

Near-well BC measurements were lower than background measurements for all four traffic volume categories. This suggests that during the study period oil and gas operations did not strongly influence BC concentrations. For example, for medium traffic volume sample points, near-well $\bar{x} = 0.89$ μg/m^3 versus background $\bar{x} = 1.37$ μg/m^3. For high traffic volume locations, the mean BC near wells was 1.12 μg/m^3 versus 1.49 μg/m^3 far from wells (mostly in urbanized areas of Pittsburgh). For three of the four traffic volumes (low, medium, and high), the lower median BC measurements near wells, relative to the background, were statistically significant, using the nonparametric Mann-Whitney U test with $\alpha = 0.05$. For very high traffic volumes, the difference was not statistically significant.

An explanation for this difference is that, on average, wells are located in rural areas with lower traffic. Even within one traffic volume classification, near-well measurement sites had fewer vehicles than background sites. For example, for medium traffic volume, mean AADT near wells was 2,164 vehicles versus 3,247 vehicles for background areas. Similarly, for high traffic volume, near-well mean AADT is 6,550 vehicles, compared to 7,560 vehicles in background areas.

This sampling campaign did not aim to quantify the increase in truck traffic associated with the development of a new well, and therefore any increase in local concentrations owing to the development and servicing of a specific well pad were not captured. However, our data suggest that the net effect of shale gas development on measured BC is small.

Even among all near-well categories, traffic volume significantly affects the shape of the sampled distribution. Cumulative distribution functions (CDFs) of these data are illustrated in Figure 5.4a. All underlying distributions of near-well measurements were distinct from each other, tested with the Kolmogorov-Smirnov test with $\alpha = 0.05$. These results indicate that traffic volume is more important than distance from natural gas sites for predicting elevated levels of BC.

Particle-bound PAH concentrations exhibited a similar spatial pattern to BC. Elevated PAH levels were more associated with traffic volume than with well proximity. PAHs have been previously shown to be strongly associated with diesel traffic (Tan et al. 2014, 2016).

Though there is often substantial diesel truck traffic to oil and gas sites, normal highway truck traffic seems to provide a larger source of PAHs. Near-well average measurements in each traffic category were lower than background (far from well) average measurements. For medium traffic volumes, near-well average particle-bound PAH concentrations were 23.4 ng/m^3 versus 28.4 ng/m^3 in background areas. Areas with high traffic volumes had higher levels of PAHs, with near-wells averaging 25.3 ng/m^3 and background averaging 44.3 ng/m^3. To account for skewness of the data, for each traffic category the nonparametric Mann-Whitney U test with $\alpha = 0.05$ was used to test for differences between the distribution medians. Low, medium, and high traffic categories had median PAH measurements significantly lower for near-well measurements. For the very high traffic volumes, median PAH measurements were significantly higher for near-well measurements. The trend in PAH as compared to AADT is the same as for BC, since the mean AADT depends only on the traffic level and near-well/background classification, which are the same for BC and PAHs.

Similarly to BC, even within all near-well categories, the traffic category affected the shape of the underlying distribution, as shown in the CDFs in Figure 5.4b. Lower traffic categories had more low measurements. All underlying distributions were distinct using $\alpha = 0.05$ for the Kolmogorov-Smirnov test. These results further support the idea that traffic intensity is more important than well proximity for understanding the spatial patterns of PAH concentrations.

In summary, unconventional oil and gas activities seem to be minor contributors to primary $PM_{2.5}$ emissions. Both inventory data (Roy et al. 2014) and measurements suggest that this is the case. Most of the $PM_{2.5}$ emissions are expected to be associated with the use of either stationary or mobile diesel engines. Even in the largely rural areas where unconventional oil and gas development occurs, any signal from increased truck traffic seems to be dwarfed by emissions from the existing, and very large, on-road vehicle fleet, as shown in Figure 5.4.

5.3.2 Secondary $PM_{2.5}$

Impacts of unconventional oil and gas emissions on secondary $PM_{2.5}$ concentrations also appear to be small. There are two primary routes by which oil and gas emissions can impact secondary $PM_{2.5}$: (1) emissions of NOx can be converted to particulate nitrate, especially in winter months when cold temperatures favor the partitioning of nitrate salts to the particle phase and (2) secondary organic aerosol (SOA) formation from oxidation of VOCs.

Roohani et al. (2017) implemented the inventory of Roy et al. (2014) in a chemical transport model. They found that unconventional oil and gas emissions increased annual-average secondary $PM_{2.5}$ by a maximum of 0.27 μg/m^3 in their eastern United States modeling domain. This small increase, which was dominated by inorganic nitrate, was insufficient to push any of the modeled areas above the EPA NAAQS.

Contributions of unconventional oil and gas emissions to SOA have received less attention but seem to be minimal. As discussed in this chapter, the VOC emissions are dominated by light hydrocarbons (e.g., butane and propane) that do not form SOA on oxidation. Bean et al. (2018) generated secondary organic PM via oxidation of evaporated hydraulic fracturing wastewater. SOA production was small, with 24 μg of SOA per ml of wastewater. When scaled to the whole of Texas, Bean et al. estimated that potential SOA mass from oxidation of evaporated wastewater was less than 10% of the primary $PM_{2.5}$ emissions from oil and gas operations.

Fann et al. (2018) modeled the impacts of oil and gas sector emissions on $PM_{2.5}$ at the national scale. The annual median $PM_{2.5}$ attributable to oil and gas operations was 0.02 μg/m^3 (90th percentile: 0.1 μg/m^3). The largest impacts were in a few states with significant oil and gas activity: Colorado, Pennsylvania, Texas, and West Virginia. They estimated that this additional $PM_{2.5}$ burden contributed to 1,000 deaths annually (~0.5 deaths per 100,000 people).

In many cases, unconventional natural gas is used to replace coal-fired electrical power plants. Thus, assessments of the impacts of oil and gas emissions on ambient $PM_{2.5}$ should also account for potential reductions associated with changes in the power sector. Pacsi et al. (2013) modeled air quality impacts of oil and gas on Texas. They found that offsetting coal-fired electricity generation with natural gas reduced secondary sulfate $PM_{2.5}$ by up to 0.7 μg/m^3. This means that in many areas of the country, the net impact of unconventional oil and gas would be a reduction in $PM_{2.5}$, especially if the produced gas is used to replace coal for electricity generation.

5.4 Ozone

Unlike NO_2 and $PM_{2.5}$, unconventional oil and gas emissions have well-documented impacts on ground-level ozone. High ozone concentrations in three western basins have been attributed to emissions from unconventional oil and gas: the Upper Green River Basin in Wyoming (Schnell et al., 2009), the Uintah Basin in Utah (Edwards et al. 2014), and the Denver-Julesburg Basin in Colorado (Cheadle et al. 2017). In all three of these areas, ozone concentrations exceeded the EPA NAAQS of 70 ppb, with concentrations in Utah exceeding 100 ppb on multiple occasions. In Wyoming and Utah, these ozone events occur when snow-covered ground leads to the development of a shallow boundary layer that traps and concentrates oil and gas emissions near the surface (Carter and Seinfeld 2012; Edwards et al. 2013, 2014).

While the high ozone events in Wyoming and Utah are easy to identify because the concentrations are high and anomalous, occurring in winter months in rural areas, unconventional oil and gas emissions can also impact ozone in urban downwind areas. For example, areas in north Texas that are impacted by emissions from the Barnett Shale have 8% higher ozone, and a slower rate of long-term ozone decrease, than nearby areas not impacted by oil and gas emissions (Ahmadi and John 2015).

Roohani et al. (2017) predicted that cities downwind of the Marcellus Shale could have up to 2.5 ppb additional ozone (8-hr summertime maximum) as a result of oil and gas emissions. Some of these ozone enhancements could impact downwind nonattainment areas in Philadelphia, New York City, and Washington, DC. Similarly, Fann et al. (2018) estimated that oil and gas emissions contributed to median 8-hr ozone increases of 0.57 ppb (90th percentile: 2.91 ppb) nationwide.

As with $PM_{2.5}$, assessing the full impact of oil and gas emissions on ozone requires accounting for the impacts of sources that are offset, including coal-fired power generation. For example, Pacsi et al. (2015) modeled ozone production in Texas. They found that the increased use of natural gas in electricity generation reduced summertime 8-hr maximum ozone by 0.6–1.3 ppb in north Texas, but that oil and gas emissions contributed to an ozone increase of 0.3–0.7 ppb in areas downwind of the Eagle Ford shale in south Texas.

5.5 VOCs: A Key Ozone Precursor

The ozone impact of oil and gas emissions arises primarily from VOC emissions. In the extreme ozone events in the Green River Basin, the shallow boundary layers (50–200 m) that develop when the ground is covered in snow lead to high concentrations of oil- and gas-emitted VOCs that subsequently contribute to rapid ozone formation (Oltmans et al. 2014). In these cases, ozone is formed very close to ground level, within 10 m of the surface (Rappenglück et al. 2014). When the ground is not snow covered, the shallow boundary layer does not form, and ozone concentrations are significantly lower.

The association between oil and gas VOC emissions and ozone formation is bolstered by receptor studies that quantified the regional impacts of oil and gas emissions on ambient VOC concentrations and OH reactivity. Petron et al. (2012) conducted measurements at the Boulder Atmospheric Observatory (BAO) in Colorado. They reported elevated concentrations of n-alkanes (e.g., propane and butane) consistent with the composition of oil and gas produced in the Denver-Julesburg Basin. N-alkane enhancements were correlated with methane, suggesting that these hydrocarbons were co-emitted from oil- and gas-related sources. Gilman et al. (2013) performed source apportionment of the BAO data and other datasets collected in Colorado. They identified an oil and natural gas source that was distinct from urban emissions and dominated by small aliphatic hydrocarbons. This source contributed to 55% of the OH reactivity, and thus could be a driver for elevated ozone concentrations.

In a subsequent study, Pétron et al. (2014) quantified emission rates of propane, *n*-butane, *i*-pentane, *n*-pentane, and benzene for the Denver-Julesburg Basin. The total emission rates of these five species summed to 25.4 tons per hour. The measured emissions were approximately double the estimate from the Colorado state emissions inventory. This echoes a trend that emission inventories for oil and gas emissions, especially for methane (Brandt et al. 2014; Marchese et al. 2015; Zimmerle et al. 2015; Omara et al. 2018) but also for VOCs, are generally biased low.

McDuffie et al. (2016) examined summertime ozone in the Colorado Front Range downwind of urban Denver. They found that oil and natural gas emitted alkanes contributed over 80% to the observed carbon mixing ratio, roughly 50% to the regional VOC OH reactivity, and approximately 20% (~3 ppb) of the regional photochemical O_3 production.

VOC measurements in Wyoming and Utah are broadly similar to those in Colorado: significant impacts of oil and gas emissions on VOC concentrations have been identified, and the VOCs are often correlated with methane. Field et al. (2015) identified two VOC factors (fugitive natural gas and condensate tank emissions) associated with oil and gas emissions in the Upper Green River Basin. High concentrations of these VOC factors were associated with the high wintertime ozone events in the basin. Likewise, Helmig et al. (2014) showed that VOC concentrations in the Uintah Basin are correlated with methane, a tracer for oil and gas emissions. They measured high concentrations of both n-alkanes and aromatics and estimated an annual flux of C_2–C_7 hydrocarbons of 194 million kg/yr in the basin.

There are fewer measurements of VOCs and VOC emissions for the Barnett Shale (Texas) and Marcellus Shale. The existing data suggest that VOC emissions in these basins might be lower than in Colorado, Wyoming, and Utah. Marrero et al. (2016) measured VOC emissions in the Barnett Shale in Texas. They measured benzene emissions (mean = 53 kg/hr) and hexane emissions (mean = 1070 kg/hr) but not propane or other light alkanes. While the benzene emissions were in line with inventory values, measured emissions of hexane and other aromatics were substantially larger than the inventory estimates.

Swarthout et al. (2015) measured spatially distributed concentrations of VOCs in the Marcellus Shale Basin in southwestern PA. While concentrations of small alkanes (e.g., propane) were higher in areas with higher densities of unconventional gas wells, the concentration enhancements were not statistically significant despite large methane increases in those areas. Overall, they concluded that oil and gas emissions were a minor source of VOCs. The difference between western (e.g., Colorado, Utah) and eastern (Pennsylvania) measurements may be the result of different resource compositions. The western states have a mixture of gas and oil resources. The Marcellus shale is dominated by natural gas, though some of the areas produce "wet" (condensate rich) natural gas.

Emissions of small alkanes in the Marcellus Shale have impacts in downwind cities. Vinciguerra et al. (2015) quantified increases in ambient ethane in Baltimore, MD, where this species went from 7% of total measured VOCs in 2010 to 15% in 2013. This contrasts with Atlanta, which is not impacted by shale gas emissions and did not have a concurrent increase in ethane.

Overall, VOC emissions from unconventional oil and gas seem to mirror the composition of the extracted resource, suggesting that the emissions are mainly fugitives from process or tank leaks. Koss et al. (2017) summarized aircraft VOC measurements downwind of several oil- and gas-producing regions in Texas, Colorado, New Mexico, and North Dakota. They observed enhancements in alkanes and aromatics, with concentrations reaching levels seen in outflows of large cities.

Recently, more evidence has emerged from measurements of VOC emissions at the process level or facility level. Warneke et al. (2014) performed mobile sampling around oil and gas facilities in the Uintah Basin, Utah. They measured elevated concentrations of components of the extracted oil and gas (e.g., n-alkanes, aromatics) and of chemicals used onsite (methanol, which is used as an anti-freeze and anti-coagulant). Further, Warneke et al. found that the composition of the emissions varied by process component. Separators, dehydrators, and pneumatic devices and pumps emitted a light VOC mixture similar in composition to natural gas, whereas condensate tanks emitted heavier VOCs.

Li et al. (2017) measured VOC emissions from four facilities in the Alberta oil sands region. As with other measurements, the VOC composition bore a resemblance to directly emitted oil and gas, with strong alkane and aromatic characteristics. The oil sand emissions seem to be enhanced in aromatics relative to gas fields (e.g., Warneke et al. 2014), as would be expected based on the composition of the extracted resource. As with the basin-scale emissions inventories, the facility-level inventory estimates for the four facilities sampled by Li et al. (2017) were biased low by a factor of 2–4.5.

Process- and facility-level measurements of methane emissions show significant skew (Omara et al. 2018) because a small percentage of "super emitter" facilities dominate the overall emissions (Zavala-Araiza et al. 2015). Further, inventories that underestimate total emissions seem to do so because of inadequate accounting for super emitters. There has been limited attention on VOC emissions at the facility level, though new evidence is emerging that these emissions are also heavily skewed. Edie et al. (2020) report facility-level emissions of aromatic hydrocarbons from 32 oil and gas production wells in the Upper Green River Basin, Wyoming. They report a mean benzene emission rate of 17.83 g/hr/facility. The emissions distribution shows a heavy tail, with the top 20% of sites contributing 71% of total benzene emissions, though this emissions distribution is less extreme than is often observed for methane emissions. Similarly, Hecobian et al. (2019) observed significant (>10 x) variability in emission rates of n-alkanes and aromatics across 46 oil and gas wells in the Denver-Julesburg and Piecance Basins. They also observed large differences in emissions based on operating mode (drilling versus fracking versus flowback).

The heavy tail is qualitatively confirmed by aerial surveys presented by Lyon et al. (2016), who used a FLIR camera to search for visible emissions from 8,000 oil and gas wells in seven basins. To first order, the sites with visible emissions plumes can be considered as super emitters. Nationwide, only 4% of sites had visible emissions, with significant interbasin variability (basinwide means ranged from 1% to 14%). Tank vents and hatches were the most common sources of emissions, making up 92% of all captured events.

We measured the VOC concentrations downwind of individual natural gas production well pads and compressor stations in the Denver-Julesburg Basin, Uintah Basin, and Marcellus Shale in 2015 and 2016. Figure 5.5 presents box-whisker plots of all VOC (in ppb) and methane (in ppm) concentrations (not background corrected) measured downwind of natural gas production well pads and compressor stations, along with background concentrations measured upwind of nearby oil and gas sources. At all sites, the methane concentrations were 10 ppb–2.7 ppm higher than the regional background, and the ethane concentrations were 1 ppb–350 ppb higher than the regional background. Enhancements in both methane and ethane indicate that our measurements were impacted by oil and gas emissions. Other VOC concentrations are generally less than 1 ppb and most of them are not significantly higher than the ambient background concentrations.

At two well sites in the Denver-Julesburg Basin, VOC concentrations were two to three orders of magnitude higher than at the other sites. We refer to these sites as high emitters,

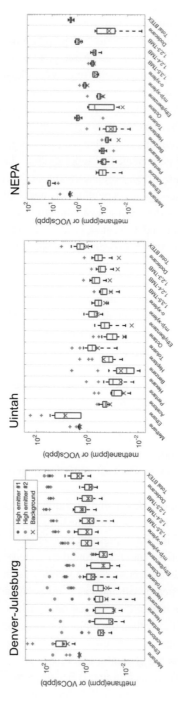

Figure 5.5 VOC (in ppb) and methane concentrations (in ppm) measured in (a) the Denver-Julesburg Basin, (b) the Uintah Basin, and (c) Northeastern PA Marcellus Shale (NEPA). Box-and-whisker plots represent distribution of measurements at gas production facilities. The red line in the middle of the box represents the median concentration; the top and the bottom of the box represent 75th and 25th percentile, respectively. The background measurements of each basin are shown as green crosses (A black and white version of this figure will appear in some formats. For the colour version, refer to the plate section.)

though local meteorological conditions probably also contributed to the high concentrations. The methane and ethane concentrations measured at high emitter #1 are similar to the median value of other sites, but VOC concentrations were significantly higher, with enrichment of heavier (C_8–C_{12}) compounds compared to lighter compounds (C_3–C_7). For the high emitter #2, the methane and lighter VOCs compounds (C_3–C_8) were significantly higher than other sites, and the heavier compounds (C_8–C_{12}) were depleted compared with the lighter compounds (C_3–C_7).

Figure 5.6 compares our measurements to data collected in 28 US cities in 1999–2005 (Baker et al. 2008) (panel a), measurements at the Boulder Atmospheric Observatory (panel b), and in the Uintah Basin (panel c). Consistent with previous studies, the concentrations of benzene (Denver-Julesburg) and toluene (Denver-Julesburg and Uintah Basins) are higher than typical urban conditions because of emissions from the oil and gas industry. In the Marcellus region, benzene and toluene concentrations near wells are lower than typical urban conditions.

In Figure 5.6(b) and (c), the VOC concentrations measured in this work are compared with the regional VOC concentrations measured previously in Boulder Atmospheric Observatory (BAO) (Gilman et al. 2013) and at a ground site in Horse Pool in Uintah County (Helmig et al. 2014), respectively. There is mixed agreement between our data and the BAO measurements (e.g., ethane is similar but other species differ by an order of magnitude). We measured lower concentrations in the Uintah than Helmig et al. (2014). This is likely because our measurements were in the spring and summer, and therefore were not affected by the snow cover-driven low mixing heights observed in the winter.

Figure 5.7 converts the VOC concentrations to emission rates using concurrent measurements of the methane emission rate from these sites; methane emissions are reported by Omara et al. (2018). As expected, emissions vary significantly across the set of sources, with order of magnitude or greater variability for most compounds.

Comparing across the basins offers insight to regional emissions characteristics. In general, the VOC emission rates measured in the Denver-Julesburg Basin and in the Uintah Basin are similar, while the VOC emission rates measured in the Marcellus Shale are slightly lower than the other two regions. Other studies have noted low VOC emissions, specifically low BTEX emissions, in the Marcellus Shale. For example, Goetz et al. (2015) were not able to measure benzene or toluene emissions from Marcellus Shale gas wells.

To summarize,

- Unconventional oil and gas sources emit VOCs. Receptor modeling and near-field measurements suggest that the emissions are dominated by fugitive emissions from leaks and tank venting.
- Since emissions are mainly from fugitives, the composition of the emissions is generally similar to the composition of the resource being extracted in each basin. This likely explains why there are higher concentrations of aromatics and heavier hydrocarbons in the western United States, where the basins contain a mixture of oil and gas wells, than in the Marcellus Shale where natural gas is the primary product.
- In general, VOC concentrations downwind of oil and gas regions are higher in Colorado, Utah, and Wyoming than in the Marcellus Shale. While that might imply higher emission

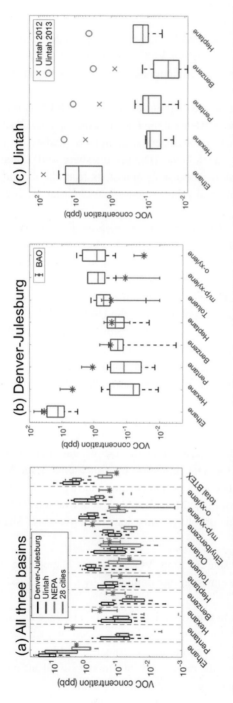

Figure 5.6 Comparison of the VOC concentration measured in this study with (a) VOC measured in 28 US cities (Baker et al. 2008), (b) VOC measured at Boulder Atmospheric Observatory (BAO) (Gilman et al. 2013), and (c) VOC measured in Horse Pool in winter 2012 and winter 2013 (Helmig et al. 2014) (A black and white version of this figure will appear in some formats. For the colour version, refer to the plate section.) Data from this study are presented using box-whisker plots, and data from previous studies are presented as symbols with standard deviation indicated by the error bars

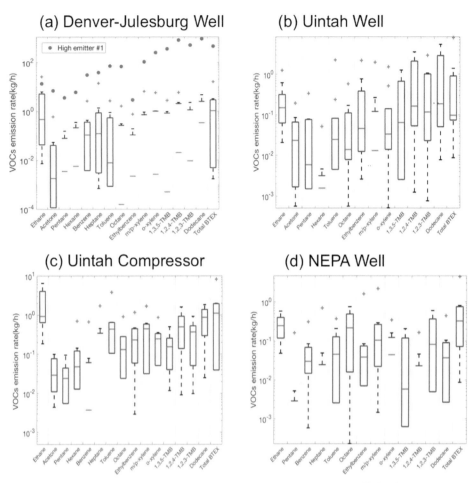

Figure 5.7 Facility-level VOC emission rates measured at (a) gas wells in the Denver-Julesburg Basin, (b) gas wells in the Uintah Basin, (c) compressor stations in the Uintah Basin, and (d) gas wells in NEPA. The emission rates of each VOC species are presented with box-whisker plots. The line in the middle of the box represents the median emission rate; the top and the bottom of the box represent 75th and 25th percentile, respectively

factors (e.g., kg-VOC/kg-gas produced), other factors may influence the measured concentrations, including well density, proximity of the measurements to the source areas, and atmospheric mixing. To date, there is insufficient data on emission rates to quantitatively state that VOC emissions are higher in one basin than another.

- The VOC emissions can contribute to elevated ozone concentrations. In general, VOC impacts on ozone have been acute in the west (Colorado, Utah, and Wyoming), contributing to very high ozone concentrations in wintertime when the ground is covered with snow. Impacts in Texas and the eastern United States are more modest, with these emissions contributing a few ppb to 8-hr maximum summertime ozone concentrations.

5.6 Hazardous Air Pollutants (HAPs) and Environmental Justice

Some of the VOCs measured downwind of oil and gas production regions, including BTEX compounds, are hazardous air pollutants. Thus, there are emerging concerns that people and communities living in areas impacted by unconventional oil and gas operations may be impacted by these HAPs. This differential or inequitable exposure, where certain communities bear higher exposures and thus higher environmental health risks, is a form of environmental injustice that may be brought on by oil and gas development (Kroepsch et al. 2019). Additionally, since emissions are not necessarily constant over time (Johnson et al. 2019), the risks may vary temporally with site age and as a function of the stage of development (Allshouse et al. 2019).

As detailed in this chapter, oil and gas operations can directly emit BTEX compounds, and there can be elevated concentrations of these and other HAPs in areas impacted by unconventional oil and gas development (Halliday et al. 2016). The downwind BTEX concentrations were 57 ppb and 110 ppb for high emitter #1 and #2 identified in Figure 5.5. These high concentrations pose a potential health risk. We calculated hazard ratios (HR) to evaluate these risks. HR is the measured VOC concentration divided by the reference concentration (RfC) from the US EPA Integrated Risk Information System (IRIS). HR greater than one indicate a potential risk of adverse health effect. As shown in Table 5.1, the HR of benzene measured at high emitter #2 and of o-xylene and trimethylbenzene measured at high emitter #1 are larger than 1, indicating people working and living on/near these high emitters may potentially have elevated risk of adverse health effects.

Table 5.1 only presents risks for two areas that we sampled, and more work is needed to better understand how oil and gas emissions impact nearby communities. The few available studies show regionally variable results, as might be expected based on variation in emissions from basin to basin (or state to state). Bunch et al. (2014) quantified community-level exposures in the Barnett Shale. They found that only one of 105 measured VOCs, 1,2-dibromoethane, a species not necessarily associated with oil and gas emissions, was above the health-based concentration threshold. Two studies focused on Colorado

Table 5.1. *Hazard Ratio (HR) of VOC measured downwind of two high emitters in Denver-Julesburg basin. Bold values indicate HR > 1*

VOC species	HR of high emitter #1	HR of high emitter #2
Hexane	0.00	0.28
Benzene	0.46	**3.43**
Toluene	0.01	0.04
Ethylbenzene	0.00	0.02
m/p-xylene	0.52	0.27
o-xylene	**1.28**	0.25
1,3,5-trimethylbenzene	**3.07**	0.00
1,2,4-trimethylbenzene	**7.08**	0.00
1,2,3-trimethylbenzene	**4.24**	0.00

(McKenzie et al., 2018; Holder et al., 2019) found elevated risks for people living or working within 500 ft of oil and gas sources, but lower risks at larger distances.

5.7 Summary, Outlook, and Research Needs

This chapter presented an overview of the impacts of unconventional oil and gas emissions on ground-level air quality. The discussion focused on three criteria pollutants defined by the U.S. EPA: NO_2, $PM_{2.5}$, and O_3. Unconventional oil and gas emissions seem to have the smallest impact on $PM_{2.5}$. The diesel engines used during well drilling and hydraulic fracturing, and trucks that service wells all emit $PM_{2.5}$, but these emissions are small compared to other sources of primary $PM_{2.5}$ emissions (e.g., all cars and trucks on the road). Oil and gas emissions can also contribute to secondary $PM_{2.5}$, primarily through formation of aerosol nitrate, but this is also a minor source.

Oil and gas sources emit NOx, which can convert to NO_2 via atmospheric chemistry. Overall, unconventional oil and gas is a minor source of NOx, especially in the Marcellus Shale, which has significant anthropogenic emissions from traffic and electricity generation. However, in the more rural areas of the basins in the western United States, NOx emissions from oil and gas can contribute, along with co-emitted VOCs, to high ozone events.

Unconventional oil and gas emissions have the largest impact on ozone concentrations. These impacts include high ozone events, especially in winter in the western United States, along with more modest contributions to summertime ozone in Texas and the eastern United States. Much of this ozone impact is associated with VOC emissions that seem to be dominated by fugitives from process and tank leaks.

Understanding the full impact of oil and gas emissions on air quality requires an assessment of what sources were offset by the extraction and use of this resource (Allen 2016). In Wyoming and Utah, where ozone concentrations increased rapidly and exceeded the NAAQS, air pollution and hence human exposures almost certainly worsened. However, in other locations, natural gas offset coal combustion for electricity generation. This trade-off was most likely a net positive for NOx, $PM_{2.5}$, and O_3.

Overall, more work is needed to understand emissions of VOCs from unconventional oil and gas resources. There has been significant work over the last decade to quantify emissions of methane and other greenhouse gases from this sector (Allen et al. 2013, 2014a, 2014b; Mitchell et al. 2015; Subramanian et al. 2015; Omara et al. 2016), but quantification of VOC emissions has lagged. Better understanding of VOC emissions is important because, as we learned with methane, measurements at a large number of facilities are needed to gain an understanding of the frequency and magnitude of emissions from super emitters and to quantify basin-to-basin variations in emissions, which in turn are needed to improve both national and statewide inventories that almost certainly underestimate emissions (Alvarez et al. 2018). Better understanding of basin-specific VOC emission rates, composition, and super emitter frequency will also enable better understanding of the risks that HAP emissions pose to communities living near unconventional oil and gas sources.

References

Adgate JL, Goldstein BD, and McKenzie LM. (2014). Potential public health hazards, exposures amd health effects from unconventional natural gas development. *Environmental Science & Technology*. 48: 8307–8320.

Ahmadi M and John K. (2015). Statistical evaluation of the impact of shale gas activities on ozone pollution in North Texas. *Science of the Total Environment*. 536: 457–467. https://doi.org/http://dx.doi.org/10.1016/j.scitotenv.2015.06.114

Allen DT. (2016). Emissions from oil and gas operations in the United States and their air quality implications. *Journal of the Air Waste Management Association*. 66: 549–575. https://doi.org/10.1080/10962247.2016.1171263

Allen DT, Pacsi AP, Sullivan DW, Zavala-Araiza D, Harrison M, Keen K, Fraser MP, Daniel Hill A, Sawyer RF, and Seinfeld JH. (2014a). Methane emissions from process equipment at natural gas production sites in the United States: Pneumatic controllers. *Environmental Science & Technology*. 49: 633–640. https://doi.org/10.1021/es5040156

Allen DT, Sullivan DW, Zavala-Araiza D, Pacsi AP, Harrison M, Keen K, Fraser MP, Daniel Hill A, Lamb BK, Sawyer RF, and Seinfeld JH. (2014b). Methane emissions from process equipment at natural gas production sites in the United States: Liquid unloadings. *Environmental Science & Technology*. 49: 641–648. https://doi.org/10.1021/es504016r

Allen DT, Torres VM, Thomas J, Sullivan DW, Harrison M, Hendler A, Herndon SC, Kolb CE, Fraser MP, Hill AD, Lamb BK, Miskimins J, Sawyer RF, and Seinfeld JH. (2013). Measurements of methane emissions at natural gas production sites in the United States. *Proceedings of the National Academy of Sciences*. 110: 17768–17773. https://doi.org/10.1073/pnas.1304880110

Allshouse WB, McKenzie LM, Barton K, Brindley S, and Adgate JL. (2019). Community noise and air pollution exposure during the development of a multi-well oil and gas pad. *Environmental Science & Technology*. 53: 7126–7135. https://doi.org/10.1021/acs.est.9b00052

Alvarez RA, Zavala-Araiza D, Lyon DR, Allen DT, Barkley ZR, Brandt AR, Davis KJ, Herndon SC, Jacob DJ, Karion A, Kort EA, Lamb BK, Lauvaux T, Maasakkers JD, Marchese AJ, Omara M, Pacala SW, Peischl J, Robinson AL, Shepson PB, Sweeney C, Townsend-Small A, Wofsy SC, and Hamburg SP. (2018). Assessment of methane emissions from the U.S. oil and gas supply chain. *Science*. 361: 186–188. https://doi.org/10.1126/science.aar7204

Anenberg SC, Henze DK, Tinney V, Kinney PL, Raich W, Fann N, Malley CS, Roman H, Lamsal L, Duncan B, Martin RV, Van Donkelaar A, Brauer M, Doherty R, Jonson JE, Davila Y, Sudo K, and Kuylenstierna JCI. (2018). Estimates of the global burden of ambient PM 2.5, Ozone, and NO2 on asthma incidence and emergency room visits. *Environmental Health Perspectives*. 126. https://doi.org/10.1289/EHP3766

Baker AK, Beyersdorf AJ, Doezema LA, Katzenstein A, Meinardi S, Simpson IJ, Blake DR, and Sherwood Rowland F. (2008). Measurements of nonmethane hydrocarbons in 28 United States cities. *Atmospheric Environment*. 42: 170–182. https://doi.org/10.1016/j.atmosenv.2007.09.007

Bean JK, Bhandari S, Bilotto A, and Hildebrandt Ruiz L. (2018). Formation of particulate matter from the oxidation of evaporated hydraulic fracturing wastewater. *Environmental Science & Technology*. 52: 4960–4968. https://doi.org/10.1021/acs.est.7b06009

Bishop GA, Schuchmann BG, and Stedman DH. (2013). Heavy-duty truck emissions in the South Coast Air Basin of California. *Environmental Science & Technology*. 47: 9523–9529. https://doi.org/10.1021/es401487b

Bolden AL, Kwiatkowski CF, and Colborn T. (2015). New look at BTEX: Are ambient levels a problem? *Environmental Science & Technology*. 49: 5261–5276. https://doi.org/10.1021/es505316f

Brandt AR, Heath GA, Kort EA, O'Sullivan F, Petron G, Jordaan SM, Tans P, Wilcox J, Gopstein AM, Arent D, Wofsy S, Brown NJ, Bradley R, Stucky GD, Eardley D, and Harriss RC. (2014). Methane leaks from North American natural gas systems. *Science*. 343: 733–735. https://doi.org/10.1126/science.1247045

Brauer M, Amann M, Burnett RT, Cohen A, Dentener F, Ezzati M, Henderson SB, Krzyzanowski M, Martin RV, Van Dingenen R, van Donkelaar A, and Thurston GD. (2012). Exposure assessment for estimation of the global burden of disease attributable to outdoor air pollution. *Environmental Science & Technology*. 46: 652–660. https://doi.org/10.1021/es2025752

Brauer M, Freedman G, Frostad J, van Donkelaar A, Martin RV, Dentener F, Dingenen R. van, Estep K, Amini H, Apte JS, Balakrishnan K, Barregard L, Broday D, Feigin V, Ghosh S, Hopke PK, Knibbs LD, Kokubo Y, Liu Y, Ma S, Morawska L, Sangrador JLT, Shaddick G, Anderson HR, Vos T, Forouzanfar MH, Burnett RT, and Cohen A. (2016). Ambient air pollution exposure estimation for the global burden of disease 2013. *Environmental Science & Technology*. 50: 79–88. https://doi.org/10.1021/acs.est.5b03709

Bunch AG, Perry CS, Abraham L, Wikoff DS, Tachovsky JA, Hixon JG, Urban JD, Harris MA, and Haws LC. (2014). Evaluation of impact of shale gas operations in the Barnett Shale region on volatile organic compounds in air and potential human health risks. *Science Of the Total Environment*. 468–469: 832–842. https://doi.org/10.1016/J.SCITOTENV.2013.08.080

Burnett RT, Pope III CA, Ezzati M, Olives C, Lim SS, Mehta S, Shin HH, Singh G, Hubbell B, Brauer M, Anderson HR, Smith KR, Balmes JR, Bruce N, Kan H, Laden F, Pruss-Ustun A, Turner MC, Gapstur SM, Diver WR, and Cohen A. (2014). An integrated risk function for estimating the global burden of disease attributable to ambient fine particulate matter exposure. *Environmental Health Perspectives*. 122: 397–403. https://doi.org/http://dx.doi.org/10.1289/ehp.1307049

Carlton AG, Little E, Moeller M, Odoyo S, and Shepson PB. (2014). The data gap: Can a lack of monitors obscure loss of Clean Air Act benefits in fracking areas? *Environmental Science & Technology*. 48: 893–894. https://doi.org/dx.doi.org/10.1021/es405672t

Carter WPL and Seinfeld JH. (2012). Winter ozone formation and VOC incremental reactivites in the Upper Green River Basin of Wyoming. *Atmospheric Environment*. 50: 255–266.

Cheadle LC, Oltmans SJ, Petron G, Schnell RC, Mattson EJ, Herndon SC, Thompson AM, Blake DR, and McClure-Begley A. (2017). Surface ozone in the Colorado northern Front Range and the influence of oil and gas development during FRAPPE/DISCOVER-AQ in summer 2014. *Elementa: Science of the Anthropocene*. 5: 61. https://doi.org/10.1525/elementa.254

Clark NA, Demers PA, Karr CJ, Koehoorn M, Lencar C, Tamburic L, Brauer M. (2010). Effect of early life exposure to air pollution on development of childhood asthma. *Environmental Health Perspectives*. 118: 284–290.

Correia AW, Pope III CA, Dockery DW, Wang Y, Ezzati M, and Dominici F. (2013). Effect of air pollution control on life expectancy in the United States: An analysis of 545 U.S. counties for the period from 2000 to 2007. *Epidemiology*, 24: 23–31. https://doi.org/10.1097/EDE.0b013e3182770237

Dallmann TR, DeMartini SJ, Kirchstetter TW, Herndon SC, Onasch TB, Wood EC, and Harley RA. (2012). On-road measurements of gas and particle phase pollutant emission factors for individual heavy-duty diesel trucks. *Environmental Science & Technology.* 46: 8511–8518. https://doi.org/10.1021/es301936c

Di Q, Wang Yan, Zanobetti A, Wang Yun, Koutrakis P, Choirat C, Dominici F, and Schwartz JD. (2017). Air pollution and mortality in the Medicare population. *New England Journal of Medicine.* 376: 2513–2522. https://doi.org/10.1056/NEJMoa1702747

Edie R, Robertson AM, Soltis J, Field RA, Snare D, Burkhart MD, and Murphy SM. (2020). Off-site flux estimates of volatile organic compounds from oil and gas production facilities using fast-response instrumentation. *Environmental Science & Technology.* 54: 1385–1394. https://doi.org/10.1021/acs.est.9b05621

Edwards PM, Young CJ, Aikin K, DeGouw JA, Dube WP, Geiger F, Gilman JB, Helmig D, Holloway JS, Kercher J, Lerner B, Martin R, McLaren R, Parrish DD, Peischl J, Roberts JM, Ryerson TB, Thornton J, Warneke C, Williams EJ, and Brown SS. (2013). Ozone photochemistry in an oil and natural gas extraction region during winter: Simulations of a snow-free season in the Uintah Basin, Utah. *Atmospheric Chemistry and Physics.* 13: 8955–8971. https://doi.org/10.5194/acp-13-8955-2013

Edwards PM, Brown SS, Roberts JM, Ahmadov R, Banta RM, de Gouw JA, Dube WP, Field RA, Flynn JH, Gilman JB, Graus M, Helmig D, Koss A, Langford AO, Lefer BL, Lerner BM, Li R, Li S-M, McKeen SA, Murphy SM, Parrish DD, Senff CJ, Soltis J, Stutz J, Sweeney C, Thompson CR, Trainer MK, Tsai C, Veres P, Washenfelder RA, Warneke C, Wild RJ, Young CJ, Yuan B, and Zamora R. (2014). High winter ozone production from carbonyl photolysis in an oil and gas basin. *Nature.* 514: 351–354. https://doi.org/10.1038/nature13767

Fann N, Baker KR, Chan EAW, Eyth A, Macpherson A, Miller E, and Snyder J. (2018). Assessing human health $PM_{2.5}$ and ozone impacts from U.S. oil and natural gas sector emissions in 2025. *Environmental Science & Technology.* 52: 8095–8103. https://doi.org/10.1021/acs.est.8b02050

Field RA, Soltis J, McCarthy MC, Murphy S, and Montague DC. (2015). Influence of oil and gas field operations on spatial and temporal distributions of atmospheric non-methane hydrocarbons and their effect on ozone formation in winter. *Atmospheric Chemistry and Physics.* 15: 3527–3542. https://doi.org/10.5194/acp-15-3527-2015

Fine PM, Sioutas C, and Solomon PA. (2008). Secondary particulate matter in the United States: Insights from the particulate matter supersites program and related studies secondary particulate matter in the United States: Insights from the particulate matter supersites program and related studies. *Journal of the Air Waste Management Association.* 58: 234–253. https://doi.org/10.3155/1047-3289.58.2.234

Gehring U, Gruzieva O, Agius RM, Beelen R, Custovic A, Cyrys J, Eeftens M, Flexeder C, Fuertes E, Heinrich J, Hoffmann B, de Jongste JC, Kerkhof M, Klumper C, Korek M, Molter A, Schultz ES, Simpson A, Sugiri D, Svartengren M, von Berg A, Wijga AH, Pershagen G, and Brunekreef B. (2013). Air Pollution Exposure and Lung Function in Children: The ESCAPE Project. *Environmental Health Perspectives.* 121: 1357–1364. https://doi.org/10.1289/ehp.1306770

Gilman JB, Lerner BM, Kuster WC, and de Gouw J. (2013). Source signature of volatile organic compounds (VOCs) from oil and natural gas operations in northeastern Colorado. *Environmental Science & Technology.* 47: 1297–1305. https://doi.org/10.1021/es304119a

Goetz JD, Floerchinger C, Fortner EC, Wormhoudt J, Massoli P, Knighton WB, Herndon SC, Kolb CE, Knipping E, Shaw SL, and DeCarlo PF. (2015). Atmospheric Emission Characterization of Marcellus Shale Natural Gas Development Sites. *Environmental Science & Technology.* 49: 7012–7020. https://doi.org/10.1021/acs.est.5b00452

Grahame TJ and Schlesinger RB. (2010). Cardiovascular health and particulate vehicular emissions: A critical evaluation of the evidence. *Air Quality and Atmospheric Health*. 3: 3–27. https://doi.org/10.1007/s11869-009-0047-x

Halliday HS, Thompson AM, Wisthaler A, Blake DR, Hornbrook RS, Mikoviny T, Müller M, Eichler P, Apel EC, and Hills AJ. (2016). Atmospheric benzene observations from oil and gas production in the Denver-Julesburg Basin in July and August 2014. *Journal of Geophysical Research*. 121: 11,055–11,074. https://doi.org/10.1002/2016JD025327

Hanninen O, Knol AB, Jantunen M, Lim T-A, Conrad A, Rappolder M, Carrer P, Fanetti A-C, Kim R, Buekers J, Torfs R, Iavarone I, Classen T, Hornberg C, Mekel OC, and Group EbW. (2014). Environmental burden of disease in Europe: Assessing Nine Risk Factors in Six Countries. *Environmental Health Perspectives*. 122: 439–446. https://doi.org/http://dx.doi.org/10.1289/ehp.1206154

Hecobian A, Clements AL, Shonkwiler KB, Zhou Y, MacDonald LP, Hilliard N, Wells BL, Bibeau B, Ham JM, Pierce JR, and Collett JL. (2019). Air toxics and other volatile organic compound emissions from unconventional oil and gas development. *Environmental Science & Technology Letters*. 6: 720–726. https://doi.org/10.1021/acs.estlett.9b00591

Helmig D, Thompson CR, Evans J, Boylan P, Hueber J, and Park J-H. (2014). Highly elevated atmospheric levels of volatile organic compounds in the Uintah Basin, Utah. *Environmental Science and Technology*. 48: 4707–4715. https://doi.org/dx.doi.org/10.1021/es405046r

Hildebrandt L, Donahue NM, and Panids SN. (2009). High formation of secondary organic aerosol from the photo-oxidation of toluene. *Atmospheric Chemistry and Physics*. 9 2973–2986.

Holder C, Hader J, Avanasi R, Hong T, Carr E, Mendez B, Wignall J, Glen G, Guelden B, and Wei Y. (2019). Evaluating potential human health risks from modeled inhalation exposures to volatile organic compounds emitted from oil and gas operations. *Journal of the Air Waste Management Association*. 69: 1503–1524. https://doi.org/10.1080/10962247.2019.1680459

Jerrett M, Burnett RT, Beckerman BS, Turner MC, Krewski D, Thurston G, Martin RV, van Donkelaar A, Hughes E, Shi Y, Gapstur SM, Thun MJ, and Pope III CA. (2013). Spatial analysis of air pollution and mortality in California. *American Journal of Respiratory Critical Care Medicine*. 188: 593–599. https://doi.org/10.1164/rccm.201303-0609OC

Jimenez JL, Canagaratna MR, Donahue NM, Prevot ASH, Zhang Q, Kroll JH, DeCarlo PF, Allan JD, Coe H, Ng NL, Aiken AC, Docherty KS, Ulbrich IM, Grieshop AP, Robinson AL, Duplissy J, Smith JD, Wilson KR, Lanz VA, Hueglin C, Sun YL, Tian J, Laaksonen A, Raatikainen T, Rautiainen J, Vaattovaara P, Ehn M, Kulmala M, Tomlinson JM, Collins DR, Cubison MJE, Dunlea J, Huffman JA, Onasch TB, Alfarra MR, Williams PI, Bower K, Kondo Y, Schneider J, Drewnick F, Borrmann S, Weimer S, Demerjian K, Salcedo D, Cottrell L, Griffin R, Takami A, Miyoshi T, Hatakeyama S, Shimono A, Sun JY, Zhang YM, Dzepina K, Kimmel JR, Sueper D, Jayne JT, Herndon SC, Trimborn AM, Williams LR, Wood EC, Middlebrook AM, Kolb CE, Baltensperger U, and Worsnop DR. (2009). Evolution of organic aerosols in the atmosphere. *Science*. 80(326): 1525–1529. https://doi.org/10.1126/science.1180353

Johnson D, Heltzel R, and Oliver D. (2019). Temporal variations in methane emissions from an unconventional well site. *ACS Omega*. 4: 3708–3715. https://doi.org/10.1021/acsomega.8b03246

Kanakidou M, Seinfeld JH, Pandis SN, Barnes I, Dentener FJ, Facchini MC, Van Dingenen R, Ervens B, Nenes A, Nielsen CJ, Swietlicki E, Putaud JP, Balkanski Y, Fuzzi S, Horth J, Moortgat GK, Winterhalter R, Myhre CEL, Tsigaridis K, Vignati E, Stephanou EG, and Wilson J. (2005). Organic aerosol and global climate modelling: A review. *Atmospheric Chemistry and Physics*. 5: 1053–1123.

Koss A, Yuan B, Warneke C, Gilman JB, Lerner BM, Veres PR, Peischl J, Eilerman S, Wild R, Brown SS, Thompson CR, Ryerson T, Hanisco T, Wolfe GM, St Clair JM, Thayer M, Keutsch FN, Murphy S, and De Gouw J. (2017). Observations of VOC emissions and photochemical products over US oil- and gas-producing regions using high-resolution H3O+ CIMS (PTR-ToF-MS). *Atmospheric Measurement Techniques*. 10: 2941–2968. https://doi.org/10.5194/amt-10-2941-2017

Kroepsch AC, Maniloff PT, Adgate JL, McKenzie LM, and Dickinson KL. (2019). Environmental Justice in Unconventional Oil and Natural Gas Drilling and Production: A Critical Review and Research Agenda. *Environmental Science & Technology*. 53: 6601–6615. https://doi.org/10.1021/acs.est.9b00209

Lefler JS, Higbee JD, Burnett RT, Ezzati M, Coleman NC, Mann DD, Marshall JD, Bechle M, Wang Y, Robinson AL, and Arden Pope C. (2019). Air pollution and mortality in a large, representative U.S. cohort: Multiple-pollutant analyses, and spatial and temporal decompositions. *Environmental Health*. 18: 101. https://doi.org/10.1186/s12940-019-0544-9

Li SM, Leithead A, Moussa SG, Liggio J, Moran MD, Wang D, Hayden K, Darlington A, Gordon M, Staebler R, Makar PA, Stroud CA, McLaren R, Liu PSK, O'Brien J, Mittermeier RL, Zhang J, Marson G, Cober SG, Wolde M, and Wentzell JJB. (2017). Differences between measured and reported volatile organic compound emissions from oil sands facilities in Alberta, Canada. *Proceedings of the National Academy of Sciences U.S.A.* 114: E3756–E3765. https://doi.org/10.1073/pnas.1617862114

Lyon DR, Alvarez RA, Zavala-Araiza D, Brandt AR, Jackson RB, and Hamburg SP. (2016). Aerial surveys of elevated hydrocarbon emissions from oil and gas production sites. *Environmental Science & Technology*. 50: 4877–4886. https://doi.org/10.1021/acs.est.6b00705

Marchese AJ, Vaughn TL, Zimmerle DJ, Martinez DM, Williams LL, Robinson AL, Mitchell AL, Subramanian R, Tkacik DS, Roscioli JR, and Herndon SC. (2015). Methane emissions from United States natural gas gathering and processing. *Environmental Science & Technology*. 49: 10718–10727. https://doi.org/10.1021/acs.est.5b02275

Marrero JE, Townsend-Small A, Lyon DR, Tsai TR, Meinardi S, and Blake DR. (2016). Estimating emissions of toxic hydrocarbons from natural gas production sites in the Barnett Shale region of Northern Texas. *Environmental Science & Technology*. 50: 10756–10764. https://doi.org/10.1021/acs.est.6b02827

May AA, Nguyen NT, Presto AA, Gordon TD, Lipsky EM, Karve M, Gutierrez A, Robertson WH, Zhang M, Brandow C, Chang O, Chen S, Cicero-Fernandez P, Dinkins L, Fuentes M, Huang S-M, Ling R, Long J, Maddox C, Massetti J, McCauley E, Miguel A, Na K, Ong R, Pang Y, Rieger P, Sax T, Truong T, Vo T, Chattopadhyay S, Maldonado H, Maricq MM, and Robinson AL. (2014). Gas- and particle-phase primary emissions from in-use, on-road gasoline and diesel vehicles. *Atmospheric Environment*. 88: 247–260. https://doi.org/10.1016/j.atmosenv.2014.01.046

McDuffie EE, Edwards PM, Gilman JB, Lerner BM, Dubé WP, Trainer M, Wolfe DE, Angevine WM, deGouw J, Williams EJ, Tevlin AG, Murphy JG, Fischer EV, McKeen S, Ryerson TB, Peischl J, Holloway JS, Aikin K, Langford AO, Senff CJ,

Alvarez RJ, Hall SR, Ullmann K, Lantz KO, and Brown SS. (2016). Influence of oil and gas emissions on summertime ozone in the Colorado Northern Front Range. *Journal of Geophysical Research: Atmospheres*. 121: 8712–8729. https://doi.org/10.1002/2016JD025265

McKenzie LM, Blair B, Hughes J, Allshouse WB, Blake NJ, Helmig D, Milmoe P, Halliday H, Blake DR, and Adgate JL. (2018). Ambient nonmethane hydrocarbon levels along Colorado's Northern Front Range: Acute and chronic health risks. *Environmental Science & Technology*. 52: 4514–4525. https://doi.org/10.1021/acs.est.7b05983

Mitchell AL, Tkacik DS, Roscioli JR, Herndon SC, Yacovitch TI, Martinez DM, Vaughn TL, Williams LL, Sullivan MR, Floerchinger C, Omara M, Subramanian R, Zimmerle D, Marchese AJ, and Robinson AL. (2015). Measurements of methane emissions from natural gas gathering facilities and processing plants: Measurement results. *Environmental Science & Technology*. 49: 3219–3227. https://doi.org/10.1021/es5052809

Moore CW, Zielinska B, Petron G, and Jackson RB. (2014). Air impacts of increased natural gas acquisition, processing, and use: A critical review. *Environmental Science & Technology*. 48: 8349–8359. https://doi.org/dx.doi.org/10.1021/es4053472

Ng NL, Kroll JH, Chan AWH, Chhabra PS, Flagan RC, and Seinfeld JH. (2007). Secondary organic aerosol formation from m-xylene, toluene, and benzene. *Atmospheric Chemistry and Physics*. 7: 3909–3922.

Oltmans S, Schnell R, Johnson B, Pétron G, Mefford T, and Neely R. (2014). Anatomy of wintertime ozone associated with oil and natural gas extraction activity in Wyoming and Utah. *Elementa: Science of the Anthropecene*. 2: 000024. https://doi.org/10.12952/journal.elementa.000024

Omara M, Sullivan MR, Li X, Subramian R, Robinson AL, and Presto AA. (2016). Methane emissions from conventional and unconventional natural gas production sites in the Marcellus Shale Basin. *Environmental Science & Technology*. 50: 2099–2107. https://doi.org/10.1021/acs.est.5b05503

Omara M, Zimmerman N, Sullivan MR, Li X, Ellis A, Cesa R, Subramanian R, Presto AA, and Robinson AL. (2018). Methane emissions from natural gas production sites in the United States: Data synthesis and national estimate. *Environmental Science & Technology*. 52: 12915–12925. https://doi.org/10.1021/acs.est.8b03535

Pacsi AP, Alhajeri NS, Zavala-Araiza D, Webster MD, and Allen DT. (2013). Regional air quality impacts of increased natural gas production and use in Texas. *Environmental Science & Technology*. 47. https://doi.org/10.1021/es3044714

Pacsi AP, Kimura Y, McGaughey G, McDonald-Buller EC, and Allen DT. (2015). Regional ozone impacts of increased natural gas use in the Texas power sector and development in the Eagle Ford shale. *Environmental Science & Technology*. 49: 3966–3973. https://doi.org/DOI: 10.1021/es5055012

Penard-Morand C, Schillinger C, Armengaud A, Debotte G, Chretien E, and Annesi-Maesano I. (2006). Assessment of schoolchildren's exposure to traffic-related air pollution in the French Six Cities using a dispersion model. *Atmospheric Environment*. 40: 2274–2287.

Petron G, Frost G, Miller BR, Hirsch AI, Montzka SA, Karion A, Trainer M, Sweeney C, Andrews AE, Miller L, Kofler J, Bar-lian A, Dlugokencky EJ, Patrick L, Moore Jr. CT, Ryerson TB, Siso C, Kolodzey W, Lang PM, Conway T, Novelli P, Masarie K, Hall B, Guenther D, Kitzis D, Miller J, Welsh D, Wolfe D, Neff W, and Tans P. (2012). Hydrocarbon emissions characterization in the Colorado Front Range: A pilot study. *Journal of Geophysical Research*. 117: D04304. https://doi.org/10.1029/2011JD016360

Pétron G, Karion A, Sweeney C, Miller BR, Montzka SA, Frost GJ, Trainer M, Tans P, Andrews A, Kofler J, Helmig D, Guenther D, Dlugokencky E, Lang P, Newberger T, Wolter S, Hall B, Novelli P, Brewer A, Conley S, Hardesty M, Banta R, White A, Noone D, Wolfe D, and Schnell R. (2014). A new look at methane and nonmethane hydrocarbon emissions from oil and natural gas operations in the Colorado Denver-Julesburg Basin. *Journal of Geophysical Research*. 119: 6836–6852. https://doi.org/10.1002/2013JD021272

Pope CA, Coleman N, Pond ZA, and Burnett RT. (2019). Fine particulate air pollution and human mortality: 25+ years of cohort studies. *Environmental Research*. 108924. https://doi.org/10.1016/j.envres.2019.108924

Pope III CA, Burnett RT, Thun MJ, Calle EE, Krewski D, Ito K, and Thurston GD. (2002). Lung cancer, cardiopulmonary mortality, and long-term exposure to fine particulate air pollution. *Journal of the American Medical Association*. 287: 1132–1141.

Pope III CA, Burnett RT, Thurston GD, Thun MJ, Calle EE, Krewski D, and Godleski JJ. (2004). Cardiovascular mortality and long-term exposure to particulate air pollution: Epidemiological evidence of general pathophysiological pathways of disease. *Circulation*. 109: 71–77.

Pope III, CA and Dockery DW. (2006). Health effects of fine particulate air pollution: lines that connect. *Journal of the Air Waste Management Association*. 56: 709–742.

Pope III CA, Ezzati M, and Dockery DW. (2009). Fine-particulate air pollution and life expectancy in the United States. *New England Journal of Medicine*. 360: 376–386.

Raaschou-Nielsen O, Andersen ZJ, Beelen R, Samoli E, Stafoggia M, Weinmayr G, Hoffmann B, Fischer P, Nieuwenhuijsen MJ, Brunekreef B, Xun WW, Katsouyanni K, Dimakopoulou K, Sommar J, Forsberg B, Modig L, Oudin A, Oftedal B, Schwarze PE, Nafstad P, De Faire U, Pedersen NL, Östenson C-G, Fratiglioni L, Penell J, Korek M, Pershagen G, Eriksen KT, Sørensen M, Tjønneland A, Ellermann T, Eeftens M, Peeters PH, Meliefste K, Wang M, Bueno-de-Mesquita B, Key TJ, de Hoogh K, Concin H, Nagel G, Vilier A, Grioni S, Krogh V, Tsai M-Y, Ricceri F, Sacerdote C, Galassi C, Migliore E, Ranzi A, Cesaroni G, Badaloni C, Forastiere F, Tamayo I, Amiano P, Dorronsoro M, Trichopoulou A, Bamia C, Vineis P, and Hoek G. (2013). Air pollution and lung cancer incidence in 17 European cohorts: Prospective analyses from the European Study of Cohorts for Air Pollution Effects (ESCAPE). *Lancet Oncology*. 14: 813–822. https://doi.org/http://dx.doi.org/10.1016/S1470-2045(13)70279-1

Rappenglück B, Ackermann L, Alvarez S, Golovko J, Buhr M, Field RA, Soltis J, Montague DC, Hauze B, Adamson S, Risch D, Wilkerson G, Bush D, Stoeckenius T, and Keslar C. (2014). Strong wintertime ozone events in the Upper Green River basin, Wyoming. *Atmospheric Chemistry and Physics*. 14: 4909–4934. https://doi.org/10.5194/acp-14-4909-2014

Roohani YH, Roy AA, Heo J, Robinson AL, and Adams PJ. (2017). Impact of natural gas development in the Marcellus and Utica shales on regional ozone and fine particulate matter levels. *Atmospheric Environment*. 155: 11–20. https://doi.org/10.1016/j.atmosenv.2017.01.001

Roy AA, Adams PJ, and Robinson AL. (2014). Air pollutant emissions from the development, production, and processing of Marcellus Shale natural gas. *Journal Of the Air Waste Management Association*. 64: 19–37. https://doi.org/10.1080/10962247.2013.826151

Schildcrout JS, Sheppard L, Lumley T, Slaughter JC, Koenig JQ, and Shapiro GG. (2006). Ambient air pollution and asthma exacerbations in children: An eight-city analysis. *American Journal of Epidemiology*. 164: 505–517.

Schnell RC, Oltmans SJ, Neely RR, Endres MS, Molenar JV, and White AB. (2009). Rapid photochemical production of ozone at high concentrations in a rural site during winter. *Nature Geoscience*. 2: 120–122. https://doi.org/10.1038/NGEO415

Seinfeld JH and Pandis SN. (2006). *Atmospheric Chemistry and Physics: From Air Pollution to Climate Change*, 2nd ed. John Wiley & Sons.

Shekarrizfard M, Valois M-F, Weichenthal S, Goldberg MS, Fallah-Shorshani M, Cavellin LD, Crouse D, Parent M-E, and Hatzopoulou M. (2018). Investigating the effects of multiple exposure measures to traffic-related air pollution on the risk of breast and prostate cancer. *Journal of Transport and Health*. 11: 34–46. https://doi.org/10.1016/J.JTH.2018.09.006

Su JG, Jerrett M, Beckerman B, Wilhelm M, Ghosh JK, and Ritz B. (2009). Predicting traffic-related air pollution in Los Angeles using a distance decay regression selection strategy. *Environmental Research*. 109: 657–670. https://doi.org/10.1016/j.envres.2009.06.001

Subramanian R, Williams LL, Vaughn TL, Zimmerle D, Roscioli JR, Herndon SC, Yacovitch TI, Floerchinger C, Tkacik DS, Mitchell AL, Sullivan MR, Dallmann TR, and Robinson AL. (2015). Methane emissions from natural gas compressor stations in the transmission and storage sector: Measurements and comparisons with the EPA greenhouse gas reporting program protocol. *Environmental Science & Technology*. 49: 3252–3261. https://doi.org/10.1021/es5060258

Swarthout RF, Russo RS, Zhou Y, Miller BM, Mitchell B, Horsman E, Lipsky E, McCabe DC, Baum E, and Sive BC. (2015). Impact of Marcellus Shale natural gas development in Southwest Pennsylvania on volatile organic compound emissions and regional air quality. *Environmental Science & Technology*. 49: 3175–3184. https://doi.org/10.1021/es504315f

Tan Y, Dallmann TR, Robinson AL, and Presto AA. (2016). Application of plume analysis to build land use regression models from mobile sampling to improve model transferability. *Atmospheric Environment*. 134: 51–60. https://doi.org/10.1016/j.atmosenv.2016.03.032

Tan Y, Lipsky EM, Saleh R, Robinson AL, and Presto AA. (2014). Characterizing the spatial variation of air pollutants and the contributions of high emitting vehicles in Pittsburgh, PA. *Environmental Science & Technology*. 48: 14186–14194. https://doi.org/10.1021/es5034074

Vinciguerra T, Yao S, Dadzie J, Chittams A, Deskins T, Ehrman S, and Dickerson RR. (2015). Regional air quality impacts of hydraulic fracturing and shale natural gas activity: Evidence from ambient VOC observations. *Atmospheric Environment*. 110: 144–150. https://doi.org/http://dx.doi.org/10.1016/j.atmosenv.2015.03.056

Warneke C, Geiger F, Edwards PM, Dube W, Petron G, Kofler J, Zahn A, Brown SS, Graus M, Gilman J, Lerner B, Peischl J, Ryerson TB, de Gouw JA, and Roberts JM. (2014). Volatile organic compound emissions from the oil and natural gas industry in the Uinta Basin, Utah: Oil and gas well pad emissions compared to ambient air composition. *Atmospheric Chemistry and Physics*. 14: 10977–10988. https://doi.org/10.5194/acp-14-10977-2014

Zavala-Araiza D, Lyon D, Alvarez RA, Palacios V, Harriss R, Lan X, Talbot R, and Hamburg, SP. (2015). Toward a functional definition of methane super-emitters: Application to natural gas production sites. *Environmental Science & Technology*. 49: 8167–8174. https://doi.org/10.1021/acs.est.5b00133

Zimmerle DJ, Williams LL, Vaughn TL, Quinn C, Subramanian R, Duggan GP, Willson B, Opsomer JD, Marchese AJ, Martinez DM, and Robinson AL. (2015). Methane emissions from the natural gas transmission and storage system in the United States. *Environmental Science & Technology*. 49: 9374–9383. https://doi.org/10.1021/acs.est.5b01669

6

Methane and Climate Change

ROBERT W. HOWARTH

6.1 Introduction

The beginning of the shale gas revolution coincided in time with the promotion of natural gas as a "bridge fuel." Not only the oil and gas industry but political leaders, including both President G. W. Bush and President Obama, argued that society could substitute natural gas for coal, thereby continuing to use fossil fuels as a bridge while reducing greenhouse gas emissions for some period of time until the world could better rely on renewable energy. Part of this premise is true: To gain the same amount of energy, fewer carbon dioxide emissions are released from burning natural gas compared to either coal or oil (Hayhoe et al. 2002). However, the continued reliance on any fossil fuel, including shale gas or conventional natural gas, is an outdated concept. The atmosphere already contains so much carbon dioxide from the fossil fuels burned to date that even a few more years of carbon dioxide emissions at current rates will lock the world into a global average of warming 1.5 degrees Celsius (2.7 degrees Fahrenheit) above preindustrial levels (Hausfather 2018). Beyond this, natural gas – including shale gas – is overwhelmingly composed of methane, and some of this methane is inevitably released to the atmosphere as the gas is developed and used. Methane is an incredibly powerful greenhouse gas – 120-fold more powerful than carbon dioxide compared mass-to-mass when both gases are in the atmosphere (IPCC 2013). Consequently, methane emissions from both shale gas and conventional natural gas can more than counteract the advantage of lower carbon dioxide emissions from using natural gas. Rather than serving as a bridge fuel, the use of gas may accelerate global warming in the next few decades.

Both methane and carbon dioxide are important drivers of global change. The radiative forcing of methane is approximately 1 watt per square meter when the indirect effects of methane are included, compared to approximately 1.66 watts per square meter for carbon dioxide, and methane has contributed roughly 25% of the warming seen over recent decades (IPCC 2013). However, the gases behave quite differently, and the climate system responds far more quickly to changes in emissions of methane compared to carbon dioxide (Shindell et al. 2012; IPCC 2018). Consequently, a reduction in methane emissions would significantly slow the rate of global warming almost immediately, while reducing carbon dioxide emissions would only slow global warming decades later.

In December 2015, the nations of the world came together in Paris under the COP21 agreement to pledge to try to keep the Earth well below 2°C from the preindustrial baseline, with the clear acknowledgment that warming to even 1.5°C poses significant risks. These risks include large social disruption caused by more extreme weather events and possible food and water shortages, as well as an increasing probability of fundamental changes in the climate system, leading to runaway catastrophic change over the long term as important thresholds are exceeded. These risks become more severe as the Earth's temperature rises above 1.5°C from the preindustrial baseline, which is predicted to occur within 10 to 20 years from now, by 2030 or 2040 (Shindell et al. 2012; IPCC 2018). Again, because of the relatively fast response of the climate to methane, reducing methane emissions can help provide a pathway to reaching the COP21 climate goal (Collins et al. 2018).

6.2 Sources of Methane

While some atmospheric methane comes from natural sources, 60% or more comes from human-controlled sources such as fossil fuels, agriculture, landfills, sewage treatment plants, and biomass burning (Kirschke et al. 2013; Begon et al. 2014). Atmospheric methane concentrations remained level for the first 10 years of the twenty-first century, but over the past decade or so, methane concentrations have been rising rapidly. Evidence from changes in the carbon-13 stable isotopic composition of atmospheric methane suggests that emissions from the natural gas industry may be the largest driver of this recent increase in atmospheric methane (Howarth 2019). Other recently gained evidence – from carbon-14 radiocarbon dating of the methane content in glacial ice laid down before the Industrial Revolution – suggests that fossil fuel emissions of methane have historically been significantly underestimated. Specifically, the ice-core studies show virtually no fossil methane before the Industrial Revolution (Petrenko et al. 2017; Hmiel et al. 2020). This means that natural emissions of fossil methane from geological seeps have always been small, far less than the 50 Tg per year assumed in many global budgets. Since we know from the carbon-14 content of atmospheric methane at the end of the twentieth century that approximately 30% of emissions were from fossil sources (fossil fuels plus natural seeps; Lassey et al. 2007), if the seep emissions are smaller, the fossil fuel emissions must be correspondingly larger than previously assumed. This larger estimate for methane emissions from fossil fuels of 50 Tg per year means that fossil fuels contribute approximately 40% more to global methane emissions than assumed in most prior budgets (Begon et al. 2014).

Some methane emissions are associated with the extraction of any fossil fuel. But for coal and petroleum products, methane is a minor contaminant, while natural gas – including shale gas – is composed overwhelmingly of methane. It therefore should not come as a surprise that some of this methane is released into the atmosphere as natural gas is developed and used. These emissions come both from leaks and from purposeful release – such as occurs, for instance, during the venting of natural gas pipelines before performing routine maintenance, or while controlling the pressure in storage tanks. Often when industry purposefully releases methane to the atmosphere, they will flare it, that is, burn the methane to convert it to carbon dioxide. However, flares sometimes do not remain lit, and unburned methane is instead emitted. A very graphic visual article published in late 2019 by the *New York Times* well documents such events (Kessel and Tabouchi 2019).

6.3 Early Estimates of Methane from Shale Gas

The first analysis of how much methane is emitted from the development of shale gas was published in 2011 (Howarth et al. 2011). Shale gas is also a form of natural gas but is composed of methane that has remained trapped in shale rock over geological timeframes, while conventional natural gas is methane that has migrated from the shale or other source rock to reservoirs where further migration is prevented by an impermeable barrier. Note that some of the older geological literature refers to any gas that originated in shale as "shale gas," whether or not it has migrated out of the shale to another reservoir. Here, and in all the literature and data on gas production, shale gas refers to the gas produced directly from a shale formation, that is, the gas that had been trapped in the shale. Gas that has migrated from the shale over geological time is considered conventional natural gas. Shale gas was not commercially exploitable until quite recently when a combination of new technologies was employed to break the trapped gas free of the shale. These technologies include high-volume hydraulic stimulation ("fracking"), high-precision directional drilling, the invention of a new stimulation fluid ("slickwater"), and the introduction of injection equipment that could generate the very high downhole pressures required to permeate large volumes of fractured shale with this new stimulant. There was virtually no shale gas development until very late in the twentieth century, and as of 2005 global shale gas production was only 31 billion cubic meters per year (US EIA 2016). Since then, the shale gas revolution has accelerated tremendously, particularly in the United States. Global production in 2015 was 435 billion cubic meters, with 89% of this in the United States and 10% in western Canada (US EIA 2016). By 2019, shale gas production in the United States alone had increased to 716 billion cubic meters (US EIA 2020-a). Today, shale gas dominates natural gas production in the United States (approximately 75% of total production is from shale), and almost two-thirds of the total global increase in natural gas production between 2005 and 2015 came from shale gas in North America (Howarth 2019).

Our 2011 analysis was the first peer-reviewed effort to estimate methane emissions from shale gas (Howarth et al. 2011). We used a full-lifecycle approach, estimating emissions during hydraulic fracturing and production at the well sites, during processing and storage, and from transportation of the gas to the consumer. We estimated that as a percentage of natural gas produced, emissions from conventional natural gas were probably in the range of 1.7% to 6.0%, and from shale gas 3.6% to 7.9%. Of this, we estimated that "downstream" emissions (during transport, storage, and distribution to the consumer) were likely to be in the range of 1.4% to 3.6% for both conventional and shale gas. We estimated "upstream" emissions (at the well site and from processing) as 0.3% to 2.4% for conventional gas and 2.2% to 4.3% for shale gas. We used the best available data but noted that these data were often poorly documented, with very little information in published, peer-reviewed papers. We therefore called for more and better measurement of emissions.

In the ten years since our paper was published, there has been an explosion of new studies, leading to a much better understanding of methane emissions from natural gas systems. Perhaps somewhat surprisingly, our original conclusions have held up remarkably well.

6.4 Recent Estimates of Methane Emissions

Table 6.1 synthesizes data from 12 recent studies that measured upstream methane emissions, with most of these from shale gas operations, from a total of 9 different gas-producing geological basins, using either aircraft or satellite remote sensing data. Estimates range from 0.2% to 40% of production. All of these studies appear to have been well designed and executed, and the variation in observed emission rates probably reflects true variation in time and space: Emission rates are probably higher in some shale gas fields than in others, and emissions in any given field probably vary over time, for instance,

Table 6.1. *Top-down estimates for upstream emissions of methane from natural gas systems, including studies based on aircraft flyovers and satellite remote-sensing data. Estimates are the percentage of the methane in natural gas that is produced.*

Aircraft data			
	Peischl et al (2013)	Los Angeles Basin, CA	17.%
	Karion et al. (2013)	Uintah shale, UT	9.0%
	Caulton et al. (2014)	Marcellus shale, PA	10.%
	Karion et al. (2015)	Barnett shale, TX	1.6%
	Peischl et al. (2015)	Marcellus shale, PA	0.2%
	Peischl et al. (2016)	Bakken shale, ND	6.3%
	Alvarez et al. (2018)	Marcellus shale, PA	0.4%
	Peischl et al. (2018)	Bakken shale, ND	5.4%
		Eagle Ford shale, TX	3.2%
		Barnett shale, TX	1.5%
		Haynesville shale, LA	1.0%
	Ren et al. (2019)	Marcellus shale, PA & WV	1.1%
Satellite data			
	Schneising et al. (2014)	Eagle Ford shale, TX	20.%*
		Bakken shale, ND	40.%*
	Zhang et al. (2020)	Permian Basin shale, NM	3.7%
	Schneising et al. (2020)	Permian Basin shale, NM	3.7%
		Appalachia (Marcellus + Utica), PA	1.2%
		Eagle Ford shale, TX	3.5%
		Bakken shale, ND	5.2%
		Anadarko shale, OK	5.8%

* Schneising et al. (2014) reported emissions as percentage of combined production of oil and gas. Here these are converted to percentage of just gas production using data on relative production of oil and gas from Schneising et al. (2020).

depending on the amount of high-volume hydraulic fracturing occurring at the time. It is noteworthy that the highest values come from the earliest studies, suggesting industry may have improved their operations over time (Schneising et al. 2020).

Table 6.2 presents estimates of the actual mass of methane emitted from each field using data from the rate of shale gas production during 2015 in individual fields (US EIA 2020-b) and the rate of emissions reported for fields from Table 6.1. It omits the very high values reported in the earlier studies shown in Table 6.1, assuming that those high emissions do not well represent emissions in more recent years. Quality data for both emissions and production in 2015 exist for only six shale gas fields, but these represent a total production of 325 billion cubic meters per year (Table 6.2), or three-quarters of the total global production of shale gas that year (Howarth 2019). Comparing the total mass of methane emitted (5.6 Tg per year) with the production for these six fields (325 billion cubic meters per year), the volume-weighted average rate of upstream emissions is 2.6% (Table 6.2). This is in the range of 2.2% to 4.3% we had estimated for upstream emissions from shale gas in our original paper (Howarth et al. 2011), and again note that this does not include the very high emissions reported by Peischl et al. (2013) and Schneising et al. (2014). Applied to the global increase in shale gas production over the period 2005–2015, this 2.6%

Table 6.2. *Shale gas production and upstream methane emissions from various major shale gas producing fields in 2015*

	Production (billion m³/yr)	% emitted upstream (with 90% CL)*	Mass emitted upstream (Tg/yr)**
Marcellus	155	2.58% (+/– 4.2%)	2.64
Eagle Ford	50	3.35% (+/– 0.21%)	1.11
Barnett	38	1.55% (+/– 0.06%)	0.39
Haynesville	36	1.0%	0.24
Permian	36	3.7% (+/– 0.0%)	0.88
Bakken	10	5.63% (+/– 0.59%)	0.37
Total for above fields	325		5.63
Volume-weighted average		2.6%***	
Global total	435		7.4****

* The values from Schneising et al. (2014) shown in Table 6.1 are considered outliers and are not included here.
** Assumes 93 percent of produced gas is methane (Schneising et al. 2020).
*** Calculated from total production and methane mass emission for six shale gas fields listed.
**** Calculated from volume-weighted average percent methane emitted.

upstream emission rate leads to an estimated increase in global methane emissions of 7.4 Tg per year (Table 6.2), or a little over 30% of the entire global increase in methane over that time period (Howarth 2019). This does not include the downstream emissions, and again, it does not include the highest emission rates shown in Table 6.1.

Fewer studies have attempted to characterize downstream emissions. Perhaps the most comprehensive effort that used top-down airplane flyovers is a study by Plant et al. (2019) that reported emissions for the urban, northeastern US seaboard from Boston south to Washington, DC. Their data indicate an emission rate of 0.8 percent of the natural gas consumption in that region (Howarth 2020). Other top-down estimates in Boston, Los Angeles, and Indianapolis give estimates that are at least this high, and up to 2.5 percent in the case of Boston (McKain et al. 2015; Lamb et al. 2016; Wunch et al. 2016). The available data suggest that we may have overestimated the downstream emissions in our 2011 study (1.4 to 3.6%), but not greatly so.

Combining an upstream emission estimate of 2.6% (volume-weighted mean from Table 6.2) with the downstream estimate of 0.8% derived from Plant et al. (2019) yields an overall emission estimate for shale gas of 3.4%, somewhat lower than our original study estimate: 3.6 to 7.9% for shale gas (Howarth et al. 2011). This 3.4% emission rate corresponds to an increase of 10 Tg per year from shale gas development between 2005 and 2015, or 40% of the entire global increase in methane emissions from all sources over that time period (Howarth 2019).

Numerous studies have used "bottom-up" approaches for estimating methane emissions from natural gas systems; that is, estimates based on evaluating individual emission sources on the ground and summing these up to get a total emission. In general, this approach gives lower emission estimates than do top-down studies such as those shown in Table 6.1 (Miller et al. 2013; Howarth 2014; Vaughn et al. 2018). There are many reasons why this might be the case. One is that the bottom-up approaches tend not to include all possible emission sources: for example, by leaving out emissions during initial well drilling, which can be high (Caulton et al. 2014). Another reason is that bottom-up measurements often require researchers to get permission to access sites controlled by natural gas operators, so that the researchers can make measurements near their operations. It seems likely that companies that are more willing to allow such access are also more careful in their operations, and perhaps emit less methane. There is always the possibility that if natural gas operators know when the emission measurements will be made, they may take particular care to reduce emissions at those times.

Alvarez et al. (2018) synthesized data from a large number of bottom-up studies of shale and conventional gas operations coordinated by the Environmental Defense Fund, and came up with estimates of 1.9% of production for upstream emissions (production, gathering-line leaks, and processing), 0.4% for downstream emissions (transmission, storage, and local distribution), and 2.3% overall. From the bottom-up data, they concluded that emissions from shale gas are no higher than from conventional gas. The US Environmental Protection Agency (US EPA) in its official greenhouse gas reporting relies exclusively on bottom-up estimates, and often uses old and outdated non-peer-reviewed studies, resulting in estimates that are even lower than those from the Environmental

Defense Fund (Miller et al. 2013; Howarth 2014; Alvarez et al. 2018; Ren et al. 2019). The US EPA too assumes no difference in emissions from shale gas and conventional gas operations. As discussed in the remainder of this chapter, the bottom-up studies may not adequately characterize differences in emissions between shale and conventional natural gas operations.

6.5 Comparing Methane Emissions from Shale and Conventional Gas

Some methane is emitted during each step of developing, processing, transporting, storing, and distributing shale gas to consumers. Many of these emission sources are similar for both conventional natural gas and shale gas, but some are greater for shale and other unconventional gas such as tight-sand formations. The most obvious differences between shale gas and conventional gas development are the higher volume of stimulation fluid that is central to developing shale gas and the much larger number of shale gas wells that are completed per unit area. A substantial amount of methane can be emitted to the atmosphere during the flowback of this fluid that immediately follows the stimulation. In Howarth et al. (2011), we summarized data indicating that two shale gas wells emitted 1.1% and 3.6% of their lifetime production of gas during the short flowback period, while two unconventional tight sand wells emitted 0.6% and 1.3% of their lifetime production total during flowback. The technology exists for industry to capture this gas and sell it to market, but to do so is expensive and slows the whole process of well completion; consequently, little of this gas was being captured, at least as of 2011 (US EPA 2011; Howarth et al. 2012; Howarth 2014). As of 2015, the US EPA regulated methane emissions during well completion, in general requiring the gas to be captured if technically possible, and flaring (burning) the gas otherwise, although with many exceptions (US EPA 2016). However, effectiveness of this regulation has not been independently determined, and as noted previously, qualitative evidence suggests unlit flares that vent unburned methane may be common (Kessel and Tabouchi 2019). Further, under the Trump administration, the EPA repeatedly took steps to end these regulations (Lavelle 2019).

Another difference in methane emissions between conventional gas and shale gas operations is less obvious. Caulton et al. (2014) observed substantial emissions of methane while wells were being drilled in the Marcellus shale region in southwestern Pennsylvania even before the drillers reached the shale. This area has a long history of fossil fuel exploitation, with development of oil, conventional gas, and coal dating back to the 1800s. The emissions during shale gas well drilling may be the result of hitting pockets of trapped methane from these earlier fossil fuel operations, which must be drilled through to reach the shale, which is much deeper underground. In such an environment, the gas industry sometimes employs "underbalanced" or negative-pressure drilling to reduce the chance of blowouts, and this could increase the emission of methane from any pockets that are encountered while drilling (Caulton et al. 2014).

The Alvarez et al. (2018) synthesis of the studies coordinated by the Environmental Defense Fund do not refer to emissions during flowback, to emissions during well drilling, or to higher emissions from producing shale gas wells (Ingraffea et al., 2020). Since they are not including these shale-specific emissions, it may have misled them to conclude that overall shale emissions are no higher than for conventional natural gas. Further, unintended emissions can occur during well completions. The well blowout at Powhatan Point, OH in March 2018 released over a period of 20 days methane equivalent to 25% of the state's total annual natural gas emissions (Pandey et al. 2019).

6.6 Evidence from Change in Carbon-13 Content of Atmospheric Methane

Another, completely independent, approach has been used to estimate the full lifecycle (upstream plus downstream) of methane emissions from shale gas: an analysis of the change in the carbon isotopic composition of atmospheric methane globally over time (Howarth 2019).

After remaining constant for the first decade of the twenty-first century (when methane concentrations in the atmosphere were constant), the carbon-13 content of methane has been decreasing since 2007 or so, coinciding with the increase in atmospheric methane concentrations. Some studies interpreted this to mean that there had been an increase in methane emissions from biogenic sources, such as animal agriculture, rather than an increase from fossil fuels (Schaefer et al. 2016; Schwietzke et al. 2016). There are many reasons to doubt this conclusion. One reason is that satellite data indicated that 30 to 60% of the global increase in methane emissions over the past decade came from the United States (Turner et al. 2016), yet the number of cows and cattle in the United States decreased by 5% to 10% over this time (USDA 2020).

The work of Schaefer et al. (2016) and Schwietzke et al. (2016) had assumed that methane emissions from biomass burning had remained constant over time. Worden et al. (2017) noted that this is not true: biomass burning had actually decreased globally as a source of methane over the past decade. Biomass burning is a relatively small contribution to global methane emissions, but the methane from this source is quite enriched in carbon-13 compared to most other emissions. Therefore, as biomass burning decreased, the carbon-13 content of atmospheric methane would be expected to decrease, a decrease earlier attributed to an increase in emissions that were depleted in carbon-13. When models are corrected for this change, Worden et al. (2017) concluded that the largest increase in emission sources since 2007 was from fossil fuels, not biogenic sources.

Methane from shale gas may have slightly less carbon-13 than does methane from conventional natural gas. This is because of fractionation, as some methane is oxidized over geological timescales during migration from shale formations to conventional gas reservoirs. Fractionation is the tendency for the oxidation reaction to slightly favor the lighter carbon-12 isotope, so that the methane that migrates ends up having a slightly higher proportion of the heavier carbon-13 isotope. Correcting for this difference in the carbon-13 content of methane results in an estimate that methane from shale gas contributed at least one-third of the total increase in global methane emissions since 2007, with total emissions from the oil and gas industry (including shale) contributing approximately two-thirds of the total global increase in methane fluxes. This corresponds to emission rates of 2.8% to 3.5%

of production for conventional natural gas, and 3.5% to 4.1% of production for shale gas (Howarth 2019). There is much uncertainty in these estimates, but they are broadly consistent with the upstream emissions volume-weighted mean of 2.6% (Table 6.2) plus estimates for downstream emissions of 0.8% or more discussed in this chapter. If anything, it would appear that my estimate based on global changes in the carbon-13 of methane over time underestimated total emissions from shale gas, but not greatly so.

This interpretation has been questioned by Milkov et al. (2020) who relied on a published database to conclude that there is no difference in the carbon-13 composition of methane from shale gas compared to conventional natural gas. However, the "shale gas" referred to in this dataset includes data on methane that has migrated from shale formations to conventional reservoirs over geological time. As noted previously, the older geological literature refers to this as shale gas, but it is not the gas that is released from shale by high-volume hydraulic fracturing with slickwater. This migrated gas would be called conventional gas in terms of production statistics, and would be expected to be enriched in carbon-13 compared to actual produced shale gas. While the dataset used in the Milkov et al. (2020) study was known at the time, it was not used in Howarth (2019) for that reason, as elucidated in Howarth (2019a). Further, the Milkov et al. (2020) approach leads to an estimate of methane emission from shale gas that is very much lower than any of the estimates presented in Table 6.1.

The use of carbon-13 and other isotopes as tracers of methane in the atmosphere and groundwater is discussed in more detail in Chapter 11 (Townsend-Small, 2022).

6.7 What Methane Emission Estimate Should Policymakers Use?

In January 2020, legislation took effect in New York State that outlines a new approach for evaluating greenhouse gas emissions. Under the new rules, New York is to account for all methane emissions associated with the use of natural gas and other fuels in the state, including those emissions that occur outside of the state's boundaries. Previously, New York, in common with all other states and most nations, only included methane emissions that occurred within its state boundaries in its greenhouse gas inventory. New York chose this new methodology, using a consumer-based approach to indicate the entire greenhouse gas consequences of different fuels, to better allow comparisons of different energy choices. The new law also requires the state to estimate methane emissions based on the best, peer-reviewed science in the literature, and not rely on the inventory estimates of the US Environmental Protection Agency.

Howarth (2020) provided guidance on how the state of New York might implement this, suggesting that total methane emissions for natural gas be calculated as 3.6% of the consumption of gas within the state. Note that 3.6% of consumption is equivalent to 3.2% of natural gas production; consumption is less than production both because some of the produced gas is emitted to the atmosphere before it reaches consumers and because the gas industry uses some of the produced gas to power their operations, including the compressors that move gas through high-pressure pipelines. The suggested factor for total emissions associated with natural gas (primarily shale gas) is at the low end of that estimated from the global carbon-13 data or compared to volume-weighted upstream methane emission rate reported in Table 6.2. A conservative value was chosen in the interest of helping to promote a consensus value for the state to use. The recommendation was based on the Alvarez et al. (2018) synthesis, using their average bottom-up estimates for upstream emissions, increasing this by 11% to better reflect the top-down estimates with which they compare their values, and using the downstream emission estimate from Plant et al. (2019). Note that in concluding that top-down estimates were 11% greater than their bottom-up estimates, Alvarez et al. (2018) did not include several of the higher top-down estimates shown in Table 6.1, including the estimates by Peischl et al. (2013), Caulton et al. (2014), Schneising et al. (2014), Zhang et al. (2020) and Schneising et al. (2020): 5 out of the 12 papers in Table 6.1.

6.8 Greenhouse Gas Footprint of Gas Compared to Coal and Petroleum

Figure 6.1 compares the greenhouse gas footprint of natural gas, coal, and petroleum products, including both the direct emissions of carbon dioxide from the burning of the fuels and the full lifecycle emissions of unburned methane associated with developing and

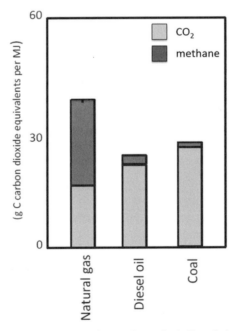

Figure 6.1 Greenhouse gas footprint of natural gas (including shale gas), diesel oil, and coal per unit of heat energy released as the fuels are burned. Direct emissions of carbon dioxide are shown in light gray. Methane emissions expressed as carbon dioxide equivalents are shown in dark gray. As discussed in the text, the methane emission rate used here for natural gas, 3.2% of production, is conservative.
Emission estimates are from Howarth (2020).

using the fuels. Methane emissions in this figure are based on the emission factors presented in Howarth (2020). To obtain the same amount of heat energy, the carbon dioxide emissions from natural gas are smaller than those from coal and petroleum, and this is the foundation of the natural gas as a bridge-fuel concept. However, when methane emissions are included (as carbon dioxide equivalents), the greenhouse gas footprint of natural gas is substantially larger than that even of coal. And as stressed earlier, methane emissions may well be greater than the values used in this figure, 3.2% of production for the natural gas estimate.

Methane emissions are converted to carbon dioxide equivalents in Figure 6.1, allowing a direct comparison with carbon dioxide emissions. Methane is a far more potent greenhouse gas, and here the methane emissions are multiplied by a factor that reflects this greater warming potential, putting both the methane and carbon dioxide emissions into the same units. This factor, known as the global warming potential, compares the warming of methane relative to carbon dioxide on average for a defined period of time after a pulse emission of both gases to the air. Figure 6.1 uses a 20-year time period for this global warming potential. This is consistent with the new climate legislation in New York state but differs from the approach used in almost all greenhouse gas inventories in other states and nations, which use a 100-year timeframe based on a recommendation from the Kyoto Protocol of 1997. The 100-year timeframe severely understates the role of methane in global warming since most of the influence of methane on the climate occurs in the first 30 years after emission, as seen in Figure 6.2 (IPCC 2013; Howarth 2020).

The original choice of 100 years by the Kyoto Protocol was arbitrary (IPCC 2013), and as we have learned more about the role of methane in global warming in the years since 1997, a growing number of researchers have called for using a 20-year timeframe, either instead of (Howarth 2014, 2020) or in addition to the 100-year approach (Ocko et al. 2017; Fesenfeld et al. 2018). Note that both carbon dioxide and methane are critical drivers of global warming, and both the shorter timeframe and longer timeframe are important to consider, as discussed briefly earlier in this chapter. However, combining methane emissions and carbon dioxide emissions into a common metric is a poor approach to communicate information on these emissions, particularly when using the 100-year global warming potential. A better approach is to separately provide information on methane and carbon dioxide, in equivalent units of carbon dioxide equivalents, using the 20-year or other shorter timeframe for methane (Howarth 2014, 2020). The long-term perspective is best characterized simply from the data on carbon dioxide emissions.

It is critically important to reduce methane emissions in a shorter timeframe in order to reduce the risk of moving past tipping points in the climate system, reduce damage to society and natural ecosystems from global warming over the coming decades, and provide the best chance of meeting the COP21 climate goals (Shindell et al. 2012; Collins et al. 2018). From this viewpoint, natural gas is far worse than coal. It is also important to note that the global warming potential approach compares the warming influence of methane to carbon dioxide based on single pulsed releases of both gases. As long as one continues to use natural gas, the global warming consequences remain worse than for coal or oil, and for at least two to three decades after natural gas is no longer used as a fuel. Given the state of

Methane and Climate Change 143

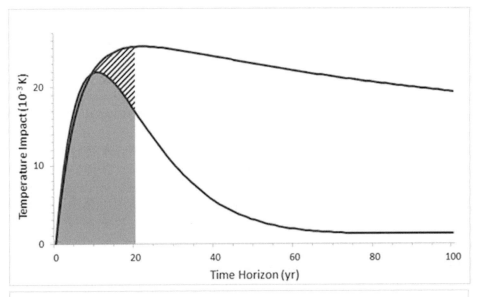

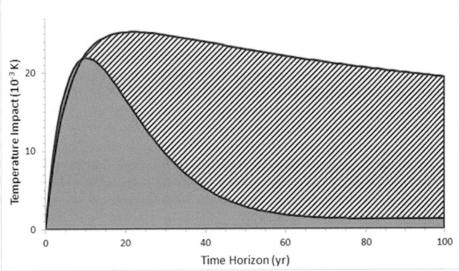

Figure 6.2 Stylized comparison of the global temperature response over time from carbon dioxide and methane emitted for a one-year pulse of each gas at time zero. Top shows area under curve integrated for the time zero to year 20 time period, with methane shown as solid gray and carbon dioxide as striped. Bottom panel is the same except through the 100-year time period. Note that the integrated area for carbon dioxide in both panels also underlies the area for methane, except for the extreme left-handed side of the curves.

Adapted from IPCC (2013) and based on the absolute global temperature change potential. Reprinted from Howarth (2020).

increasing climate disruption in 2020, the continued use of natural gas would be a bridge to disaster.

The comparison of footprints presented in Figure 6.1 is based on the generation of heat. How this heat energy is used matters in the comparison of fuels. For instance, electric power plants powered by natural gas are often (but not always) more efficient than those powered by coal, which tend to be older. On the other hand, internal combustion engines in cars and trucks have a lower efficiency when powered by natural gas than when powered by petroleum products (Alvarez et al. 2012).

Including efficiency concerns, methane emissions from natural gas must be less than 3.2% of consumption in order for natural gas to have a lower greenhouse gas impact than coal for generating electricity, and methane emissions from natural gas must be less than 1% of consumption in order for natural gas to be preferred over diesel fuel for use in long distance, heavy trucks (Alvarez et al. 2012). Hong and Howarth (2016) demonstrated that greenhouse gas emissions from using natural gas water heaters in homes are greater than from using high efficiency heat pumps, with the electricity to power the heat pumps coming from either natural gas or coal, if methane emissions from natural gas are greater than 0.8% of consumption. It makes no sense to use natural gas as a transportation fuel, and the use of natural gas for heating – which is the largest use of gas globally and in the United States – should be phased out as quickly as possible. Even with an increase in electricity production needed for heat pumps, and even if this electricity were to come from fossil fuels (until electricity generation becomes 100% renewable), greenhouse gas emissions are reduced by switching away from natural gas as the energy source for heating buildings and hot water (Hong and Howarth 2016).

6.9 Conclusions

A large and growing body of evidence indicates that methane emissions from the development and use of shale gas are substantial, probably in the range of 3.4% of production based on the most recent top-down estimates for upstream and downstream emissions, such as shown in Table 6.1. Such emissions give shale gas a large greenhouse gas footprint, greater than that of coal or other fossil fuels when emissions are considered on a 20-year timeframe after an emission. Atmospheric methane has been rising rapidly over the past decade, after having been stable to the first decade of the twenty-first century. Given a full lifecycle emission rate of 3.4% of production, shale gas is responsible for 40% of the total global increase in atmospheric methane from all sources since 2005. This increase makes it far more difficult to meet the COP21 target of keeping the Earth well below 2° C from the pre-industrial baseline.

Acknowledgments

The author thanks Roxanne Marino, Stan Ridley, John Stolz, and Tony Ingraffea for their comments and suggested edits on earlier version of this chapter. Preparation of this chapter was supported by a grant from the Park Foundation and by an endowment given to Cornell University by David R. Atkinson.

References

Alvarez RA, Pacala SW, Winebrake JJ, Chameides WL, and Hamburg SP. (2012). Greater focus needed on methane leakage from natural gas infrastructure. *Proceedings of the National Academy of Sciences.* 109: 6435–6440, doi:10.1073/pnas.1202407109

Alvarez RA, Zavalao-Araiza D, Lyon DR, Allen DT, Barkley ZR, Brandt AR, Davis KJ, Herndon SC, Jacob DJ, Karion A, Korts EA, Lamb BK, Lauvaux T, Maasakkers JD, Marchese AJ, Omara M, Pacala JW, Peischl J, Robinson AJ, Shepson PB, Sweeney C, Townsend-Small A, Wofsy SC, and Hamburg SP. (2018). Assessment of methane emissions from the U.S. oil and gas supply chain. *Science.* 361: 186–188, doi:10.1126/science.aar7204

Begon M, Howarth RW, and Townsend C. (2014). *Essentials of Ecology*, 4th Edition. Wiley. ISBN-13: 978-0470909133

Caulton DR, Shepson PD, Santoro RL, Sparks JP, Howarth RW, Ingraffea A, Camaliza MO, Sweeney C, Karion A, Davis KJ, Stirm BH, Montzka SA, and Miller B. (2014). Toward a better understanding and quantification of methane emissions from shale gas development. *Proceedings of the National Academy of Sciences.* 111: 6237–6242, doi:10.1073/pnas.1316546111

Collins WJ, Webber CP, Cox PM, Huntingford C, Lowe J, Sitch S, Chadburn SE, Comyn-Platt E, Harper AB, Hayman G, and Powell T. (2018). Increased importance of methane reduction for a 1.5 degree target. *Environmental Research Letters.* 13: 054003, doi:10.1088/1748-9326/aab89c

Fesenfeld LP, Schmidt TS, and Schrode A. (2018). Climate policy for short- and long-lived pollutants. *Nature Climate Change.* 8: 933–936, doi:10.1038/s41558-018-0328-1

Hausfather Z. (2018). Analysis: How much 'carbon budget' is left to limit global warming to 1.5C? *Carbon Brief* www.carbonbrief.org/analysis-how-much-carbon-budget-is-left-to-limit-global-warming-to-1-5c

Hayhoe K, Kheshgi HS, Jain AK, Wuebbles DJ. (2002). Substitution of natural gas for coal: Climatic effects of utility sector emissions. *Climatic Change.* 54: 107–139

Hmiel B, Petrenko VV, Dyonisius MN et al. (2020). Preindustrial $^{14}CH_4$ indicates greater anthropogenic fossil CH_4 emissions. *Nature.* 578: 409–412, doi:10.1038/s41586-020-1991-8

Hong B and Howarth RW. (2016). Greenhouse gas emissions from domestic hot water: heat pumps compared to most commonly used systems. *Energy Science & Engineering.* 4: 123–133, doi:10.1002/ese3.112

Howarth RW. (2014). A bridge to nowhere: Methane emissions and the greenhouse gas footprint of natural gas. *Energy Science & Engineering.* 2: 47–60, doi:10.1002/ese3.35

Howarth RW. (2019). Ideas and perspectives: Is shale gas a major driver of recent increase in global atmospheric methane? *Biogeosciences.* 16: 3033–3046, doi:10.5194/bg-16-3033-2019

Howarth RW. (2019a). Interactive comment on "Is shale gas a major driver of recent increase in global atmospheric methane" by Robert W. Howarth et al. Biogeosciences Discussion, doi.org/10.5194/bg-2019-131-AC3

Howarth RW. (2020). Methane emissions from fossil fuels: Exploring recent changes in greenhouse-gas reporting requirements for the State of New York. *Journal of Integrative Environmental Sciences*, doi.org/10.1080/1943815X.2020.1789666

Howarth RW, Santoro R, and Ingraffea A. (2011). Methane and the greenhouse gas footprint of natural gas from shale formations. *Climatic Change Letters*. 106: 679–690, doi:10.1007/s10584-011-0061-5

Howarth RW, Santoro R, and Ingraffea A. (2012). Venting and leakage of methane from shale gas development: Reply to Cathles et al. *Climatic Change*. 113: 537–549, doi:10.1007/s10584-012-0401-0

Ingraffea AR, Wawrzynek PA, Santoro R, Wells M. (2020). Reported methane emissions from active oil and gas wells in Pennsylvania, 2014–2018. *Environmental Science & Technology*. 54: 5783–5789, doi:10.1021/acs.est.0c00863

Intergovernmental Panel on Climate Change (IPCC). (2013). Climate Change 2013: The Physical Science Basis. Contribution of Working Group I to the Fifth Assessment Report of the Intergovernmental Panel on Climate Change,. www.ipcc.ch/report/ar5/wg1/

Intergovernmental Panel on Climate Change (IPCC). (2018). Global warming of 1.5°C. An IPCC Special Report on the impacts of global warming of 1.5°C above pre-industrial levels and related global greenhouse gas emission pathways, in the context of strengthening the global response to the threat of climate change, sustainable development, and efforts to eradicate poverty. Cambridge, United Kingdom and New York, NY, USA, Cambridge University Press.

Karion A, Sweeney C, Pétron G, Frost G, Hardesty RM, Kofler J, Miller BR, Newberger T, Wolter S, Banta R, and Brewer A. (2013). Methane emissions estimate from airborne measurements over a western United States natural gas field. *Geophysical Research Letters*. 40: 4393–4397, doi:10.1002/grl.50811, 2013.

Karion A, Sweeney C, Kort EA, Shepson PB, Brewer A, Cambaliza M, et al. (2015). Aircraft-based estimate of total methane emissions from the Barnett Shale region. *Environmental Science & Technology*. 49: 8124–8131, doi:10.1021/acs.est.5b00217

Kessel JM and Tabuchi H. (2019). It's a vast, invisible climate menace: We made it visible. *New York Times*, December 12, 2019, www.nytimes.com/interactive/2019/12/12/climate/texas-methane-super-emitters.html

Kirschke S, Bousquest P, Ciais P, Saunois M, Canadell J, Dlugokencky EJ, Beramaschi P, Beergmann D, Blake D, et al. (2013). Three decades of global methane sources and sinks. *Nature Geosciences*. 6: 813–823, doi:10.1038/ngeo1955

Lamb BK, Cambaliza M, Davis K, Edburg S, Ferrara T, Floerchinger C, Heimburger A, Herndon S, Lauvaux T, Lavoie T, Lyon D, Miles N, Prasad K, Richardson S, Roscioli J, Salmon O, Shepson P, Stirm B, and Whetstone J. (2016). Direct and indirect measurements and modeling of methane emissions in Indianapolis, Indiana. *Environmental Science & Technology*. 50: 8910–8917, doi:10.1021/acs.est.6b01198

Lassey KR, Etheridge DM, Lowe DC, Smith AM, and Ferretti DF. (2007). Centennial evolution of the atmospheric methane budget: What do the carbon isotopes tell us? *Atmospheric Chemistry and Physics*. 7: 2119–2139, doi:10.5194/acp-7-2119-2007

Lavelle M. (2019). Trump EPA tries again to roll back methane rules for oil and gas industry. *Inside Climate News*, August 30, 2019. https://insideclimatenews.org/news/29082019/methane-regulation-oil-gas-storage-pipelines-epa-rollback-trump-wheeler

McKain K, Down A, Raciti S, Budney J, Hutyra LR, Floerchinger C, Herndon SC, Nehrkorn T, Zahniser M, Jackson R, Phillips N, and Wofsy S. (2015). Methane emissions from natural gas infrastructure and use in the urban region of Boston, Massachusetts. *Proceedings of the National Academy of Sciences*. 112: 1941–1946, doi:10.1073/pnas.1416261112

Milkov AV, Schwietzke S, Allen G, Sherwood OA, and Etiope G. (2020). Using global isotopic data to constrain the role of shale gas production in recent increases in atmospheric methane. *Scientific Reports*. 10: 4199.

Miller SM, Wofsy SC, Michalak AM, Kort EA, Andrews AE, Biraud SC, Dlugokencky EJ, Janusz Eluszkiewicz J, Fischer ML, Janssens-Maenhout G, Miller BR, Miller JB, Montzka SA, Nehrkorn T, and Sweeney C. (2013). Anthropogenic emissions of methane in the United States. *Proceedings of the National Academy of Sciences.* 110: 20018–20022, doi:10.1073/pnas.1314392110

Ocko IB, Hamburg SP, Jacob DJ, Keith DW, Keohane NO et al. (2017). Unmask temporal trade-offs in climate policy debates. *Science.* 356: 492–493, doi:10.1126/science.aaj2350

Pandey S, Gautam R, Houweling S, van der Gon HD, Sadavarte P, Borsdorff T, Hasekamp O, Landgraf J, Tol P, van Kempen T, Hoogeveen R, van Hees R, Hamburg SP, Maasakkers JD, and Aben Ilse. (2019). Satellite observations reveal extreme methane leakage from a natural gas well blowout. *Proceedings of the National Academy of Sciences.* 116: 26376–26381. doi.org/10.1073/pnas.1908712116

Peischl J, Ryerson T, Brioude J, Aikin K, Andrews A, Atlas E et al. (2013). Quantifying sources of methane using light alkanes in the Los Angeles basin. *California. Journal of Geophysical Research: Atmospheres.* 118: 4974–4990, doi:10.1002/jgrd.50413

Peischl J, Ryerson T, Aikin K, de Gouw J, Gilman J, Holloway J et al. (2015). Quantifying atmospheric methane emissions from the Haynesville, Fayetteville, and northeastern Marcellus Shale gas production regions. *Journal of Geophysical Research: Atmospheres.* 120: 2119–2139, doi:10.1002/2014JD022697

Peischl J, Karion A, Sweeney C, Kort E, Smith M, Brandt A et al. (2016). Quantifying atmospheric methane emissions from oil and natural gas production in the Bakken Shale region of North Dakota. *Journal of Geophysical Research: Atmospheres.* 121: 6101–6111, doi:10.1002/2015JD024631

Peischl J, Eilerman S, Neuman J, Aikin K, de Gouw J, Gilman J et al. (2018). Quantifying methane and ethane emissions to the atmosphere from central and western U.S. oil and natural gas production regions. *Journal of Geophysical Research: Atmospheres.* 123: 7725–7740, doi:10.1029/2018JD028622

Petrenko V, Smith A, Schaefer H et al. (2017). Minimal geological methane emissions during the Younger Dryas–Preboreal abrupt warming event. *Nature.* 548: 443–446, doi:10.1038/nature23316

Plant G, Kort EA, Floerchinger C, Gvakharia A, Vimont I, and Sweeney C. (2019). Large fugitive methane emissions from urban centers along the US east coast. *Geophysical Research Letters.* 46: 8500–8507, doi:10.1029/2019GL082635

Ren X, Hall D, Vinciguerra T, Benish S, Stratton P, Ahn D et al. (2019). Methane emissions from the Marcellus Shale in Southwestern Pennsylvania and northern West Virginia based on airborne measurements. *Journal of Geophysical Research: Atmospheres.* 124: 1862–1878, doi:10.1029/2018JD029690

Schaefer H, Mikaloff-Fletcher S, Veid C. Lassey K, Brailsford G, Bromley T, Dlubokenck E, Michel S, Miller J, Levin I, Lowe D, Martin R, Vaugn B, and White J. (2016). A 21st century shift from fossil-fuel to biogenic methane emissions indicated by $^{13}CH_4$. *Science.* 352: 80–84, doi:10.1126/science.aad2705.

Schwietzke S, Sherwood O, Bruhwiler L, Miller J, Etiiope G, Dlugokencky E, Michel S, Arling V, Vaughn B, White J, and Tans P. (2016). Upward revision of global fossil fuel methane emissions based on isotope database. *Nature.* 538: 88–91, doi:10.1038/nature19797

Schneising O, Burrows JP, Dickerson RR, Buchwitz M, Reuter M, and Bovensmann H. (2014). Remote sensing of fugitive emissions from oil and gas production in North American tight geological formations. *Earth's Future.* 2: 548–558, doi:10.1002/2014EF000265

Schneising O, Buchwitz M, Reuter M, Vanselow S, Bovensmann H, and Burrows JP. (2020). Remote sensing of methane leakage from natural gas and petroleum systems revisited. *Atmospheric Chemistry and Physics*. 20: 9169–9183.

Shindell D, Kuylenstierna JC, Vignati E, van Dingenen R, Amann M, Klimont Z, Anenberg SC, Muller N, Janssens-Maenhout G, Raes R, Schwartz J, Falvegi G, Pozzoli L, Kupiainent K, Höglund-Isaksson L, Emberson L, Streets D, Ramanathan V, Kicks K, Oanh NT, Milly G, Williams M, Demkine V, and Fowler D. (2012). Simultaneously mitigating near-term climate change and improving human health and food security. *Science*. 335: 183–189, doi:10.1126/science.1210026

Townsend-Small A. (2022). Isotopes as tracers of atmospheric and groundwater methane sources. In Stolz JF, Griffin WM, and Bain DJ (eds.) *Environmental Impacts from the Development of Unconventional Oil and Gas Reserves*. Cambridge University Press.

Turner AJ, Jacob DJ, Benmergui J, Wofsy SC, Maasakker JD, Butz A, Haekamp O, and Biraud SC. (2016). A large increase in US methane emissions over the past decade inferred from satellite data and surface observations, *Geophysical Research Letters*. 43: 2218–2224, doi:10.1002/2016GL067987

US Department of Agriculture (USDA). (2020). *Cattle Inventory*. National Agricultural Statistics Service, US Department of Agriculture. www.nass.usda.gov/Surveys/Guide_to_NASS_Surveys/Cattle_Inventory/ downloaded August 28, 2020.

US Energy Information Administration (US EIA). (2016). Shale gas production drives world natural gas production growth. Energy Information Administration, U.S. Department of Energy. www.eia.gov/todayinenergy/detail.php?id=27512, downloaded September 13, 2018.

US EIA. (2020a). How much shale gas is produced in the United States? Energy Information Agency, U.S. Department of Energy. www.eia.gov/tools/faqs/faq.php?id=907&t=8, downloaded November 13, 2020.

US EIA. (2020b). Natural Gas: Dry Shale Gas Production Estimates by Play. Energy Information Agency, U.S. Department of Energy. www.eia.gov/naturalgas/data.php, downloaded September 9, 2020.

US EPA. (2011). Regulatory Impact Analysis: Proposed New Source Performance Standards and Amendments to the National Emissions Standards for Hazardous Air Pollutants for the Oil and Gas Industry. July 2011. U.S. Environmental Protection Agency, Office of Air and Radiation.

US EPA. (2016). Oil and Natural Gas Sector: Emission Standards for New, Reconstructed, and Modified Sources. U.S. Environmental Protection Agency, final rule, 40 CFR Part 60, EPA–HQ–OAR–2010–0505; FRL–9944–75– OAR, RIN 2060–AS30. Federal Register 81 (#107): 35824-35942. www.govinfo.gov/content/pkg/FR-2016-06-03/pdf/2016-11971.pdf

Vaughn TL, Bella CS, Picering CK, Schwietzke S, Heath GA, Pétron G, Zimmerle DJ, Schnell RC, and Nummedal D. (2018). Temporal variability largely explains top-down/bottom-up difference in methane emission estimates from a natural gas production region. *Proceedings of the National Academy of Sciences*. 115: 11712–11717, doi:10.1073/pnas.1805687115

Worden J, Bloom A, Pandey S, Jiang Z, Worden H, Walter T, Houweling S, and Röckmann T. (2017). Reduced biomass burning emissions reconcile conflicting estimates of the post-2006 atmospheric methane budget. *Nature Communications*. 8: 2227, doi:10.1038/s41467-017-02246-0, 2017.

Wunch D, Toon G, Hedelius J, Vizenor N, Roehl C, Saad K, Blavier J, Blake D, and Wennberg P. (2016). Quantifying the loss of processed natural gas within California's South Coast Air Basin using long-term measurements of the ethane and methane. *Atmospheric Chemistry and Physics*. 16: 14091–14105, doi:10.5194/acp-16-14091-2016

Zhang Y, Gautam R, Pandey S, Omara M, Maasakkers J, Sadavarte P, Lyon D et al. (2020). Quantifying methane emissions from the largest oil-producing basin in the United States from space. *Science Advances*. 6(17): eaaz5120 doi: 10.1126/sciadv.aaz5120

7

Water Usage and Management

JESSICA M. WILSON AND JEANNE M. VANBRIESEN

7.1 Introduction

Fossil fuel extraction and utilization has significant effects on the water cycle. Water is used at multiple stages in the lifecycle of energy production, and various wastewaters are produced during extraction and energy generation from fossil fuels. As noted elsewhere in this book, shale oil and shale gas production have recently emerged and expanded as energy sources worldwide, and the interactions of large-scale operations to develop these resources with the hydrologic cycle are of concern. Figure 7.1 is an overview of the role of water within the development of oil and gas resources, including five stages or activities that involve potential interactions with water resources.

Extensive prior literature and summary reports exist on this topic. The United States Environmental Protection Agency (EPA) conducted a study on the impacts of hydraulic fracturing on drinking water resources in the United States (US EPA 2015) that was reviewed in a lengthy report by the EPA Science Advisory Board (USEPA Science Advisory Board 2016). The USGS has also studied this issue with reports in 2009 and 2013 (Soeder and Kappel 2009; Kappel et al. 2013). The National Academy of Sciences, Engineering, and Medicine conducted several workshop-based studies on health impacts (Institute of Medicine 2014), risk and risk governance (National Research Council 2014), and opportunities and challenges for innovation in water management (National Academies of Sciences 2017) for shale development. In addition, many nongovernmental organizations have published reports on the potential environmental challenges associated with oil and gas development. For example, the Natural Resources Defense Council released a report in 2012 that included a review of hydraulic fracturing wastewater production and treatment options (VanBriesen and Hammer 2012). The Pacific Institute released a report in 2012 summarizing key issues and synthesizing recent research (Ross and Luu 2012). The American Water Works Association published a report on potential impacts on drinking water resources (AWWA 2013; Koplos et al. 2014) and the National Groundwater Association released a report looking at proximity of oil and gas development to water wells (NGWA 2013)

These reports provide extensive discussion of water use and wastewater production across the lifecycle of hydraulic fracturing. In addition to these summary reports, primary research and literature citations related to water resources and hydraulic fracturing have expanded significantly over the past several decades. Figure 7.2 shows the amount of

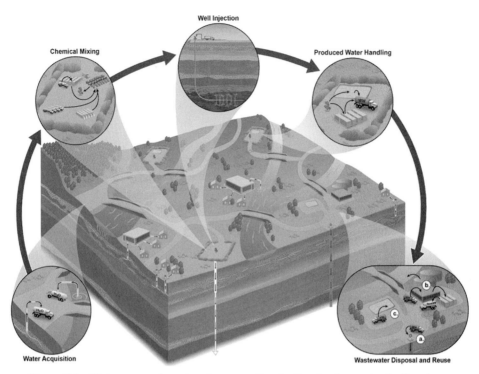

Figure 7.1 Water use and wastewater generation in the development of oil and gas resources. Specific activities in the "Wastewater disposal and reuse" inset are: (a) disposal via injection well, (b) wastewater treatment with reuse or discharge, and (c) evaporation or percolation pit disposal (US EPA 2015) (A black and white version of this figure will appear in some formats. For the colour version, refer to the plate section.)

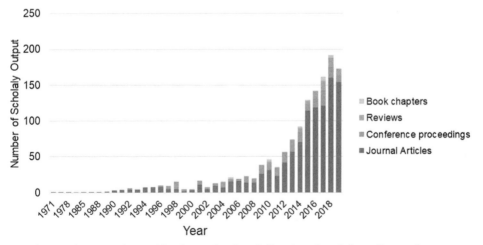

Figure 7.2 Increasing publications related to "oil and gas" and "water" over the past 50 years. (A black and white version of this figure will appear in some formats. For the colour version, refer to the plate section.)

Figure courtesy Xiaoju (Julie) Chen

scholarly output over 50 years based on a search in Web of Science (WoS). The search was based on the combinations of the terms: "oil and gas," or "oil & gas," or "O&G," or "hydraulic fracking," or "Marcellus Shale," and the terms "produced water," or "water use/using/usage," or "brine," or "formation water." A total of 1,385 records were identified, including 1,153 journal articles, 203 conference proceedings, 75 reviews, 20 book chapters, and 29 other records. The earliest publication on this topic was in 1971; the amount of scholarly output grew rapidly after 2010. Mentions in the general press have also increased significantly, with overview articles appearing in *National Geographic*, *Forbes*, and *The New York Times Magazine* (Griswold 2011; National Geographic 2013; Nunez 2013, 2015; Stone 2017; Greenstone 2018).

The present chapter does not attempt to summarize this extensive prior work, but rather provides a holistic overview of the major aspects of water use and wastewater production with national context and examples from our home state of Pennsylvania, which has seen extensive development of the Marcellus shale formation in the past decade.

7.2 Water Resource Acquisition and Water Use

Hydraulic fracturing requires significant quantities of water, and most is acquired locally to reduce transportation costs. Water can be extracted from streams and rivers, pumped from groundwater sources, or purchased from existing water treatment plants. This water use may contribute to global water stress in some areas. Rosa et al. (2018) report that 31–44% of the world's shale deposits are located in areas where water stress might be exacerbated by extraction.

In the United States, national data for water withdrawals does not differentiate among different types of water used for mining; however, analyses have considered water use in specific states or regions. Nicot and Scanlon (2012) estimated water use for shale gas production in Texas for 2011 at 100 million m^3. Murray (2013) estimated water use in Oklahoma in 2011 at 16.2 million m^3. Mitchell et al. (2013) estimated water withdrawals for Marcellus Shale production in the upper Ohio River basin (in Pennsylvania) in 2011–2012 as 10.4 million m^3/year. In these three studies, this water use represented less than 1% of statewide withdrawals. In drier regions or those with less water use for other activities, the fraction of water used for shale development can be more significant. For example, Lin et al. (2018) report from 2008 to 2014 that the annual total industrial water use for Bakken shale oil development (in North Dakota) ranged from 0.5 and 10% of statewide total consumptive water use. This analysis is consistent with a report from the state of North Dakota that water used for shale gas development in 2012 was about 4% of statewide consumptive use (North Daoka State Water Commission 2019). Nicot and Scanlon (2012) reported water consumption for the Eagle Ford shale play represents 5% of total consumption in 2008 in the relevant counties (which are sparsely populated), and they projected an increase to 89% of total consumption when the gas play is at peak production. While most assessments have considered regional effects (e.g., state level or across a shale play), small-scale water withdrawals from headwaters streams may be more

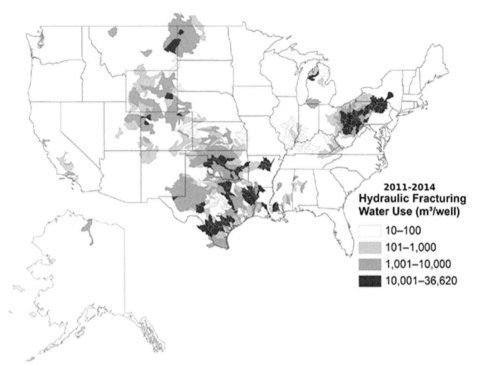

Figure 7.3 Hydraulic fracturing water use per well. Figure shows average of the data (ISH Energy, 2014) within 1st–99th percentile of water volumes used per well from Jan. 2011 to Aug. 2014 in watershed (8 HUCs) with at least two wells reporting water use.
(Reproduced with permission from Gallegos et al. 2015)

impactful (Maloney et al. 2018) Thus, regionally, water withdrawals for oil and gas development are generally small compared with other uses; however, local effects may be of concern, because of significant competing needs, low water availability, or extraction from higher order streams (e.g., small headwaters locations).

In addition to differences in total consumption across the shale development region, there is significant variation in the amount of water used per well (and per unit of energy associated with the extracted resource) based on different geological characteristics of the shale plays. Figure 7.3 shows water use per well across multiple active oil and gas development regions in the United States by watershed (Gallegos et al. 2015). Water use intensity is highest in regions with more horizontal wells (e.g., the Marcellus and Utica shales in Pennsylvania, West Virginia, and eastern Ohio) and lower in regions with lower water use vertical wells (Gallegos et al. 2015). Differences are also seen in oil and gas development in the same plays; Scanlon et al. (2014) report significantly higher water use for Eagle Ford Oil than for Eagle Ford Gas. The water intensity, as measured by volume of freshwater consumed per unit of energy in the fuel produced, also shows wide variability. Kuwayama et al. (2015) report shale gas water intensity values from the literature ranging from 1 to 28 gal/MMBtu (mean of 5), and shale oil water intensity values in the literature ranging from 1.6 to 21.7 gal MMBtu (mean of 8.2).

Water use for unconventional oil and gas development is increasing on a per well basis, and in most regions, water intensity is also increasing. Kondash et al. (2018) report water use per well increased by 770% from 2011 to 2016. In all basins except Marcellus, water intensity also increased. Kondash et al. (2018) caution that while water withdrawals for unconventional oil and gas development are a small fraction of total use annually, cumulative water removed from the hydrologic cycle through this process could have long-term impacts.

Few studies have investigated the type of water being sourced, which can be groundwater, surface water, recycled water, or even treated potable water from public supplies with excess capacity. Mitchell et al. (2013) investigated sources of water used for hydraulic fracturing in the Upper Ohio Basin (in Pennsylvania). In this water rich region, most withdrawals are from surface water and public supply (which is predominately surface water); an estimated 85% of water was sourced directly or indirectly from surface waters. In drier regions, groundwater use is more significant (e.g., the Eagle Ford play relies on the Carrizo-Wilcox aquifer) (Nicot and Scanlon 2012). Scanlon et al. (2014) examined the sources of water for production in the Barnett Shale play, and they concluded surface and groundwater were used in approximately equal amounts.

7.3 Flowback and Produced Water Quantity

Following hydraulic fracturing, some water returns to the surface (initially known as flowback). Produced water returns to the surface during oil or gas production. In general, the amount of water returned to the surface is much less than the amount used for the hydraulic fracturing operation as water is held in the formation. Volumes of produced water vary significantly across different shale plays and can vary significantly over time as new exploration and development ramps up in an area or older wells end production and close. Veil (2020) reports on produced water volumes and management practices for 2017, and compares with data previously reported by Clark and Veil for Argonne National Laboratory for 2007 and Veil for the Ground Water Research and Education Foundation for 2012 (Clark and Veil 2009; Veil 2015). In all three years, the top nine states are Texas, California, Oklahoma, Wyoming, Kansas, Louisiana, Alaska, New Mexico, and Colorado. Eleven states showed significant increases in produced water from 2007 to 2017 (>25% rise), including five states with >100% increase (Pennsylvania, Ohio, West Virginia, North Dakota, and Illinois). States showing significant decreases from 2007 to 2017 in produced water (>25% decline) include Tennessee, Arizona, New York, Kentucky, Michigan, Alabama, Mississippi, and Wyoming. Figure 7.4 shows the volumes reported for the three studies; the inset focuses on those states not in the top 10 produced water generators.

The quantity of produced water also varies significant by wells in different formations. The Eagle Ford and Haynesville plays report high volumes of produced water per well, while the Marcellus production is much lower. Kondash et al. (2017) considered six major unconventional oil and gas formations in the United States, and they report median volume ranges for flowback and produced water from 1.7 to 14.3 million L per well over the first

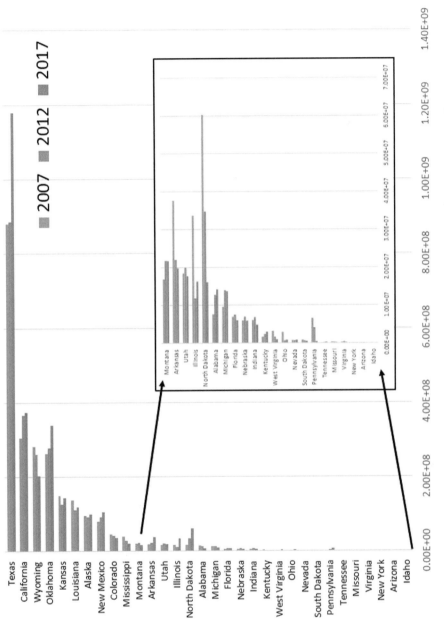

Figure 7.4 Volume of produced water by state for 2007, 2012, and 2017. Boxed inset focuses on those states not in the top 10 produced water generators (note x-axis scale). Data from Clark and Veil (2009); Veil (2015); Veil (2020) (A black and white version of this figure will appear in some formats. For the colour version, refer to the plate section.)

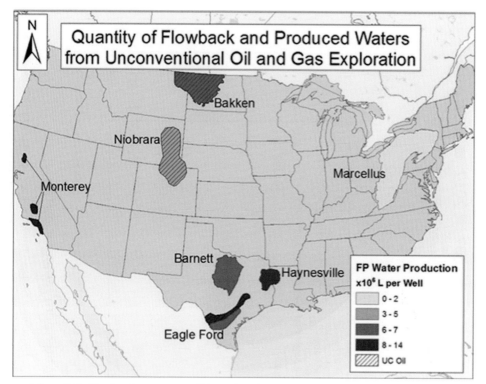

Figure 7.5 Produced water per well in various US shale plays.
(Reproduced with permission from Kondash et al. 2017)

five years of production. Figure 7.5 shows these estimates of produced water per well for major gas plays in the United States. Kondash et al. (2017) report that the median produced water volumes for higher producing reserves (e.g., Eagle Ford) is 2.3 million L per well, while a lower producing reserve (e.g. Marcellus Shale) is 1.7 million L per well. Further, Kondash et al. (2017) report that in all regions except the Marcellus, produced water volumes per well increased from 2011 to 2016. There is significant variability in the amount of produced water compared with the original injection water, but the ratio of year 1 produced water to initial load water increased from 2011 to 2014 before decreasing through 2016.

While Pennsylvania did not report high produced water volumes compared with other states (Figure 7.4), and the Marcellus Shale formation produces less water per well than other resources (Figure 7.5), unconventional natural gas production has risen steadily, from 2.8×10^{10} cubic meters in 2011 to over 1.7×10^{11} cubic meters in 2018; with 2018 representing the largest volume of natural gas on record that has been produced in Pennsylvania in a single year (PA DEP 2018a). With this increased natural gas production comes increased volumes of produced water; as shown in the inset to Figure 7.4. Figure 7.6 shows the volumes of conventional and unconventional produced water in Pennsylvania as reported to the PADEP from 2010 to 2018 (PA DEP 2018b).

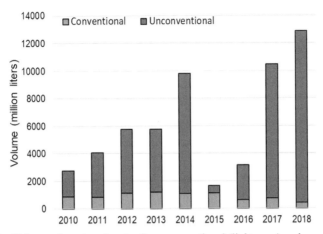

Figure 7.6 Volume of produced water from conventional (light gray) and unconventional (dark gray) natural gas development in Pennsylvania

The produced water volumes shown in Figure 7.6 should be considered "high estimates," as recent work by Lutz et al. (2013) and Veil (2020) suggest that the PA DEP database contains records that are effectively replicates, leading to overreporting of volumes of produced water when using the data as reported. In order to compare the current work with our prior work, we are using the same methodology as reported in Wilson and VanBriesen (2012) and VanBriesen et al. (2014).

In 2006, the total volume of produced water was reported as 1.0×10^6 ML, which is considered baseline for conventional produced water since this was prior to significant development in the Marcellus shale formation. Since 2006, the total volume of produced water in Pennsylvania increased steadily to 12.8×10^6 ML in 2018. Unconventional produced water has steadily increased from 1.8×10^6 ML in 2010 to 12.4×10^6 ML 2018, while conventional sources have remained relatively constant. The years 2015 and 2016 are potential outliers, probably owing to reporting errors. For comparison, this analysis indicates total produced water volume in 2012 is 5.8×10^6 ML, while Veil (2015) reports 4.0×10^6 ML (difference of 35%). For 2017, this analysis reports a total produced water volume or 10.4×10^6 ML, while Veil (2020) reports 8.8×10^6 ML (difference of 17%). In conclusion, Figure 7.6 probably represents an over-estimation of produced water volumes for Pennsylvania because of unidentified replicates in the dataset, but the overall trend of increasing unconventional produced water values from 2010 to 2018 is consistent with other reports.

7.4 Produced Water Quality

Produced water contains residual chemicals that were originally added to enable the hydraulic fracturing operations, as well as chemicals that dissolved from the formation itself (most notably salts). Some components of produced water are potentially harmful

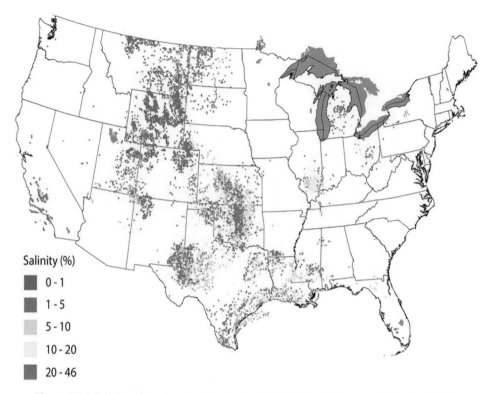

Figure 7.7 Salinity of produced waters in the United States (Allison and Mandler 2018) (A black and white version of this figure will appear in some formats. For the colour version, refer to the plate section.)

environmental or human health pollutants including salts, organic hydrocarbons (e.g., oil and grease), inorganic and organic additives, and naturally occurring radioactive material (NORM). These pollutants can be toxic to humans and aquatic life, radioactive, or corrosive. They can damage ecosystem health or they can interact with disinfectants at drinking water plants to form cancer-causing byproducts.

Both conventional and unconventional produced water is elevated in total dissolved solids (TDS) or salts, when compared with natural waters. Figure 7.7 shows national data for produced water salinity. For comparison, seawater is around 3.5% salinity, and drinking water is typically below 0.05% salinity. Salinity includes a variety of dissolved chemicals, for example, sodium, calcium, chloride, bromide, and carbonates. Chloride is particularly concerning for aquatic ecosystems, and bromide causes concerns for drinking water utilities.

Figure 7.8 shows concentration ranges of TDS, chloride, and bromide for conventional gas well produced water in Pennsylvania and unconventional Marcellus Shale produced water (Hayes 2009). Again, for comparison, seawater is between 34,000 and 38,000 mg/L, while freshwater typically has a TDS less than 1,500 mg/L. Both unconventional and conventional produced water are an order of magnitude saltier (in both TDS and chloride) than seawater. Produced water in Pennsylvania is also enriched in bromide, with an elevated bromide to chloride ratio compared with seawater.

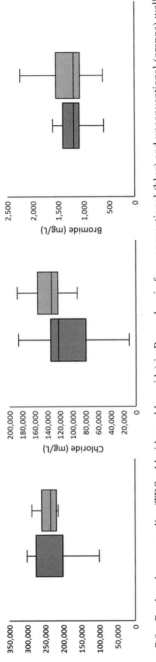

Figure 7.8 Produced water quality (TDS, chloride, and bromide) in Pennsylvania from conventional (blue) and unconventional (orange) wells. (A black and white version of this figure will appear in some formats. For the colour version, refer to the plate section.) Data from Hayes (2009)

When salts are released to surface waters, the resultant changes in water quality can affect aquatic life and alter the ecology of the surface water systems (Nielsen et al. 2003; Kaushal et al. 2005). Increases in salinity up to 1,000 mg/L can have lethal and sublethal effects on aquatic plants and invertebrates (Hart et al. 1991), and chronic concentrations of chloride as low as 250 mg/L have been recognized as harmful to freshwater life (US EPA 1988; Nagpal et al. 2003; Government of Canada 2007). Elevated chloride concentrations are also associated with changes in mortality and reproduction of aquatic plants and animals (Hart et al. 1991; Bidwell et al. 1995; Eaton et al. 1999) and altered community composition of plants and animals (Warwick and Bailey 1997). Even at lower salinity (<10% saline), the structure of microbial communities is altered (Elshahed et al. 2004) and denitrification can be inhibited (Groffman et al. 1995).

When wastewaters with elevated salt concentrations enter source waters used for drinking water plants, effects range from increased corrosion of lead or copper (AWWA Research Foundation 1996; Hong and Macauley 1998), increased scaling and sedimentation (Muylwyk et al. 2014), and poor taste and odor (Young et al. 1996). Bromide is a particular challenge, as even at very low concentrations, it alters drinking water treatability, increasing the formation of carcinogenic disinfection by-products (Hellergrossman et al. 1993; Cowman and Singer 1996; Chang et al. 2001; Liang and Singer 2003; Ates et al. 2007; Richardson et al. 2007).

Produced water from other shale plays in the United States are also elevated in salts (see Figure 7.7), although many are not as high as those found in the Marcellus Shale. Median reported TDS concentrations for other areas include concentrations of 20,000 mg/L for the Denver-Julesburg Basin in northeastern Colorado, 60,000 mg/L for the Barnett play in Texas, and 140,000 mg/L for the Permian Basin in Texas/New Mexico (Rosenblum et al. 2017).

In addition to inorganic salts, produced waters also contain organic compounds commonly found with petroleum and natural gas, including oil and grease; benzene, toluene, ethylbenzene, and xylene (BTEX); and polycyclic aromatic hydrocarbons (PAHs). A study of produced water from the Marcellus Shale showed that these compounds range in concentration from undetectable to greater than 5,000 μg/L for BTEX, indicating high variability in the wastewater that will require treatment (Hayes 2009). Benzene, in particular, occurs naturally in the shale formation but can also be present in fracturing additives, increasing the likelihood it will return in the produced water. Exposure to many of these petroleum hydrocarbons can have direct effects on human health, but as a result of the diversity of possible hydrocarbons, range of concentrations found in produced water, and different chemical characteristics (e.g., solubility), it is difficult to predict the effect of produced water discharged to the environment.

In addition to naturally occurring chemicals from the formation, produced water also contains residual chemicals that were used in the hydraulic fracturing process. Hydraulic fracturing fluid is a mixture of chemicals that is used to maintain the efficiency of well production, including corrosion inhibitors, friction reducers, proppants, etc. The makeup of this fluid varies from site to site and is based on an evaluation of well conditions and evolving industry practices. Concern has been raised about some of these additives with several of those used by industry being identified as known carcinogens or hazardous air pollutants (e.g., 2-butoxyethanol, benzene, naphthalene).

Finally, shale gas produced water also typically contains naturally occurring radioactive material (NORM) at levels elevated from background (NYS DEC 1999). The most abundant types of NORM found in produced water from the Marcellus Shale formation are radium-226 and radium-228, which are produced from radioactive decay of uranium and thorium present in the shale formation. Concentrations of NORM in Marcellus Shale produced water ranged from nondetect to 18,000 pCi/L (Hayes 2009).

7.5 Produced Water Management

Management and disposal of produced water is a significant challenge. Figure 7.9 is a schematic of options, which include, produced water minimization, reuse and recycling, treatment, disposal, and beneficial reuse. Flowback water is more frequently reused, while offsite treatment and disposal are more common for produced water.

The dominant management choice for produced water has been underground injection for enhanced oil recovery, or for disposal. Changes in management between 2007 and 2012 (Veil 2015) show increases in disposal vs EOR for injection and increases in alternative options, including surface discharge, evaporation, and offsite commercial disposal (see Figure 7.10). In many states with abundant capacity for deep well injection, continued use

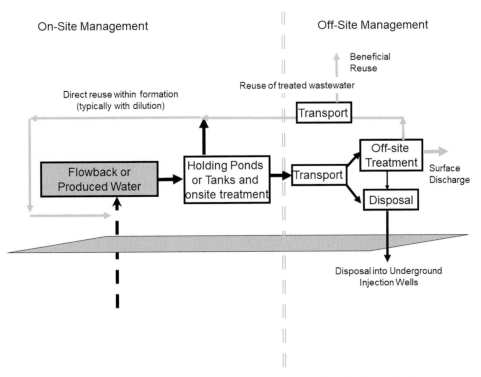

Figure 7.9   Onsite and offsite management of produced water from shale gas development. Modified VanBriesen and Hammer (2012)

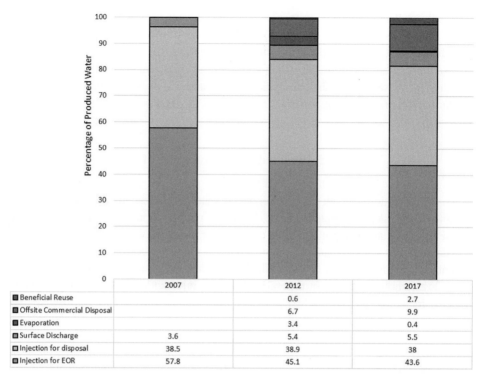

Figure 7.10 Produced water management practices for 2007, 2012, and 2017. (A black and white version of this figure will appear in some formats. For the colour version, refer to the plate section.)

Data from Clark and Veil (2009); Veil (2015); Veil (2020)

of this disposal method is not controversial; however, in some areas, injection of produced water has been linked to increases in earthquakes (Horton 2012; Ellsworth 2013; Walsh and Zoback 2015). From 2012 to 2017, underground injection was still the dominant management option, and remained relatively constant. Offsite commercial disposal and beneficial reuse increased from 2012 to 2017, although they represent a small percentage of the overall management options (<13%).

Despite injection being the dominant produced water management option, some subsurface formations are not suited to injection, or injection wells have not been built and permitted. In these locations, alternative disposal methods are needed. Pennsylvania is an example of one such location. As the volume of produced water requiring management in Pennsylvania increased rapidly after 2010 (see Figure 7.6), and disposal wells were not available in many locations, offsite commercial disposal rose, including disposal at centralized waste treatment (CWT) facilities as well as at publicly owned treatment works (POTWs) designed to treat household wastewater. These facilities provided partial treatment but were not designed to remove salts and thus discharged treated wastewater with elevated TDS. Figure 7.11 shows the changing management options in Pennsylvania for unconventional produced water from the Marcellus Shale from 2010 to 2012. Volumes of produced water going to POTWs and CWTs in 2010 are not large; however, during low

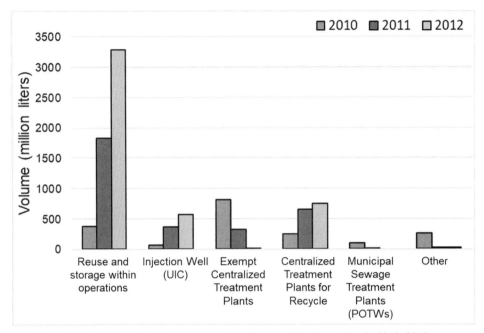

Figure 7.11 Produced water management options used in Pennsylvania 2010–2012. Data from PA DEP (2018)

flow conditions in rivers in southwestern Pennsylvania, these discharges were implicated in significant water quality issues for downstream drinking water providers (discussed further in Section 7.6). In response to these significant concerns, the PA DEP requested that unconventional wastewater not be sent to surface-discharging facilities after May 2011. The significant drops in POTWs and CWTs receiving produced water in 2012 and the concomitant increases in volumes being stored for reuse or sent to centralized treatment for recycle are associated with this policy change.

7.6 Environmental Effects of Produced Water Management

The different management options for unconventional produced water can have various environmental impacts. A major concern with produced water management strategies is the potential negative effects on drinking water resources. These may include accidental releases during transport, discharge of wastewaters from CWTs or POTWs with inadequate treatment, migration of wastewaters that have been applied to land (e.g., for dust and ice control), or leakage from onsite storage pits. Significant research has been conducted on the potential effects of hydraulic fracturing wastewater on water resources and it has been suggested that the most significant effect of unconventional produced water on drinking water resources is through discharge of partially treated wastewater (Kuwayama et al. 2015). As noted in Section 7.5, this wastewater is elevated in certain constituents (e.g., bromide) that may adversely impact drinking water resources.

High TDS is one of the most obvious concerns with hydraulic fracturing wastewater entering drinking water resources. For the Marcellus shale formation, as previously discussed, there is limited availability of underground injection wells for wastewater disposal, leading to a higher rate of treated wastewater discharge to surface water sources. Most drinking water treatment plants are not designed to address high concentrations of TDS (including bromide), resulting in drinking water with poor taste and odor and increased formation of brominated disinfection by-products (States et al. 2013; Wang et al. 2017). In Pennsylvania, prior to May 2011, shale gas associated produced water could be disposed of at publicly owned treatment works (POTWs) that provided minimal treatment of these high TDS wastewaters. Several studies found that wastewater from these WWTPs contained elevated concentrations of constituents associated with produced water, including TDS, chloride, and bromide (Ferrar et al. 2013; Warner et al, 2013). Warner et al. (2013) found that effluent discharge from treatment plants contained chloride concentrations elevated 6,000-fold, and bromide elevated 12,000-fold above stream background levels.

In response to the elevated concentrations of concern to drinking water resources, the Pennsylvania Department of Environmental Protection requested that drillers voluntarily cease delivering shale gas associated produced water to surface-discharging plants (PA DEP 2011). Compliance with this request significantly decreased the use of discharging plants throughout the industry, and only 20% of oil and gas produced water was treated at such plants in 2011, and no produced water was delivered to POTWs in 2012 (Wilson and VanBriesen 2012). Instead, increasing amounts of produced water was recycled and reused within operations, or treated offsite and recycled (see Figure 7.11).

Work conducted by Wilson and VanBriesen (2013) in the Monongahela River Basin from 2009 to 2012 showed that constituent concentrations and resulting anion loads changed as a result of these management changes. As part of this work, chloride and bromide concentrations were measured at multiple locations in the Monongahela River; Figure 7.12 shows estimated chloride and bromide loads for one sampling location. Flow data was obtained from United States Geological Survey (USGS) gages. It is expected that flow will have an important effect on constituent concentrations, as higher flows driven by rainfall lead to more dilution, and lower flows lead to higher concentrations. The loads were calculated using the average (or estimated) daily flow data and the measured chloride and bromide concentrations. The flow is seasonal, with higher flows in the winter and spring than in the fall and summer. It is expected that the highest constituent loads will be observed during periods of high flow and that, during periods of similar flow, the constituent loadings will be similar. From Figure 7.12, there is significant temporal variability in the loads for bromide and chloride. For bromide, the highest loads are observed in March and June 2011. The high bromide load in March 2011 is expected as the flow was quite high during that time. Despite the lower flow in June 2011, the bromide load is still elevated, indicating that flow alone cannot account for the elevated load. During periods of similar low flow (summer 2010 and summer 2012), the bromide loadings are significantly different. The bromide load is statistically significantly higher in 2010 than in 2012. Thus, despite having similar flows in summer 2010 and summer 2012, the bromide loadings are significantly different, indicating a change to bromide load in the system from 2010 to 2012. In contrast, chloride loading is highest during the late winter and early

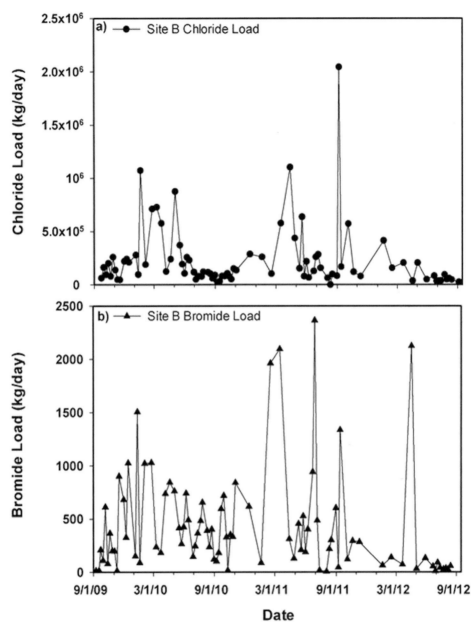

Figure 7.12 Observed changes in chloride (a) and bromide (b) loads in the Monongahela River during 2009–2012.
Data from Wilson and VanBriesen (2013)

spring, probably because of increased runoff to the river from road salt applied for road deicing. However, when bromide load is high during the summer 2010 and 2011, the chloride load is generally low. These differences suggest that the sources of bromide and chloride loads to the basin are different and that the change in produced water management in mid-2011 resulted in a decrease in overall bromide loading to the basin.

Besides salts, unconventional oil and gas produced water often contains naturally occurring radioactive material (NORM) at elevated levels of concern (NYS DEC 1999). Radium-226 and radium-228 have been found at the highest concentrations in produced water. A study by Warner et al. (2013) found that large quantities of unconventional produced water with high loads of radium were discharged from two CWT plants, leading to elevated Ra-226 and Ra-228 levels in stream sediments immediately downstream (200 times greater than activities measured in upstream and background sediments) (Warner et al. 2013). Other pollutants present in unconventional produced water that may have an impact on water resources include organics (BOD, COD), suspended solids, nutrients, oil and grease, and metals. Some of these pollutants are typically removed during conventional wastewater treatment (e.g., suspended solids), while others (e.g., metals) are more likely to pass through treatment plants and enter receiving waters.

Another management option that may have an effect on surface water resources is land application of oil and gas produced waters for deicing or dust suppression, which is permissible in thirteen states in the United States. The high salt content of the O&G wastewater is effective in retaining road moisture for suppressing dust or lowering freezing points for deicing. However, wastewater contaminants may threaten environmental and public health by leaching into surface or groundwater, accumulating around roads, modifying adjacent soil chemistry, or migrating in air and dust. The PA DEP (2016) investigated the potential for radium, a known carcinogen, to accumulate around roads treated with O&G wastewaters. Large variabilities in radium concentrations measured in untreated and treated roads led to inconclusive results. Wastewater pollutants that migrate to water resources could have toxic effects on fish, macroinvertebrates, amphibians, and other salinity-intolerant species. A laboratory study by Tasker et al. (2018) found that nearly all contaminants in O&G wastewaters were leached from road aggregate when applied with synthetic rainwater (chloride, bromide, sodium, magnesium, calcium, and strontium). Iron and lead were retained in the road aggregate after rainwater leaching. For NORM, it was found that radium would accumulate in roads following spreading events but probably at concentrations below the regulatory standards. Leaching experiments conducted with radium showed that 45% of radium applied to a road aggregate leached out after one application, and increased following multiple applications, eventually leveling off at a maximum concentration of 2 pCi/g. The radium not retained in the road aggregate will run off into ditches or underlying soils, where the behavior in the environment is uncertain, as numerous environmental processes influence the mobility of radium (e.g., pH, salinity) (Tasker et al. 2018). Other studies have found cases of well water contamination that were linked to road brining activities (Bair and Digel 1990). Additionally, a study in Ohio found increased chloride concentrations (above the EPA public drinking water standards) in groundwater underlying roads where brine spreading was practiced (Eckstein 2011).

With concerns over produced water discharge to surface waters, other management options may be preferable, but may lead to different environmental effects. With deep well injection, produced water is often partially treated via sedimentation and filtration where suspended solids or other material that may clog the well are removed. A study conducted by the EPA of disposal to Class I underground injection wells (wells where industrial or municipal wastewater is injected below the lowest underground drinking water source) showed that the regulations are adequately protective of human and environmental health (US EPA 2001). Studies conducted prior to the establishment of these regulations demonstrate four significant cases of fluid migration at hazardous waste wells (Clark et al. 2006).

Finally, residuals of oil and gas produced water must also be considered in terms of management options that might have environmental impacts. Residuals are the concentrations of brines and solids containing the chemicals from produced waters, and are typically disposed of at injection wells or landfills, or put to beneficial reuse (VanBriesen and Hammer 2012). The chemicals present in these residuals are present at higher concentrations than in the original produced waters, and this must be considered prior to selection of a management option.

7.7 Conclusion

As unconventional natural gas production grows, selection of management options for water acquisition and wastewater management that do not lead to negative environmental impacts must be prioritized. Research discussed here has shown that the volumes of produced water requiring management are growing significantly in many states, and the most significant environmental impact is probably from surface water disposal of produced water. Underground injection of produced water remains the most commonly used management option. States with limited or no underground injection wells, such as Pennsylvania, have started increasing the amounts of produced water being recycled or reused, resulting in fewer impacts on drinking water resources.

References

Allison E and Mandler B. (2018). *Petroleum and the Environment.* American Geosciences Institute. https://www.americangeosciences.org/critical-issues/petroleum-environment

Ates N, Yetis U, and Kitis M. (2007). Effects of bromide ion and natural organic matter fractions on the formation and speciation of chlorination by-products. *Journal of Environmental Engineering.* **133**: 947–954.

AWWA. (2013). Water and hydraulic fracturing. Available at www.spe.org/jpt/print/archives/2010/12/10Hydraulic.pdf. [Accessed February 24, 2020].

AWWA Research Foundation. (1996). *Internal Corrosion of Water Distribution Systems.* Second Ed., Denver.

Bair ES and Digel RK. (1990). Subsurface transport of inorganic and organic solutes from experimental road spreading of oil-field brine. *Ground Water Monitoring & Remediation.* **10**: 94–105. Available at: http://doi.wiley.com/10.1111/j.1745-6592.1990.tb00008.x.

Bidwell JR, Farris JL, and Cherry DS. (1995). Comparative response of the zebra mussel, Dreissena polymorpha, and the Asian clam, Corbicula fluminea, to DGH/QUAT, a nonoxidizing molluscicide. *Aquatic Toxicology.* **33**: 183–200.

Chang EE, Lin YP, and Chiang PC. (2001). Effects of bromide on the formation of THMs and HAAs. *Chemosphere.* **43**: 1029–1034.

Clark CE and Veil JA (2009). *Produced Water Volumes and Management Practices in the United States.* U.S. Department of Energy.

Clark JE, Bonura DK, and Vorhees RF. (2006). An overview of injection well history in the United States of America. In Tsang C-F and Apps JA (eds.) *Underground Injection: Science and Technology.* Elsevier, pp. 3–12.

Cowman GA and Singer PC. (1996). Effect of bromide ion on haloacetic acid speciation resulting from chlorination and chloramination of aquatic humic substances. *Environmental Science & Technology.* **30**: 16–24.

Eaton LJ, Hoyle J, and King A. (1999). Effect of deicing salt on lowbush blueberry flowering and yield. *Canadian Journal of Plant Science.* **79**: 125–128.

Eckstein Y. (2011). Is use of oil-field brine as a dust-abating agent really benign? Tracing the source and flowpath of contamination by oil brine in a shallow phreatic aquifer. *Environmental Earth Science.* **63**: 201–214.

Ellsworth WL (2013) Injection-induced earthquakes. *Science.* **341**(6142). Available at https://doi.org/10.1126/science.1225942 [Accessed March 12, 2020].

Elshahed MS, Najar FZ, Roe BA, Oren A, Dewers TA, and Krumholz LR. (2004). Survey of archaeal diversity reveals an abundance of halophilic Archaea in a low-salt, sulfide- and sulfur-rich spring. *Applied Environmental Microbiology.* **70**: 2230–2239.

Ferrar KJ, Michanowicz DR, Christen CL, Mulcahy N, Malone SL, and Sharma RK. (2013). Assessment of effluent contaminants from three facilities discharging Marcellus shale wastewater to surface waters in Pennsylvania. *Environmental Science & Technology.* **47**(7): 3472–3481.

Gallegos TJ, Varela BA, Haines SS, and Engle MA. (2015). Hydraulic fracturing water use variability in the United States and potential environmental implications. *Water Resources Research.* **51**: 5839–5845.

Government of Canada. (2007). Priority substances list assessment report for road salts. Available at www.canada.ca/en/health-canada/services/environmental-workplace-health/reports-publications/environmental-contaminants/canadian-environmental-protection-act-1999-priority-substances-list-assessment-report-road-salts.html [Accessed February 11, 2020].

Greenstone M. (2018). Fracking Has Its Costs And Benefits: The Trick Is Balancing Them. *Forbes.* Available at www.forbes.com/sites/ucenergy/2018/02/20/fracking-has-its-costs-and-benefits-the-trick-is-balancing-them/#1075fe8e19b4 [Accessed February 24, 2020].

Griswold E. (2011). The Fracturing of Pennsylvania. *New York Times.* Available at www.nytimes.com/2011/11/20/magazine/fracking-amwell-township.html [Accessed February 24, 2020].

Groffman PM, Gold AJ, and Howard G. (1995). Hydrologic tracer effects on soil microbial activities. *Soil Science Society of America Journal.* **59**: 478–481.

Hart BT, Bailey P, Edwards R, Hortle K, James K, McMahon A, Meredith C, and Swadling K. (1991). A review of the salt sensitivity of the Australian freshwater biota. *Hydrobiologia.* **210**: 105–144.

Hayes T. (2009). *Sampling and Analysis of Water Streams Associated with the Development of Marcellus Shale gas.* Prepared for Marcellus Shale Coalition.

Available at https://edx.netl.doe.gov/dataset/sampling-and-analysis-of-water-streams-associated-with-the-development-of-marcellus-shale-gas/resource/4a092e1c-f824-4ecf-8562-0556cd52e353/download/MSCommission-Report.pdf.

Hellergrossman L, Manka J, Limonirelis B, and Rebhun M. (1993). Formation and distribution of haloacetic acids, THM, and TOX in chlorination of bromide-rich lake water. *Water Research*. **27**: 1323–1331.

Hong PKA and Macauley Y-Y. (1998). Corrosion and leaching of copper tubing exposed to chlorinated drinking water. *Water, Air, & Soil Pollution*. **108**: 457–471.

Horton S (2012) Disposal of hydrofracking waste fluid by injection into subsurface aquifers triggers earthquake swarm in central Arkansas with potential for damaging earthquake. *Seismology Research Letters*. **83**: 250–260.

Institute of Medicine. (2014). *Health Impact Assessment of Shale Gas Extraction*. National Academies Press.

Kappel WM, Williams JH, and Szabo Z. (2013). *Water Resources and Shale Gas/Oil Production in the Appalachian Basin-Critical Issues and Evolving Developments*. U.S. Geological Survey Open-File Report 2013-1137. 12p. Available at: https://pubs.usgs.gov/of/2013/1137/pdf/ofr2013-1137.pdf

Kaushal S, Groffman PM, Likens GE, Belt KT, Stack WP, Kelly VR, Band LE, and Fisher GT. (2005). Increased salinization of fresh water in the northeastern United States. *Proceedings of the National Acadamy of Sciences*. **102**: 13517–13520.

Kondash AJ, Albright E, and Vengosh A. (2017). Quantity of flowback and produced waters from unconventional oil and gas exploration. *Science of the Total Environment*. **574**: 314–321.

Kondash A. J. Lauer NE, and Vengosh A. (2018). The intensification of the water footprint of hydraulic fracturing. *Science Advances*. **4**(8).

Koplos J. Tuccillo ME, and Ranalli B. (2014). Hydraulic fracturing overview: How, where, and its role in oil and gas. *Journal of the American Water Works Association*. **106**: 38–46.

Kuwayama Y, Olmstead S, and Krupnick A. (2015). Water quality and quantity impacts of hydraulic fracturing. *Current Sustainable/Renewable Energy Reports*. **2**: 17–24.

Liang L and Singer PC. (2003). Factors influencing the formation and relative distribution of haloacetic acids and trihalomethanes in drinking water. *Environmental Science & Technology*. **37**: 2920–2928.

Lin Z, Lin T, Lim SH, Hove MH, and Schuh WM. (2018). Impacts of Bakken shale oil development on regional water uses and supply. *Journal of the American Water Resources Association*. **54**: 225–239.

Lutz BD, Lewis AN, and Doyle MW. (2013). Generation, transport, and disposal of wastewater associated with Marcellus Shale gas development. *Water Resources Research*. **49**: 647–656.

Maloney KO, Young JA, Faulkner SP, Hailegiorgis A, Slonecker ET, and Milheim LE. (2018). A detailed risk assessment of shale gas development on headwater streams in the Pennsylvania portion of the Upper Susquehanna River Basin, U.S.A. *Science of the Total Environment*. **610–611**: 154–166.

Mitchell AL, Small M, and Casman EA. (2013). Surface water withdrawals for Marcellus shale gas development: Performance of alternative regulatory approaches in the upper Ohio river basin. *Environmental Science & Technology*. **47**: 12669–12678.

Murray KE. (2013). State-scale perspective on water use and production associated with oil and gas operations, Oklahoma, U.S. *Environmental Science & Technology*. **47**: 4918–4925.

Muylwyk Q, Sandvig A, and Snoeyink V. (2014). Developing corrosion control for drinking water systems. *Opflow*. **40**: 24–27.

Nagpal NK, Levy SA, MacDonald DD, and Ministry of Environment Canada B. C. (2003). Water Quality: Ambient water quality guidelines for choride – overview report. Available at www.env.gov.bc.ca/wat/wq/BCguidelines/chloride/chloride.html.

National Academies of Sciences. (2017). *Flowback and Produced Waters: Opportunities and Challenges for Innovation: Proceedings of a Workshop*. National Academies Press.

National Geographic. (2013). How hydraulic fracturing works. *National Geographic Magazine* Available at www.nationalgeographic.org/media/how-hydraulic-fracturing-works/ [Accessed February 24, 2020].

National Research Council. (2014). *Risks and Risk Governance in Shale Gas Development*. National Academies Press.

NGWA. (2013). Water wells in proximity to natural gas or oil development. Available at www.ntllabs.com [Accessed February 24, 2020].

Nicot JP and Scanlon BR. (2012). Water use for shale-gas production in Texas, U.S. *Environmental Science & Technology*. **46**: 3580–3586.

Nielsen DL, Brock MA, Rees GN, and Baldwin DS. (2003). Effects of increasing salinity on freshwater ecosystems in Australia. *Australian Journal of Botany*. **51**: 655–665. Available at www.publish.csiro.au/paper/BT02115.htm.

North Dakota State Water Commission. (2019). North Dakota Fracking & Water Facts. Available at www.swc.nd.gov/pdfs/fracking_water_use.pdf [Accessed February 24, 2020].

Nunez C. (2015). Fracking, quakes, and drinking water: Your questions answered. *National Geographic Magazine* Available at www.nationalgeographic.com/news/energy/2015/07/150723-fracking-questions-answered/#close [Accessed February 24, 2020].

Nunez C. (2013). How has fracking changed our future? *National Geographic Magazine* Available at: www.nationalgeographic.com/environment/energy/great-energy-challenge/big-energy-question/how-has-fracking-changed-our-future/#close [Accessed February 24, 2020].

NYS DEC. (1999). An Investigation of Naturally Occurring Radioactive Materials (NORM) in Oil and Gas Wells in New York State. Available at www.dec.ny.gov/docs/materials_minerals_pdf/normrpt.pdf.

PA DEP. (2018a). 2018 Oil and Gas Annual Report. Available at www.depgis.state.pa.us/2018OilGasAnnualReport/index.html [Accessed January 2, 2020].

PA DEP. (2018b). Oil and gas reports. Available at www.dep.pa.gov/DataandTools/Reports/Oil and Gas Reports/Pages/default.aspx [Accessed January 2, 2020].

PA DEP. (2016). www.elibrary.dep.state.pa.us/dsweb/Get/Document-112658/Pennsylvania%20Department%20of%20Environmental%20Protection%20TENORM%20Study%20Report%20Rev%201.pdf

PA DEP. (2011). Proposed Rulemaking [25 PA. CODE CH. 95] Wastewater treatment requirements [39 Pa.B. 6467]. Available at www.pabulletin.com/secure/data/vol39/39-45/2065.html.

Richardson SD, Plewa MJ, Wagner ED, Schoeny R, and DeMarini DM. (2007). Occurrence, genotoxicity, and carcinogenicity of regulated and emerging disinfection by-products in drinking water: A review and roadmap for research. *Mutation Research*. **636**: 178–242.

Rosa L, Rulli MC, Davis KF, and D'Odorico P. (2018). The water-energy nexus of hydraulic fracturing: A global hydrologic analysis for shale oil and gas extraction. *Earth's Future*. **6**: 745–756.

Rosenblum J, Nelson AW, Ruyle B, Schultz MK, Ryan JN, and Linden KG. (2017). Temporal characterization of flowback and produced water quality from a hydraulically fractured oil and gas well. *Science of the Total Environment*. **596–597**: 369–377.

Ross N and Luu P. (2012). Hydraulic fracturing and water resources: Separating the frack from the fiction. Available at www.pacinst.orgphone:510.251.1600Facsimile:510.251.2203 [Accessed February 24, 2020].

Scanlon BR, Reedy RC, and Nicot JP. (2014). Comparison of water use for hydraulic fracturing for unconventional oil and gas versus conventional oil. *Environmental Science & Technology*. **48**: 12386–12393.

Soeder DJ and Kappel WM. (2009). Water resources and natural gas production from the Marcellus Shale. Available at http://geology.com/articles/marcellus-shale.shtml [Accessed February 18, 2020].

States S, Cyprych G, Stoner M, Wydra F, Kuchta J, Monnell J, and Casson L. (2013). Brominated THMs in drinking water: A possible link to Marcellus Shale and other wastewaters. *Journal of the American Water Works Association*. **105**: E432–E448. Available at www.awwa.org/publications/journal-awwa/abstract/articleid/38156568.aspx [Accessed June 8, 2017].

Stone J. (2017). Fracking is dangerous to your health: Here's why. *Forbes*. Available at www.forbes.com/sites/judystone/2017/02/23/fracking-is-dangerous-to-your-health-heres-why/#45a60fd75945 [Accessed February 24, 2020].

Tasker TL, Burgos WD, Piotrowski P, Castillo-Meza L, Blewett TA, Ganow KB, Stallworth A, Delompré PLM, Goss GG, Fowler LB, Vanden Heuvel JP, Dorman F, and Warner NR. (2018). Environmental and human health impacts of spreading oil and gas wastewater on roads. *Environmental Science & Technology*. **52**: 7081–7091.

US EPA. (1988). *Ambient Aquatic Life Water Quality Criteria for Chloride*. Washington, DC

US EPA. (2015). *Assessment of the Potential Impacts of Hydraulic Fracturing for Oil and Gas on Drinking Water Resources*. Washington, DC.

US EPA. (2001). Class I underground injection control program: Study of the risks associated with class I underground injection wells. Available at www.epa.gov/safewater [Accessed January 2, 2020].

US EPA Science Advisory Board. (2016). SAB review of the EPA's draft assessment of the potential impacts of hydraulic fracturing for oil and gas on drinking water resources.

VanBriesen JM and Hammer R. (2012). *In Fracking's Wake: New Rules Are Needed to Protect our Health and Environment from Contaminated Wastewater*. NRDC.

VanBriesen JM, Wilson JM, and Wang Y. (2014). Management of produced water in Pennsylvania: 2010–2012. In *Proceedings of the 2014 Shale Energy Engineering Conference*. Pittsburgh, pp. 1–7.

Veil J. (2015). U.S. Produced Water Volumes and Management Practices in 2012. Available at www.gwpc.org/sites/default/files/Produced Water Report 2014-GWPC_0.pdf [Accessed February 24, 2020].

Veil J. (2020). U.S. Produced Water Volumes and Management Practices in 2017. Available at www.veilenvironmental.com/publications/pw/pw_report_2017_final.pdf [Accessed February 24, 2020]

Walsh FR and Zoback MD. (2015). Oklahoma's recent earthquakes and saltwater disposal. *Science Advances*. **1**: e1500195.

Wang Y, Small MJ, and VanBriesen JM. (2017). Assessing the risk associated with increasing bromide in drinking water sources in the Monongahela River. *Pennsylvania Journal of Environmental Engineering*. **143**: 04016089.

Warner NR, Christie CA, Jackson RB, and Vengosh A. (2013). Impacts of shale gas wastewater disposal on water quality in Western Pennsylvania. *Environmental Science & Technology*. **47**: 11849–11857.

Warwick NWM and Bailey PCE. (1997). The effect of increasing salinity on the growth and ion content of three non-halophytic wetland macrophytes. *Aquatic Botany*. **58**: 73–88.

Wilson JM and VanBriesen JM. (2012). Oil and gas produced water management and surface drinking water sources in Pennsylvania. *Environmental Practice* **14**: 288–300.

Wilson JM and VanBriesen JM. (2013). Source water changes and energy extraction activities in the Monongahela River, 2009–2012. *Environmental Science & Technology*. **47**: 12575–12582.

Young WF, Horth H, Crane R, Ogden T, and Arnott M. (1996). Taste and odour threshold concentrations of potential potable water contaminants. *Water Research*. **30**: 331–340.

8

Seismicity Induced by the Development of Unconventional Oil and Gas Resources

DAVID W. EATON

8.1 Introduction

Induced seismicity refers to earthquakes attributed to human activities (Eaton 2018). Seismic events can be induced by a variety of energy technologies, including injection or withdrawal of fluids into the subsurface, or impoundment of water behind a dam (NRC 2013). In the case of unconventional resource development, the main triggering processes are massive waste-brine disposal (e.g., Ellsworth 2013; Weingarten et al. 2015; Walsh & Zoback 2015) and hydraulic fracturing (e.g., Holland 2013; Clark et al. 2014; Bao & Eaton 2016). Of the various environmental risks associated with unconventional oil and gas development, induced seismicity is remarkable for the unusually high level of public anxiety it has evoked; this may reflect, in part, public perceptions about hydraulic fracturing and earthquakes (Rubinstein and Mahani 2015) and/or heightened public sensitivity to anthropogenic risks compared with natural risks (CCA 2014).

During the last decade, a significant surge of induced seismicity linked to unconventional oil and gas development[1] has led to considerable new research and a proliferation of publications on this topic, as highlighted in a number of recent review articles (Foulger et al. 2018; Keranen and Weingarten 2018; Kang et al. 2019; Atkinson et al. 2020; Schultz et al. 2020). However, this area of research is not particularly new, since the scientific basis for injection-induced seismicity has been understood, at least in part, since the 1960s. A causal link between fluid injection and induced seismicity was initially conjectured based on a sequence of earthquakes that took place from 1962 to 1968 near Denver, Colorado (Healy et al. 1968). During this time, a deep well at the Rocky Mountain Arsenal was being used to dispose large volumes of chemical fluid waste into fractured crystalline basement rocks at a depth of 3.7 km. Low-magnitude seismicity occurred throughout the injection operations, but the largest earthquake (4.85 M_W) resulted in minor damage and occurred more than a year after injection had ceased (Herrmann et al. 1981). These observations prompted an experiment at the Rangely conventional oil field in Colorado from 1969 to 1974, which involved multiple cycles of raising and lowering the reservoir pressure to observe the seismicity response (Raleigh et al. 1976). This experiment resulted in a pattern of onset and cessation of seismicity, validating the basic principles of effective stress and confirming that a fault can be activated by increasing the *in situ* pore pressure.

International attention to injection-induced seismicity has been renewed by a dramatic increase in the level of seismicity in Oklahoma and other parts of the United States starting in about 2008 (Ellsworth et al. 2013; Keranen et al. 2013; Ellsworth et al. 2015), and over the following years by sequences of small-to-moderate earthquakes associated with hydraulic fracturing operations in shale basins. Induced seismicity linked to unconventional oil and gas development has occurred around the world (Figure 8.1), including western Canada (BC OGC 2012; 2014; Schultz et al. 2017), the UK (Clarke et al. 2014), United States (Holland 2013; Friberg et al. 2014; Skoumal et al. 2015), China (Lei et al. 2019), and Colombia (Gómez-Alba et al. 2020). To date, the largest induced events have been the September 2016 5.8 M_W Pawnee earthquake in Oklahoma (McGarr and Barbour 2017), triggered by saltwater disposal (SWD), and the December 2018 5.2 M_W earthquake in the Sichuan Basin (Lei et al. 2019), triggered by hydraulic fracturing (HF). These induced events have led to economic losses, injuries, and – in the case of the Sichuan event – several fatalities (Lei et al. 2019). Fortunately, induced seismicity in Oklahoma is abating in response to reduced injection rates (Langenbruch and Zoback 2016) and there are large parts of North America where HF-induced seismicity has not been observed, despite extensive development of unconventional oil and gas resources in low-permeability formations (Van der Baan and Calixto 2017). The underlying physical basis for the high variability in geological susceptibility to HF-induced seismicity is currently the subject of ongoing research (Schultz et al. 2016; Shah and Keller 2017; Eaton and Schultz 2018; Pawley et al. 2018).

From geomechanical and engineering perspectives, HF and SWD are entirely different processes (Rubinstein and Mahani 2015). HF involves the injection of engineered fluids under high pressure (above the fracture gradient), designed to create new tensile fractures and/or activate existing fracture systems within a relatively impermeable rockmass. At a given location such as a well pad, the HF process typically occurs on a timescale of days or weeks in numerous individual stages of fluid injection at discrete segments along a horizontal wellbore (Eaton 2018). After completion of a HF program, injected fluids are flowed back to the wellhead prior to production of oil or gas. During the flowback process, injected hydraulic-fracturing fluid is generally only partially recovered (Rubinstein and Mahani 2015). By contrast, in the case of SWD, larger volumes of fluid are injected into a well over a typical timescale of months or years. By design, the injection zone has a high permeability so that, for a given depth, the fluids can be injected at significantly lower pressures (below the fracture gradient) than in the case of HF. Consequently, new tensile fractures are not formed during SWD operations and, at a single well, the net volume of injected fluid is significantly greater than for HF (Rubinstein and Mahani 2015).

The primary objective of this chapter is to introduce the basic underlying principles of injection-induced seismicity associated with both HF and SWD during unconventional oil and gas development. The basic physical principles provide the basis for a discussion of the space-time evolution of induced seismicity within three case studies from North America that highlight regulatory measures and subsurface response to SWD, HF, and a combination of both of these industrial processes. Several important research themes are covered, including a review of evidence that induced events always occur on preexisting faults, rather than generating new faults; tools for screening induced-seismicity risk based on fault-slip potential; and proposed methods to forecast the maximum magnitude of an

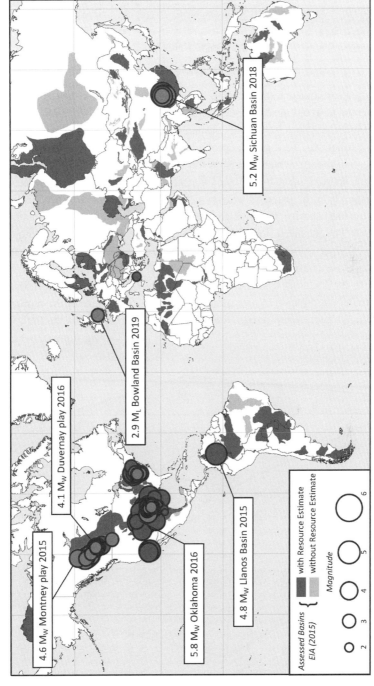

Figure 8.1 Locations of seismicity induced by saltwater disposal and hydraulic fracturing. (A black and white version of this figure will appear in some formats. For the colour version, refer to the plate section.)

induced seismicity sequence. Finally, the benefits and limitations are considered for traffic-light protocols (TLPs), a reactive-control approach that have been implemented in some areas to manage risks of injection-induced seismicity (Eaton 2018).

8.2 Basic Principles

8.2.1 Seismology

Because certain earthquake terms have become part of the modern vernacular, defining key terms is essential to avoid misconceptions and to ensure a full understanding of underlying processes. Figure 8.2 is a schematic drawing that illustrates some key terms used to describe an earthquake. To a good approximation, most earthquakes are well represented using dislocation theory, which describes the shear displacement across a fault: that is, a surface (often approximated by a plane) that accommodates displacement between rock masses on either side. In general, the *displacement* (or *rupture*) is initially localized within a region of the fault surface, from which it spreads out across all or part of the fault during an earthquake. The idealized point of rupture initiation is known as the earthquake *focus*, while the *hypocenter* is the estimated location of the focus. The *epicenter* is the point at the Earth's surface vertically above the hypocenter (but the epicenter does not necessarily correspond to the location with the highest intensity of ground shaking). Since rupture velocity is finite, the duration of an earthquake depends on the size of the rupture area (A). When earthquake rupture terminates, the net displacement (d) across the fault is spatially variable and falls to zero at the edge of the rupture area. During rupture, the earthquake

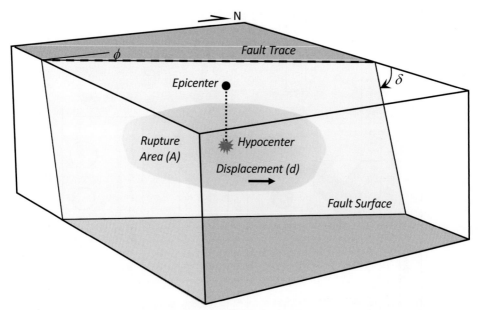

Figure 8.2 Schematic diagram illustrating basic earthquake terminology (A black and white version of this figure will appear in some formats. For the colour version, refer to the plate section.)

radiates elastic waves including P waves (compressional waves) and S waves (shear waves), which propagate through the Earth to seismograph stations at local, regional, and teleseismic (>1,000 km) distances.

A number of basic geometrical parameters are also used to describe faulting (Figure 8.2). For example, the intersection of a fault plane with the Earth's surface is known as the *fault trace*. The clockwise angle ϕ between geographic north and the fault trace is known as the *strike*,[2] while the angle δ between the fault plane and the horizontal is known as the *dip*. Together with *rake*, an angle that defines the direction of slip on a fault plane, these parameters can be used to specify the *focal mechanism* – a graphical representation of the faulting mechanism of an earthquake, discussed later in this chapter.

Given the underlying complexities of earthquake rupture processes, a description of the size of an earthquake is approximate at best. Nevertheless, a useful basic measure of earthquake size is given by the scalar seismic moment (Kanamori and Anderson 1975),

$$M_0 = G\bar{d}A \tag{8.1}$$

where G is the *rigidity* (or *shear modulus*) and $\bar{d}$ is the estimated average displacement calculated over the rupture area, A. The rigidity is region-dependent, with typical values for crustal earthquakes in the range of $1 - 4 \times 10^{10}$ Pa.

Earthquake *magnitude* is a quantitative measure of the size of an earthquake based on seismograph recordings, typically expressed using a logarithmic scale. The moment magnitude (M_W) scale, now considered definitive, was introduced by Hanks and Kanamori (1979) and is given by

$$M_W = \tfrac{2}{3} \log_{10} M_0 - 6.03 \tag{8.2}$$

where M_0 is expressed here in units of N-m. Since the quantities required to calculate M_0 (Eq. 8.1) are usually not readily available in the immediate aftermath of an earthquake, a number of other scales are widely used for rapid estimation of earthquake magnitude based directly on measurements of ground motion at a set of seismograph stations. An example of such a scale is the local magnitude (M_L) for southern California developed by Richter (1935). This scale can be written as (Shearer 2019)

$$M_L = \log_{10} A + 2.56 \log_{10} \Delta - 1.67 \tag{8.3}$$

where A is the peak amplitude of ground motion from the radiated waves, measured in units of μm on a (now archaic) Wood-Anderson seismograph, and Δ is the *epicentral distance* (distance from the epicenter to the seismograph station) in units of km. Although they are approximately equivalent within certain magnitude ranges, it is important to avoid conflating different magnitude scales.

Earthquake occurrences are inherently unpredictable, but for a given region it has long been recognized that, over time, the magnitude–frequency relationship of earthquakes is reasonably approximated by the *Gutenberg-Richter* (G-R) formula (Ishimoto and Iida 1939; Gutenberg and Richter 1944)

$$\log_{10} N = a - bM, \quad M \geq M_C \tag{8.4}$$

In this equation, $N(M)$ is the total number of earthquakes of magnitude $\geq M$, M_C is a low-magnitude cutoff based on catalog completeness, and a and b are parameters that describe, respectively, the earthquake productivity and relative size distribution of events (Eaton 2018). For tectonic earthquakes, b value typically ranges between 0.8 and 1.2, whereas for events induced by hydraulic fracturing, higher b values are common (Eaton and Maghsoudi 2015). In such cases of unusually high b value ($b > 1.5$), the magnitude distribution is thus more heavily weighted to smaller events than is typical for tectonic earthquakes.

Figure 8.3 illustrates the G-R relationship for three regions where there has been notable induced seismicity owing to unconventional oil and gas development, as discussed in the case studies section in this chapter. In each graph, red symbols show the cumulative seismicity distribution (i.e., total number of events of magnitude $\geq M$), while the G-R relationship corresponding to the best-fitting parameters (a, b, M_C) is shown by the solid line. The maximum likelihood estimate of b and its 95% confidence bounds were obtained using the method of Aki (1965), while the magnitude cutoff (M_C) and productivity (a) parameters were obtained using a grid search using the method of least-squares to minimize the misfit in the semilogarithmic domain. The bar graphs show the differential (noncumulative) distribution, which provides an independent check on the model fit since the differential distribution is noisier but should have the same slope as the cumulative distribution. It should be cautioned that these examples are presented for purposes of illustration, since the calculations use nondeclustered seismicity catalogs (see Knopoff 2000), as well as a mix of M_W and M_L (identified in the respective seismicity catalogs as the preferred magnitudes) rather than a uniform magnitude scale such as M_W.

8.2.2 Geomechanics

Crustal deformation, including earthquakes, represents a response to changes in the ambient stress field within the Earth. The stress at a position **x** within a medium can be fully represented using the stress tensor,

$$\boldsymbol{\sigma}(\mathbf{x}) = \begin{bmatrix} \sigma_{11} & \sigma_{12} & \sigma_{13} \\ \sigma_{21} & \sigma_{22} & \sigma_{23} \\ \sigma_{31} & \sigma_{32} & \sigma_{33} \end{bmatrix} \quad (8.5)$$

where each tensor element σ_{ij} denotes the force per unit area acting in the $\hat{x}_j$ direction on a surface normal to $\hat{x}_i$. The diagonal elements of the stress tensor are known as normal stresses[3] (σ_n) and the off-diagonal elements are known as shear stresses (τ). In general, any stress tensor can be rotated into a principal coordinate system, defined by three orthogonal stress axes, which transforms the tensor into a diagonal form. After transformation into this coordinate system, the shear elements of the stress tensor vanish; the remaining diagonal (normal stress) elements are known as the principal stresses.

Stresses caused by Earth's gravitational field mean that, of the three principal stress axes, one is usually very close to vertical. It is therefore common to represent the three principal

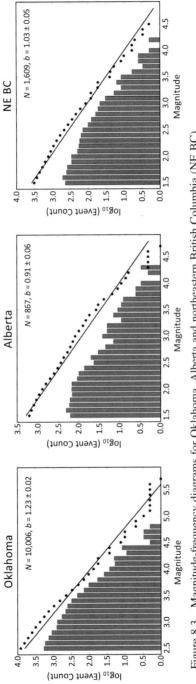

Figure 8.3 Magnitude-frequency diagrams for Oklahoma, Alberta and northeastern British Columbia (NE BC)

stresses using the vertical stress (S_V), maximum horizontal stress (S_{Hmax}), and minimum horizontal stress (S_{hmin}) (Zoback 2010). In this representation, in order to fully characterize the stress tensor at a point in the Earth, it is sufficient to specify the magnitudes of the three principal stresses and the orientation of one of the principal horizontal axes. For example, Figure 8.4 shows measured orientations of S_{Hmax} axes (vectors) compiled by the World Stress Map project (Heidbach et al. 2016; 2018). The color of each vector symbol indicates the inferred stress regime, based on a faulting classification scheme introduced by Anderson (1951). As shown in Figure 8.4 (inset), according to this scheme the magnitudes of the principal stresses are ordered as follows:

- Normal (extensional) faulting: $S_V > S_{Hmax} > S_{hmin}$
- Strike-slip faulting: $S_{Hmax} > S_V > S_{hmin}$
- Reverse (compressional) faulting: $S_{Hmax} > S_{hmin} > S_V$

Overlain on the stress map are *focal-mechanism* (or *beachball*) diagrams for representative induced earthquakes, showing one or more examples for each of these faulting types. Simply put, focal-mechanism diagrams are lower-hemispheric projections, centered on the earthquake hypocenter, indicating the polarity of first motion for P waves radiated from an earthquake source (Eaton 2018). These diagrams are characterized by distinct quadrants, within which the observed P-wave first motion is either outward (denoted by solid fill) or inward (no fill). Curvilinear boundaries between the quadrants represent lower-hemispheric projections of planar surfaces, known as nodal planes. One of the nodal planes is the fault surface. Each fault category in Anderson's classification scheme can be recognized based on a characteristic focal mechanism fill pattern, as shown by beachballs symbols with different colors in Figure 8.4. In the case of strike-slip faults, the nodal planes are close to vertical and four quadrants are evident in the lower hemisphere. Otherwise only three of the four quadrants are evident, where the central region has no fill for normal faults and is filled in the case of reverse faults. The orientation of the fill pattern in a focal mechanism diagram also depends on the fault strike direction.

Of the three principal stresses, S_V is the most straightforward to calculate from first principles as a function of depth, z. Within the Earth's crust (i.e., at depths up to 10's km), the vertical stress can be calculated using

$$S_V(z) = \int_0^z \rho(z)g \, dz' \qquad (8.6)$$

where ρ is density and $g = 9.81$ m/s^2 is the acceleration of gravity. At typical depths of induced earthquakes in the Earth's shallow crust ($z < 10$ km), fluids play a particularly important role in determining the stress regime. Biot (1941) developed a comprehensive constitutive model, known as *poroelasticity*, for two-phase materials consisting of a solid elastic frame with a network of fluid-filled pore spaces. Subject to a number of assumptions (Eaton 2018), poroelastic theory leads to a commonly used model for *effective normal stress* (σ'_n),

$$\sigma'_n = \sigma_n - \alpha P \qquad (8.7)$$

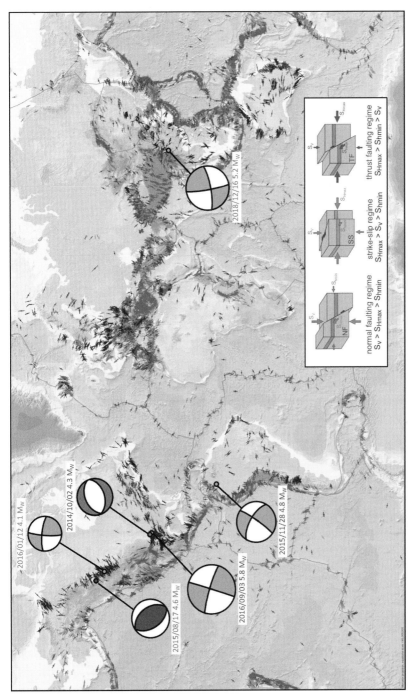

Figure 8.4 World stress map (Heidbach 2016) showing direction of SHmax, stress regime and method of stress determination, with overlay showing selected focal mechanisms for injection-induced events (A black and white version of this figure will appear in some formats. For the colour version, refer to the plate section.)

181

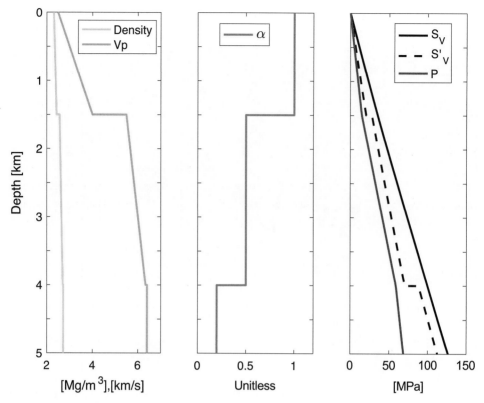

Figure 8.5 Graphs illustrating the calculation of a 1-D effective vertical stress (S'_V) profile for a hypothetical sedimentary basin. Symbols are defined in the text. (A black and white version of this figure will appear in some formats. For the colour version, refer to the plate section.)

where P denotes pore pressure and the Biot parameter, α, provides a measure of the degree of coupling between pore fluid and the elastic rock frame ($0 \leq \alpha \leq 1$). For porous rocks in sedimentary basins that have experienced a relatively simple history of subsidence, the pore pressure is typically close to hydrostatic, such that $P(z) \sim 0.01z$ [MPa], where z is expressed in meters. In other cases, where thick impermeable rock layers impede pore-pressure equilibration, and a basin has experienced episodes of subsidence and uplift (and/or hydrocarbon generation), it is common for the vertical gradient of pore pressure to locally exceed hydrostatic values. Such pore overpressure (e.g., Eaton and Schultz 2018) results in reduced effective stress.

The concepts of stress and poroelasticity are illustrated in Figure 8.5. The left panel in this diagram shows a graph of a hypothetical three-layered structure for P-wave velocity (V_P) and density (ρ). The P-wave velocity profile is representative of a sedimentary basin with a younger sedimentary succession in the upper 1.5 km, unconformably overlying an older sedimentary succession that, in turn, rests unconformably on crystalline basement below 4.0 km. The depth gradient of V_P within the sedimentary layers is indicative of gradually diminishing porosity resulting from increasing age and consolidation of the rock layers, while ρ is calculated based on V_P using the Nafe-Drake relationship (Ludwig et al. 1971). In addition, a uniform value of α

is assumed for each layer, as shown in the central panel. In this example a decrease in α with depth, as shown in Figure 8.8, is indicative of generally diminishing dependence of effective stress on pore pressure with increasing depth. Finally, pore pressure (P) is assumed to be continuous, with depth-varying gradient that reflects hydrostatic conditions in the top layer and overpressured conditions in the deeper sedimentary succession. On the other hand, the discontinuous nature of α leads to stress layering. Such layering of stresses can influence the distribution of seismicity (e.g., Roche and van der Baan 2015).

A Mohr diagram (Parry 2004) provides a convenient tool to depict the state of stress in terms of effective normal-stress (σ'_n) versus total shear-stress ($|\tau|$). In 3-D space, a Mohr diagram contains three semi-circles, each of which is bisected by the σ'_n axis, with intersection points for each circle along the σ'_n axis at two of the effective principal stress points (Figure 8.6). For a surface of arbitrary orientation, such as a fault or fracture, the traction vector (force per unit area acting on the surface) is given by

$$\mathbf{T} = \sigma \hat{n} \tag{8.8}$$

where $\hat{n}$ is the normal to the surface. When the traction, $\mathbf{T}$, is resolved into normal and shear stress components, it transpires that the values are bounded by (i.e., they lie on or within) the largest diameter semicircle and are also bounded by (i.e., they lie on or outside) the two smaller semicircles (Figure 8.6c).

Although many researchers assume a simplified poroelastic pressure effect, based on Eqn. 8.7 with α = 1, the use of a constitutive model for poroelasticity results in a more complex situation in which a change in pore pressure also changes the deviatoric stress (the difference between maximum and minimum principal stress) in a stress-regime dependent manner (Eaton 2018). In particular, in reverse and normal faulting regimes, deviatoric stress increases or decreases, respectively, with increasing pore pressure, whereas in a strike-slip regime there is no change in deviatoric stress with increasing pore pressure (Lavrov 2016).

Representative Mohr diagrams are shown in Figure 8.6 for three depths (1.0 km, 2.5 km, and 5.0 km), based on depth-dependent stress conditions depicted in Figure 8.5. Here, it is further assumed that $S_{hmin} = 0.6\ S_V$ and $S_{Hmax} = 1.6\ S_V$; considering Anderson's (1951) fault classification scheme, this ordering of principal stresses ($S_{Hmax} > S_V > S_{hmin}$) falls within the same strike-slip faulting regime as the ones observed in most of the largest induced-seismic events shown in Figure 8.4. These Mohr diagrams also contain a line that represents the Mohr-Coulomb failure criterion,

$$|\tau| = C + \mu \sigma'_N \tag{8.9}$$

where C denotes cohesion and μ is the coefficient of static friction. In the context of fault activation, the Mohr-Coulomb criterion distinguishes stress regimes that are frictionally stable from those that are frictionally unstable. For the sake of illustration, parameters $C = 5$ MPa and $\mu = 0.6$ are used for the Mohr-Coulomb criterion plotted in Figure 8.6. It should be noted that while this value of friction coefficient falls within a range that is typical for faults, it is more common for a cohesionless state ($C = 0$) to be used for active faults (Zoback 2010). However, faults that are activated during fluid injection are generally located in areas that were seismically quiescent prior to industrial activities (e.g., Ellsworth

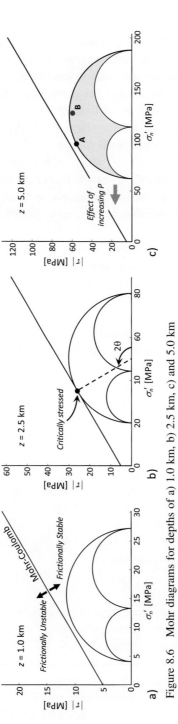

Figure 8.6 Mohr diagrams for depths of a) 1.0 km, b) 2.5 km, c) and 5.0 km

et al. 2015; Bao and Eaton 2016). Experimental data and theoretical models show that faults regain cohesive strength, owing to long-term fluid-rock interactions after the most recent earthquake (Tenthorey and Cox 2006); hence, ancient faults that have not slipped for millennia are likely to have nonnegligible cohesion.

In Figure 8.6b, the outer Mohr circle is approximately tangent to the Mohr-Coulomb criterion at $z = 2.5$ km. At the point of tangency, an *optimally oriented* fault is in a *critically stressed* state. The normal to an optimally oriented fault makes an angle $\theta = 90° - \tan^{-1}\mu$ with respect to the maximum principal stress axis, as shown in Figure 8.6b. In a strike-slip regime, where fault planes are typically vertical, in the case of $\mu = 0.6$ this means that an optimally oriented fault plane makes an angle of ~ 31° with respect to the S_{Hmax} direction. The term critically stressed means a state of incipient failure (Townend and Zoback 2000). In accordance with Byerlee's law, in a critically stressed state the crust is in a state of brittle failure equilibrium wherein the differential stress is limited by the frictional strength of well-oriented, preexisting faults (Zoback and Harjes 1997). In the presence of nonnegligible fault cohesion, C, under this condition the maximum (σ_1) and minimum (σ_3) principal stresses are related by (Eaton 2018):

$$\sigma_1 = \left[\sqrt{\mu^2 + 1} + \mu\right]^2 \sigma_3 + 2C\left[\sqrt{\mu^2 + 1} + \mu\right] \tag{8.10}$$

For the example stress regime considered here, at $z = 1.0$ km and $z = 5.0$ km the stress states lie entirely within the frictionally stable regime (Figure 8.6). However, based on Eqn. 8.7 it is clear that a sufficient increase in pore pressure could destabilize the stress regime at these depths, by transforming the Mohr circles to lower values of σ'_n until they intersect the Mohr-Coulomb failure line.

Laboratory measurements of frictional characteristics of faults show that, under varying slip conditions, friction exhibits dynamic behavior that depends on the state (past history) and rate (velocity) of sliding (see reviews by Marone 1998; Scholz 1998). In many cases, an increase in sliding velocity results in an increase in dynamic friction. Such velocity-strengthening behavior inhibits dynamic rupture (earthquakes) and results in stable sliding, known as fault creep. Conversely, under certain rheological conditions, after an initial transient increase in friction an increase in sliding velocity can produce a net reduction in the dynamic friction. This velocity-weakening behavior occurs over a characteristic slip-weakening (earthquake nucleation) distance and leads to dynamic earthquake rupture. Kohli and Zoback (2013) measured friction parameters for rock samples from unconventional resource plays in the United States and found that those with high total organic carbon (TOC) plus clay content have velocity-strengthening characteristics. In contrast, Kohli and Zoback (2013) reported that samples with lower TOC + clay (<33%) exhibit velocity-weakening and therefore are more prone to dynamic rupture.

8.2.3 Activation Mechanisms

Understanding the underlying physical mechanisms of fault activation resulting from HF and SWD is important for the development of science-informed strategies for risk assessment and mitigation. Although it is abundantly clear that an increase to the *in situ* pore pressure can unclamp a preexisting fault and bring it to a frictionally unstable condition, several other mechanisms for fault reactivation have been proposed and are considered here (Figure 8.7). Two such models, direct pore-pressure effect and a change in fault loading conditions, have been variously ascribed to induced seismicity from SWD and/or HF. A third model, aseismic creep, has recently been proposed as a mechanism for HF-induced seismicity (Eyre et al. 2019b).

As noted previously, the potential to induce earthquakes by the direct pore-pressure effect has been recognized since the time of the Denver earthquakes in the 1960s (Healy et al. 1968) and the subsequent experiment at the Rangely field in Colorado (Raleigh et al. 1976). Observation and modeling of seismicity at 8.9 km depth, induced by small-scale brine injection at the KTB borehole, indicates that critically stressed, permeable faults are present at this depth and the required pressure perturbation for fault activation is <1 MPa (Zoback and Harjes 1997). Based on hydrodynamic modeling coupled with observations of an expanding seismicity front centered on several high-volume SWD wells, Keranen et al. (2014) concluded that a pore-pressure increase of as little as 0.07 MPa was sufficient to initiate induced seismicity in central Oklahoma. This result implies that basement faults in this area were exceedingly close to brittle failure prior to the onset of massive wastewater injection.

Shapiro et al. (1997; 2003) demonstrated that an expanding seismicity front can be explained by linear diffusion of pore pressure within a poroelastic medium. According to this model, the triggering front of injection-induced seismicity propagates outward like a diffusive process and can be used to characterize the fluid-transport properties of reservoirs (Shapiro et al. 2002). A nonlinear diffusion model was introduced by Shapiro and Dinske (2009) to provide a better fit to observations during hydraulic fracturing. In southern Kansas, Peterie et al. (2018) showed that pressure diffusion from multiple high-volume SWD wells could explain fault activation to distances of up to 90 km.

In the case of HF, the direct pore-pressure effect seems intuitively plausible, at least in close proximity to the hydraulic fracture, owing to the use of high injection pressures that exceed the fracture gradient.[4] Fluid leak off from hydraulic fractures extends the stimulated region of a reservoir where elevated pore-pressure could induce slip (Warpinski et al. 2013). However, at distances of more than a few 10's m away from hydraulic fractures, the timescales of diffusion are significantly longer than required to explain the time lag of observed seismicity from HF (Atkinson et al. 2016). In particular, observation of induced events, during or shortly after (days) the end of HF operations and at distances of up to several km (e.g., Bao and Eaton 2016; Kozłowska et al. 2018), is difficult to reconcile with limited extent of fluid leak off and extremely low matrix permeability of shale reservoirs, which is typically on the order of microDarcies $\left(1 \text{ microDarcy} = 10^{-18} \text{ m}^2\right)$.

These potential shortcomings of the pore-pressure diffusion model have provided motivation to consider other possible mechanisms for HF-induced seismicity. One possibility is the presence of enhanced permeability pathways, owing to preexisting networks of natural fractures, which would allow elevated pore pressures to propagate more rapidly in an otherwise low-permeability medium. Unconventional reservoirs commonly contain

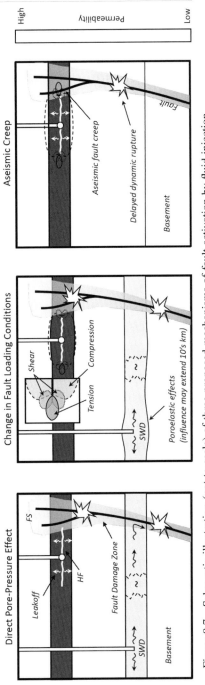

Figure 8.7 Schematic illustration (not to scale) of three proposed mechanisms of fault activation by fluid injection

natural fracture networks, which have been documented using analog outcrops, well logs, cores, and microseismicity (MacKay et al. 2018). Moreover, fracture density is expected to increase within fault damage zones, a volume of deformed rocks around a fault that results from the initiation, propagation, and interaction of slip processes (Kim et al. 2004).

A change in fault loading conditions, the second mechanism presented in Figure 8.7, could occur without any change in pore pressure within a fault core. For example, one of the primary mechanisms for propagation of hydraulic fractures is stress perturbations that concentrate within the process zone near a fracture tip (Yao 2012). Based on analytical expressions for internal deformation in a half space (Okada 1992), the stress pattern around tensile fracture results in lobate patterns of shear stress and tensile stress at the tip and compression orthogonal to the fracture (Eaton 2018). For hydraulic fracturing in proximity to a fault, these elastostatic stress changes in the surrounding medium could be significantly larger than the pore-pressure effect owing to leak off (Kettlety et al. 2019; 2020). Poroelastic stress transfer is another potentially important factor; for example, by modeling hydraulic fracturing within a fully coupled poroelastic numerical simulation, Deng et al. (2016) showed that stress changes owing to poroelastic coupling could explain the timing and location of induced seismic events at distances of up to several km from HF operations.

In the context of a stationary point injection source such as a SWD well, Segall and Lu (2015) considered the effects of poroelastic coupling of stress with pore pressure, along with time-dependent earthquake nucleation predicted by the rate-state friction model of Dieterich (1994). Their analysis indicates that, for a given magnitude, seismicity rate depends on the time-varying volume of perturbed crust as well as the size distribution of fault segments. They also found that seismicity rate decays following an initial peak and that large events are not unexpected after a well is shut in. Goebel et al. (2017) investigated a 2016 earthquake sequence in Oklahoma that was manifested as two linear seismicity clusters located from 10 to 40 km from the nearest, high-rate SWD well. Using a semi-analytical approach, they concluded that poroelastically induced stress changes provide a plausible mechanism for changing fault loading conditions at large distances. This potential long-range influence of poroelastic effects is depicted in the central panel in Figure 8.7.

Finally, a model for seismicity triggering by aseismic creep during HF operations, depicted in the right panel of Figure 8.7, was proposed by Eyre et al. (2019b). This model was motivated by several recent studies. Experimental observations of Guglielmi et al. (2015), involving a small-scale fluid injection into carbonate faults at a depth of ~ 280 m in southeastern France, provided direct measurements of aseismic fault slip and diffuse microseismicity that accompanied fluid injection. Bhattacharya and Viesca (2019) developed a physical model based on these experimental results and showed that stress changes induced by fault creep extend beyond the fluid-pressurized region, outpacing pore-fluid migration. Eyre et al. (2019a) reported high-resolution locations for microseismicity during HF operations prior to a January 2016 HF-induced event in central Alberta, measuring 4.1 M_W. Their study showed that the induced earthquake nucleated within a massive carbonate, located about 400m above the HF treatment zone. Eyre et al. (2019b) used rate-state friction parameters inferred from well-log derived TOC plus clay values as the basis for numerical simulations. These confirmed that the 4.1 M_W induced event could be explained by precursory aseismic fault creep within the TOC-rich unconventional reservoir. According to the model results, aseismic slip loaded other parts of the fault, above and below the reservoir, and, in this case, ultimately triggered dynamic rupture within the massive carbonate. Since it is common for HF to occur in rocks that have a high TOC and clay composition, Eyre et al. (2019b) argued that aseismic fault creep, driven by either a direct pore-pressure increase or a change in fault loading conditions, could be a relatively common occurrence during HF.

8.3 Case Studies

The concepts discussed in Section 8.2 are illustrated in this section, based on previous studies of induced seismicity in three areas of North America. The first area, covering most of Oklahoma and parts of neighboring states, effectively represents "ground zero" in terms of scientific recognition and analysis of seismicity triggered by widespread saltwater injection associated with development of unconventional oil and gas resources. In the past few years, there has been growing recognition of HF-induced seismicity in parts of Oklahoma. The second area is located near Fox Creek in central Alberta, Canada and is a simpler case, inasmuch as it represents an unconventional play (Duvernay) in a relatively early phase of development. This is also one of the first regions in the world where induced seismicity hazard has been closely linked to hydraulic fracturing alone. The third area considered here is in northeastern British Columbia, Canada. This area includes a more mature unconventional play (Montney) that exhibits a mixture of seismicity induced by both HF and SWD.

8.3.1 Central Oklahoma

Starting in about 2008, central Oklahoma began to experience an unprecedented surge of induced seismicity within a region of former relative seismic quiescence (Ellsworth 2013). Over the next decade, the main region of seismicity expanded northward into southern Kansas, with isolated clusters of seismic activity in the Dallas Fort-Worth and Snyder areas in northern Texas (Figure 8.8). At its peak, Oklahoma surpassed California when measured in terms of the rate of M > 3 earthquakes per unit area (Keranen et al. 2014), but the activity rate has declined markedly since 2016 (Langenbruch and Zoback 2016). Notable earthquakes during this time include the 5.7 Mw 5 November 2011 Prague (Keranen et al. 2013), 5.1 Mw 13 February 2016 Fairview (Yeck et al. 2016), 5.8 Mw 3 September 2016 Pawnee (Barbour et al. 2017), and 5.0 Mw 7 November 2016 Cushing earthquakes (McNamara et al. 2015). Based on analysis of surface waves, earthquakes in Oklahoma are predominantly represented by strike-slip focal mechanisms on fault planes that are well oriented for slip in the contemporary stress field (Herrmann 2020).

The dramatic growth in SWD operations arose because some of the major unconventional resource plays in this area, such as the Mississippi Lime play, are dewatering plays: they produce large volumes of saline water that cannot be re-injected into

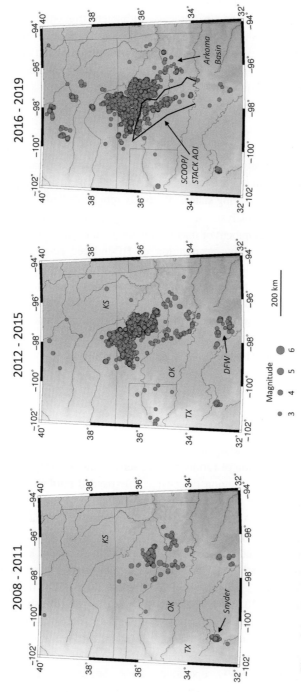

Figure 8.8 $M \geq 2.5$ earthquakes in Oklahoma (OK) and adjacent states including Texas (TX) and Kansas (KS) (A black and white version of this figure will appear in some formats. For the colour version, refer to the plate section.)

the producing formation owing to low reservoir permeability (Eaton 2018). Most of the wells in de-watering plays produce substantially more water than oil; hence, reducing the volume of co-produced water requires a reduction in the volume of produced oil (Rubinstein and Mahani 2015). Owing to the high salinity of the co-produced formation water (40,000–300,000 ppm), the industry practice has been to inject the waste brine into a permeable formation that is not in hydrological contact with potable groundwater (Oklahoma Produced Water Working Group 2017). The Cambrian-Ordovician Arbuckle Group beneath Oklahoma and Kansas, as well as the stratigraphically equivalent Ellenburger Formation in Texas, have been widely used for disposal operations because of their great lateral extent as well as high permeability and porosity. These units unconformably overlie Precambrian crystalline basement and appear to be in hydraulic communication with basement faults, which are thought to host the majority of induced earthquakes (Walsh and Zoback 2015).

Precise relocation of hypocenters of Oklahoma earthquakes from May 2013 to November 2014, using data from publicly available stations supplemented by industry operated seismograph networks, has revealed a pattern of seismicity within distinct linear trends up to 20 km in length (Schoenball and Ellsworth 2017). In areas of denser station coverage within this dataset, the resolved strike and dip of fault planes are in good agreement with focal mechanisms derived from analysis of surface waves (Schoenball and Ellsworth 2017). These planar features most likely represent reactivated basement faults, yet it is striking that many do not correlate with mapped faults: despite a significant effort to compile and characterize geological faults in Oklahoma from a variety of published sources (Marsh and Holland 2016). As discussed later, this apparent lack of a strong correlation between mapped and inferred reactivated faults provides a strong indication of the challenges that exist for induced seismicity risk assessment.

Most of the induced seismicity in this area during the last 12 years is attributable to SWD. However, hydraulic fracturing is key to a number of unconventional oil and gas plays in this region that are generally associated with a petroleum system sourced by the Late Devonian Woodford shale (Cardott 2012). Of particular note are the STACK play (Sooner Trend, Anadarko, Canadian, and Kingfisher) and the SCOOP play (South Central Oklahoma Oil Province), in addition to dry gas that is produced farther east in the Arkoma basin (Schultz et al. 2020). From 2013 to 2020, there have been at least 12,000 HF wells completed in Oklahoma (www.fracfocus.org). The first published cases of anomalous induced seismicity near oil and gas wells with active hydraulic fracturing operations in Oklahoma (Holland 2013; Darold et al. 2014) were followed by more extensive studies that identified hundreds of HF wells that correlated to seismicity up to 3.5 M_L (Skoumal et al. 2018). In 2016 the Oklahoma Corporation Commission (OCC), the state regulator for oil and gas wells in Oklahoma, issued seismicity guidelines for operators working in the SCOOP/STACK area of interest (Figure 8.8) (OCC 2018). This oversight by the regulator has resulted in a catalog of well completions and potentially associated seismicity (Shemata et al. 2019), in which the largest induced event to date occurred on July 25, 2019 (3.9 M_L, 3.6 Mw). This analysis indicates that 7.7% of the HF wells in Oklahoma have associated induced seismicity of magnitude greater than 2, a substantially higher association rate than has been observed in unconventional plays in western Canada (Ghofrani and Atkinson 2020).

8.3.2 Alberta Duvernay Formation

The Duvernay Formation, a fine-grained, organic-rich carbonaceous of Late Devonian age, is the primary source rock for a number of conventional hydrocarbon pools in the Western Canada Sedimentary Basin (Switzer et al. 1994). The Duvernay is coeval with, and was deposited adjacent to, extensive carbonate reef complexes of the lower Leduc Formation (Stoakes 1980). Through a process of tectonic heredity, these reef complexes are interpreted to have formed along paleo-bathymetric highs associated with Precambrian basement structures (Eaton et al. 1995) that divide the Duvernay play into distinct sub-basins. The Duvernay Formation is emerging as a major resource play in North America (Rokosh et al. 2012), with thermal maturity that varies considerably throughout the basin. The Kaybob region near Fox Creek, Alberta is particularly attractive as it falls generally within the gas condensate window, a product that is locally in high demand to be used as diluent for transporting bitumen from the oil sands in northern Alberta (Schultz et al. 2020). Development of the Duvernay resource play began in 2010 and, to date, more than 1,000 HF wells have been completed (Schultz and Pawley 2019).

The first reported cases of induced seismicity associated with the Duvernay play occurred within the Kaybob area in 2013 (Schultz et al. 2015), and as unconventional oil and gas development continued, several seismicity clusters became evident in this area (Figure 8.9). Public attention was drawn to this area as it was among the first unconventional plays to generate HF-induced earthquakes above M_L 4.0; examples include induced events in January 2015 (Bao and Eaton 2016), January 2016 (Eyre et al. 2019a), June 2016 (Schultz et al. 2017), and March 2019 (Schultz and Wang 2020). Starting in 2015 the Alberta Energy Regulator (AER), the provincial regulator for oil and gas wells, introduced new requirements for pre-completion risk assessment, seismicity monitoring, and mitigation in the Kaybob area. These measures were expanded in 2019 to encompass other parts of the province.

Similar to the pattern for induced seismicity in Oklahoma, focal mechanisms of induced events in the Duvernay play consistently show right-lateral strike-slip motion on north–south faults that are generally well oriented for slip in the regional stress field (Bao and Eaton 2016; Wang et al. 2016; 2018; Zhang et al. 2019). Detailed analysis indicates that induced events in the Duvernay play often nucleate within sedimentary layers rather than in the Precambrian basement (Poulin et al. 2019). Collaboration with industry operators has enabled access to local seismograph arrays and 3-D seismic data, which has been instrumental for imaging and characterizing reactivated fault systems in this area (e.g., Bao and Eaton 2016; Eaton et al. 2018; Eyre et al. 2019b). These are interpreted as basement-rooted transtensional fault systems (Wang et al. 2018) that extend into Devonian strata above the Duvernay (Eyre et al. 2019a) and are characterized by negative flower structures that span a width of nearly 10 km (Eaton et al. 2018). In one case, foreshock microseismicity observed using a dense local geophone array appears to highlight precursory fault-activation processes leading up to the mainshock (Eyre et al. 2019a).

The Duvernay play is geographically extensive, but induced seismicity tends to be highly localized within certain areas that appear to be unusually prone to fault activation

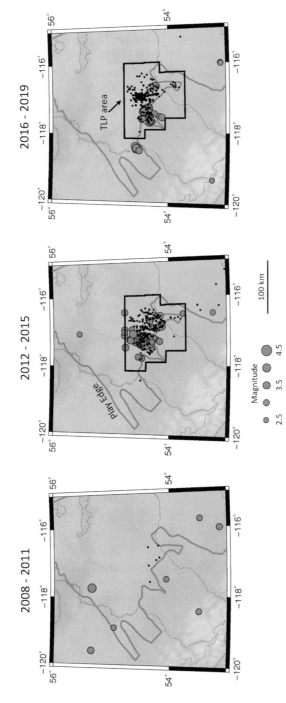

Figure 8.9 $M \geq 2.5$ earthquakes (red circles) in central Alberta, Canada (A black and white version of this figure will appear in some formats. For the colour version, refer to the plate section.)

193

(Pawley et al. 2018; Schultz et al. 2018). The most obvious indication of such geological susceptibility – the presence of critically stressed faults – has proved challenging to map, owing to the subtle seismic expression of near-vertical strike slip faults with predominantly lateral offset that is difficult to discern in a horizontally layered sedimentary sequence (Eaton et al. 2018). There is, however, a clearer seismic signature of interpreted lateral offsets of deeper, Mid-Devonian fluvial channel systems (Weir et al. 2018). A number of other factors have also been proposed to explain this apparent localization of geological susceptibility to induced earthquakes, including proximity to coeval or preexisting reef complexes (Schultz et al. 2016) and areas of significant pore overpressure (Eaton and Schultz 2018). Once geological susceptibility is accounted for using a spatiotemporal association filter, statistical analysis shows that the productivity of induced earthquakes in the Duvernay play scales linearly with stimulation volume (Schultz et al. 2018).

8.3.3 British Columbia Montney Formation

The Montney Formation is a siltstone-dominated clastic unit of Lower Triassic age that extends in the subsurface across the border between British Columbia (BC) and Alberta. It was deposited in a marine, open-shelf environment and forms a westward-thickening wedge that exceeds 300m in thickness adjacent to the Rocky Mountains thrust belt (Edwards et al. 1994). Like target formations in the SCOOP/STACK plays other than the Woodford shale (Schultz et al. 2020), the Montney Formation was charged with hydrocarbons from other source units (Ducros et al. 2017). In BC, many parts of the Montney play have a structurally deformed substrate that includes a buried foreland extension of the north-northwest trending Rocky Mountains thrust belt (Hosseini and Eaton 2018), and/or the east-west axis of the Carboniferous Fort St. John graben complex and Permian Peace River embayment (Barclay et al. 1990). Prior to 2005, conventional oil and gas development focused on localized, high permeability sandstone units within elongate pools, while subsequent development employed horizontal drilling and HF methods to develop this unconventional resource play (BC OGC 2014a). Based on industry reporting to regulators in BC and Alberta, since 2008 more than 14,000 HF wells have been drilled and completed in the Montney play.

Starting in 2013, a sharp increase in seismicity was evident in the BC part of the Montney play (Figure 8.10). Over a period from August 2013 to October 2014, an investigation by the BC Oil and Gas Commission (BC OGC), the provincial regulator, showed that 231 seismic events ranging from 2.4–4.4 M_L could be attributed to unconventional oil and gas development in the Montney play. Based on a spatiotemporal association filter, 38 of these events were linked to wastewater disposal, while 193 were inferred to be induced by HF operations (BC OGC 2014b). There are two prominent zones of seismicity in the BC Montney play, each with distinct characteristics (Figure 8.10). North of 56°N and west of 121°W, the northern Montney area overlies, in part, a buried foreland extension of the Rocky Mountains thrust belt. This area is characterized by a G-R *b*-value of ~1.0–1.3 and a time lag of up to two days following the end of HF operations

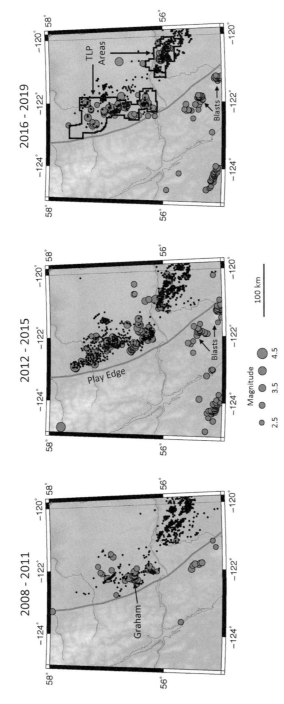

Figure 8.10 $M \geq 2.5$ earthquakes (red circles) in northeastern British Columbia (BC), Canada (A black and white version of this figure will appear in some formats. For the colour version, refer to the plate section.)

195

(Babaie Mahani et al. 2017). To the south and east, the Kiskatinaw area overlies part of the Fort St. John graben (Yu et al. 2019) and has a G-R b-value of ~1.5 (Igonin et al. 2018) with seismicity predominantly occurring during or immediately following injection (BC OGC 2014a). Notable HF-induced events include a 4.6 M_W earthquake in the northern Montney area in August 2015 (Babaie Mahani et al. 2017) and a 4.5 M_W earthquake in the Kiskatinaw area in November 2018 (Babaie Mahani et al. 2019). In both regions, induced earthquakes tend to show a mixture of reverse and strike-slip focal mechanisms (Babaie Mahani et al. 2020), suggesting that this area straddles the boundary between strike-slip and reverse faulting regimes.

Some clusters of induced seismicity in the Montney region have been directly linked to SWD, the most prominent of which is the Graham cluster in the northern Montney (BC OGC 2014b). The injection interval for the Graham disposal well was located in the Mississippian Debolt Formation, a 200-m thick fractured carbonate unit with porosity of up to 12–15% (Hosseini and Eaton 2018). SWD operations commenced in 2001 and continued until July 2015, when the injection switched to a shallower zone. Associated seismicity of magnitude greater than 2.5 commenced in 2003 and continued until 2016 (Hosseini and Eaton 2018), including a M_W 4.0 event in September 2010 (BC OGC 2014b).

8.4 Discussion

8.4.1 Regulatory Measures: Traffic Light Protocols and Exclusion Zones

To manage the risks of induced seismicity, regulatory measures have been implemented in many jurisdictions. One of the most widely used approaches is known as a traffic-light protocol (TLP), which forms one component within a comprehensive risk-management strategy that also includes monitoring (feedbacks) and hazard assessment (planning). A TLP is a reactive-control measure with multiple discrete response thresholds, such as "green," "yellow," and "red" light conditions, wherein each threshold imposes specific actions designed to mitigate the associated risk based on observable criteria (Trutnevyte and Wiemer 2017; Eaton 2018). Ideally, a TLP makes use of a dedicated seismograph array with real-time processing (Bommer et al. 2006); in practice, however, monitoring is often provided by sparse, regional networks. In the case of HF-induced seismicity, TLP thresholds are usually defined in terms of M_L, although more elaborate thresholds that include ground-motion parameters (e.g., peak ground acceleration) and public response (number of complaints) have been used for managing induced seismicity risks of engineered geothermal systems (e.g., Häring et al. 2008). As elaborated later, the range of TLP thresholds has varied by jurisdiction and changed over time.

In BC, the thresholds and regulatory measures have been revised over time in response to increasing knowledge of induced seismicity. In 2015, the BC OGC made enhancements to their Drilling and Production Regulation that required suspension of HF and SWD operations in the event of an induced earthquake of M_L 4.0 or greater within 5 km of operations. As a further regulatory response, a regional seismograph network was established through a public–private partnership, resulting in improved seismicity monitoring of the Montney play (Kao et al. 2018). In 2016, an amendment to well permit conditions added additional reporting requirements for peak ground motion. In 2018, the regulator issued a Special Project Order for the Kiskatinaw region (southern TLP area in Figure 8.10) that reduced the magnitude threshold for suspension of operations to 3.0 M_L and imposed additional conditions that included notifying nearby residents, a seismic hazard pre-assessment, and deploying an accelerometer within three km of the well pad. In Alberta, the AER issued a subsurface order in 2015 that established TLP requirements for the Kaybob-Duvernay region (Figure 8.9), with magnitude thresholds set at 2.0 and 4.0 M_L (Shipman et al. 2018). Under this TLP, activation of the yellow-light condition – occurrence of an induced event above 2.0 M_L within 5 km of HF operations – requires the operator to report the event to the regulator and to implement a mitigation plan. This plan may include a reduction in injection rate, a reduction in injection pressure, or skipping of HF stages (Shipman et al. 2018). Activation of the red-light condition at 4.0 M_L under this TLP requires a suspension of HF operations until a site-specific mitigation plan is developed and approved by the regulator. In response to a M_W 4.0 event in March 2019 that was widely felt in the nearby city of Red Deer, Alberta (Schultz and Wang 2020), the AER established a new TLP in a different part of the Duvernay play with a red light at 3.0 M_L (AER 2109). In Oklahoma, the OCC implemented mandatory submission of HF Notices in 2016 and subsequently established seismicity guidelines for operators working within the SCOOP/STACK area of interest (Figure 8.8), with TLP magnitude thresholds of 2.0, 2.5, 3.0, and 3.5 M_L (Shemeta et al. 2019). These magnitude thresholds require the following actions: implementation of a mitigation plan (2.0 M_L); pause completion operations for a minimum of six hours and initiate a technical conference with the OCC to discuss mitigation procedure (2.5 M_L); receive permission from the OCC to resume HF (3.0 M_L); and immediately suspend all HF and meet in person with OCC to determine under what circumstances the operator can safely resume (3.5 M_L).

Regulatory measures have also been established to manage induced seismicity risks arising from SWD. Starting in 2013, the OCC started to impose new requirements for reduction in saltwater injection volumes into the Arbuckle Group and Precambrian basement, or plugging back wells, in certain areas of the state. These measures, in addition to reduced oil production linked to a drop in world oil price, have led to an overall abatement in the level of seismicity in the state (Langenbruch and Zoback 2016). In BC, permit conditions applied to deep well disposal of produced water limit the increase in reservoir pressure to no more than 20% above the initial reservoir pressure prior to operations, which is confirmed by annual reservoir pressure testing and reporting (BC OGC 2017). As well, the OGC requires monthly statements that included wellhead injection pressure and injected volumes. These measures, together with monitoring low-level, initial seismic events, are reported to have been successful in reducing this source of induced seismicity in the Montney play (M. Gaucher and S. Venables, pers. comm., 2019).

In the UK a TLP for hydraulic fracturing has been established with the lowest threshold at 0.5 M_L (Clarke et al. 2019). To a certain extent, these differences in TLP thresholds reflect a risk-based (as opposed to a strictly hazard-based) management approach; risk is the product of hazard and exposure (e.g., Eaton 2018) and therefore factors in variability in exposure between remote and populated areas. Comparisons between different jurisdictions are further complicated by substantial differences in terms of public level of tolerance and the minimum magnitude that can be felt by local residents. The implementation of an exclusion zone for hydraulic fracturing in the vicinity of critical infrastructure, such as dam facilities (e.g., AER 2019) represents an even stricter risk-management approach to avoid high-consequence events. Nevertheless, the vast difference in TLP thresholds across different jurisdictions, especially in the case of the UK, raises questions about the underlying basis on which these protocols are constructed. For example, TLPs implicitly assume that larger events are preceded by smaller precursory events that will activate warning levels. Kao et al. (2018) provided corroborating evidence, documenting three cases where red-light events in the Kaybob-Duvernay area of Alberta were preceded by yellow light events. Similarly, Mahani et al. (2017) documented a number of $2 < M_L < 4$ events within 5 km of the 2015 4.6 M_W event in the Montney play. However, Van der Elst et al. (2016) argued that the timing of the largest event within a seismicity sequence is random; consequently, the largest event could, in principle, occur first. Furthermore, Eyre et al. (2019b) developed a model in which anomalous induced seismicity is preceded by aseismic slip processes, suggesting that observations of slow deformation could be useful. TLPs are also based on a further premise that stopping injection is an effective measure to prevent levels of seismicity from escalating. However, in many cases, seismicity is observed to continue after operations have ceased, with the largest events sometimes occurring after shut-in (Häring et al. 2008; Bao and Eaton 2016; Kao et al. 2018; Clarke et al. 2019).

8.4.2 Fault Slip Potential

Many of the TLPs that have been established by regulators to manage induced-seismicity risk from hydraulic fracturing include a requirement for operators to perform a risk assessment prior to HF operations. This risk assessment could incorporate information about historical or recent seismicity, or information about known faults. However, there are a number of challenges in carrying out a quantitative risk assessment. For example, since induced seismicity has been documented in areas of previous seismic quiescence (e.g., the Kaybob-Duvernay region in Alberta, Schultz et al. 2017), the historical seismicity record may have limited value. In addition, preexisting faults may have a subtle seismic expression (e.g., Eaton et al. 2018), and even where faults have been mapped it is common for induced events to occur on unmapped faults (e.g., Schoenball and Ellsworth 2017). In areas where information is available about preexisting faults, a quantitative screening tool has been developed known as fault-slip potential (FSP). This tool provides a probabilistic approach to estimate the likelihood of inducing slip on a given fault, based on prior knowledge of the distribution of various parameters that control the risk (Walsh et al. 2017). Various approaches based on FSP have been applied for risk screening in Oklahoma (Schoenball et al. 2018), Alberta (Zhang and Van der Baan 2019), and the Fort Worth Basin (Hennings et al. 2019).

To illustrate this approach, a modified FSP algorithm has been applied to faults A and B, whose stress states are plotted in Figure 8.6. Like the original version of FSP (Walsh et al.,

2017), the approach here uses a Monte Carlo method to assess the *a posteriori* likelihood of slip on a fault, based on a set of assumed parameters. Unlike Walsh et al. (2017), several additional parameters are considered here, namely the fault cohesion (*C*) and the Biot parameter (α). The complete set of nine input parameters used here, along with their assumed prior distributions, are shown in Figure 8.11. Based on these distributions, the probability of slip is plotted in Figure 8.12 as a function of the nominal pore-pressure increase required to exceed the Mohr-Coulomb criterion (Eqn. 8.9). The probability curves were calculated using 10,000 Monte Carlo iterations and are smooth, rather than step-like, owing to the range of input parameters. Since fault A is much closer to failure (Figure 8.6), the probability of slip approaches unity at a much lower level of pore-pressure increases compared to fault B, as expected. Since the stress states vary with depth, the FSP analysis is similarly depth dependent as discussed by Zhang and van der Baan (2019).

The right panel in Figure 8.11 shows the FSP parameter sensitivity for fault B in the form of a tornado plot, expressed in terms of the range of nominal pore pressure required to induce slip. To produce this plot, each parameter was varied separately within its assumed range of deviation, whilst other parameters were unperturbed. This type of diagram is

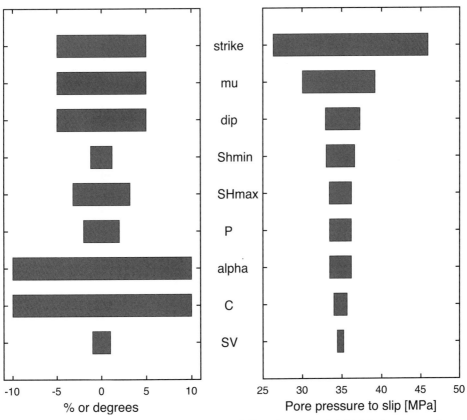

Figure 8.11 Variability in input parameters (left) and tornado plot (right) showing sensitivity of fault B (Figure 9.6) to input parameters for Monte Carlo fault slip potential analysis

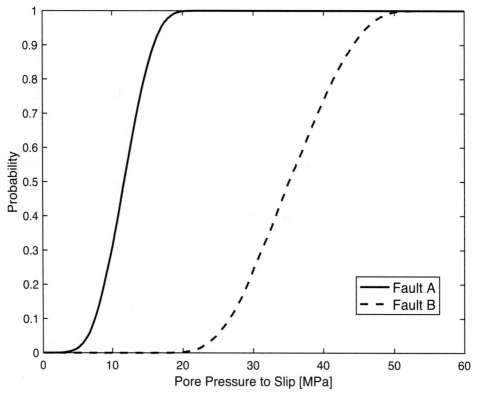

Figure 8.12 Probability of nominal pore-pressure increase required to exceed Mohr-Coulomb criterion on faults A and B (Figure 9.6), at 5.0 km depth

useful to highlight where to focus data gathering efforts for assessment to reduce uncertainties in determining the probability of fault slip (Walsh et al. 2017). Based on the assumed parameter distributions here, the greatest sensitivity is seen for the fault strike direction and the static friction coefficient, μ, whereas the lowest sensitivity is found for the vertical principal stress (S_V) and the fault cohesion.

8.4.3 Maximum Magnitude

Operational forecasting of the maximum magnitude of an induced event is a challenging but important task for earthquake risk analysis (Eaton 2018; Eaton and Igonin 2018) as well as real-time mitigation of induced seismicity (Clarke et al. 2019). As outlined here, a number of authors have proposed functional limits on the maximum magnitude (M_{max}) that are based on certain operational parameters such as the net injected fluid volume.

Shapiro et al. (2011) proposed a purely geometrical condition, by postulating that for injection-induced earthquakes the maximum fault rupture area is contained entirely within a subsurface volume, known as the stimulated volume, that is defined by an expanding triggering front based on microseismicity. Since seismic moment scales with the rupture area (Eqn. 8.1), this model places an upper bound on event magnitude. By fitting an ellipsoid to the stimulated volume, they showed empirically that the distribution of observed maximum magnitudes scales with the logarithm of L_{min}, the minor axis of the ellipsoid. This geometrical condition has important implications for TLPs, since it implies that real-time tracking of the stimulated volume of an expanding cloud of microseismicity could be used to estimate M_{max} (Eaton and Igonin 2018). However, as discussed previously, this approach assumes that large events are preceded by detectable foreshock activity, which neglects the possibility of fault loading owing to aseismic slip (Eyre et al. 2019b).

Building on an extensive body of previous work, McGarr (2014) proposed an upper bound on induced seismic moment (M_0^{max}), given by

$$M_0^{max} = G \cdot V \tag{8.11}$$

where G is the shear modulus of the medium, typically about 30 GPa, and V is the net injected fluid volume. This formulation arises from a set of simplifying assumptions, including the coefficient of static friction is $\mu = 0.6$; the medium is fully saturated and in a state of incipient failure; the pressure increase is given by the product of the bulk modulus with the normalized increase in fluid volume; and, the magnitude-frequency distribution is described by the G-R relationship (Eqn. 8.4) with $b = 1$. Although this relationship implies that induced-seismicity risk can be managed easily by limiting the net injected volume (Eaton and Igonin 2018), in the case of hydraulic fracturing a seismic moment exceeding most yellow light thresholds occurs after a typical single stage (Bao and Eaton 2016).

For a seismicity sequence with N events, whose statistical behavior is Poissonian and whose magnitude-recurrence statistics follow the G-R relationship with a magnitude of completeness M_c, Van der Elst et al. (2016) showed that a naïve statistical estimate of the most probable maximum magnitude is given by

$$M_{max} = M_c + \frac{1}{b} \log_{10} N \tag{8.12}$$

By combining this result with a seismotectonic parameter known as the seismogenic index (Shapiro et al. 2010),

$$\Sigma = \log_{10} N - \log_{10} V + b M_c \tag{8.13}$$

Van der Elst et al. (2016) showed that the maximum magnitude can be expressed as

$$M_{max} = \frac{1}{b}(\Sigma + \log_{10} V) \tag{8.14}$$

They also showed that this naïve estimate of maximum magnitude is statistically biased because it is retrospective; an unbiased estimate of M_{max} requires an additive correction of $\Delta M \sim 0.4$ magnitude units. Eqn. 8.14 implies that the maximum expected seismic moment scales with net injected volume as $M_0^{max} \propto V^{2/3b}$.

Galis et al. (2017) used a different approach to derive a similar relationship for maximum seismic moment for rupture that spontaneously arrests within the limits of the perturbed part of the fault, similar to the model of Shapiro et al. (2011). Galis et al. (2017)

also invoked medium assumptions similar to McGarr (2014) and applied Griffith's fracture-mechanics principle for energy balance, equating the elastostatic energy release rate owing to crack growth to the fracture energy dissipated by unit of crack growth. In the case of arrested rupture, their analysis resulted in a semi-analytical relationship similar to Van der Elst et al. (2016), with $M_0^{max} \propto V^{7/3}$. They also noted that under certain conditions rupture may continue beyond the fluid-stimulated part of the fault, leading to the concept of runaway rupture (Atkinson et al. 2020). In the case of runaway rupture, the maximum magnitude is ultimately limited by the maximum slip surface area on a preexisting fault (Van der Elst et al. 2016; Galis et al. 2017).

Figure 8.13 shows published examples of injection-induced events compiled by McGarr (2014) and Atkinson et al. (2016), along with the relationship between M_0^{max} and V that is predicted by Eqn. 8.11 (McGarr 2014) and Eqn. 8.14 (Van der Elst et al. 2016). In addition, the 95% confidence region is shown based on the method of Van der Elst et al. (2016). The theoretical relationship of Galis et al. (2017) is not shown in Figure 8.13, because it has the same slope as the relationship of Van der Elst et al. (2016) with $b = 1$, as used here. There are a number of published examples of induced events that plot above the moment cap predicted by Eqn. 8.11, whereas all of the events lie within the 95% confidence limits of the statistical model. Several events plot below the 95% confidence interval, but that is not

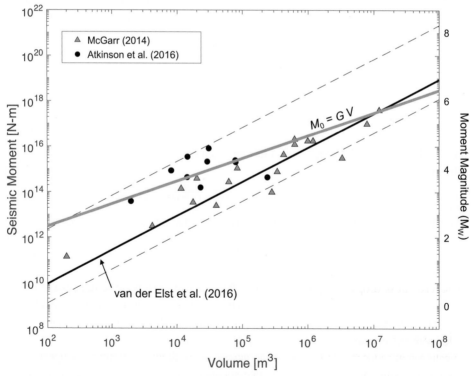

Figure 8.13 Maximum seismic moment (M_0) versus injected volume (V) for some reported cases of large-volume waste brine disposal and hydraulic-fracturing induced seismicity

problematic because the formula predicts an upper bound and magnitude. Various formulas for maximum magnitude such as these have been used in real-time TLPs (Clarke et al. 2019; Kwiatek et al. 2019).

8.4.4 Preexisting Faults

All three activation mechanisms illustrated in Figure 8.7 assume that induced events nucleate on a preexisting fault; this assumption is effectively treated as axiomatic by seismologists who study induced earthquakes. Yet, shear-failure in the lithosphere reflects a continuum of behavior from frictional slip on a preexisting fault to failure of intact rock in the brittle to brittle-plastic regimes (Ohnaka et al. 1997). Indeed, all currently active fault systems came into existence at some point in the geologic past, usually in response to changes in tectonic forces. One could therefore reasonably ask whether induced events *always* occur on a preexisting fault or, alternatively, if it is possible that new crack surfaces could form within an intact rockmass as a result of stimulation. The answer to this question will determine strategies for site evaluation and therefore carries important ramifications for induced-seismicity risk assessment.

Studies of acoustic emissions recorded during shear failure of intact rock, including samples prepared from granite (Ponomarev et al. 1997) and shale (Jia et al. 2018) materials, show that formation of new fracture surfaces owing to coalescence of microfractures is accompanied by brittle failure events with a complex pattern. Although the pattern of seismic activity is complex, the statistical properties and focal mechanisms are broadly similar to those observed in earthquake fault systems (Ponomarev et al. 1997). These experimental processes have also been documented in the field, where macroscope and microscopic analysis of rock fabrics within the process zone of small brittle faults reveal structures that are consistent with formation within the altered stress fields of propagating shear fracture tips (Vermilye and Scholz 1998). However, results obtained with a testing apparatus designed to measure stress changes associated with failure of intact rock under confining conditions at seismogenic depths shows that the associated stress drops are of the same order as the unconfined compressive strength of the rock (i.e., 100's MPa) (Ohnaka et al. 1997.

Such stress drops are significantly higher than most stress drop measurements determined using seismological observations of stress drops for induced earthquakes. For example, a recent comprehensive study of induced earthquakes in western Canada yielded average estimated stress drop values of 7.5 ± 0.5 MPa (Holmgren et al. 2019). However, observed stress drops for approximately co-located events at the Soultz-sous-Forêts geothermal site in France exhibit variations by a factor of 300 (Lengliné et al. 2014). This variability may result from transition from stable to unstable slip on the imaged asperities (Lengliné et al. 2014). Taken together, this stress-drop evidence suggests that most injection-induced earthquakes indeed occur on preexisting fault surfaces; however, the creation of new fracture surfaces – possibly associated with nonnegligible fault cohesion owing to cohesive strength from long-term fluid-rock interactions (Tenthorey and Cox

2006) – cannot be ruled out for preexisting faults that have been inactive on a geological timescale.

8.5 Conclusions

Among the environmental risks associated with unconventional oil and gas development, induced seismicity is remarkable for the unusually high level of public anxiety it has evoked. During the development of unconventional resources, seismicity has been triggered by various mechanisms, including large-volume disposal of produced brines and hydraulic fracturing operations. Examples of significant induced events have been documented in North America, Europe, Asia, and South America, but it is clear that areas prone to induced seismicity are highly localized. In addition, the vast majority of HF and SWD operations have no associated anomalous induced seismicity.

This chapter provides a basic seismological and geomechanical background for induced-seismicity processes. An understanding of the physical mechanisms of fault activation is key for understanding the associated risks and developing a science-informed mitigation strategy. While the basic framework of reduction of effective stress owing to increased pore pressure has been understood for over half a century, observations of induced earthquakes indicate that this is only part of the story. There is evidence that elastostatic effects, poroelastic coupling between pore fluid pressure and rock stresses, and fault loading by aseismic slip may be important. These concepts have been incorporated into the design of regulatory measures such as traffic-light protocols for induced seismicity. Variations in how these measures are implemented in different jurisdictions are a reflection of how the level exposure modifies the risk. Research is ongoing to develop new tools for risk assessment and hazard mitigation, including how they can be applied to forecast the maximum magnitude.

Acknowledgments

This research was supported by the Canada First Research Excellence Fund through its support of the Global Research Initiative in Sustainable Low Carbon Unconventional Resources, as well as by NSERC Discovery Grant RGPIN/03823-2017. Data products for this study were accessed through the Human-Induced Earthquake Database (HiQuake), (www.inducedearthquakes.org). The Generic Mapping Tools (GMT) were used in the preparation of some figures (Wessel et al. 2019).

Notes

1 Oil and gas extracted using techniques other than conventional production, including the use of hydraulic fracturing applied to low-permeability formations and disposal of large volumes of fluid for de-watering plays.
2 To avoid ambiguity, strike direction is measured using the right-hand rule, meaning that the fault dips toward the observer's right while facing along the direction of strike.

3 In this chapter, the sign convention for normal stress is that positive stresses are compressive. The opposite sign convention is also common.
4 The depth-normalized fluid pressure required to initiate a tensile fracture.

References

AER (Alberta Energy Regulator). (2019). Subsurface Order No. 6. www.aer.ca/documents/orders/subsurface-orders/SO6.pdf

Aki K. (1965). Maximum likelihood estimate of b in the formula log $N = a - bM$ and its confidence limits. *Bulletin of the Earthquake Research Institute*. 43(1): 237–239.

Anderson EM. (1951). *The Dynamics of Faulting and Dyke Formation with Applications to Britain*. Hafner Pub. Co.

Atkinson GM, Eaton DW, Ghofrani H, Walker D, Cheadle B, Schultz R, Shcherbakov R, Tiampo K, Gu J, Harrington RM, and Liu Y. (2016). Hydraulic fracturing and seismicity in the western Canada sedimentary basin. *Seismological Research Letters*. 87(3): 631–647.

Atkinson GM, Eaton DW, Igonin N. (2020). Developments in understanding seismicity triggered by hydraulic fracturing. *Nature Reviews Earth & Environment*. 1(1): 264–277.

Babaie Mahani AB, Schultz R, Kao H, Walker D, Johnson J, and Salas C. (2017). Fluid injection and seismic activity in the northern Montney play, British Columbia, Canada, with special reference to the 17 August 2015 Mw 4.6 induced earthquake. *Bulletin of the Seismological Society of America*. 107(2): 542–552.

Babaie Mahani A, Kao H, Atkinson GM, Assatourians K, Addo K, and Liu, Y. (2019). Ground-motion characteristics of the 30 November 2018 injection-induced earthquake sequence in Northeast British Columbia, Canada. *Seismological Research Letters*. 90(4): 1457–1467.

Babaie Mahani A, Esfahani F, Kao H, Gaucher M, Hayes M, Visser R, and Venables S. (2020). A systematic study of earthquake source mechanism and regional stress field in the Southern Montney Unconventional Play of northeast British Columbia, Canada. *Seismological Research Letters*. 91(1): 195–206.

Bao X, and Eaton DW. (2016). Fault activation by hydraulic fracturing in western Canada. *Science*. 354(6318): 1406–1409.

Barbour AJ, Norbeck JH, and Rubinstein JL. (2017). The effects of varying injection rates in Osage County, Oklahoma, on the 2016 Mw 5.8 Pawnee earthquake. *Seismological Research Letters*. 88(4): 1040–1053.

Barclay JE, Krause FF, Campbell RI, and Utting J. (1990). Dynamic casting and growth faulting: Dawson Creek graben complex, Carboniferous-Permian Peace River embayment, western Canada. *Bulletin of Canadian Petroleum Geology*. 38(1): 115–145.

BC OGC (British Columbia Oil and Gas Commission). (2014a). Montney Formation Play Atlas NEBC. www.bcogc.ca/node/8131/download

BC OGC (British Columbia Oil and Gas Commission). (2014b). Investigation of observed seismicity in the Montney trend. www.bcogc.ca/sites/default/files/documentation/technical-reports/investigation-observed-seismicity-montney-trend.pdf

BC OGC (British Columbia Oil and Gas Commission). (2017). Application guideline for deep well disposal of produced water. www.bcogc.ca/application-guideline-deep-well-disposal-produced-water-non-hazardous-waste.

Bhattacharya P and Viesca RC. (2019). Fluid-induced aseismic fault slip outpaces pore-fluid migration. *Science*. 364(6439): 464–468.

Biot MA. (1941). General theory of three-dimensional consolidation. *Journal of Applied Physics*. 12(2): 155–164.

Bommer JJ, Oates S, Cepeda JM, Lindholm C, Bird J, Torres R, Marroquín G, and Rivas J. (2006). Control of hazard due to seismicity induced by a hot fractured rock geothermal project. *Engineering Geology*. 83(4): 287–306.

Cardott BJ. (2012). Thermal maturity of Woodford Shale gas and oil plays, Oklahoma, USA. *International Journal of Coal Geology*. 103(1): 109–119.

CCA (Council of Canadian Academies). (2014). *Environmental Impacts of Shale Gas Extraction in Canada*. Council of Canadian Academies.

Clarke H, Eisner L, Styles P, and Turner P. (2014). Felt seismicity associated with shale gas hydraulic fracturing: The first documented example in Europe. *Geophysical Research Letters*. 41(23): 8308–8314.

Clarke H, Verdon JP, Kettlety T, Baird AF, and Kendall JM. (2019). Real-time imaging, forecasting, and management of human-induced seismicity at Preston New Road, Lancashire, England. *Seismological Research Letters*. 90(5): 1902–1915.

Darold A, Holland AA, Chen C, and Youngblood A. (2014). *Preliminary Analysis of Seismicity near Eagleton 1–29, Carter County, July 2014*. Oklahoma Geological Survey Open File Report, OF2–2014.

Deng K, Liu Y, and Harrington RM. (2016). Poroelastic stress triggering of the December 2013 Crooked Lake, Alberta, induced seismicity sequence. *Geophysical Research Letters*. 43(16): 8482–8491.

Dieterich J. (1994). A constitutive law for rate of earthquake production and its application to earthquake clustering. *Journal of Geophysical Research: Solid Earth*. 99(B2): 2601–2618.

Ducros M, Sassi W, Vially R, Euzen T, and Crombez V. (2017). 2-D basin modeling of the Western Canada Sedimentary Basin across the Montney-Doig system: Implications for hydrocarbon migration pathways and unconventional resources potential. In AbuAli MA Moretti I, and Bolås HMN (eds.) *Memoir 114: Petroleum Systems Analysis: Case Studies, Tulsa, OK*. American Association of Petroleum Geologists.

Dziewonski AM, Chou TA, and Woodhouse JH. (1981). Determination of earthquake source parameters from waveform data for studies of global and regional seismicity. *Journal of Geophysical Research: Solid Earth*. 86(B4): 2825–2852.

Eaton DW. (2018). *Passive Seismic Monitoring of Induced Seismicity: Fundamental Principles and Application to Energy Technologies*. Cambridge University Press.

Eaton DW and Igonin N. (2018). What controls the maximum magnitude of injection-induced earthquakes? *The Leading Edge*. 37(2): 135–140.

Eaton DW anf Maghsoudi S. (2015). 2b… or not 2b? Interpreting magnitude distributions from microseismic catalogs. *First Break*. 33(10): 79–86.

Eaton DW and Schultz R. (2018). Increased likelihood of induced seismicity in highly overpressured shale formations. *Geophysical Journal International*. 214(1): 751–757.

Eaton DW, Igonin N, Poulin A, Weir R, Zhang H, Pellegrino S, and Rodriguez G. (2018). Induced seismicity characterization during hydraulic-fracture monitoring with a shallow-wellbore geophone array and broadband sensors. *Seismological Research Letters*. 89(5): 1641–1651.

Eaton DW, Milkereit B, Ross GM, Kanasewich ER, Geis W, Edwards DJ, Kelsch L, and Varsek J. (1995). Lithoprobe basin-scale seismic profiling in central Alberta: Influence of basement on the sedimentary cover. *Bulletin of Canadian Petroleum Geology*. 43(1): 65–77.

Edwards DE, Barclay J, Gibson D, Kvill G, and Halton E. (1994). Triassic strata of the Western Canada Sedimentary Basin. In Mossop GD and Shetsen I (eds.) *Geological*

Atlas of the Western Canada Sedimentary Basin. Calgary, AB: Canadian Society of Petroleum Geologists and Alberta Research Council.

EIA (Energy Information Administration). (2015). World Shale Resource Assessments, www.eia.gov/analysis/studies/worldshalegas.

Ekström G, Nettles M, and Dziewoński AM. (2012). The global CMT project 2004–2010: Centroid-moment tensors for 13,017 earthquakes. *Physics of the Earth and Planetary Interiors*. 200(1): 1–9.

Ellsworth WL. (2013). Injection-induced earthquakes. *Science*: 341(6142), 1225942, https://doi.org/10.1126/science.1225942.

Ellsworth WL, Llenos AL, McGarr AF, Michael AJ, Rubinstein JL, Mueller CS, Petersen MD, and Calais E. (2015). Increasing seismicity in the US midcontinent: Implications for earthquake hazard. *The Leading Edge*. 34(6): 618–626.

Eyre TS, Eaton DW, Zecevic M, D'Amico D, and Kolos D. (2019a). Microseismicity reveals fault activation before M_W 4.1 hydraulic-fracturing induced earthquake. *Geophysical Journal International*. 218(1): 534–546.

Eyre TS, Eaton DW, Garagash DI, Zecevic M, Venieri M, Weir R, and Lawton DC. (2019b). The role of aseismic slip in hydraulic fracturing–induced seismicity. *Science Advances*. 5(8): eaav7172. https://doi.org/10.1126/sciadv.aav7172

Foulger GR, Wilson M, Gluyas J, Julian BR, and Davies R. (2018). Global review of human-induced earthquakes. *Earth-Science Reviews*. 178(1), 438–514.

Friberg PA, Besana-Ostman GM, and Dricker I. (2014). Characterization of an earthquake sequence triggered by hydraulic fracturing in Harrison County, Ohio. *Seismological Research Letters*. 85(6), 1295–1307.

Frohlich C, DeShon H, Stump B, Hayward C, Hornbach M, and Walter JI. (2016). A historical review of induced earthquakes in Texas. *Seismological Research Letters*. 87(4): 1022–1038.

Galis M, Ampuero JP, Mai PM, and Cappa F. (2017). Induced seismicity provides insight into why earthquake ruptures stop. *Science Advances*. 3(12): eaap7528, https://doi.org/10.1126/sciadv.aap7528.

Gan W and Frohlich C. (2013). Gas injection may have triggered earthquakes in the Cogdell oil field, Texas. *Proceedings of the National Academy of Sciences*. 110(47): 18786–18791.

Ghofrani H and Atkinson GM. (2020). Activation rate of seismicity for hydraulic fracture wells in the Western Canada Sedimentary Basin. *Bulletin of the Seismological Society of America*, in press.

Goebel THW, Weingarten M, Chen X, Haffener J, and Brodsky EE. (2017). The 2016 Mw5. 1 Fairview, Oklahoma earthquakes: Evidence for long-range poroelastic triggering at > 40 km from fluid disposal wells. *Earth and Planetary Science Letters*. 472(1): 50–61.

Gómez-Alba S, Vargas C, and Zang A. (2020). Evidencing the relationship between injected volume of water and maximum expected magnitude during the Puerto Gaitán (Colombia) earthquake sequence from 2013 to 2015. *Geophysical Journal International*. 220(1): 335–344.

Guglielmi Y, Cappa F, Avouac JP, Henry P, and Elsworth D. (2015). Seismicity triggered by fluid injection–induced aseismic slip. *Science*. 348(6240): 1224–1226.

Gutenberg, B. and Richter CF. (1944). Frequency of earthquakes in California. *Bulletin of the Seismological Society of America*. 34(4): 185–188.

Hanks TC and Kanamori H. (1979). A moment magnitude scale. *Journal of Geophysical Research: Solid Earth*. 84(B5): 2348–2350.

Häring MO, Schanz U, Ladner F, and Dyer BC. (2008). Characterisation of the Basel 1 enhanced geothermal system. *Geothermics*. 37(5): 469–495.

Healy JH, Rubey WW, Griggs DT, and Raleigh CB. (1968). The Denver earth-quakes. *Science*. 161(3848): 1301–1310.

Heidbach O, Rajabi M, Reiter K, and Ziegler M. (2016): World Stress Map Database Release 2016. GFZ Data Services, https://doi.org/10.5880/WSM.2016.001.

Heidbach O, Rajabi M, Cui X, Fuchs K, Müller B, Reinecker J, Reiter K, Tingay M, Wenzel F, Xie F, and Ziegler MO. (2018). The World Stress Map database release 2016: Crustal stress pattern across scales. *Tectonophysics*. 744(1): 484-498.

Hennings PH, Lund Snee JE, Osmond JL, DeShon HR, Dommisse R, Horne E, Lemons C, and Zoback MD (2019). Injection-induced seismicity and fault-slip potential in the Fort Worth Basin, Texas. *Bulletin of the Seismological Society of America*. 109(5): 1615–1634.

Herrmann RB.(2020). North America Moment Tensors. www.eas.slu.edu/eqc/eqc_mt/MECH.NA/.

Herrmann RB, Park SK, and Wang CY. (1981). The Denver earthquakes of 1967–1968. *Bulletin of the Seismological Society of America*. 71(3), 731–745.

Herrmann RB, Benz H, and Ammon CJ. (2011). Monitoring the earthquake source process in North America. *Bulletin of the Seismological Society of America*. 101(6): 2609–2625.

Holland A. (2013). Earthquakes triggered by hydraulic fracturing in south-central Oklahoma. *Bulletin of the Seismological Society of America*. 103(3): 1784–1792.

Holmgren JM, Atkinson GM, and Ghofrani H. (2019). Stress drops and directivity of induced earthquakes in the Western Canada Sedimentary Basin. *Bulletin of the Seismological Society of America*. 109(5): 1635–1652.

Hosseini BK and Eaton DW. (2018). Fluid flow and thermal modeling for tracking induced seismicity near the Graham disposal well, British Columbia, Canada. SEG Technical Program Expanded Abstracts 2018 (pp. 4987–4991). Society of Exploration Geophysicists, https://doi.org/10.1190/segam2018-2996360.1.

Igonin N, Zecevic M, and Eaton DW. (2018). Bilinear magnitude-frequency distributions and characteristic earthquakes during hydraulic fracturing. *Geophysical Research Letters*. 45(23): 12866–12874.

Ishimoto M and Iida K. (1939). Observations of earthquakes registered with the microseismograph constructed recently. *Bulletin of the Earthquake Research Institute*. 17(1): 443–478.

Jia SQ, Wong RCK, Eaton DW, and Eyre TS. (2018). Investigating fracture growth and source mechanisms in shale using acoustic emission technique. In *52nd US Rock Mechanics/Geomechanics Symposium*. American Rock Mechanics Association, ARMA-2018-136.

Kanamori H and Anderson DL. (1975). Theoretical basis of some empirical relations in seismology. *Bulletin of the Seismological Society of America*. 65(5): 1073–1095.

Kang JQ, Zhu JB, and Zhao J. (2019). A review of mechanisms of induced earthquakes: from a view of rock mechanics. *Geomechanics and Geophysics for Geo-Energy and Geo-Resources*. 5(2): 171–196.

Kao H, Visser R, Smith B, and Venables S. (2018). Performance assessment of the induced seismicity traffic light protocol for northeastern British Columbia and western Alberta. *The Leading Edge*. 37(2): 117–126.

Keranen KM and Weingarten M. (2018). Induced seismicity. *Annual Review of Earth and Planetary Sciences*. 46(1): 149–174.

Keranen KM, Savage HM, Abers GA, and Cochran ES. (2013). Potentially induced earthquakes in Oklahoma, USA: Links between wastewater injection and the 2011 Mw 5.7 earthquake sequence. *Geology*. 41(6): 699–702.

Keranen KM, Weingarten M, Abers GA, Bekins BA, and Ge S. (2014). Sharp increase in central Oklahoma seismicity since 2008 induced by massive wastewater injection. *Science*. 345, 448–451.

Kettlety T, Verdon JP, Werner MJ, and Kendall JM. (2020). Stress transfer from opening hydraulic fractures controls the distribution of induced seismicity. *Journal of Geophysical Research: Solid Earth*. 125: e2019JB018794. https://doi.org/10.1029/2019JB018794

Kettlety T, Verdon JP, Werner MJ, Kendall JM, and Budge J. (2019). Investigating the role of elastostatic stress transfer during hydraulic fracturing-induced fault activation. *Geophysical Journal International*. 217(2): 1200–1216.

Kim YS, Peacock DC, and Sanderson DJ. (2004). Fault damage zones. *Journal of Structural Geology*. 26(3): 503–517.

Knopoff L. (2000). The magnitude distribution of declustered earthquakes in Southern California. *Proceedings of the National Academy of Sciences*. 97(22): 11880–11884.

Kohli AH and Zoback MD. (2013). Frictional properties of shale reservoir rocks. *Journal of Geophysical Research: Solid Earth*. 118(9): 5109–5125.

Kozłowska M, Brudzinski MR, Friberg P, Skoumal RJ, Baxter ND, and Currie BS. (2018). Maturity of nearby faults influences seismic hazard from hydraulic fracturing. *Proceedings of the National Academy of Sciences*. 115(8): E1720–E1729.

Kwiatek G, Saarno T, Ader T, Bluemle F, Bohnhoff M, Chendorain M, Dresen G, Heikkinen P, Kukkonen I, Leary P, and Leonhardt M. (2019). Controlling fluid-induced seismicity during a 6.1-km-deep geothermal stimulation in Finland. *Science Advances*. 5(5): eaav7224, https://doi.org/10.1126/sciadv.aav7224.

Langenbruch C and Zoback MD. (2016). How will induced seismicity in Oklahoma respond to decreased saltwater injection rates? *Science Advances*. 2(11): e1601542. https://doi.org/10.1126/sciadv.1601542.

Lavrov A. (2016). Dynamics of stresses and fractures in reservoir and cap rock under production and injection. *Energy Procedia*. 86(1): 381–390.

Lei X, Wang Z, and Su J. (2019). The December 2018 ML 5.7 and January 2019 ML 5.3 Earthquakes in South Sichuan Basin Induced by Shale Gas Hydraulic Fracturing. *Seismological Research Letters*. 90(3): 1099–1110.

Lengliné O, Lamourette L, Vivin L, Cuenot N, and Schmittbuhl J. (2014). Fluid-induced earthquakes with variable stress drop. *Journal of Geophysical Research: Solid Earth*. 119(12): 8900–8913.

Ludwig J, Nafe J, and Drake C. (1971). Seismic refraction. In Maxwell AE (ed.) *The Sea*, Vol. 4. Wiley, pp. 53–84.

MacKay MK, Eaton DW, Pedersen PK, and Clarkson CR. (2018). Integration of outcrop, subsurface, and microseismic interpretation for rock-mass characterization: An example from the Duvernay Formation, Western Canada. *Interpretation* 6(4): T919–T936.

Mahani AB, Schultz R, Kao H, Walker D, Johnson J, and Salas C. (2017). Fluid injection and seismic activity in the northern Montney play, British Columbia, Canada, with special reference to the 17 August 2015 M_W 4.6 induced earthquake. *Bulletin of the Seismological Society of America*. 107(2): 542–552.

Marone C. (1998). Laboratory-derived friction laws and their application to seismic faulting. *Annual Review of Earth and Planetary Sciences*. 26(1): 643–696.

Marsh S and Holland A. (2016). *Comprehensive Fault Database and Interpretive Fault Map of Oklahoma*. Oklahoma Geological Survey Open-File Rep. OF2–2016.

McGarr A. (2014). Maximum magnitude earthquakes induced by fluid injection. *Journal of Geophysical Research: Solid Earth*. 119(2): 1008–1019.

McGarr A and Barbour AJ. (2017). Wastewater disposal and the earthquake sequences during 2016 near Fairview, Pawnee, and Cushing, Oklahoma. *Geophysical Research Letters*. 44(18): 9330–9336.

McNamara DE, Hayes GP, Benz HM, Williams RA, McMahon ND, Aster RC, Holland A, Sickbert T, Herrmann R, Briggs R, and Smoczyk G. (2015). Reactivated faulting near Cushing, Oklahoma: Increased potential for a triggered earthquake in an area of United States strategic infrastructure. *Geophysical Research Letters*. 42(20): 8328–8332.

NRC (National Research Council). (2013). *Induced Seismicity Potential in Energy Technologies*. National Academies Press.

Ohnaka M, Akatsu M, Mochizuki H, Odedra A, Tagashira F, and Yamamoto Y. (1997). A constitutive law for the shear failure of rock under lithospheric conditions. *Tectonophysics*. 277(1–3): 1–27.

Okada Y. (1992). Internal deformation due to shear and tensile faults in a half-space. *Bulletin of the Seismological Society of America*. 82(2): 1018–1040.

OCC (Oklahoma Corporation Commission). (2018). Moving forward: New protocol to further address seismicity in state's largest oil and gas play, www.occeweb.com/og/02-27-18PROTOCOL.pdf.

Oklahoma Produced Water Working Group. 2017. Oklahoma Water for 2060: Produced Water Reuse and Recycling. www.owrb.ok.gov/2060/pwwg.php.

Parry RH. (2004). *Mohr Circles, Stress Paths and Geotechnics*, 2nd Edition, CRC Press.

Pawley S, Schultz R, Playter T, Corlett H, Shipman T, Lyster S, and Hauck T. (2018). The geological susceptibility of induced earthquakes in the Duvernay play. *Geophysical Research Letters*. 45(4): 1786–1793.

Peterie SL, Miller RD, Intfen JW, and Gonzales JB. (2018). Earthquakes in Kansas induced by extremely far-field pressure diffusion. *Geophysical Research Letters*. 45(3): 1395–1401.

Ponomarev AV, Zavyalov AD, Smirnov VB, and Lockner DA. (1997). Physical modeling of the formation and evolution of seismically active fault zones. *Tectonophysics*. 277 (1–3). 57–81.

Poulin A, Weir R, Eaton D, Igonin N, Chen Y, Lines L, and Lawton D. (2019). Focal-time analysis: A new method for stratigraphic depth control of microseismicity and induced seismic events. *Geophysics*. 84(6): KS173-KS182.

Raleigh CB, Healy JH, and Bredehoeft JD. (1976). An experiment in earthquake control at Rangely, Colorado. *Science*. 191(4233): 1230–1237.

Richter CF. (1935). An instrumental earthquake magnitude scale. *Bulletin of the Seismological Society of America*: 25(1): 1–32.

Roche V and Van der Baan M. (2015). The role of lithological layering and pore pressure on fluid-induced microseismicity. *Journal of Geophysical Research: Solid Earth*. 120 (2): 923–943.

Rokosh CD, Lyster S, Anderson SDA, Beaton AP, Berhane H, Brazzoni T, Chen D, Cheng Y, Mack T, Pana C, and Pawlowicz JG. (2012). *Summary of Alberta's Shale- and Siltstone-Hosted Hydrocarbon Resource Potential*. Energy Resources Conservation Board, ERCB/AGS Open File Report, 2012-06.

Rubinstein JL and Mahani AB. (2015). Myths and facts on wastewater injection, hydraulic fracturing, enhanced oil recovery, and induced seismicity. *Seismological Research Letters*. 86(4): 1060–1067.

Schoenball M and Ellsworth WL. (2017). Waveform-relocated earthquake catalog for Oklahoma and southern Kansas illuminates the regional fault network. *Seismological Research Letters*. 88(5): 1252–1258.

Schoenball M, Walsh FR, Weingarten M, and Ellsworth WL. (2018). How faults wake up: The Guthrie-Langston, Oklahoma earthquakes. *The Leading Edge*. 37(2): 100–106.

Scholz CH. (1998). Earthquakes and friction laws. *Nature*. 391(6662): 37–42.

Schultz R. and Pawley S. (2019). *Induced Earthquakes Geological Susceptibility Model for the Duvernay Formation, Central Alberta: Version 2*. AER/AGS Open File Report 2019-02.

Schultz R. and Wang R. (2020). Newly emerging cases of hydraulic fracturing induced seismicity in the Duvernay East Shale Basin. *Tectonophysics*. 228393, https://doi.org/10.1016/j.tecto.2020.228393.

Schultz R, Stern V, Novakovic M, Atkinson G, and Gu YJ. (2015). Hydraulic fracturing and the Crooked Lake Sequences: Insights gleaned from regional seismic networks. *Geophysical Research Letters*. 42(8): 2750–2758.

Schultz R, Atkinson G, Eaton DW, Gu YJ, and Kao H. (2018). Hydraulic fracturing volume is associated with induced earthquake productivity in the Duvernay play. *Science*. 359(6373): 304–308.

Schultz R, Wang R, Gu YJ, Haug K, and Atkinson G. (2017). A seismological overview of the induced earthquakes in the Duvernay play near Fox Creek, Alberta. *Journal of Geophysical Research: Solid Earth*. 122(1): 492–505.

Schultz R, Skoumal RJ, Brudzinski MR, Eaton DW, Baptie B, and Ellsworth WL. (2020). Hydraulic Fracturing Induced Seismicity. *Reviews of Geophysics*, submitted.

Schultz R, Corlett H, Haug K, Kocon K, MacCormack K, Stern V, and Shipman T. (2016). Linking fossil reefs with earthquakes: Geologic insight to where induced seismicity occurs in Alberta. *Geophysical Research Letters*. 43(6), 2534–2542.

Segall P and Lu S. (2015). Injection-induced seismicity: Poroelastic and earthquake nucleation effects. *Journal of Geophysical Research: Solid Earth*. 120(7): 5082–5103.

Shah AK and Keller GR. (2017). Geologic influence on induced seismicity: Constraints from potential field data in Oklahoma. *Geophysical Research Letters*. 44(1): 152–161.

Shapiro SA and Dinske C. (2009). Fluid-induced seismicity: Pressure diffusion and hydraulic fracturing. *Geophysical Prospecting*. 57(2): 301–310.

Shapiro SA, Huenges E, and Borm G. (1997). Estimating the crust permeability from fluid-injection-induced seismic emission at the KTB site. *Geophysical Journal International*. 131(2): F15–F18.

Shapiro SA, Dinske C, Langenbruch C, and Wenzel F. (2010). Seismogenic index and magnitude probability of earthquakes induced during reservoir fluid stimulations. *The Leading Edge*. 29(3): 304–309.

Shapiro SA, Krüger OS, Dinske C, and Langenbruch C. (2011). Magnitudes of induced earthquakes and geometric scales of fluid-stimulated rock volumes. *Geophysics*. 76(6): WC55–WC63.

Shapiro SA, Patzig R, Rothert E, and Rindschwentner J. (2003). Triggering of seismicity by pore-pressure perturbations: Permeability-related signatures of the phenomenon. *Pure and Applied Geophysics*. 160(5–6): 1051–1066.

Shapiro SA, Rothert E, Rath V, and Rindschwentner J. (2002). Characterization of fluid transport properties of reservoirs using induced microseismicity. *Geophysics*. 67(1): 212–220.

Shearer PM. (2019). *Introduction to Seismology*, 2nd Edition. Cambridge University Press.

Shemeta JE, Brooks CE, and Lord CC. (2019). Well Stimulation Seismicity in Oklahoma: Cataloging Earthquakes Related to Hydraulic Fracturing. Unconventional Resources

Technology Conference (URTEC), https://doi.org/10.15530/AP-URTEC-2019-198283.

Shipman T, MacDonald R, and Byrnes T. (2018). Experiences and learnings from induced seismicity regulation in Alberta. *Interpretation*. 6(2): SE15–SE21.

Skoumal RJ, Brudzinski MR, and Currie BS. (2015). Earthquakes induced by hydraulic fracturing in Poland Township, Ohio. *Bulletin of the Seismological Society of America*. 105(1): 189–197.

Skoumal RJ, Ries R, Brudzinski MR, Barbour AJ, and Currie BS. (2018). Earthquakes induced by hydraulic fracturing are pervasive in Oklahoma. *Journal of Geophysical Research: Solid Earth*. 123(12): 10918–10935.

Stoakes FA. (1980). Nature and control of shale basin fill and its effect on reef growth and termination: upper Devonian Duvernay and Ireton formations of Alberta, Canada, *Bulletin of Canadian Petroleum Geology*. 28(3): 345–410.

Switzer SB, Holland WG, Christie DS, Graf GC, Hedinger AS, McAuley RJ, Wierzbicki RA, Packard JJ, Mossop GD, and Shetsen I. (1994). Devonian Woodbend-Winterburn strata of the Western Canada Sedimentary Basin. In Mossop GD and Shetsen I (eds.) *Geological Atlas of the Western Canada Sedimentary Basin*. Canadian Society of Petroleum Geologists and Alberta Research Council, pp. 165–202.

Tenthorey E and Cox SF. (2006). Cohesive strengthening of fault zones during the interseismic period: An experimental study. *Journal of Geophysical Research: Solid Earth*. 111: B09202, https://doi:10.1029/2005JB004122.

Townend J and Zoback MD. (2000). How faulting keeps the crust strong. *Geology*. 28(5): 399–402.

Trutnevyte E and Wiemer S. (2017). Tailor-made risk governance for induced seismicity of geothermal energy projects: An application to Switzerland. *Geothermics*. 65: 295–312.

Van der Baan M and Calixto F. (2017). Human-induced seismicity and large-scale hydrocarbon production in the USA and Canada. *Geochemistry, Geophysics, Geosystems*. 18(7): 2467–2485.

Van der Elst NJ, Page MT, Weiser DA, Goebel TH, and Hosseini SM. (2016). Induced earthquake magnitudes are as large as (statistically) expected. *Journal of Geophysical Research: Solid Earth*. 121(6): 4575–4590.

Vermilye JM and Scholz CH. (1998). The process zone: A microstructural view of fault growth. *Journal of Geophysical Research: Solid Earth*. 103(B6): 12223–12237.

Walsh FR and Zoback MD. (2015). Oklahoma's recent earthquakes and saltwater disposal. *Science Advances*. 1(5): e1500195. https://doi.org/10.1126/sciadv.1500195

Walsh FRI, Zoback MD, Pais D, Weingartern M, and Tyrell T. (2017). FSP 1.0: A Program for Probabilistic Estimation of Fault Slip Potential Resulting from Fluid Injection, scits.stanford.edu/software.

Wang R, Gu YJ, Schultz R, and Chen Y. (2018). Faults and non-double-couple components for induced earthquakes. *Geophysical Research Letters*. 45(17): 8966–8975.

Wang R, Gu YJ, Schultz,R, Kim A, and Atkinson G. (2016). Source analysis of a potential hydraulic-fracturing-induced earthquake near Fox Creek, Alberta. *Geophysical Research Letters*. 43(2): 564–573.

Wang R, Gu YJ, Schultz R, Zhang M, and Kim A. (2017). Source characteristics and geological implications of the January 2016 induced earthquake swarm near Crooked Lake, Alberta. *Geophysical Journal International*. 210(2): 979–988.

Warpinski NR, Mayerhofer M, Agarwal K, and Du J. (2013). Hydraulic-fracture geomechanics and microseismic-source mechanisms. *SPE Journal*. 18(04): 766–780.

Weingarten M, Ge S, Godt JW, Bekins BA, and Rubinstein JL. (2015). High-rate injection is associated with the increase in US mid-continent seismicity. *Science.* 348(6241): 1336–1340.

Weir RM, Eaton DW, Lines LR, Lawton DC, and Ekpo E. (2018). Inversion and interpretation of seismic-derived rock properties in the Duvernay play. *Interpretation.* 6(2): SE1–SE14.

Wessel P, Luis JF, Uieda L, Scharroo R, Wobbe F, Smith WHF, and Tian D. (2019). The Generic Mapping Tools version 6. *Geochemistry, Geophysics, Geosystems.* 20(1): 5556–5564.

Wilson MP, Foulger GR, Gluyas JG, Davies RJ, and Julian BR. (2017). HiQuake: The human-induced earthquake database. *Seismological Research Letters.* 88(6): 1560–1565.

Yao Y. (2012). Linear elastic and cohesive fracture analysis to model hydraulic fracture in brittle and ductile rocks. *Rock Mechanics and Rock Engineering.* 45(3): 375–387.

Yeck WL, Weingarten M, Benz HM, McNamara DE, Bergman EA, Herrmann RB, Rubinstein JL, and Earle PS. (2016). Far-field pressurization likely caused one of the largest injection induced earthquakes by reactivating a large pre-existing basement fault structure. *Geophysical Research Letters.* 43(19): 10–198.

Yu H, Harrington RM, Liu Y, and Wang B. (2019). Induced Seismicity Driven by Fluid Diffusion Revealed by a Near-Field Hydraulic Stimulation Monitoring Array in the Montney Basin, British Columbia. *Journal of Geophysical Research: Solid Earth.* 124(5): 4694–4709.

Zhang, J. and Van der Baan M. (2019). Depth-dependent fault slip potential. In *SEG Technical Program Expanded Abstracts 2019* (pp. 3016–3020), Society of Exploration Geophysicists, https://doi.org/10.1190/segam2019–3214231.1.

Zhang H, Eaton DW, Rodriguez G, and Jia SQ. (2019). Source-mechanism analysis and stress inversion for hydraulic-fracturing-induced event sequences near Fox Creek, Alberta. *Bulletin of the Seismological Society of America.* 109(2): 636–651.

Zoback MD. (2010). *Reservoir Geomechanics*, 1st edition, Cambridge University Press.

Zoback MD and Harjes HP. (1997). Injection-induced earthquakes and crustal stress at 9 km depth at the KTB deep drilling site, Germany. *Journal of Geophysical Research: Solid Earth.* 102(B8): 18477–18491.

9

Naturally Occurring Radioactive Material (NORM)

NATHANIEL R. WARNER, MOSES A. AJEMIGBITSE, KATHARINA PANKRATZ, AND BONNIE MCDEVITT

9.1 Discovery and Fundamentals of Radioactivity

As the building block of all matter and life, the atom enraptured the minds of great thinkers long before the discovery of nuclear radiation in the late nineteenth and early twentieth century. Wilhelm Röntgen discovered a new form of radiation in 1895, which he coined "x-rays," denoting their mysterious, unknown origins (Röntgen 1970). Henri Becquerel later realized that uranium was responsible for this new radiation, earning the name "uranic rays." Around the same time, Marie Curie used an electrometer (an early ionization chamber) and identified other substances, such as thorium, that could emit these rays. Using her quantitative approach, Marie was able to determine that pitchblende, an ore of uranium, emitted more radiation than could be attributed to its uranium content, leading to the discovery of two new elements, polonium and radium. At this point the word *radioactive* was coined to describe the scientific phenomena whereby substances spontaneously produce radiation without external stimuli. In 1899, Ernest Rutherford built on Curie's work and discovered uranium radiation consisted of multiple types of radiation, which laid the foundation for the discovery of heavy, positively charged particles termed alpha and lighter, negatively charged particles termed beta radiation. In 1900, Paul Villard identified yet a third radiation, which became known as gamma rays. These three types of radiations would later be identified as the helium nuclei (alpha), electrons (beta), and high energy photons (gamma) that can be released from the nucleus of an atom.

In 1902, Rutherford and Soddy introduced the concept of the radioactive series: the idea that a radioactive element transforms into another radioactive element, which itself decays to yet another radioactive element, and so on until a stable element is formed. This discovery would lead to the law of radioactive decay,

$$N = N_0 e^{-\lambda t} \tag{9.1}$$

Where N is the number of radioactive atoms present at time t; N_0 is the number of atoms present at $t = 0$; and λ is the probability for any particular atom to decay per unit time. Radioactive decay occurs because of an imbalance in the number of protons and neutrons, leading to an unstable nucleus where electrostatic repulsion overcomes the strong nuclear force. Generally, elements with atomic numbers greater than 82 (Pb) have some isotopes (elements with the same number of protons but different number of neutrons) that are naturally radioactive.

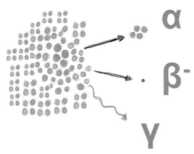

Figure 9.1 Three types of radioactivity that originate in the nucleus of an atom; alpha, a particle with two protons and two neutrons, beta, a negatively charged particle that changes a neutron to a proton, and gamma an electromagnetic energy that originates from the nucleus (A black and white version of this figure will appear in some formats. For the colour version, refer to the plate section.)

Alpha decay, being the release of a helium atom, results in the loss of two protons and two neutrons from the nucleus of the emitting atom (Figure 9.1). The atomic number is reduced by 2 and the mass number is reduced by 4, creating a new "daughter" element. The force of alpha particle ejection from the nucleus produces movement of the daughter element, a process known as alpha recoil. The alpha particle, with both atomic mass and kinetic energy, interacts strongly with matter (through ionization) over very short distances. In this manner, alpha particles are highly damaging to the local environment in which they are emitted but relatively easy to terminate; a thin piece of paper is sufficient. However, when the source of alpha radiation is ingested, the alpha particles dissipate high energy into surrounding tissues, causing damage to cells.

Beta decay involves the transformation of a neutron into a proton and an electron. The electron is emitted from the nucleus, resulting in a new element with one more proton (Figure 9.1). The radiated electron, being relatively lightweight, interacts with matter to a lesser degree (also through ionization), penetrating further into materials than alpha radiation. The vast majority of times the emission of an alpha or beta particle results in an excited nucleus (i.e., one in which the protons or neutrons are above the ground state), and a gamma ray is also emitted.

Gamma rays are short wavelength, high energy photons (and are a form of electromagnetic radiation) that penetrate much farther into materials than both alpha and beta radiation (Figure 9.1). Being electromagnetic in nature, gamma rays react with matter through three processes: the Photoelectric Effect, the Compton Effect, and Pair Production. These processes involve the absorption of the gamma ray energy by an electron in the affected atom. This adsorption can be complete (Photoelectric Effect and Pair Production) or partial (Compton Effect). These processes cause the intensity of gamma radiation to decrease with increasing penetration depth and are found to be directly proportional to this depth (Hubbell 1982; Bolivar et al. 1997; Gilmore 2008).

9.2 Naturally Occurring Radioactive Decay

Naturally occurring radioactive material (NORM) is found in low concentrations in the environment and includes elements uranium, thorium, radium, and radon. The most commonly reported type of oil and gas-related NORM is the element radium ($_{88}$Ra), which is an alkaline earth metal with 88 protons exhibiting similar chemical behavior to calcium, barium, strontium, and magnesium (Ca, Ba, Sr, and Mg). Radium is relatively soluble over a wide range of pH and redox conditions, exhibits one aqueous oxidation state (+2) in environmental conditions, and is not easily complexed (Kirby and Salutsky 1964). The low hydration energy, and thus low hydrated radius, of radium results in high selectivity for ion exchange processes in clays or ion exchange resins relative to other cations (e.g., Ca, Ba, Sr) (US EIA 2014).

Radium is not vital to living organisms, and its radioactivity and the radioactivity of its daughter products create adverse health effects when incorporated into biochemical processes (Kirby and Salutsky 1964). Radium is known to cause lymphoma, bone cancer, and leukemia owing to the uptake of the radium ion into animal bones and calcium-rich tissues where it then decays (United Nations 2011; Carvalho et al. 2014). Radium's mobility, bioavailability, and toxicity in the environment depend on its phase as an aqueous, sorbed, or co-precipitated ion (International Atomic Energy Agency 2014). Radium is often the most important element for the consideration of NORM associated with oil and gas (O&G) extraction, despite its presence in small concentrations relative to other ions. For example, radium is often reported at concentrations of 10^{-12} g/L compared to 10^{-3} g/L for other cations such as Ca, Ba, and Na.

While there are over 30 isotopes of radium, there are four common naturally occurring isotopes, ^{223}Ra, ^{224}Ra, ^{226}Ra, and ^{228}Ra. The relatively short half-lives of ^{223}Ra ($t_{1/2}$ =11.4 days) and ^{224}Ra ($t_{1/2}$ = 3.6 days) preclude their measurement and/or discussion in most studies of NORM. Instead, the two most persistent radioisotopes, ^{226}Ra ($t_{1/2}$ = 1600 years) and ^{228}Ra ($t_{1/2}$ = 5.75 years) are often referred to as "total radium = ^{226}Ra + ^{228}Ra" in NORM evaluations. ^{226}Ra and ^{228}Ra are sourced from the radioactive decay of parent ^{238}U and ^{232}Th (respectively) (Figure 9.2). ^{228}Ra is sourced directly from alpha decay of parent ^{232}Th, whereas ^{238}U first decays through ^{234}Th, ^{234}Pa, and ^{230}Th. ^{226}Ra then decays to radon-222 by alpha decay (Eq. 9.2). Radon-222 is a short-lived daughter product of radium decay and is a radioactive gas known to cause lung cancer (National Research Council 1999). During decay of radium-226 to radon-222 the nucleus releases an alpha particle and a characteristic gamma ray at 186.2 keV (Eq. 9.2). Radon-222 is in turn radioactive and gives rise to multiple daughter products, including multiple isotopes of Po, Bi, and Pb until the ^{226}Ra (^{238}U) series terminates with the stable isotope of ^{206}Pb (Figure 9.2).

$$^{226}_{88}Ra \rightarrow \, ^{222}_{86}Rn + \, ^{4}_{2}He^{2+} + \gamma + h\nu \qquad (9.2)$$

Radium-228, on the other hand, decays by beta decay to ^{228}Ac (Eq. 9.3) with a half-life of 5.75 years. Gamma radiation is also released, at a different energy level, 911.2 keV, compared to the ^{226}Ra decay. Actinium-228 is also radioactive, beginning a decay chain for ^{228}Ra (Eq. 9.3) that terminates in stable isotope of lead, ^{208}Pb.

$$^{228}_{88}Ra \rightarrow \, ^{228}_{89}Ac + e^- + \gamma + h\nu \qquad (9.3)$$

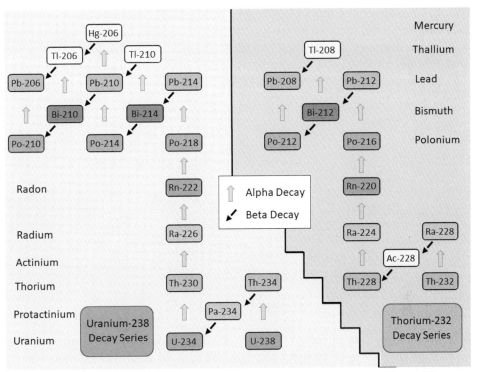

Figure 9.2 ^{226}Ra and ^{228}Ra decay chains showing the primordial parent nuclides, ^{238}U and ^{232}Th. (A black and white version of this figure will appear in some formats. For the colour version, refer to the plate section.)
Adapted from Nelson et al. (2015)

The relatively short half-lives of ^{226}Ra and ^{228}Ra (1600 and 5.8 years) reveal that they are only persistently formed because of the presence of their primordial parent nuclides ^{238}U and ^{232}Th, whose half-lives are on the order of billions of years. The long half-lives of ^{238}U and ^{232}Th allow for daughter products such as radium to maintain constant activity because they are continually supplied/produced by the parents at a rate that equals the decay rate. This is termed secular equilibrium. Therefore, naturally occurring radium exists with the occurrence of uranium and thorium in ores of geologic formations, because radium atoms, if left alone, would decay to radon within about 10 half-lives, ~16,000 years (radium-226) or ~60 years (radium-228). This does not imply that radium, uranium, and thorium are always co-located. Observed deviations in the secular equilibrium between the parent and daughter nuclide (i.e., the balance between production and decay) indicate that radium will separate from its parents because of differences in chemical properties (e.g., solubility), which are discussed in more detail in Section 9.3.

The variations in half-lives of the decay series for ^{226}Ra and ^{228}Ra introduces a complexity when measuring material that contains these elements, particularly when they have been separated from their parent rock material. Following the law of radioactive decay

(Eq. 9.1), a solution to a series of differential equations allows calculation of activity of each daughter in the decay series and can help determine when the decay series is not in equilibrium (Bateman 1910). These equations are fundamental to understanding and predicting the behavior of radioactive decay chains, particularly when disequilibrium in the chain is observed and geologic and/or anthropogenic processes lead to breaks in the chain.

$$A_n = \lambda_n N_n \tag{9.4}$$

$$N_n = c_1 e^{-\lambda_1 t} + c_2 e^{-\lambda_2 t} + c_3 e^{-\lambda_3 t} + \ldots + c_n e^{-\lambda_n t} \tag{9.5}$$

$$c_1 = \frac{\lambda_1 \lambda_2 \ldots \lambda_{n-1}}{(\lambda_2 - \lambda_1)(\lambda_3 - \lambda_1)(\lambda_n - \lambda_1)} N_1^0 \tag{9.6}$$

$$\frac{\lambda_1 \lambda_2 \ldots \lambda_{n-1}}{(\lambda_1 - \lambda_2)(\lambda_3 - \lambda_2)(\lambda_n - \lambda_2)} N_1^0 \tag{9.7}$$

$$c_n = \frac{\lambda_1 \lambda_2 \ldots \lambda_{n-1}}{(\lambda_1 - \lambda_n)(\lambda_2 - \lambda_n)(\lambda_{n-1} - \lambda_n)} N_1^0 \tag{9.8}$$

Where A_n is the activity of the nth element in the decay series, λ_n is the probability of an atom of the nth element to decay per unit time, and N_n is the number of radioactive atoms of the nth element present at time t; N_1^0 is the number of atoms of the first element in the decay series present at $t = 0$.

The units used to describe the amount of radioactivity in a sample are typically presented in terms of radioactive decays per second (dps), also known as activity. The SI unit for radioactivity is the Becquerel (Bq), which is equal to a single decay per second. In the United States the standard unit is the Curie (for Marie Curie), which represents 3.7×10^{10} radioactive dps. One pCi is therefore equivalent to 27.027 Bq. The Curie is a value that represents the number of decays that occur in 1 gram of radium per second. A gram of radium is very large relative to values observed in most environmental systems, so typically activity values are reported in pico-Curie (pCi) [1×10^{-12}] per liter or per kilogram. For radium this means that the activity of 1 pCi is equivalent to the mass of 1 picogram, but the same would not be true for measuring the mass of uranium or thorium and extrapolating the number of decays per second (i.e., 1 picogram of U $\neq$ 1 picocurie U). Note also that the unit prefix pico (10^{-12}) is much smaller than that used for other major or trace elements, such as milligrams (10^{-3}), micrograms (10^{-6}), or even nanograms (10^{-9}).

Units of activity are often normalized to mass (solid) or volume (liquid). Most states and the US EPA regulate ^{226}Ra for O&G produced water disposal to surface water at 60 pCi or 2.22 Bq for each liter (L) of water. The drinking water standard in the United States is set much lower at 0.185 Bq/L (5 pCi/L) for combined ^{226}Ra + ^{228}Ra (US Environmental Protection Agency 1976; US Nuclear Regulatory Commission 2017). For comparison, high salinity produced waters typically contain a median total dissolved salts (TDS) concentration of 250,000 mg/L and median ^{226}Ra concentration of around 111 Bq/L (3,000 pCi/L) (https://energy.usgs.gov/) (Blondes et al. 2018). However, other produced waters have much lower TDS that can range from 4,000 to 10,000 mg/L with a much lower reported ^{226}Ra concentration (e.g., 3.14 Bq/L (~85 pCi/L) in the Niobrara formation produced water) (Rowan et al. 2011a; Rosenblum et al. 2017).

9.3 Generation of NORM in Formation Water

9.3.1 Transfer of Radium to Pore Spaces

The recognition of NORM (from radium) associated with oil and gas fields started nearly at the same time (1904) as the discovery of radioactivity itself (see Section 9.1). But the high activity did not associate with the oil and gas. Instead, while operators aim to extract oil and gas from the ground large volumes of water, termed coproduced or produced water, also come to the surface. The produced water is often in much larger volumes than the oil or gas that is extracted. In the United States, O&G extraction generates 21 billion barrels (3.3 trillion liters) of oil and gas produced water from nearly one million active O&G wells each year (Clark and Veil 2015). A portion of the water injected into the formation to fracture the rock returns to the surface with the oil/gas and is sometimes referred to as *flowback*, but more commonly it is also termed *produced water*. The portion of the injected water that returns to the surface as produced water varies greatly between formations (Kondash et al. 2017). Even when freshwater is injected, billions of barrels of produced water can contain thousands of pCi/L of radium, but often negligible amounts of U and Th (Fisher 1998). Produced water from unconventional wells appear to have higher activities of radium than conventional wells (Landis et al. 2018) and increase with time following fracturing (Rowan et al. 2011a).

Three mechanisms, solubility, alpha recoil, and cation exchange, can help explain abundant activities of radium while U and Th concentrations are negligible. First, under reducing conditions such as those found at the depth of typical oil and gas wells, radium is soluble and concentrates in pore fluids, while U and Th are insoluble and will remain associated with mineral and/or adsorbed phases. Indeed, adequate concentrations of U and Th must be present and remain in the host rock formations because radium would decay away to radon in about 16,000 (^{226}Ra) and 60 years (^{228}Ra) without the support (Langmuir and Melchior 1985; Sturchio et al. 2001).

Second, radium is transferred out of the solid rocks and into the aqueous phase via alpha recoil. Alpha recoil is a phenomenon wherein the transfer of kinetic energy following alpha decay is great enough to move the Ra daughter nuclei, which causes damage to the crystalline structure of minerals that host the U and Th. This damage creates a path for liquid infiltration through which the soluble radium escapes from the rock into the pore fluids (Fleischer and Raabe 1978; Kraemer and Reid 1984; Wiegand and Sebastian 2002).

In oil and gas formations this means that radium atoms tend to recoil out of mineral phases that host the U and Th and accumulate in the surrounding pore fluids.

Once Ra^{2+} is present in the aqueous fluids it could be removed from solution if either sorbed to cation exchange sites in clays, organic matter, and oxides (Nathwani and Phillips 1979; Krishnaswami et al. 1982; Shao et al. 2009; Alhajji et al. 2016) or coprecipitated with Ba or Ca in sulfates or carbonates. However, both removal mechanisms, exchange and coprecipitation, are limited in produced waters, which keeps radium activities in produced water high. Cation exchange sites for radium are limited because of competition from other ions (e.g., Ca, Mg, Na), which are often at concentrations orders of magnitude higher than radium. For example, the produced water of the Marcellus Shale is typical of conventional produced water throughout the Appalachian Basin: dominated by chloride and rich in metal content, primarily Na and Ca; and yielding high salinities, up to 400,000 mg/L, among the highest in mineral waters (Dresel and Rose 2010; Rowan et al. 2015). This high salinity is thought to originate from formation water that evaporated beyond halite precipitation (Rowan et al. 2015; Stewart et al. 2015). Because of the competition for cation exchange sites from Ca^{2+}, Mg^{2+}, and Na^+, increasing salinity in water desorbs radium (Webster et al. 1995). Exchange sites within the formation that could host divalent cations such as Ra^{2+} are instead occupied by Ca^{2+} and Mg^{2+}(Rowan et al. 2011b).

In an experimental leaching procedure of Marcellus Formation core material, Landis et al. (2018) demonstrated that the radium leaching responded in particular to concentration of calcium and in fact the $^{226}Ra/^{228}Ra$ also increased with calcium. The change in ratios was explained because the solids contained radium in two distinct phases. First a mineral phase hosted labile radium with moderate $^{226}Ra/^{228}Ra$, while a second organic phase hosted exchangeable radium with much higher $^{226}Ra/^{228}Ra$. High concentrations of calcium were necessary to mobilize radium at activities approaching activities and ratios similar to those observed in Marcellus produced water.

Coprecipitation with barium or strontium in radiobarite $(Ba,RaSO_4)$ or radiocelestite $(Sr, RaSO_4)$ is limited in reducing conditions; therefore barite precipitation does not control radium mobility in oil and gas formations at depth (Bloch and Key 1953; Lowson 1985; Wiegand and Sebastian 2002). Note that the four processes (solubility, alpha recoil, cation exchange, and coprecipitation) described earlier lead to higher radium in pore fluids under reducing conditions with high salinity regardless of the type of reservoir (conventional high permeability oil and gas reservoirs or unconventional low permeability organic-rich shale source rocks).

The presence of Uranium and thorium in oil and gas formations ultimately leads to the presence of ^{226}Ra and ^{228}Ra in produced water. The relative concentrations of ^{226}Ra and ^{228}Ra in the produced water can be explained by the concentrations of parent U and Th in the surrounding rock. Because ^{226}Ra is ultimately sourced from parent ^{238}U and ^{228}Ra from ^{232}Th, the ratio of U/Th in the source rock will dictate the $^{226}Ra/^{228}Ra$ in the produced water (Rowan et al. 2015). For example, one of the largest unconventional formations, the Marcellus Shale, often contains high concentrations of organic matter (the source material to generate the oil and gas). Organic matter often concentrates U content, and therefore the water produced from the Marcellus Formation yields ^{226}Ra at higher concentrations than

^{228}Ra and leads to a high ^{226}Ra/^{228}Ra ratio (Rowan et al. 2011b). The Marcellus Shale formation has median concentrations of ~148 Bq/L (4,000 pCi/L) ^{226}Ra and ~37 Bq/L (1000 pCi/L) ^{228}Ra (Rowan et al. 2011b) in produced waters. The combination of high ^{226}Ra activities – typically > 74 Bq/L (2,000 pCi/L) – and low ^{228}Ra/^{226}Ra values – less than 0.3 – serves as a tool for identifying the Marcellus produced waters in the natural environment relative to produced water from many conventional formations where the ^{226}Ra/^{228}Ra ratio is closer to the source rocks and produced water contains ^{226}Ra/^{228}Ra ratios of ~1. In northeastern Poland, a study demonstrated that hydraulic fracturing fluids contained <2 Bq/L before injection but returned as produced water/flowback fluid with an average of 42 Bq/L Ra 226, and 20 Bq/L of Ra-228 (Jodlowski et al. 2017).

9.3.2 Solids and NORM: Drill Cuttings, Proppant, and Drilling Mud

The solid fraction of material that comes to the surface during drilling and stimulation comprises the shale drill cuttings from the vertical and horizontal portions of the wellbore, the spent proppant, and spent drilling mud (Barry and Klima 2013). NORM values for drill cuttings, proppant, and drilling mud are not widely available for unconventional formations. However, the Pennsylvania Department of Environmental Protection did conduct a survey of solids associated with the Marcellus Formation. Among these solids, the radium activities of the vertical drill cuttings and drilling mud are very low, typically ~104 Bq/kg (2.8 pCi/g) ^{226}Ra and ~37 Bq/kg (1 pCi/g) ^{228}Ra, being of low U/Th content (Eitrheim et al. 2016; PA DEP 2016). The spent proppant contains more radium than its pristine counterparts (PA DEP 2017), but the activities are also low, with ^{226}Ra activities ranging from 6 Bq/kg (0.17pCi/g) – 13.2 Bq/kg (0.358 pCi/g) (PA DEP 2016). It is the horizontal clay and organic-rich shale fragments that have greater radium content, averaging ~185 Bq/kg (5 pCi/g) ^{226}Ra and 23.3 Bq/kg (0.63 pCi/g) ^{228}Ra, owing to their higher U/Th content (Rowan et al. 2011b; Eitrheim et al. 2016; PA DEP 2017). In northeastern Poland, similar values were reported for waste solids with ranges of 15–415 Bq/kg Ra-226 and <10 to 516 Bq/kg Ra-228, typical of average soils (Jodlowski et al. 2017).

In 2011 in Pennsylvania, O&G development of the Marcellus generated ~800,000 tons of drill cuttings, and ~15,000 tons of flowback fracturing sands (Maloney and Yoxtheimer 2012). The management and disposal of these co-products continues to be a concern, with regards to their high salts, metals, and NORM content. It is important to note that the solid scale that forms on equipment in contact with produced water (sometimes referred to as technologically enhanced, naturally occurring radioactive material, TENORM) is often barite mineral and can contain thousands of pCi per gram of radium, and represents an area of concern for exposure. All the solids have typically been landfilled in Subtitle D RCRA hazardous waste landfills. (Maloney and Yoxtheimer 2012; Veil 2015; US EPA 2018a, 2018b). The disposal of the solids in hazardous waste landfills has largely gone without scrutiny. However, a study by the PA DEP reported that there was no significant difference in the chemical compositions of leachate from a landfill accepting oil and gas waste solids when compared to one that had not accepted any such material, suggesting that this practice is without additional environmental concern borne from the origin of the solids (PA DEP 2016).

In Pennsylvania, centralized waste treatment facilities process produced water with the addition of sodium sulfate to stimulate precipitation of Ra into a mixed barite sludge. While the high TDS produced water is discharged to surface water with limited radium activity, the sludge often greatly exceeds 259 Bq/kg, which can be a great cost to treatment facilities to attenuate sludge activities below allowable detection for landfill disposal. For example, in 2017, 7,000 tons of this radioactive sludge from treatment facilities were landfilled in Pennsylvania (Ajemigbitse et al. 2019), with ^{226}Ra activities in the landfilled sludge reportedly ranged from 0.11 Bq/g to 17.8 Bq/g (Zhang et al. 2015b; Pennsylvania Department of Environmental Protection 2016). The Ohio solid and hazardous waste landfill code for radioactive materials, ORC § 3734.02 (P)(2), states that technologically enhanced naturally occurring radioactive material (TENORM) should not exceed 185 Bq/kg (5 pCi/g) above natural background levels (set at 2 pCi/g unless otherwise measured). This value is similar to the action level 40 CFR 192 for ^{226}Ra in surface sediments not to exceed 185 Bq/kg (5 pCi/g) at uranium and thorium mill tailing sites.

9.3.3 Salinity as an Estimate for Radium Activity

Data on radium/NORM in produced water across the United States is limited given the number of O&G wells in operation. The United States Geological Survey (USGS) produced water database contains over 108,000 chloride data, which includes both conventional and unconventional wells but has only 746 values for radium-226, the most commonly reported isotope of radium (Figure 9.3a). These radium values are not distributed across all formations; instead the majority of values are reported in the Appalachian Basin. Of the data reported, radium-226 activities vary widely, where produced water from O&G wells in Texas and the Appalachian Basin appear to contain higher radium-226 than produced water in Wyoming or California. Mean values of radium in produced water appear to vary based on the geologic basin and not necessarily whether the formation was conventional or unconventional. This is also true of major element chemistry, which in some cases requires specific chemical fingerprinting tools to determine if the origin of the produced water was a conventional well or an unconventional well that was hydraulically fractured.

Based on work from O&G wells in Texas, radium values in produced water can be estimated in some cases based on the Cl or TDS content of the produced water as proposed by Fisher (1998). In some formations in Texas there was a strong correlation ($r^2 > 0.7$) between TDS and radium in produced water from select formations. However, in other formations the correlation was negligible. If we assume that the lack of correlation was because of smaller sample size, would the larger USGS produced water database show a similar trend? Cation exchange competition (see Section 9.2) is one of the factors controlling the radium activity observed in produced water. Therefore, TDS or Cl could indicate greater cation exchange competition and allow a *very generalized* estimate of radium

Naturally Occurring Radioactive Material

Figure 9.3a Map of the continental United States with published NORM data for produced water. Note the wide range in values reported in (pCi/L) but also the lack of publicly available data for many oil and gas basins. Most data is sourced from the USGS PW database and Fisher 1998 (A black and white version of this figure will appear in some formats. For the colour version, refer to the plate section.)

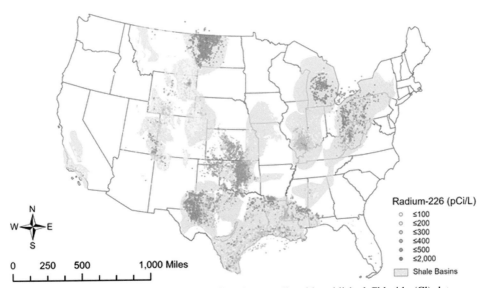

Figure 9.3b There are over 100,000 oil and gas wells with published Chloride (Cl) data available in the USGS PW database. Using the relationship between Cl and radium-226 described by Fisher (1998) for PW from Texas, an estimate of the radium 226 in other basins can be calculated (A black and white version of this figure will appear in some formats. For the colour version, refer to the plate section.)

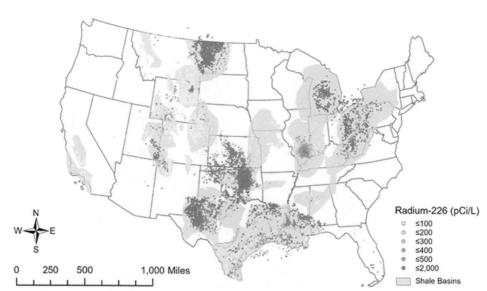

Figure 9.3c There are over 100,000 oil and gas wells with published TDS data available in the USGS PW database. Using the relationship between TDS and radium-226 described by Fisher (1998) for PW from Texas, an estimate of the radium 226 in other basins can be calculated (A black and white version of this figure will appear in some formats. For the colour version, refer to the plate section.)

activity across O&G basins with little to no radium data. Here, with the larger set of data for Cl and TDS reported in the USGS produced water database, and the correlations proposed by Fisher (1998), a correlation ($r^2 > 0.7$) is observed between Cl or TDS and radium-226 (n = 746). The relationships between salinity and radium can help estimate radium values for other formations based on either TDS (Eq. 9.9) or chloride concentrations (Eq. 9.10) as follows.

$$\text{Log}(\text{Ra} - 226) = \left(1.3536 * \text{Log}(\text{TDS})\right) - 4.4513 \tag{9.9}$$

$$\text{Log}(\text{Ra} - 226) = \left(0.6521 * \text{Log}(\text{Cl})\right) - 0.8448 \tag{9.10}$$

These regressions, applied to the larger database that contains either TDS or chloride for over ~100,000 O&G wells, create the maps of *estimated radium* in geologic plays displayed in Figure 9.3b and 9.3c. It is important to note that within each O&G basin and even within formations there will be variations for which this broad estimate does not account. However, the maps indicate that there are many formations with estimated high radium and therefore more measurements of radium in produced water should be obtained and made publicly available.

9.4 Disposal of Large Volumes of Produced Water

What happens once produced water, often with elevated radium activity, is brought to the surface? In the United States, states often regulate the disposal of produced water. The top

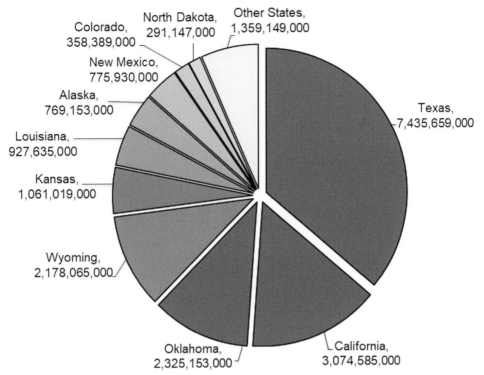

Figure 9.4 Top 10 states of oil and gas produced water volume generation (in barrels/year for 2012) (A black and white version of this figure will appear in some formats. For the colour version, refer to the plate section.)

four O&G produced water-generating US states are Texas, California, Oklahoma, and Wyoming (Figure 9.4). While Texas is the leading producer of both US oil and gas, Oklahoma, California, and Wyoming are not; however, they still contribute to high levels of generated produced water. These states contain some of the older conventional O&G fields that generate increasing volumes of produced water relative to O&G with ratios upward from 50 to 1,000 (Scanlon et al. 2014). The average produced water to O&G ratio in the United States is lower, from 7 to 10 (Guerra et al. 2011). Produced water generation and the quality of the produced water (e.g., the presence of NORM) can determine the economic feasibility of drilling owing to disposal-tailored treatment, which can range from 1–15 U.S. dollars per m^3 of produced water (Dolan et al. 2018). Thus, the less treatment required particularly for NORM, the longer O&G wells can operate profitably.

Of the 21 billion barrels of produced water generated in the United States annually, approximately 84% is reinjected into formations either for disposal via Class II UIC or for enhanced oil recovery (EOR) processes (Clark and Veil 2015) (Figure 9.5). While disposal via wells is commonly practiced, it continues to be under scrutiny because of reports of injection-induced seismicity (Ellsworth 2013). In some states, such as Pennsylvania, the number of disposal wells has been limited by regulatory bodies (in 2016 PA had eight active wells while OH had over 200) (PA DEP, 2016). For such states, produced water

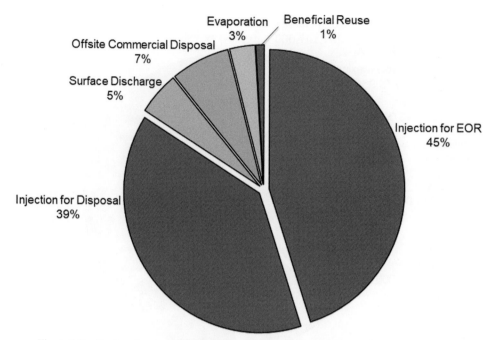

Figure 9.5 Produced water management for 2007 (based on Clark and Veil 2015) (A black and white version of this figure will appear in some formats. For the colour version, refer to the plate section.)

must be transported to other states for disposal via injection incurring high transportation costs, necessitating the need for a different management and disposal practice in those areas (Maloney and Yoxtheimer 2012). Second to injection, the largest volumes of PW (16%) are disposed via surface discharge to freshwater streams, evaporated, and/or concentrated, or in some cases (1%), beneficial reuse (Figure 9.6). Disposal to streams often takes place after minimal treatment, while beneficial reuse can include management options such as spreading on roads for dust suppression, irrigation, or resource recovery. The US Environmental Protection Agency (US EPA), US Geological Survey (USGS), and US Department of Energy (US DOE) are currently seeking alternative uses for O&G produced water for beneficial use, as commonly practiced in the western United States (Kondash et al. 2018; McDevitt et al. 2018, 2020b, 2021; US EPA 2018b; McLaughlin et al. 2020).

9.5 Environmental Transport and Fate Following Disposal to Surface Water

Because produced water is a complex wastewater (upward of 10–300 g/L TDS) containing high concentrations of hydrocarbons, trace and heavy metals, NORM, and significantly elevated concentrations of other alkaline earth metals such as Ba and Sr, surface water disposal to streams poses significant challenges (Wilson and Vanbriesen 2012; Vengosh et al. 2014a; Akob et al. 2016; Cozzarelli et al. 2016; Geeza et al. 2018; Kassotis et al. 2016; Burgos et al. 2017; Lauer et al. 2018; K. Van Sice et al. 2018). Disposal via surface discharges from centralized treatment plants is permitted by the National Pollutant

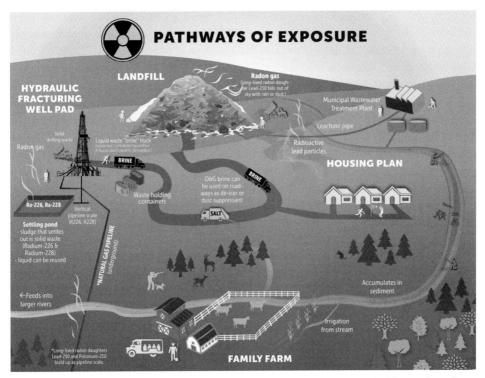

Figure 9.6 Depiction of NORM management and disposal. (A black and white version of this figure will appear in some formats. For the colour version, refer to the plate section.)
Reprinted with permission from Center for Coal Field Justice

Discharge Elimination System (NPDES) under the Clean Water Act of 1972 (USEPA 2018b). The effluent limits for these discharges are permitted based on either the technology available in these plants or the water quality of the receiving water bodies (USEPA 2018b). However, these facilities were unable to sufficiently treat contaminant concentrations to nontoxic levels. As a result, surface waters have been exposed to radium, threatening the health of the receiving streams and their wildlife (Warner et al. 2013b; Vengosh et al. 2014b; Burgos et al. 2017; Geeza et al. 2018; Lauer et al. 2018; K. Van Sice et al. 2018). Once discharged, radium from produced water often associates with suspended particles and other precipitating ions and accumulates in streambed sediments. According to 40 CFR 192, the action level for ^{226}Ra in the top 15 cm of surface soils in inactive uranium and thorium processing sites should not exceed 185 Bq/kg (5 pCi/g) above background concentrations and not exceed 555 Bq/kg (15 pCi/g) above background in any 15 cm layer below the surface layer in any 100 square meter area (www.ecfr.gov).

9.5.1 Eastern US Surface Disposal Case Studies: Pennsylvania

Studies of CWTs (centralized O&G waste treatment facilities) in Pennsylvania examined radium accumulation downstream of facilities located on Black Lick Creek that discharged between 150–250 million L of treated produced water to surface water annually from 2002–2011 (Burgos et al. 2017). Combined ^{226}Ra + ^{228}Ra accumulation in sediments 200 times background levels were noted at the discharge of a facility that discharged treated brine with mean ^{226}Ra concentration of 1.97 Bq/L. Despite relatively low radium activity in the fluids disposed to the surface water, the sediments downstream of the discharge contained radium concentrations 1.5 times background levels up to 31 km downstream from the discharge three years later (Warner et al., 2013a; Burgos et al. 2017). Blacklick Creek was listed as an impaired stream because previous coal mining resulted in AMD discharges to the stream. Much of the radium at the discharge was associated with fine-grained radiobarite that precipitated when the high TDS and Ba-rich discharge was diluted and mixed with high sulfate Blacklick creek. Radium-sulfate minerals accumulated at the discharge were easily transported fine-grained sediments and mobilized to areas with lower radium concentrations and anoxic conditions downstream. The reduction of these radium-sulfate minerals by sulfate-reducing bacteria can solubilize the Ra, Sr, and Ba, thereby posing a potential risk for radium bioavailability (Phillips 2001; Renock et al. 2016; Ouyang et al. 2017).

Iron and manganese oxides (HFO and HMO, respectively) are also commonly associated with AMD (Cravotta 2008b) and have been shown to readily sorb Ra (International Atomic Energy Agency 2014; Sajih et al. 2014; K. Van Sice et al. 2018). At Blacklick Creek, Ra sequestration mechanisms could be adequately modeled, for the first time, by adding a mass of sorbent equivalent to 100 mg/L HMO and 1000 mg/L HFO, indicating the higher sorption potential for the HMO compared to the HFO (K. Van Sice et al. 2018). In the PHREEQC model developed for Van Sice et al. (2018) in dilute downstream conditions where $Ba_xSr_{1-x}SO_4$ coprecipitation of Ra was not thermodynamically favorable, HMO sorption of the Ra was modeled to reduce Ra activities in Blacklick Creek by 40%, Ba by 95% and Sr by 75%. Because of the effectiveness of HMO as a sorbent for Ra, the US EPA has allowed small treatment systems to utilize preformed HMO with filtration as a verified Ra and arsenic removal prior to discharge.

Studies that focused on other CWT facilities in Pennsylvania also showed elevated radium activities at the discharges with mean radium values in sediments from 740 to 7400 Bq/kg, with much of the activity attributed to conventional O&G brines typically produced from the Appalachian Basin (Lauer et al. 2018).

9.5.2 Western US Surface Disposal Case Studies: Wyoming, North Dakota, and California

Agricultural beneficial use in the western United States is permitted through the National Pollutant Discharge Elimination System (NPDES) under 40 CFR § 435 Subpart E for irrigation, livestock watering, and wildlife propagation. While there remain effluent limits in place for pollutants, the regulations for consistent self-reporting Discharge Monitoring Reports (DMR) are relatively limited and variable by state and discharge.

There are approximately 600 NPDES-permitted produced water discharges in Wyoming for beneficial use, though there have been limited studies completed on the environmental and human health impacts of this practice (Ramirez,2002; McDevitt et al. 2020b, 2021; McLaughlin et al. 2020). From NPDES permits for Wyoming, there are only standards for NORM of ^{226}Ra (60 pCi/L). McDevitt et al. (2018) demonstrated O&G produced water discharges had low total Ra concentrations around 0.44 Bq/L (well below the permitted limit) but Ra accumulations in the sediments below the outfall were significantly higher than background sediment Ra activities, upward of approximately 3,700 Bq/kg at the discharge compared to 74 Bq/kg background (McDevitt et al. 2018). Surprisingly, downstream of several NPDES O&G discharges in Wyoming, radium was shown to accumulate in carbonate mineral phases. Radium sequestration by partitioning into sulfate solid solutions (i.e., mineral precipitates that contain both Ba and Sr in the crystal lattice with SO_4) involving barite ($BaSO_4$) and celestite ($SrSO_4$) is widely reported; however, incorporation of radium into calcium carbonate ($CaCO_3$), strontianite ($SrCO_3$), witherite ($BaCO_3$), and other carbonate phases is less well established (Langmuir and Melchior 1985; Tesoriero and Pankow 1996; Curti 1999; Grandia et al. 2008; Vinson et al. 2012; Vinograd et al. 2013, 2018a, 2018b; He et al. 2014; Kondash et al. 2014a; Sajih et al. 2014; Vengosh et al. 2014a; Zhang et al. 2014, 2015b; Brandt et al. 2015). Because cations with crystal ionic radii larger than Ca^{2+} can fit well in the orthorhombic structure, aragonite may take up significant amounts of Sr^{2+}, Pb^{2+}, Ba^{2+}, and, presumably, Ra^{2+}, compared to rhombohedral calcite, which has low potential for equilibrium partitioning of Sr and Ba (Back and Hanshaw 1970; Hanshaw and Back 1979; Tesoriero and Pankow 1996; Curti 1999; Glynn 2000). Additional carbonate phases, including dolomite, ankerite, magnesite, and siderite, exhibit potential for adsorption of low concentrations of Ra^{2+} and Ba^{2+} (Jones et al. 2011). Carbonates generally dissolve more readily than celestite and barite, suggesting potential for radium re-mobilization to the water column in response to episodic dilution and acidification (Dove and Platt 1996; McDevitt et al. 2018). Created wetlands from O&G produced water in Wyoming were observed to accumulate radium in vegetation upward of 1,110 Bq/kg and in bird bones up to 37 Bq/kg, indicating the potential for bioaccumulation of O&G-derived pollutants through the aquatic food chain (Ramirez 2002).

Studies by two independent teams showed that a 2015 O&G pipeline fluid spill in Williston Basin, North Dakota resulted in sediments containing combined ^{226}Ra + ^{228}Ra activities 15–100 times background; even though the pipeline fluid concentration was relatively low at ~0.33 Bq/L. The accumulation in sediment indicated the potential for long-term impact, as transport of radium-enriched sediments, and subsequent release into the aqueous phase, served as a secondary contaminant source to the local ecosystem (Cozzarelli et al. 2016; Lauer et al., 2016).

California's Central Valley Water Board has been practicing beneficial use for crop irrigation for over 30 years. In 2017, 16% of the total 1.9 billion barrels of produced water generated in California's Central Valley were blended with surface and groundwater and irrigated 90,000 acres of cropland (Boards 2019). In a greenhouse irrigation study, spring wheat that was irrigated with 10% and 50% O&G produced water exhibited negative morphological and physiological effects; grain yields decreased by 70–100% compared to a tap water control (Sedlacko et al. 2019). In Kern County, California, the use of low salinity produced water for irrigation did not lead to higher radium activities than irrigation with local groundwater (Kondash et al. 2020). However, when utilizing O&G produced water for crop irrigation, soil vulnerability to sodification and salinization from boron and sodium toxicity was found (Kondash et al. 2020); whereas, clay soil structure disintegrates over time after continuing inputs of high salinity irrigation water (Burkhardt et al. 2015; Echchelh et al. 2019).

9.5.3 Disposal via Road Spreading

At the time of publication, Tasker et al. (2018) identified 13 US states that utilized conventional O&G wastewater for dust suppression and deicing, targeting high TDS brines enriched in Ca and Mg (Tasker et al. 2018; Stallworth et al. 2019). The lower the value of fluid SAR (sodium absorption ratio) and the higher the value of TDS for a dust suppressant fluid, the lower the predicted dust generated after application (Stallworth et al. 2019). NORM, lead, and arsenic removal from O&G produced water is essential when considering use as a dust suppressant owing to association with small particulate matter that is easily transported into human lungs (Tasker et al. 2018; Stallworth et al. 2019). Conventional and unconventional produced waters share similar inorganic chemistry (Blondes et al. 2018). Michigan did not differentiate between the allowable use of unconventional versus conventional produced water on roads until 2012 (Goodman 2017). In a regulatory survey of a subset of states that allowed road spreading, only oil–water separation was required prior to spreading, or no treatment (Goodman 2017). Ra was not monitored in brines of states surveyed, and New York and North Dakota were the only states to monitor lead and arsenic concentrations. Over 60 million L of conventional produced water was road spread for dust suppression in 2012 in Pennsylvania (Tasker et al. 2018). If road maintenance is a desired treatment outcome for produced waters, operators can utilize the malleable Sr/Ba ratios for secondary treatment that maximizes Ra removal. However, further risk assessments should be conducted to ensure public health.

9.6 Novel Treatment Techniques for Resource Recovery and Sequestration

The oil and gas industry is interested in increasing its sustainability and decreasing costs through beneficial reuse and resource recovery of produced water (Sekhran et al., 2017; Alnuaim, 2018). (US EPA 2018a, b). Coonrod et al. (2020) proposed the closing of the hydraulic fracturing water cycle by encouraging "fit for purpose," flexible, and low-cost treatment technologies for industry reuse of unconventional produced water followed by recycling for alternative beneficial uses (i.e., agriculture, road deicing, and dust suppression). Secondary treatment for the sequestration of Ra and scalants, Ba and Sr, is necessary to provide a flexible treatment for either surface disposal or recycling in hydraulic fracturing processes.

A limited volume of produced water is currently treated to create beneficial co-products. Some O&G produced waters with high Na and Cl are used as a feed stock for NaCl salt generation and are commercially available for road deicing in winter months and for pool maintenance (Dorich 2017). Lithium recovery from the produced water is also of interest, being in high demand for the manufacture of batteries (Schaller et al. 2014; Paranthaman et al. 2017; Swain 2017; Flexer et al. 2018; Jang and Chung 2018; MGX Minerals Inc. 2019). However, the elimination of NORM in these products has not been documented and may prove to be a limiting factor in resource recovery and reuse.

The current treatment and handling of radium in produced water results in the generation of hundreds of tons of low-level radioactive sludge that require substantial transport and disposal costs. The lack of radium-specific treatment technologies likely results in the loss of valuable feedstock for crystallization of marketable salts ($CaCl_2$, NaCl, and $BaSO_4$). In light of the potential risks to human and environmental health surrounding current management and treatment practices, and the desire for beneficial reuse of the co-products through resource recovery, there is an apparent need for new treatment techniques for produced water that can quantify removal of NORM. Such treatment techniques would reduce the high volumes of radioactive treatment solids requiring specialized disposal. In addition, a new treatment protocol for the solid co-products that recovers a portion of the raw materials and ultimately diminishes the volume of landfill waste is highly desirable.

9.6.1 Radium Specific Treatment of Clay and Sand Products from O&G Wastes

While some effort has been applied to beneficial use of fluids or the dissolved salts in fluids, there been has been little effort documented to recover resources in the solid co-products, despite their high volume and presence of potentially useful materials such as fracturing proppant and clay minerals (Maloney and Yoxtheimer 2012). The handling and treatment of hydraulic fracturing solid co-products also presents challenges because of the presence of NORM.

There has been little research on reclaiming the raw materials found in the solids, such as clays and sand. Sand plays a prominent role in the fracturing process and fracking a well can require as much as 5,000 tons of sand as proppant. Clay is also an important component of the drilling process. The clay and sand are therefore valuable resources that should be of interest for conservation and management; however, these materials are generally disposed in municipal waste landfills (RCRA Subtitle D). Novel treatment of solid co-products for resource recovery and NORM management can be achieved by treating O&G residual wastes with hydroacoustic cavitation that is able to reclaim marketable sand (proppant sand) with an estimated worth of $50k–$70k/year (when sold at a fraction of the price of freshly mined silica sand). A study of hydroacoustic cavitation yielded primarily cleaned proppant sand in one flow stream; and clays and fine particles in another. Consequently, the separation of particle sizes also affected radium distribution: the sand grains had low radium activities, as lows as 0.207 Bq/g (5.6 pCi/g). In contrast, the clays had elevated radium activities, as high as 1.85–3.7 Bq/g (50–100 pCi/g). The reclaimed proppant sand could potentially be reused as hydraulic fracturing proppant and reduce the volume or mass of radium wastes that are disposed in landfills. The reclaimed sand, along with re-claimable clay, could reduce waste volumes by 50%, which represents a yearly savings of $200,000 for facilities handling ~5,000 tons/year of residual solid waste. This could represent a significant savings to facilities handling O&G waste, as much as $100,000–$300,000 per year. Disposing the radium-enriched salts and organics downhole will mitigate radium release to the surface (Ajemigbitse et al. 2019). Reclamation of solids could result in the reduction of radioactivity disposed in landfills, mitigating the risk of radioactive exposure and contamination.

Clays are materials that are often used to remediate metal-contaminated wastes. Radium removal from highly saline produced water with a synthetic clay mineral possessing high Ra selectivity and high charge for radium removal could be a potential novel treatment technique that requires further study. This synthetic clay, Na-4-mica, presents itself as an ideal candidate for radium removal as its interlayer can collapse on complete substitution, hence sequestering radium and mitigating release and environmental exposure. While Na-4-mica can remove radium at a range of pH and salinities, Ba presents a significant competition for adsorption sites (Ajemigbitse et al. 2019).

9.6.2 Radium Specific Treatment of O&G Wastes with Acid Mine Drainage

Owing to the proximity of O&G extraction activities and abandoned mine drainage (AMD) (estimated to contribute a total waste stream of 8,633 million L/day) (Growitz et al.,1985; Cravotta 2008a; Curtright and Giglio 2012) in Pennsylvania, use of AMD in place of local freshwater withdrawals poses a unique opportunity to repurpose AMD as fracturing fluid that otherwise pollutes local waterways (He et al. 2014; Kondash et al. 2014a; Wang et al. 2018; McDevitt et al. 2020a). However, AMD contains high concentrations of sulfate, and sequestration of alkaline earth metals from O&G wastewater including Ba, Sr, and Ra by sulfate mineral co-precipitation into barite ($BaSO_4$) and radiobarite $(Ba,Ra)SO_4$ is widely reported as a source of scaling in O&G wells and is also commonly utilized to treat produced water prior to surface water disposal (Langmuir and Melchior 1985; Grandia et al. 2008; Vinograd et al. 2013, 2018b; He et al. 2014; Kondash et al. 2014b; Vengosh et al. 2014a; Zhang et al. 2014, 2015b; Brandt et al. 2015; McDevitt et al. 2020a). Less commonly reported but also effective in Ra and Sr sequestration is celestite ($SrSO_4$) and radiocelestite $(Sr,Ra)SO_4$ precipitation (Langmuir and Melchior 1985; He et al. 2014; K. Van Sice et al. 2018; Vinograd et al. 2018a). However, instead of pursuing AMD as an alternative source to freshwater for injection as hydraulic fracturing, previous studies have explored the potential use of co-treatment of hydraulic fracturing fluid with AMD that otherwise pollutes local waterways (He et al. 2013, 2016; Kondash et al. 2014a; Cavazza 2016). Depending on the ultimate treatment outcome for the produced water volumes, it has been shown that the Sr/Ba molar ratio of the initial fluid mix can be adjusted for either maximum Ra removal into a smaller mass of solids (increasing the Sr/Ba) or adjusted for maximum removal of problematic borehole scalants Ba, Sr SO_4, and Ra in a larger mass of solids (decreasing the Sr/Ba) (McDevitt et al. 2020a).

9.7 Accurate Measurements in Complex Matrices

The management of O&G wastes necessitates the need for accurate and reliable radium measurements. While there are many methods to quantify radium in liquids, these methods have been shown to be unreliable when performed on produced waters (Nelson et al. 2014; Tasker et al. 2019). This is exacerbated by the fact that the recommended US EPA methods for measuring radium are based on freshwater methods and do not consider the high-TDS concentrations of the produced waters or require pre-concentration steps that have been shown to have poor radium recovery (EPA 2005; Nelson et al. 2014; Zhang et al.,2015a). While gamma spectroscopy remains the recommended method for measuring radium, radium measurements of produced waters and O&G solids have been shown to have high variability, even when the same samples are measured by different labs using similar methods (Tasker et al. 2019). There has been little attempt made to address the causes of such high variability or to propose methods to account for and correct this variance, especially in the context of O&G products.

9.7.1 Accounting for TDS in Liquids with Gamma Spectroscopy

The high total dissolved solids (TDS) concentration of the produced waters introduces difficulties when measuring ^{226}Ra by recommended US EPA methods that were specifically developed several decades ago for drinking water. Radium measurements in high TDS fluids from oil and gas extraction can have unfavorable precision and accuracy, in part because these high concentrations of dissolved salts incur attenuation. While other techniques for measuring radium in these high-TDS fluids have since been developed, these newer techniques often require extensive and complicated pre-concentration steps; thus, they require extensive analytical chemistry skills, utilize hazardous chemicals like hydrofluoric acid, demand long holding times or measurement times, and require high sample volumes.

Accurately measuring radium in brines is a challenge because of the variability and often high TDS and organic content in both the raw material and the co-products. TDS concentrations and composition greatly influence gamma spectroscopy radium measurements of high-TDS produced waters. Indeed, radium activities can be underestimated by up to 40% if the attenuation offered by the high-TDS fluid environment is ignored. However, rapid and accurate gamma spectroscopy radium measurements of high-TDS produced water can be achieved by direct measurement of ^{226}Ra at 186 keV when the attenuation caused by TDS concentration and composition are accounted for, including the deconvolution of interference in peak signals at the 186 keV from thorium (Ajemigbitse et al. 2019).

In order to reduce measurement duration and to maintain or improve accuracy while measuring NORM with a well-detector gamma spectrometer, a method to rapidly assess both ^{226}Ra and ^{228}Ra and to account for the self-attenuation of gamma rays in high-TDS O&G fluids is necessary. Comparisons between a NaCl-only and a multi-cation-chloride synthetic brine spiked with known amounts of ^{226}Ra and ^{228}Ra indicated that both the TDS concentration and the type of TDS (i.e., Na only vs. Na-Mg-Ba-Ca-Sr) influenced self-attenuation in well-detector gamma spectroscopy; thus, variations in brines (and NORM) across regions requires personalized corrections for this TDS-influenced self-attenuation. Radium activities can be underestimated if the correction is not applied. Using a NaCl-only brine to match the matrix of high-TDS oil and gas brines is inadequate to produce accurate measurements: rather, the full set of cations should be included (Ajemigbitse et al. 2019).

9.7.2 Rapid Measurement with Liquid Scintillation Counting

While gamma spectroscopy is the recommended method, it remains expensive, inaccessible for many O&G operators, and in many cases, the analysis requires long waiting times. Therefore, there is a need for new rapid and accurate methods for quantifying radium in co-products (prior to and after treatment), beneficially recovered resources, and any treatment wastes generated as a result of these efforts. There are rapid methods for quantifying radium in produced water through the application of alpha-beta discrimination coupled with spectrum analysis in liquid scintillation counting (LSC). Liquid scintillation counting is a method for quantifying the amount of radioactive material of alpha and beta decay in an aqueous sample. LSC is a common analysis technique for measuring radioactivity, but it is seldom applied to measurements of NORM because of the high TDS and interference of organic matrices. Ajemigbitse et al. 2019 developed a new rapid method for radium measurement of brines and liquid co-products that uses a relatively simple sample preparation yet produces high radium recoveries of > 90%. This new LSC method could yield an R^2 of 0.92 when compared to high-accuracy gamma spectroscopy. The reduced sample preparation steps, low cost, and rapid analysis present this as a method ideal for field appraisal prior to comprehensive radiochemical analysis. The rapid and inexpensive analysis is probably a hurdle that must be overcome to reach commercial beneficial reuse of products.

Samples were prepared for analysis by evaporating the fluid and re-suspending the evaporate with acidified distilled deionized water prior to liquid scintillation counting for one hour. This protocol yielded radium recoveries $\geq$ 93%. Per this protocol, the alpha and beta spectra of ^{226}Ra and its daughters were computationally separated by alpha-beta discrimination and spectrum deconvolution. The minimum detectable activity of ^{226}Ra was 0.33 Bq/L (9.0 pCi/L) when the counting time was 60 minutes and the sample volume was 4 mL. Nine produced waters of varying TDS and radium concentrations from the Marcellus Shale Formation were analyzed by this method and compared with gamma spectroscopy; these yielded comparable results with an R^2 of 0.92. The reduced sample preparation steps, low cost, and rapid analysis position this as a well-suited protocol for field-appraisal and screening, when compared to comprehensive radiochemical analysis. We believe that for a given produced water region, routine and local liquid scintillation analyses can be compared and calibrated with infrequent gamma spec analyses, so as to yield a near-real time protocol for monitoring ^{226}Ra levels during hydrofracturing operations. We present this as a pragmatic and efficient protocol for monitoring ^{226}Ra when produced water samples host low levels of ^{228}Ra: since the progeny of ^{228}Ra can significantly confound the LSC analyses.

9.7.3 Measurements of Solids for Co-Products

The radium content of the treatment waste solids and beneficial products must be accurately determined when evaluating the efficacy and social validity of such treatments. A rapid and accurate method of radioactivity determination will allow waste treatment facilities to make treatment decisions more efficiently. Gamma spectroscopy is often recommended for determining radium activities in O&G solids. However, this method can produce a range of reported activities for the same sample; and can be impacted by the composition/mineralogy of the solids, which influence the attenuation of the gamma decay energy: with denser sediments incurring greater degrees of attenuation. For small volumes of NORM, the sample density and sample volume in the measuring vial must be accurately measured. Corrections are relevant to a wide range of solid samples and sediment densities that may be encountered during treatment and management of oil and gas solids, including clays, environmental sediment samples, sand grains, and precipitated salts. These corrections can also be applied for situations where low volumes of material are present, as in bench scale studies, thereby rendering this technique applicable to a wider range of scenarios. Therefore, self-attenuation must be accounted for with an empirical technique when accurately measuring radium, otherwise radium measurements are found to be inaccurate, sometimes by as much as 50% (Ajemigbitse et al. 2019). While these techniques are often applied by commercial, academic, and government laboratories, interlab comparisons of NORM solids show that variations in measurement techniques can lead to a broad range of reported values (Tasker et al. 2019).

Accurate measurements of radium in solid co-products are often achieved by incubating samples for 21 days prior to analysis of daughter decay products of radium. Unfortunately, these measurements can be inaccurate by up to ~50% when sample density and specific volumes are ignored in calibration of gamma spectrometry. However, rapid and accurate gamma spectroscopy radium measurements of oil and gas solids can be achieved by direct measurement of ^{226}Ra at 186 keV when ^{235}U interference is removed and the attenuation caused by sample density and volume are accounted for (Landis et al. 2018; Ajemigbitse et al. 2019).

Accurate radium measurements in solid co-products also require an empirical technique to account for sample attenuation in well-detector gamma spectroscopy.

9.8 Summary

Unconventional extraction is predicted to increase cumulative water use 20-fold in gas plays and 13-fold in oil plays from 2018–2030 (Kondash et al. 2018). While data on NORM from various O&G basins throughout the USA is limited, elevated activities are reported in many areas. Treatment and disposal of produced water to local freshwater streams (even when radium activities are decreased to permitted values) results in the accumulation of radium in stream sediments that is transported downstream. This radium could be bioavailable and represents an area of potential concern for exposure for humans and aquatic life.

References

Ajemigbitse MA, Cannon FS, Klima MS, Furness JC, Wunz C, and Warner NR. (2019). Raw material recovery from hydraulic fracturing residual solid waste with implications for sustainability and radioactive waste disposal. *Environmental Science Process. Impacts.* 21: 308–323. https://doi.org/10.1039/c8em00248g

Akob DM, Mumford AC, Orem W, Engle MA, Klinges JG, Kent DB, and Cozzarelli IM. (2016). Wastewater Disposal from Unconventional Oil and Gas Development Degrades Stream Quality at a West Virginia Injection Facility. *Environmental Science & Technology.* 50: 5517–5525. https://doi.org/10.1021/acs.est.6b00428

Alhajji E, Al-Masri MS, Khalily H, Naoum BE, Khalil HS, and Nashawati A. (2016). A Study on Sorption of 226Ra on Different Clay Matrices. *Bulletin of Environmental Contaminants and Toxicology.* 97: 255–260. https://doi.org/10.1007/s00128–016-1852-1

Alnuaim S.. (2018). *What Does Sustainability Mean for Oil and Gas?* Soc. Pet. Eng.

Back W, Hanshaw BB. (1970). Comparison of chemical hydrogeology of the carbonate peninsulas of Florida and Yucatan. *Journal of Hydrology.* 10: 330–368. https://doi.org/10.1016/0022-1694(70)90222-2

Barry B and Klima MS. (2013). Characterization of Marcellus Shale natural gas well drill cuttings. *Jouranl of Unconventional Oil and Gas Resources.* 1–2: 9–17. https://doi.org/10.1016/J.JUOGR.2013.05.003

Bateman H. (1910). The solution of a system of differential equations occurring in the theory of radio-active transformations. *Cambridge Philosophical Society.* 15: 423–427.

Blewett TA, Delompré PLM, Glover CN, and Goss GG. (2018). Physical immobility as a sensitive indicator of hydraulic fracturing fluid toxicity towards Daphnia magna. *Science of the Total Environment.* 635: 639–643. https://doi.org/10.1016/j.scitotenv.2018.04.165

Blewett TA, Weinrauch AM, Delompré PLM, and Goss GG. (2017). The effect of hydraulic flowback and produced water on gill morphology, oxidative stress and antioxidant response in rainbow trout (Oncorhynchus mykiss). *Scientific Reports.* 7: 46582. https://doi.org/10.1038/srep46582

Bloch S and Key RM. (1953). Modes of Formation of Anomalously High Radioactivity in Oil-Field Brines.

Blondes MS, Gans KD, Engle MA, Kharaka YK, Reidy ME, Saraswathula V, Thordsen JJ, Rowan EL, and Morrissey EA. (2018). U.S. Geological Survey National Produced Waters Geochemical Database (ver. 2.3, January 2018) [WWW Document]. U.S. Geol. Surv. data release. https://doi.org/doi.org/10.5066/F7J964W8

Boards CW. (2019). Fact Sheet [WWW Document]. URL www.waterboards.ca.gov/rwqcb5/water_issues/oil.411"/>_fields/food_safety/data/fact_sheet/of_foodsafety_fact_sheet.pdf

Bolivar JP, García-León M, and García-Tenorio R. (1997). On self-attenuation corrections in gamma-ray spectrometry. *Applied Radiation and Isotopes.* 48: 1125–1126. https://doi.org/10.1016/S0969–8043(97)00034-1

Brandt F, Curti E, Klinkenberg M, Rozov K, Bosbach D. (2015). Replacement of barite by a (Ba,Ra)SO4 solid solution at close-to-equilibrium conditions: A combined experimental and theoretical study. *Geochimica et Cosmochimica Acta*. 155: 1–15. https://doi.org/10.1016/j.gca.2015.01.016

Burgos WD, Castillo-Meza L, Tasker TL, Geeza TJ, Drohan PJ, Liu X, Landis JD, Blotevogel J, McLaughlin M, Borch T, and Warner NR. (2017). Watershed-scale impacts from surface water disposal of oil and gas wastewater in Western Pennsylvania. *Environmental Science & Technology*. 51: 8851–8860. https://doi.org/10.1021/acs.est.7b01696

Burkhardt A, Gawde A, Cantrell CL, Baxter HL, Joyce BL, Stewart CN, and Zheljazkov VD. (2015). Effects of produced water on soil characteristics, plant biomass, and secondary metabolites. *Journal of Environmental Quality*. 44: 1938–1947. https://doi.org/10.2134/jeq2015.06.0299

Carvalho F, Chambers D, Fesenko S, Moore WS, Porcelli D, Vandenhoven H, and Yankovich T. (2014). *Environmental Pathways and Corresponding Models: The Environmental Behaviour of Radium*, Revised Edition. Vienna.

Cavazza M. (2016). Reducing freshwater consumption in the marcellus shale play by recycling flowback with acid mine drainage, in: Proceedings – SPE Annual Technical Conference and Exhibition. https://doi.org/10.2118/184499-stu

Clark C and Veil J. (2015). U.S. Produced water volumes and management practices. *Groundwater Protection Council*. 119.

Coleman JL, Milici RC, Cook TA, Charpentier RR, Kirschbaum M, Klett TR, Pollastro RM, and Schenk CJ. (2011). Assessment of Undiscovered Oil and Gas Resources of the Devonian Marcellus Shale of the Appalachian Basin Province, 2011.

Considine T, D. Robert Watson P, E. Seth Blumsack P.. (2010). The economic impacts of the Pennsylvania Marcellus Shale natural gas play: An update.

Coonrod CL, Yin YB, Hanna T, Atkinson A, Alvarez PJJ, Tekavec TN, Reynolds MA, and Wong MS. (2020). Fit-for-purpose treatment goals for produced waters in shale oil and gas fields. *Water Research*. 173: 115467. https://doi.org/10.1016/j.watres.2020.115467

Cozzarelli IM, Skalak KJ, Kent DB, Engle MA, Benthem A, Mumford AC, Haase K, Farag A, Harper D, Nagel SC, Iwanowicz LR, Orem WH, Akob DM, Jaeschke JB, Galloway J, Kohler M, Stoliker DL, and Jolly GD. (2016). Environmental signatures and effects of an oil and gas wastewater spill in the Williston Basin, North Dakota. *Science of the Total Environment*. 579: 1781–1793. https://doi.org/10.1016/j.scitotenv.2016.11.157

Cravotta CA. (2008a). Dissolved metals and associated constituents in abandoned coal-mine discharges, Pennsylvania, USA. Part 1: Constituent quantities and correlations. *Applied Geochemistry*. 23: 166–202. https://doi.org/10.1016/j.apgeochem.2007.10.011

Cravotta CA. (2008b). Dissolved metals and associated constituents in abandoned coal-mine discharges, Pennsylvania, USA. Part 2: Geochemical controls on constituent concentrations. *Applied Geochemistry*. 23: 203–226. https://doi.org/10.1016/j.apgeochem.2007.10.003

Curti E. (1999). Coprecipitation of radionuclides with calcite: Estimation of partition coefficients based on a review of laboratory investigations and geochemical data. *Applied Geochemistry*. 14: 433–445. https://doi.org/10.1016/S0883-2927(98)00065-1

Curtright AE and Giglio K. (2012). Coal Mine Drainage for Marcellus Shale Natural Gas Extraction, Proceedings and Recommendations from a Roundtable on Feasibility and Challenges.

Dolan FC, Cath TY, and Hogue TS. (2018). Assessing the feasibility of using produced water for irrigation in Colorado. *Science of the Total Environment*. 640–641: 619–628. https://doi.org/10.1016/j.scitotenv.2018.05.200

Dorich A. (2017). Eureka Resources recovers vital resources from flowback and production wastewater. [WWW Document]. Energy Min. Int. URL www.emi-magazine.com/sections/profiles/1460-eureka-resources

Dove PM and Platt FM. (1996). Compatible real-time rates of mineral dissolution by Atomic Force Microscopy (AFM). *Chemical Geology*. 127: 331–338. https://doi.org/10.1016/0009-2541(95)00127-1

Dresel P and Rose A. (2010). Chemistry and origin of oil and gas well brines in western Pennsylvania. *Pennsylvania Geological Survey*. 48. https://doi.org/Open-File Report OFOG 1001.0

Dunne EJ. (2017). *Flowback and Produced Waters*. National Academies Press. https://doi.org/10.17226/24620

Echchelh A, Hess T, Sakrabani R, de Paz JM, and Visconti F. (2019). Assessing the environmental sustainability of irrigation with oil and gas produced water in drylands. *Agricultural Water Management*. 223. https://doi.org/10.1016/j.agwat.2019.105694

Eitrheim ES, May D, Forbes TZ, and Nelson AW. (2016). Disequilibrium of Naturally Occurring Radioactive Materials (NORM) in drill cuttings from a horizontal drilling operation. *Environmental Science & Technology Letters*. 3: 425–429. https://doi.org/10.1021/acs.estlett.6b00439

Ellsworth WL. (2013). Injection-Induced Earthquakes. *Science*. 341: 1225942. https://doi.org/10.1126/SCIENCE.1225942

Fisher RS (1998) Geologic and geochemical controls on naturally occurring radioactive materials (NORM) in produced water from oil, gas, and geothermal operations. *Environmental Geosciences*. 5: 139–150.

Fleischer RL and Raabe OG. (1978). Recoiling alpha-emitting nuclei. Mechanisms for uranium-series disequilibrium. *Geochimica et Cosmochimica Acta*. 42: 973–978. https://doi.org/10.1016/0016-7037(78)90286-7

Flexer V, Fernando Baspineiro C, and Galli CI. (2018). Lithium recovery from brines: A vital raw material for green energies with a potential environmental impact in its mining and processing. *Science of the Total Environment*. 639: 1188–1204. https://doi.org/10.1016/j.scitotenv.2018.05.223

Geeza TJ, Gillikin DP, McDevitt B, Van Sice K, and Warner NR. (2018). Accumulation of Marcellus Formation oil and gas wastewater metals in freshwater mussel shells. *Environmental Science & Technology*. 52: 10883–10892. https://doi.org/10.1021/acs.est.8b02727

Gilmore GR. (2008). *Practical Gamma-Ray Spectrometry*. John Wiley & Sons, Ltd. https://doi.org/10.1002/9780470861981

Glynn P. (2000). Solid-solution solubilities and thermodynamics: Sulfates, *Carbonates and Halides: Reviews in Mineralogy and Geochemistry*. 40: 481–511. https://doi.org/10.2138/rmg.2000.40.10

Goodman C. (2017). Beneficial use of produced water for roadspreading: perspectives for Colorado policymakers. Denver. https://doi.org/10.1017/CBO9781107415324.004

Grandia F, Merino J, andBruno J. (2008). Assessment of the radium-barium co-precipitation and its potentialinfluence on the solubility of Ra in the near-field TR-08-07, 52.

Gregory KB, Vidic RD, and Dzombak DA. (2011). Water management challenges associated with the production of shale gas by hydraulic fracturing. *Elements*. 7: 181–186. https://doi.org/10.2113/gselements.7.3.181

Growitz BDJ, Reed LA, and Beard MM. (1985). Reconnaissance of Mine Drainage in the Coal Fields of Eastern Pennsylvania. Harrisburg. USGS 83-4274 https://doi.org/10.3133/wri834274

Guerra K, Dahm K, and Dundorf S. (2011). Oil and gas produced water management and beneficial use in the Western United States. Denver. https://doi.org/www.usbr.gov/pmts/water/publications/reports.html

Hanshaw BB and Back W. (1979). Major geochemical processes in the evolution of carbonate-aquifer systems. *Developments in Water Science*. 12: 287–312. https://doi.org/10.1016/S0167-5648(09)70022-X

Harkness JS, Dwyer GS, Warner NR, Parker KM, Mitch WA, and Vengosh A. (2015). Iodide, bromide, and ammonium in hydraulic fracturing and oil and gas wastewaters: Environmental implications. *Environmental Science & Technology*. 49: 1955–1963. https://doi.org/10.1021/es504654n

He C, Zhang T, and Vidic RD. (2013). Use of Abandoned Mine Drainage for the Development of Unconventional Gas Resources. *Disruptive Science and Technology*. 1: 169–176. https://doi.org/10.1089/dst.2013.0014

He C, Zhang T, and Vidic RD. (2016). Co-treatment of abandoned mine drainage and Marcellus Shale flowback water for use in hydraulic fracturing. *Water Research*. 104: 425–431. https://doi.org/10.1016/j.watres.2016.08.030

He C, Li M, Liu W, Barbot E, and Vidic RD. (2014). Kinetics and equilibrium of barium and strontium sulfate formation in Marcellus Shale flowback water. *Journal of Environmental Engineering*. 140: B4014001-1–9. https://doi.org/10.1061/(ASCE)EE.1943-7870.0000807

Hubbell JH. (1982). Photon mass attenuation and energy-absorption coefficients. *The International Journal of Applied Radiation and Isotopes*. 33: 1269–1290. https://doi.org/10.1016/0020-708X(82)90248-4

International Atomic Energy Agency. (2014). *The Environmental Behaviour of Radium*, Revised Edition. Vienna.

Jang Y and Chung E. (2018). Adsorption of lithium from shale gas produced water using titanium based adsorbent. *Industrial & Engineering Chemistry Research*. 57: 8381–8387. https://doi.org/10.1021/acs.iecr.8b00805

Jodłowski P, Macuda J, Nowak J, and Nguyen Dinh C. (2017) Radioactivity in wastes generated from shale gas exploration and production: North-Eastern Poland. *Journal of Environmental Radioactivity*. 175–176: 34–38. doi: 10.1016/j.jenvrad.2017.04.006.

Jones MJ, Butchins LJ, Charnock JM, Pattrick RAD, Small JS, Vaughan DJ, Wincott PL, and Livens FR. (2011). Reactions of radium and barium with the surfaces of carbonate minerals. *Applied Geochemistry*. 26: 1231–1238. https://doi.org/10.1016/j.apgeochem.2011.04.012

Kassotis CD, Iwanowicz LR, Akob DM, Cozzarelli IM, Mumford AC, Orem WH, and Nagel SC. (2016). Endocrine disrupting activities of surface water associated with a West Virginia oil and gas industry wastewater disposal site. *Science of the Total Environment*. 557–558: 901–910. https://doi.org/10.1016/j.scitotenv.2016.03.113

Kerr RA. (2010). Energy: Natural gas from shale bursts onto the scene. *Science*. 328: 1624–1626. https://doi.org/10.1126/science.328.5986.1624

Kirby HW and Salutsky ML. (1964). *The Radiochemistry of Radium*. NAS-NRC Nucl. Sci. Ser.

Kondash AJ, Albright E, and Vengosh A. (2017). Quantity of flowback and produced waters from unconventional oil and gas exploration. *Science of the Total Environment*. 574: 314–321. https://doi.org/10.1016/j.scitotenv.2016.09.069

Kondash AJ, Lauer NE, and Vengosh A. (2018). The intensification of the water footprint of hydraulic fracturing. *Science Advances*. 4. https://doi.org/10.1126/sciadv.aar5982

Kondash AJ, Warner NR, Lahav O, and Vengosh, A. (2014a). Radium and barium removal through blending hydraulic fracturing fluids with acid mine drainage. *Environmental Science & Technology*. 48: 1334–1342. https://doi.org/10.1021/es403852h

Kondash AJ, Warner NR, Lahav O, and Vengosh, A. (2014b). Radium and barium removal through blending hydraulic fracturing fluids with acid mine drainage. *Environmental Science & Technology*. 48: 1334–1342. https://doi.org/10.1021/es403852h

Kondash AJ, Redmon J, Lambertini L, Feinstein L, Weinthal E, Cabrales ., and Vengosh A. (2020). The impact of using low-saline oilfield produced water for irrigation on water and soil quality in California, Science of The Total Environment, Volume 733, 139392.

Kraemer TF and Reid DF. (1984). The occurrence and behavior of radium in saline formation water of the U.S. Gulf Coast region. *Chemical Geology*. 46: 153–174. https://doi.org/10.1016/0009-2541(84)90186-4

Krishnaswami S, Graustein WC, Turekian KK, and Dowd JF. (1982). Radium, thorium and radioactive lead isotopes in groundwaters: Application to the in situ determination of adsorption-desorption rate constants and retardation factors. *Water Resources Research*. 18: 1663–1675. https://doi.org/10.1029/WR018i006p01663

Landis JD, Sharma M, Renock D, and Niu D. (2018). Rapid desorption of radium isotopes from black shale during hydraulic fracturing. 1. Source phases that control the release of Ra from Marcellus Shale, Chemical Geology, Volume 496, 2018, Pages 1-13.

Langmuir D and Melchior D. (1985). The geochemistry of Ca, Sr, Ba and Ra sulfates in some deep brines from the Palo Duro Basin, Texas. *Geochimica et Cosmochimica Acta*. 49: 2423–2432. https://doi.org/10.1016/0016-7037(85)90242-X

Lauer NE, Harkness JS, and Vengosh A. (2016). Brine Spills Associated with Unconventional Oil Development in North Dakota. Environ. Sci. Technol. acs.est.5b06349. https://doi.org/10.1021/acs.est.5b06349

Lauer NE, Warner NR, and Vengosh A. (2018). Sources of radium accumulation in stream sediments near disposal sites in Pennsylvania: implications for disposal of conventional oil and gas wastewater. *Environmental Science & Technology*. 52: 955–962. https://doi.org/10.1021/acs.est.7b04952

Lowson RT. (1985). The thermochemistry of radium. *Thermochimica Acta*. 91: 185–212. https://doi.org/10.1016/0040-6031(85)85214-X

Lutz BD, Lewis AN, and Doyle MW (2013). Generation, transport, and disposal of wastewater associated with Marcellus Shale gas development. *Water Resources Research*. 49: 647–656. https://doi.org/10.1002/wrcr.20096

Lyons W, Plisga G, and Lorenz M. (2015). *Standard Handbook of Petroleum and Natural Gas Engineering*, 3rd ed. Gulf Professional Publishing.

Lyons WC, Pilsga GJ, and Lorenz, MD. (2016). *Standard Handbook of petroleum and Natural Gas Engineering*. Gulf Professional Publishing.

Maloney JA. (1937a). *Radium-Nature's Oddest Child Pt. 4*. 157, 212–215. https://doi.org/10.2307/26070913

Maloney JA. (1937b). *Radium-Nature's Oddest Child*. 157: 18–20. https://doi.org/10.2307/24997429

Maloney KO and Yoxtheimer DA. (2012). Production and disposal of waste materials from gas and oil extraction from the Marcellus Shale play in Pennsylvania. *Environmental Practice*. 14: 278–287. https://doi.org/10.1017/S146604661200035X

McDevitt B, Cavazza M, Beam R, Cavazza ED, Burgos W, Li L, and Warner NR. (2020a). Maximum removal efficiency of barium, strontium, radium, and sulfate with optimum AMD-Marcellus flowback mixing ratios for beneficial use in the Northern Appalachian Basin. *Environmental Science & Technology*. 54: 4829–4839. https://doi.org/10.1021/acs.est.9b07072

McDevitt B, McLaughlin MC, Vinson DS, Geeza TJ, Blotevogel J, Borch T, and Warner NR. (2020b). Isotopic and element ratios fingerprint salinization impact from beneficial use of oil and gas produced water in the Western U.S. *Science of the Total Environment*. 716. https://doi.org/10.1016/j.scitotenv.2020.137006

McDevitt B, Mclaughlin M, Cravotta CA, Ajemigbitse MA, Sice KJ, Van Blotevogel J, Borch T, and Warner NR. (2018). Emerging investigator series: radium accumulation in carbonate river sediments at oil and gas produced water discharges: implications for beneficial use as disposal management. *Environmental Science: Processes & Impacts*. 21: 324–338. https://doi.org/10.1039/c8em00336j

McDevitt, B., McLaughlin, M. C., Blotevogel, J., Borch, T., & Warner, N. R. (2021). Oil & gas produced water retention ponds as potential passive treatment for radium removal and beneficial reuse. *Environmental Science: Processes & Impacts*. 23(3), 501–518.

McLaughlin M, Borch T, McDevitt B, Warner NR, and Blotevogel J. (2020). Water quality assessment downstream of oil and gas produced water discharges intended for beneficial reuse in arid regions. *Science of the Total Environment*. 713: 136607.

McLaughlin MC, Blotevogel J, Watson RA, Schell B, Blewett TA, Folkerts EJ, Goss GG, Truong L, Tanguay RL, Lucas J, and Borch T. (2020). Mutagenicity assessment downstream of oil and gas produced water discharges intended for agricultural beneficial reuse. *Science of the Total Environment*. 715: 136944. https://doi.org/10.1016/j.scitotenv.2020.136944

MGX Minerals Inc. (2019). MGX Minerals and Eureka Resources Announce Joint Venture to Recover Lithium from Produced Water in Eastern United States [WWW Document]. PW Newswire.

Montgomery CT and Smith MB. (2010). Hydraulic fracturing: History of an enduring technology. *Journal of Petroleum Technology*. 62: 26–40. https://doi.org/10.2118/1210-0026-JPT

Nathwani JS and Phillips CR. (1979). Adsorption of 226Ra by soils in the presence of Ca2 + ions. Specific adsorption (II). *Chemosphere*. 8: 293–299. https://doi.org/10.1016/0045-6535(79)90112-7

National Research Council. (1999). Health Effects of Exposure to Radon. National Academies Press. https://doi.org/10.17226/5499

Nelson AW May D, Knight AW, Eitrheim ES, Mehrhoff M, Shannon R, Litman R, Schultz MK. (2014). Matrix Complications in the Determination of Radium Levels in Hydraulic Fracturing Flowback Water from Marcellus Shale. *Environmental Science & Technology Letters*. 1: 204–208. https://doi.org/10.1021/ez5000379

Nelson AW, Eitrheim ES, Knight AW, May D, Mehrhoff MA, Shannon R, Litman R, Burnett WC, Forbes TZ, and Schultz MK. (2015). Understanding the radioactive ingrowth and decay of naturally occurring radioactive materials in the environment: An analysis of produced fluids from the marcellus shale. *Environ. Health Perspect*. https://doi.org/10.1289/ehp.1408855

Ouyang B, Akob DM, Dunlap D, and Renock D. (2017). Microbially mediated barite dissolution in anoxic brines. *Applied Geochemistry*. 76: 51–59. https://doi.org/10.1016/j.apgeochem.2016.11.008

Paranthaman MP, Li L, Luo J, Hoke T, Ucar H, Moyer BA, and Harrison S. (2017). Recovery of lithium from geothermal brine with lithium–aluminum layered double hydroxide chloride sorbents. *Environmental Science & Technology*. 51: 13481–13486. https://doi.org/10.1021/acs.est.7b03464

Parker KM, Zeng T, Harkness J, Vengosh A, and Mitch WA (2014). Enhanced formation of disinfection byproducts in shale gas wastewater-impacted drinking water supplies. *Environmental Science & Technology*. 48, 11161–11169. https://doi.org/10.1021/es5028184

Paukert Vankeuren AN, Hakala JA, Jarvis K, and Moore JE. (2017). Mineral reactions in shale gas reservoirs: Barite scale formation from reusing produced water as hydraulic fracturing fluid. *Environmental Science & Technology*. 51: 9391–9402. https://doi.org/10.1021/acs.est.7b01979

Pedersen KS, Christensen PL, and Shaikh JA. (2015). *Phase Behavior of Petroleum Reservoir Fluids*. CRC Press.

PA DEP. (2016). 2016 Oil and Gas Annual Report. https://gis.dep.pa.gov/oilgasannualreport/index.html

Pennsylvania Department of Environmental Protection. (2016). Technologically Enhanced Naturally Occurring Radioactive Materials (TENORM) Study Report. www.depgreenport.state.pa.us/elibrary/getdocument?docid=5815&docname=01 pennsylvania department of environmental protection tenorm study report rev 1.pdf

Phillips EJP, Landa ER, Kraemer T, and Zielinski R. (2001). Sulfate-reducing bacteria release barium and radium from naturally occurring radioactive material in oil-field barite. *Geomicrobiology Journal*. 18: 167–182. https://doi.org/10.1080/01490450120549

Pichtel J. (2016). Oil and gas production wastewater: Soil contamination and pollution prevention. *Applied Environmental Soil Science*. 2016: 1–24. https://doi.org/10.1155/2016/2707989

Ramirez PJ. (2002). *Oil Field Produced Water Discharges into Wetlands in Wyoming*. US Fish Wildl. Serv. Report.

Renock D, Landis JD, and Sharma M. (2016). Reductive weathering of black shale and release of barium during hydraulic fracturing. *Applied Geochemistry*. 65: 73–86. https://doi.org/10.1016/j.apgeochem.2015.11.001

Röntgen WC. (1970). On a new kind of rays. By W.C. Rontgen. Translated by Arthur Stanton from the Sitzungsberichte der Würzburger Physic-medic. Gesellschaft, 1895. Nature, January 23, 1896. *Radiography*. 36: 185–188.

Rosenblum J, Nelson AW, Ruyle B, Schultz MK, Ryan JN, Linden KG. (2017). Temporal characterization of flowback and produced water quality from a hydraulically fractured oil and gas well. *Science of the Total Environment*. 596–597: 369–377. https://doi.org/10.1016/j.scitotenv.2017.03.294

Rowan EL, Engle Ma, Kirby CS, Kraemer TF. (2011a). *Radium Content of Oil- and Gas-Field Produced Waters in the Northern Appalachian Basin (USA): Summary and Discussion of Data*. USGS Sci. Investig. Rep. 38 pp.

Rowan EL, Engle Ma, Kirby CS, Kraemer TF. (2011b). *Radium Content of Oil- and Gas-Field Produced Waters in the Northern Appalachian Basin (USA): Summary and Discussion of Data*. USGS Sci. Investig. Rep. 38 pp.

Rowan EL, Engle MA, Kraemer TF, Schroeder KT, Hammack RW, and Doughten MW. (2015). Geochemical and isotopic evolution of water produced from Middle

Devonian Marcellus shale gas wells, Appalachian basin, *Pennsylvania: American Association of Petroleum Geological Bulletin*. 99: 181–206. https://doi.org/10.1306/07071413146

Rutherford E. (1899). VIII. Uranium radiation and the electrical conduction produced by it. London, Edinburgh, *Dublin Philosophical Magazine and Journal of Science*. 47: 109–163. https://doi.org/10.1080/14786449908621245

Sajih M, Bryan ND, Livens FR, Vaughan DJ, Descostes M, Phrommavanh V, Nos J, and Morris K. (2014). Adsorption of radium and barium on goethite and ferrihydrite: A kinetic and surface complexation modelling study. *Geochimica et Cosmochimica Acta*. 146: 150–163. https://doi.org/10.1016/j.gca.2014.10.008

Scanlon BR, Reedy RC, and Nicot JP. (2014). Will water scarcity in semiarid regions limit hydraulic fracturing of shale plays? *Environmental Research Letters*. 9. https://doi.org/10.1088/1748-9326/9/12/124011

Schaller J, Headley T, Prigent S, and Breuer R. (2014). Potential mining of lithium, beryllium and strontium from oilfield wastewater after enrichment in constructed wetlands and ponds. *Science of the Total Environment*. 493: 910–913. https://doi.org/10.1016/j.scitotenv.2014.06.097

Sedlacko EM, Jahn CE, Heuberger AL, Sindt NM, Miller HM, Borch T, Blaine AC, Cath TY, and Higgins CP. (2019). Potential for beneficial reuse of oil-and-gas-derived produced water in agriculture: Physiological and morphological responses in spring wheat (*Triticum aestivum*). *Environmental Toxicology and Chemistry*. 4449. https://doi.org/10.1002/etc.4449

Sekhran N, Sheahan B, and Sullivan B. (2017). *Mapping the oil and gas industry to the SDGs: An Atlas*. New York.

Shao H, Kulik DA, Berner U, Kosakowski G, and Kolditz O. (2009). Modeling the competition between solid solution formation and cation exchange on the retardation of aqueous radium in an idealized bentonite column. *Geochemical Journal*. 43: e37-e42. https://doi.org/10.2343/geochemj.1.0069

Soddy F. (1913). Intra-atomic Charge. *Nature*. 92: 399–400.

Stallworth A.M., Chase, E.H., Burgos, W.D., Warner, N.R. (2019). *Laboratory Method to Assess Efficacy of Dust Supressants for Dirt and Gravel Roads, in: Transportation Research Record*. Transportation Research Board.

Stewart BW, Chapman EC, Capo RC, Johnson JD, Graney JR, Kirby CS, and Schroeder KT. (2015). Origin of brines, salts and carbonate from shales of the Marcellus Formation: Evidence from geochemical and Sr isotope study of sequentially extracted fluids. *Applied Geochemistry*. 60: 78–88. https://doi.org/10.1016/j.apgeochem.2015.01.004

Swain B. (2017). Recovery and recycling of lithium: A review. *Separation and Purification Technology*. 172: 388–403. https://doi.org/10.1016/j.seppur.2016.08.031

Tasker TL, Burgos WD, Ajemigbitse MA, Lauer NE, Gusa AV, Kuatbek M, May D, Landis JD, Alessi DS, Johnsen AM, Kaste JM, Headrick KL, Wilke FDH, McNeal M, Engle M, Jubb AM, Vidic RD, Vengosh A, and Warner NR. (2019). Accuracy of methods for reporting inorganic element concentrations and radioactivity in oil and gas wastewaters from the Appalachian Basin, U.S. based on an inter-laboratory comparison. *Environmental Science Processes & Impacts*. 21: 224–241. https://doi.org/10.1039/C8EM00359A

Tasker TL, Burgos WD, Piotrowski P, Castillo-Meza L, Blewett TA, Ganow KB, Stallworth A, Delompré PLM, Goss GG, Fowler LB, Vanden Heuvel JP, Dorman F, and Warner NR. (2018). Environmental and human health impacts of spreading oil and gas wastewater on roads. *Environmental Science & Technology.* 52: 7081–7091. https://doi.org/10.1021/acs.est.8b00716

Tesoriero AJ and Pankow JF. (1996). Solid solution partitioning of Sr2+, Ba2+, and Cd2+ to calcite. *Geochimica et Cosmochimica Acta.* 60: 1053–1063. https://doi.org/10.1016/0016-7037(95)00449-1

US Energy Information Administration (US EIA). (2015). Top 100 U.S. Oil and Gas Fields. www.eia.gov/naturalgas/crudeoilreserves/top100/pdf/top100.pdf

US EIA. (2018). Annual energy outlook. www.eia.gov/outlooks/aeo/section_issues.php

US EIA. (2019). Appalachia Region Drilling Productivity Report. www.eia.gov/petroleum/drilling/pdf/appalachia.pdf

US Environmental Protection Agency (US EPA). (1976). *National Interim Primary Drinking Water Regulations.* Washington, DC.

US EPA. (2005). A System's Guide to the Management of Radioactive Residuals from Drinking Water. Epa 816-R-05-004.

US EPA. (2018a). *Detailed Study of the Centralized Waste Treatment Point Source Category for Facilities Managing Oil and Gas Extraction Wastes.* Washington, DC.

US EPA. (2018b). *Study of Oil and Gas Extraction Wastewater Management.* www.epa.gov/eg/study-oil.428"/>-and-gas-extraction-wastewater-management

US EPA (2019). *Study of Oil and Gas Extraction Wastewater Management Under the Clean Water Act EPA-821-R19–001.* Washington, DC.

United Nations. (2011). *Sources and Effects of Ionizing Radiation United Nations Scientific Committee on the Effects of Atomic Radiation.* New York.

US Nuclear Regulatory Commission. (2017). Appendix B to Part 20 Annual Limits on Intake (ALIs) and Derived Air Concentrations (DACs) of Radionuclides for Occupational Exposure; Effluent Concentrations; Concentrations for Release to Sewerage.

Van Sice K, Cravotta CA, McDevitt B, Tasker TL, Landis JD, Puhr J, and Warner NR. (2018). Radium attenuation and mobilization in stream sediments following oil and gas wastewater disposal in western Pennsylvania. *Applied Geochemistry.* 98: 393–403. https://doi.org/10.1016/j.apgeochem.2018.10.011

Veil J. (2015). U.S. Produced Water Volumes and Management Practices in 2012.

Vengosh A, Jackson RB, Warner N, Darrah TH, and Kondash A. (2014a). A critical review of the risks to water resources from unconventional shale gas development and hydraulic fracturing in the United States. *Environmental Science & Technology.* 48: 8334–8348. https://doi.org/10.1021/es405118y

Vengosh A, Jackson RB, Warner N, Darrah TH, and Kondash A. (2014b). A Critical Review of the Risks to Water Resources from Unconventional Shale Gas Development and Hydraulic Fracturing in the United States. *Environmental Science & Technology.* 48: 8334–8348. https://doi.org/10.1021/es405118y

Vinograd VL, Brandt F, Rozov K, Klinkenberg M, Refson K, Winkler B, and Bosbach D. (2013). Solid-aqueous equilibrium in the BaSO4-RaSO4-H2O system: First-principles calculations and a thermodynamic assessment. *Geochimica et Cosmochimica Acta.* 122: 398–417. https://doi.org/10.1016/j.gca.2013.08.028

Vinograd VL, Kulik DA, Brandt F, Klinkenberg M, Weber J, Winkler B, and Bosbach D. (2018a). Thermodynamics of the solid solution – Aqueous solution system (Ba,Sr,Ra)SO4+ H2O: I. The effect of strontium content on radium uptake by barite. *Applied Geochemistry.* 89: 59–74. https://doi.org/10.1016/j.apgeochem.2017.11.009

Vinograd VL, Kulik DA, Brandt F, Klinkenberg M, Weber J, Winkler B, and Bosbach D. (2018b). Thermodynamics of the solid solution – Aqueous solution system (Ba,Sr,Ra)SO4+ H2O: II. Radium retention in barite-type minerals at elevated temperatures.

Applied Geochemistry. 93: 190–208. https://doi.org/10.1016/j.apgeochem.2017.10.019

Vinson DS, Lundy JR, Dwyer GS, and Vengosh A. (2012). Implications of carbonate-like geochemical signatures in a sandstone aquifer: Radium and strontium isotopes in the Cambrian Jordan aquifer (Minnesota, USA). *Chemical Geology.* 334: 280–294. https://doi.org/10.1016/j.chemgeo.2012.10.030

Wang Y, Tavakkoli S, Khanna V, Vidic RD, and Gilbertson LM. (2018). Life cycle impact and benefit trade-offs of a produced water and abandoned mine drainage cotreatment process. *Environmental Science & Technology.* 52: 13995–14005. https://doi.org/10.1021/acs.est.8b03773

Warner NR, Christie CA, Jackson RB, and Vengosh A. (2013a). Impacts of shale gas wastewater disposal on water quality in Western Pennsylvania. *Environmental Science & Technology.* 47: 11849–11857. https://doi.org/10.1021/es402165b

Warner NR, Christie CA, Jackson RB, and Vengosh A. (2013b). Impacts of shale gas wastewater disposal on water quality in Western Pennsylvania. *Environmental Science & Technology.* 47: 11849–11857. https://doi.org/10.1021/es402165b

Webster IT, Hancock GJ, and Murray AS. (1995). Modelling the effect of salinity on radium desorption from sediments. *Geochimica et Cosmochimica Acta.* 59: 2469–2476. https://doi.org/10.1016/0016-7037(95)00141-7

Wiegand JW and Sebastian F. (2002). Origin of radium in high-mineralised waters. Technol. Enhanc. Nat. Radiat. (TENR II) - Proc. an Int. Symp. Held Rio Janeiro; IAEA - TECDOC - 1271 107–111.

Wilson JM and Vanbriesen JM. (2012). Oil and gas produced water management and surface Pennsylvania. *Environmental Practice.* 14: 288–301.

Zhang T, Gregory K, Hammack RW, and Vidic RD. (2014) Co-precipitation of radium with barium and strontium sulfate and its impact on the fate of radium during treatment of produced water from unconventional gas extraction. *Environmental Science & Technology.* 48: 4596–4603. https://doi.org/10.1021/es405168b

Zhang T, Bain D, Hammack R, and Vidic RD. (2015a). Analysis of Radium-226 in High Salinity Wastewater from Unconventional Gas Extraction by Inductively Coupled Plasma-Mass Spectrometry. *Environmental Science & Technology.* 49: 2969–2976. https://doi.org/10.1021/es504656q

Zhang T, Hammack RW, and Vidic RD. (2015b). Fate of radium in Marcellus Shale flowback water impoundments and assessment of associated health risks. *Environmental Science & Technology.* 49: 9347–9354. https://doi.org/10.1021/acs.est.5b01393

10

Metal Isotope Signatures as Tracers for Unconventional Oil and Gas Fluids

BRIAN W. STEWART AND ROSEMARY C. CAPO

10.1 Introduction

Both conventional and unconventional oil and gas extraction can result in the release of salts, metals, methane, organic compounds, and NORM into air and water resources. Drilling, hydraulic fracturing, and hydrocarbon production generate waste fluids related to well construction and completion (flowback water) and production (produced water). A number of tracers have been deployed to identify the source and potential environmental impact of fugitive methane and organic and radioactive contaminants associated with flowback and produced water and drilling mud (e.g., Vengosh et al. 2014; Phan et al. 2015; McIntosh et al. 2019 and references therein). Subsurface migration of fluid with high total dissolved solids (TDS) into adjacent or overlying aquifers is also of particular concern (Soeder et al. 2014). Produced waters with TDS values as high as 345,000 mg/L have been reported from unconventional wells (Barbot et al. 2013), consisting of inorganic constituents such as sodium (Na), calcium (Ca), chloride (Cl), bromine (Br), and other dissolved metals including barium (Ba), strontium (Sr), and lithium (Li) (Hayes 2009; Chapman et al. 2012; Haluszczak et al. 2013).

Identification of contaminants directly related to onshore hydraulic fracturing can be complicated if their background levels are already high. For example, the Appalachian Plateau of the northeastern United States is a region with sensitive ecosystems and populations dependent on both aquifers and rivers for drinking water and for agriculture and fishing. Unconventional shale gas and natural gas liquids are being produced in the region amid active and legacy conventional oil and gas drilling, coal mining, and industrial activities. Extensive baseline data are critical to the early, correct identification of the source of a pollution event, and sensitive tracers are needed that can accurately discriminate between contaminants even when dilute (Brantley et al. 2014; Soeder et al. 2014).

The natural isotopic composition of metals in surface and groundwater, precipitation, soil and rock has been used for decades to identify the sources of inorganic contaminants and understand the processes involved in their mobility in the environment. In addition to complementing traditional stable isotope systems (e.g., H, C, O, N, S), metal isotope tracers can be used to fingerprint contaminant endmembers, and aid in the identification of brine sources and quantification of mixing of saline waters with shallower groundwater. They have also been applied to better understand processes such as exchange reactions, evaporation, redox changes, and mineral dissolution and precipitation that can result in the release or retention of contaminants, or adversely affect wells during oil and gas extraction. In this paper we will focus on three isotope systems, Sr, Li, and Ba, which have been applied to studies related to unconventional oil and gas extraction. More comprehensive discussions of other metal and nontraditional isotope systems can be found in reviews by Johnson et al. (2004), Weiss et al. (2008), Bullen and Eisenhauer (2009), Porcelli and Baskaran (2012), Bullen (2014), and Wiederhold (2015). For a discussion of isotopes in low molecular weight hydrocarbons, see Chapter 11 in this volume.

10.2 Background

Isotopes of an element have the same number of protons in the nucleus but differ in the number of neutrons. While differing numbers of neutrons have only a minor effect on the geochemical behavior of an element, the isotopic composition of an element (the relative abundances of different isotopes) dissolved in a fluid sample such as groundwater, river water, or oil and gas brines can be indicative of the sources of those elements, and the processes involved in fluid mixing and water–rock interactions. Variations in isotopic composition can be caused by radioactive decay, by mass-dependent fractionation processes, or, in some cases, by mass-independent processes. Recent advances in measurement techniques, including analysis by multi-collector inductively coupled plasma mass spectrometers (MC-ICPMS), have opened application of a multitude of isotope systems to geological and environmental processes.

10.2.1 Isotope Fractionation

Although isotopes of an element are chemically similar, a variety of physical processes can separate isotopes of an element owing to differences in mass, and resultant vibrational energy; this is referred to as mass-dependent fractionation. The vibrational energy of an atom can affect its complexation and bonding environment, as well as its rates of diffusion and ion exchange. Typically, light elements are affected more strongly than heavy elements as a result of the greater relative differences in mass between the isotopes (e.g., ~16% for ^{6}L and ^{7}Li *vs.* ~1% for ^{87}Sr and ^{86}Sr). Processes that can fractionate isotopes include mineral precipitation, dissolution, diffusion, changes in aqueous complexation, evaporation, and biological processes that directly involve an isotope system. As a rule, mass-dependent isotope fractionation is more pronounced at low temperatures than at high temperatures, making isotope fractionation a useful probe for processes occurring at or near the Earth's surface, and a potential temperature probe. Although measurable isotope fractionation can result from effects other than mass differences, such as photochemical reactions in the upper atmosphere (e.g., S isotopes; Eiler et al. 2014; Ono 2017) or nuclear volume effects (e.g., Hg, Tl, U isotopes; Schauble 2007), this type of mass-independent fractionation is not significant for the three metal isotope systems discussed here (Wiederhold 2015).

10.2.2 Radiogenic Isotopes

Some metal isotopes originate from the radioactive decay of an unstable parent isotope. For example, Sr has four naturally occurring stable isotopes (atomic mass 84, 86, 87, 88). Variations in the relative proportions of Sr isotopes in nature occur because an isotope of rubidium, ^{87}Rb, decays (with a half-life of 49.6 billion years; Rotenberg et al. 2012) to produce a stable (nonradioactive) daughter isotope of strontium, ^{87}Sr. This isotope is referred to as *radiogenic*. Strontium isotope compositions are generally expressed as the ratio of ^{87}Sr relative to ^{86}Sr. Because the mass-dependent fractionation of Sr from natural or analytical causes is corrected for by normalization to known values for the nonradiogenic isotopes of Sr, the ^{87}Sr/^{86}Sr ratio reflects only the addition of radiogenic ^{87}Sr to the sample, which is a function of its geologic history. Therefore the ^{87}Sr/^{86}Sr ratio of a water sample reflects the provenance of the Sr (e.g., water-rock interactions, exchange reactions on clay, fluid mixing, etc.; Capo et al. 1998). Other radiogenic isotope systems commonly applied to geological problems include the parent–daughter pairs ^{147}Sm-^{144}Nd, ^{176}Lu-^{176}Hf, and ^{187}Re-^{187}Os, and U/Th-Pb (see Banner 2004; Faure and Mensing 2005). Because the parent and daughter elements of these systems are typically present at very low levels in brines and formation waters, these isotope systems have not been extensively applied as tracers for produced waters. The disequilibrium daughter products of the U-Th decay chain (e.g., ^{226}Ra, ^{228}Ra) are outside the scope of this paper but have also been applied to unconventional shale gas systems. See Chapter 9 in this volume and papers by Rowan et al. (2011; 2015), Warner et al. (2012; 2013), Skalak et al. (2014), Lauer et al. (2016; 2018), and Tasker et al. (2020).

10.2.3 Nomenclature

While isotope compositions can be reported directly as isotope ratios, differences between labs and instruments make interlaboratory comparisons of direct isotope ratios difficult. Therefore, isotope compositions are often reported as isotope ratios normalized to a reference standard or as values relative to a reference standard (δ notation); for isotopes a and b of element X, where a is (by convention) the heavier or the radiogenic isotope,

$$\delta^a X(\text{\textperthousand}) = 10^3 \left[\frac{\left(^a X / ^b X\right)_{sample}}{\left(^a X / ^b X\right)_{standard}} - 1 \right]$$

Positive values indicate enrichment of the heavy isotope relative to the standard, which by definition has a δX value of 0, and negative values indicate enrichment in the light isotope. Values of δ are generally expressed as parts per thousand, or per mil (‰). For some isotope systems (e.g., Sr and Nd), even smaller variations in isotope ratios are measurable and significant. In these cases, a 10^4 multiplication factor is sometimes used and isotope ratios are expressed as ε values (DePaolo and Wasserburg 1976). For example, in low-temperature and environmental isotope studies, the $^{87}Sr/^{86}Sr$ ratio can be normalized to that of seawater (which is globally uniform) and expressed as

$$\varepsilon_{Sr} = 10^4 \left[\frac{(^{87}Sr/^{86}Sr)_{sample}}{(^{87}Sr/^{86}Sr)_{seawater}} - 1 \right]$$

(Andersson et al. 1992). Typical analytical uncertainty for the $^{87}Sr/^{86}Sr$ ratio measured by MC-ICPMS or TIMS is <0.00001 (2σ); this translates to an ε_{Sr} of <0.14. In the isotope mixing equations discussed in Section 10.3, isotope ratios, δ values, and ε values can all be used to express the isotope ratio terms.

10.3 Framework for Using Geochemical and Isotopic Tracking Tools

In order to use isotopes or element concentrations for tracking leakage of produced water, several conditions must be met: (1) The uncontaminated water and contaminant must have distinct chemical/isotopic signatures; (2) the element or isotope in the contaminating fluid must be at high enough levels to measurably affect the uncontaminated fluid; and (3) the element concentration or isotope ratio should not be affected by other processes during or after mixing. Contamination of a fresh surface water or ground water by produced water can be conceptualized as a mixing problem, where the dissolved constituents of the freshwater are altered by addition of a contaminant fluid. For a flowing stream or moving ground water system, quantities are expressed as fluxes; thus, the concentration of an element in a flowing water system subject to contamination from produced water depends on the flux of water in the freshwater system (J_{FW}), the flux of produced water leaking into the system (J_{PW}), and the concentration of the chemical element in each (C^i_{FW}, C^i_{PW}). The concentration in the contaminated water (C^i_{CW}) is given by

$$C^i_{CW} = \frac{J_{FW} C^i_{FW} + J_{PW} C^i_{PW}}{J_{FW} + J_{PW}} \tag{10.1}$$

The use of Eq. (10.1) can be illustrated in the case of a shallow groundwater subject to possible contamination from multiple produced water sources. An example study site is provided in southwestern Pennsylvania (Greene County), where hydraulically fractured deep Middle Devonian Marcellus Shale gas wells are located in close proximity to overlying conventional Upper Devonian gas and oil wells, and wellbores from both of these intercept shallow ground water (Hammack et al. 2013). We use data from this site reported by Kolesar Kohl et al. (2014) to illustrate the use of mixing equations.

Using concentrations of major cations (Na, Ca, Mg, Sr, Ba) as tracers for potential contamination of shallow groundwater from the two potential sources of produced water (Marcellus Shale wells and Upper Devonian conventional wells), a series of mixing curves can be generated for each source (Figure 10.1). Here, the element concentration of the mixture is normalized to the groundwater concentration to show the relative enrichment of each element; the greater the difference in concentration, the more sensitive the potential tracer can be. For all cases shown here, the elements are greatly enriched in the produced water relative to uncontaminated ground water, and addition of produced water causes a rapid change in concentration. An upward shift in the total dissolved solid (TDS) load of the groundwater would be a clear indication of contamination from a produced water source, either through a leaky wellbore or from infiltration from a surface source such as a storage pond, drilling pad spill, or from application of produced water brine for road dust control (e.g., Soeder et al. 2014; Tasker et al. 2018).

In order to determine which produced water source (in this case unconventional vs. conventional) is causing the contamination, it is necessary to look in detail at the chemistry of the contaminated water for signatures of a particular source. For example, analysis of Figure 10.1 suggests that rapid relative increases in Ba or Sr relative to Na could be indicative of a Marcellus produced water source as opposed to conventional Upper Devonian produced waters. For this reason, it can be advantageous to use element ratios

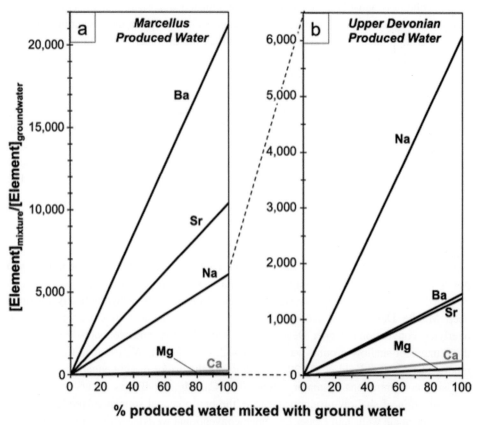

Figure 10.1 Simple major element mixing curves for the cases of mixing produced water from (a) hydraulically fractured Marcellus Shale gas wells and (b) Upper Devonian conventional wells with adjacent shallow ground water. The enrichment of each element relative to the uncontaminated ground water is plotted as a function of the mixing ratio. All data are from a site in Greene County, Pennsylvania, reported by Kolesar Kohl et al. (2014)

rather than elemental concentrations to identify specific sources of contamination (Engle and Rowan 2013). Element ratios can be incorporated into a mixing equation as follows,

$$R_{CW}^{ij} = \frac{J_{FW} C_{FW}^{j} R_{FW}^{ij} + J_{PW} C_{PW}^{j} R_{PW}^{ij}}{J_{FW} C_{FW}^{j} + J_{PW} C_{PW}^{j}} \tag{10.2}$$

where R_{CW}^{ij} is the ratio of elements i and j (C_{CW}^{i}/C_{CW}^{j}) in the contaminated water and other terms and subscripts are the same as in Eq. (10.1). We note that either mass or molar concentrations can be used, but it is critical for authors to specify which when presenting mixing models, to allow comparison of datasets. The covariance of Ca/Sr with Ca/Mg has been suggested as optimal major element ratios that can distinguish between Marcellus and Upper Devonian produced water in the Appalachian Basin (Tisherman and Bain 2019). Applying these ratios to the Greene County dataset, a clear separation between values for the Marcellus Shale produced water, Upper Devonian produced water, and groundwater from the site is observed (Figure 10.2a). All three sets of data fall along the same linear trend (Figure 10.2a), which would seem to preclude their use for distinguishing between the two potential contamination sources. However, because the absolute concentrations of Ca, Sr, and Mg vary tremendously among the different endmembers, mixing lines show significant curvature (Figure 10.2b). This shows that discrimination between the two contaminants could be identified, if at least 0.1% of either Marcellus or Upper Devonian produced water is present in the groundwater.

Johnson et al. (2015) suggested that the ratio (Ba+Sr)/Mg could sensitively distinguish Marcellus produced water from other high TDS sources, including road salt; examination of Figure 10.1 suggests that this could also be used to differentiate Marcellus and Upper Devonian produced waters. A potential complicating factor for using element concentrations and element ratios is the potential for nonconservative behavior during or after a contamination event. For example, Ba introduced by contamination could precipitate out of groundwater as the mineral barite ($BaSO_4$) if sulfate is present (which is common in acid mine drainage-impacted waters of the Appalachian Basin). Precipitation of carbonate minerals can fractionate Ca from other alkaline earth elements. This problem can potentially be avoided by using isotope systems that are not subject to mass fractionation as tracers for produced water contamination (Stewart et al. 1998). For an isotope ratio δ^i of element i, the mixing equation can be rewritten as follows,

$$\delta_{CW}^{i} = \frac{J_{FW} C_{FW}^{i} \delta_{FW}^{i} + J_{PW} C_{PW}^{i} \delta_{PW}^{i}}{J_{FW} C_{FW}^{i} + J_{PW} C_{PW}^{i}} \tag{10.3}$$

Other terms and subscripts are the same as in Eq. (10.1) and (10.2). From this equation, the flux ratio of the contaminating fluid to the flowing surface or groundwater can be calculated as,

$$\frac{J_{PW}}{J_{FW}} = \frac{C_{FW}^{i} (\delta_{CW}^{i} - \delta_{FW}^{i})}{C_{PW}^{i} (\delta_{PW}^{i} - \delta_{CW}^{i})} \tag{10.4}$$

This corrects an error in the transcription of eq. 2 of Chapman et al. (2012), although the correct form of the equation was used for all calculations and figures in that report.

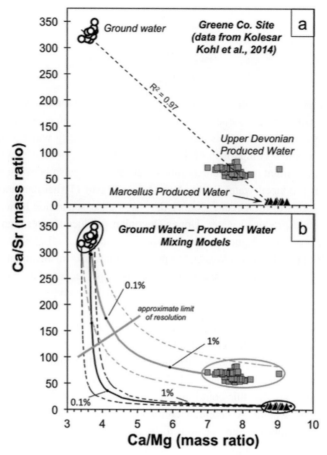

Figure 10.2 Ca/Sr vs. Ca/Mg of groundwater and conventional (Upper Devonian) and unconventional (Marcellus) produced water from a single site in Greene County, Pennsylvania. (a) The three data clusters define a linear trend. (b) Mixing curves between the groundwater and each of the produced water types. Labeled percentages on the curve are the % produced water needed to move the mixture to the indicated point on the curve. The limit of resolution is the approximate zone above which the two produced waters cannot be differentiated from each other in the mixture.
Data are from Kolesar Kohl et al. (2014)

Inspection of Eq. (10.4) shows that the flux ratio of the contaminant to uncontaminated water does not require knowledge of the concentration of element i in the contaminated fluid. Thus, even if an unknown fraction of element i is removed from the fluid by precipitation or adsorption after (or downstream from) contamination, the extent of contamination can still be quantified, as long as the endmember concentrations are known and the isotope ratio can be measured. In addition, this allows detection of past contamination events that might be preserved in proxy records such as freshwater mussel shells (e.g., Geeza et al. 2018).

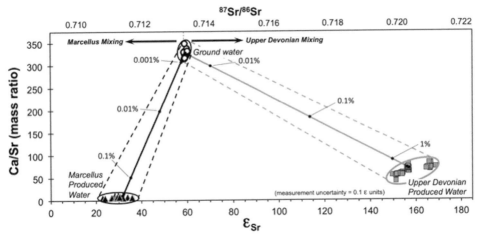

Figure 10.3 Strontium isotope mixing diagram for groundwater contamination with either Marcellus or Upper Devonian produced water for the Greene County, Pennsylvania site. The labeled percentages on the mixing lines indicate the % of each produced water needed to shift the mixture to the indicated point. Because the isotopic compositions of the two produced water sources fall on opposite sides of that of the groundwater, the direction of mixing (black arrows) allows this method to effectively discriminate between sources.
Data are from Kolesar Kohl et al. (2014)

The utility of applying an isotope system in the case of produced water contamination depends on the difference between the isotope composition of the endmembers, the relative isotopic variability within each of the endmembers, and the concentration of the element in the contaminant and freshwater source. In general, the isotope tracer will be more sensitive if the produced water contaminant is enriched in the isotope element relative to the uncontaminated water. Returning to the example of the Greene County site, Kolesar Kohl et al. (2014) measured strontium isotope compositions of the endmembers and found that there was a significant difference between the isotopic composition (expressed as $^{87}Sr/^{86}Sr$ or ε_{Sr}) of the groundwater and each of the potential contaminants (Marcellus produce water and Upper Devonian produced water). Moreover, the very high concentrations of Sr in the produced waters make this system extremely sensitive to small amounts of contamination. A mixing diagram is shown in Figure 10.3, in which the element ratio Ca/Sr is combined with the Sr isotope ratio. As indicated, contamination from the Upper Devonian produced waters would increase the $^{87}Sr/^{86}Sr$ of the groundwater, while contamination from Marcellus Shale produced waters would decrease it. Because both sources are extremely high in Sr, very small amounts of contamination (~0.001%) are detectable, and the source of contamination can be determined at levels of <0.01%. In this case, the isotope ratio is an order of magnitude more sensitive than the element ratios presented in Figure 10.2 in determining the source of contamination. Because the concentration of Sr is so high in the potential produced water contaminants, and the shifts in Sr isotope ratio of the two contaminants move in opposite directions, the $^{87}Sr/^{86}Sr$ becomes more sensitive than even those elements showing extreme contrast with the uncontaminated groundwater

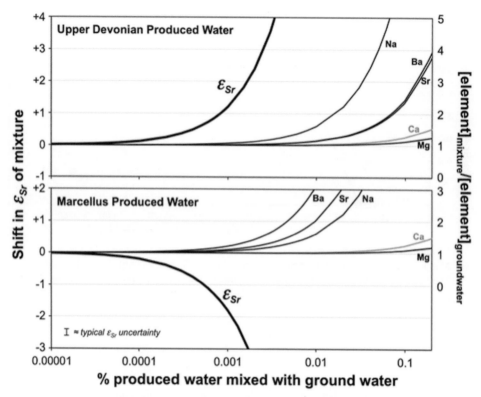

Figure 10.4 Comparison of the sensitivity of individual element concentrations and Sr isotope ratios to mixing of groundwater with Upper Devonian and Marcellus produced water at the Greene County, Pennsylvania site. Major element concentrations in the groundwater increase significantly with contamination from either produced water source, and could be detectable with, for example, a TDS meter. The Sr isotope data provide information about which produced water source is the contaminant, even at very low levels of contamination (relative to measurement uncertainty).
Data are from Kolesar Kohl et al. (2014)

such as Na and Ba (Figure 10.4), while still easily delineating between the two produced water endmembers.

10.4 Application of Isotopes as Tracers of Produced Water Contamination and Origin

10.4.1 Sr Isotopes

As discussed previously, Sr isotopes can be a sensitive tool to identify and quantify produced water contamination of freshwater resources. The Sr isotope ratio has long been used to discriminate between oilfield brines and yield information about the chemical evolution of saline waters associated with oil- and gas-producing rock formations (Chaudhuri and Clauer 1993). It can also distinguish between groundwaters from different aquifers, which, in an area of hydrocarbon extraction, could be indicative of issues with wellbore construction or incomplete aquifer isolation. Naftz et al. (1997) used Sr isotope compositions to show that the salinity increase in the freshwater Navajo Aquifer near an oil field in Utah was probably caused by mixing with water from an adjacent saline aquifer rather than from oilfield produced waters or re-injected fluids. In an area of coalbed methane extraction in the Powder River Basin, Wyoming, $^{87}Sr/^{86}Sr$ ratios in coal-bed groundwaters and associated sandstone aquifers reflect residence time, interaction with radiogenic Sr in coal, and structural features such as faults, and were used in some areas to discriminate between sources and identify communication between aquifers (Frost et al. 2002; Campbell et al. 2008). Strontium isotopes have also been used to track coalbed methane produced water discharged to the surface, where it can increase soil salinization (Brinck and Frost 2007).

10.4.1.1 Sr Isotope Applications to Hydraulic Fracturing in the Appalachian Basin

The widespread use of hydraulic fracturing beginning in the mid-2000s in the Appalachian Basin (Zagorski et al. 2012), a region already impacted by more than a century of hydrocarbon extraction and coal mining, necessitated expanded use of tracers to identify the source(s) of spills, leaks, and unauthorized disposal of flowback and produced water in the numerous surface and groundwater systems. Chapman et al. (2012) showed that Sr isotopes could effectively differentiate between produced water from the hydraulically fractured Marcellus Shale and that from overlying conventional Upper Devonian to Lower Mississippian oil and gas wells in many parts of the basin, as well as from AMD, another source of surface and groundwater contamination (Figure 10.5). Warner et al. (2012) used $^{87}Sr/^{86}Sr$ ratios along with isotopes of H, O, and Ra to suggest that deep Marcellus brines were hydraulically connected to shallower groundwater aquifers in the northeastern part of the Marcellus Shale gas play. A detailed Sr isotope study of a single site containing both hydraulically fractured Marcellus Shale gas wells and conventional Upper Devonian/Lower Mississippian wells in southwestern Pennsylvania (Kolesar Kohl et al. 2014) showed no evidence of fluid migration between reservoirs or from either reservoir to shallow groundwater over about two years of hydrocarbon extraction and water sampling. Strontium isotopes would be a sensitive tracer for long-term monitoring of the overlying groundwater systems over the lifetime of the hydraulically fractured wells, particularly to detect corrosion and leaking from the well casing. The geographic pattern of Marcellus Shale and Upper Devonian produced water $^{87}Sr/^{86}Sr$ values (Chapman et al. 2012; Osborn et al. 2012; Warner et al. 2012; Capo et al. 2014; Kolesar Kohl et al. 2014) suggests that these units have remained hydrologically separated from each other, potentially for millions of years, in the central to southwestern portions of the Appalachian Basin. Recent Sr isotope data from produced waters of the underlying Ordovician hydraulically fractured Utica/Pt. Pleasant Shale (Tasker et al. 2020) yielded $^{87}Sr/^{86}Sr$ ratios very similar to those of Marcellus produced water, indicating a possible shared formation water source or history.

Increased wastewater production associated with unconventional hydrocarbon production can affect the effluent discharged from wastewater treatment plants into surface waters.

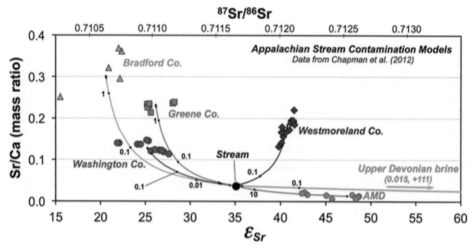

Figure 10.5 Mixing diagram for hypothetical Appalachian Basin stream water contaminated by either Marcellus Shale produced water (values shown here for four different counties in Pennsylvania), Upper Devonian conventional well produced water (endmember values far off scale), or acid mine drainage from SW Pennsylvania. The numbers indicated on the mixing curves are the percent of contaminant added to the stream water. Note that even in this worst-case scenario where the stream isotope composition falls wholly within the Marcellus field, it is still possible to differentiate among contamination sources. (A black and white version of this figure will appear in some formats. For the colour version, refer to the plate section.)
Data are from Chapman et al. (2012)

Burgos et al. (2017) concluded from $^{87}Sr/^{86}Sr$ and $^{226}Ra/^{228}Ra$ ratios that elevated Sr and Ra in water entering Conemaugh River Lake in western Pennsylvania probably originated from upstream water treatment plants that processed wastewater from Marcellus Shale gas production. Geeza et al. (2018) used $^{87}Sr/^{86}Sr$ ratios measured in calcium carbonate shells of freshwater mussels in Marcellus Shale-impacted stream waters to demonstrate that a secular isotopic record of Marcellus wastewater disposal can be preserved. Other biogenic precipitates (e.g., fish otoliths; Brandt et al. 2018) could potentially be used as records of oil and gas inputs to freshwater systems.

Because hydraulic fracturing involves injection of large volumes of water into the target formation, the origin of dissolved solids in the returning water has been an important research question. Before well operators recognized the benefit of using flowback and produced water to hydraulically fracture new wells, freshwater was typically used. The resulting flowback had a characteristic pattern of decreasing flow rate and increasing TDS concentration over the first few days, typically reaching a plateau after 1–2 months of production (Hayes 2009). Strontium concentrations and isotope ratios followed a similar trend (Capo et al. 2014), although in some cases the $^{87}Sr/^{86}Sr$ continued to increase slightly even after the first year (Figure 10.6). Capo et al. (2014) suggested that the late-stage produced water represented formation water within or adjacent to the Marcellus Shale that slowly migrated into the well via hydraulic fractures. This was supported by stable isotope

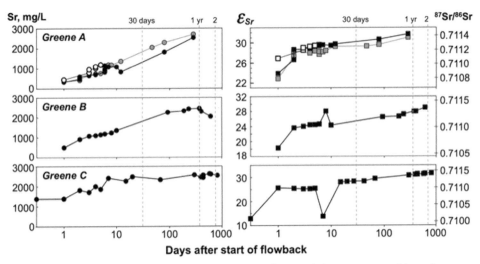

Figure 10.6 Time series data for Sr concentrations and isotope compositions from hydraulically fractured Marcellus Shale gas wells from three locations in Greene County, southwestern Pennsylvania.
Data are from Chapman et al. (2012) and Capo et al. (2014)

studies of Rowan et al. (2015), and geochemical and Sr isotopic investigations of leached Marcellus Shale cuttings and core material (Stewart et al. 2015). The initial low $^{87}Sr/^{86}Sr$ ratios of flowback water (Figure 10.6) could be caused by dissolution of low-$^{87}Sr/^{86}Sr$ concrete from the wellbore (Stewart et al. 2015) or calcium carbonate minerals from within the Marcellus Shale (Phan et al. 2020) during injection of HCl in the early stages of hydraulic fracturing. On the other hand, Balashov et al. (2015) suggested that Sr in Marcellus Shale flowback waters is derived from a combination of interaction with radiogenic Sr in clay minerals in the shale and Silurian brine trapped within the shale. In either case, the $^{87}Sr/^{86}Sr$ of the produced water after the initial flowback phase appears to be controlled primarily by preexisting formation fluids or by interaction with Sr in the rock units, so the long-term value of the produced water Sr isotope ratio is independent of the frac fluid source (i.e., freshwater vs. recycled flowback water).

10.4.1.2 Other Sr Isotope Applications to Unconventional Oil and Gas Well Waters

Since the early work on Sr isotopes in hydraulically fractured Appalachian Basin produced water (Chapman et al. 2012; Warner et al. 2012), Sr isotopes have been increasingly applied to track surface and groundwater contamination in unconventional oil and gas fields around the world. The oil-producing Bakken Shale within the Williston Basin straddles Saskatchewan, Montana, and North Dakota, and is one of the largest oil resources in the United States (Pollastro et al. 2010). Although vertical wells were drilled in the 1950s, oil production has increased rapidly with the introduction of unconventional drilling techniques. This has raised concerns about over-extraction of freshwater for hydraulic fracturing, potential migration of drilling-related or saline fluids into shallower aquifers, and spills that result in ground and surface water contamination (Shrestha et al. 2017 and references therein).

Strontium isotopes were used to determine that hydrocarbon-related brines were the source of high TDS contamination at three study sites within the Prairie Pothole region that overlaps the Williston Basin (Preston et al. 2014). Neymark et al. (2018) found that although groundwater $^{87}Sr/^{86}Sr$ values overlapped that of Bakken brines at some sites, the addition of nontraditional stable $\delta^{88}Sr$ values allowed discrimination between surface and groundwaters and produced waters. This was integrated with $^{234}U/^{238}U$ ratios in order to better discriminate between different uncontaminated groundwater end members and to detect potential pollution from agricultural phosphates.

Strontium can also be a tracer of water–rock interactions that can occur as hydraulic fracturing fluid or saline formation brines interact with shale minerals. In the Kangan gas field in Iran, Bagheri et al. (2014) concluded, based on Sr isotope signatures, that saline (TDS up to 60,000 mg/L) produced waters likely originated from within the gas reservoir and not from the underlying aquifer. In China, Zheng et al. (2017) report that Sr isotopes are sensitive tracers for contamination by hydraulic fracturing flowback fluids in the northern Qaidam Basin, particularly when combined with boron isotopes. Ni et al. (2018) distinguished between conventional and unconventional produced waters in the Sichuan Basin by integrating Sr, B, and Ra isotopic signatures. Huang et al. (2020) reported that using a combination of Sr and B isotope compositions allowed detection of shallow groundwater contamination by produced water from the Fuling gas field, even at levels of 0.05%.

10.4.2 Li Isotopes

Lithium, with two stable isotopes, ^6Li and ^7Li, is the lightest of the alkali metals, and commonly substitutes for Mg or Fe in clay minerals such as smectites (Chan et al. 1992; Huh et al. 1998). Lithium isotope variations are generally expressed using delta notation,

$$\delta^7 Li(‰) = 10^3 \left[\frac{(^7Li/^6Li)_{sample}}{(^7Li/^6Li)_{RM\ 8545}} - 1 \right]$$

where NIST RM 8545 is a standard reference material equivalent to the older (now depleted) NBS standard L-SVEC. The Li isotopic composition of produced waters from hydrocarbon-bearing units has been used to decipher the geochemical evolution of deep saline groundwaters (Chan et al. 2002; Millot et al. 2011). Some oil and gas brines contain as much as 100–500 mg/L Li (Collins 1978), and efforts are underway to recover lithium from produced waters from unconventional Marcellus and Utica wells (Bloomberg.com 2019; Eureka Resources 2019). Lithium isotopes have been used to determine that some brines, including hypersaline brines associated with Canadian Shield rocks, originated from evaporation of seawater (Bottomley et al. 2003). The Li isotopic compositions of deep saline groundwaters also reflect water–mineral interactions. Incorporation into clay minerals favors ^6Li (Chan et al. 1994; Zhang et al. 1998), leaving the remaining fluid enriched in ^7Li. Subsequent diagenetic alteration of clay minerals (e.g., smectite to illite) during burial diagenesis could release isotopically light Li into porewaters, resulting in a lower δ^7Li.

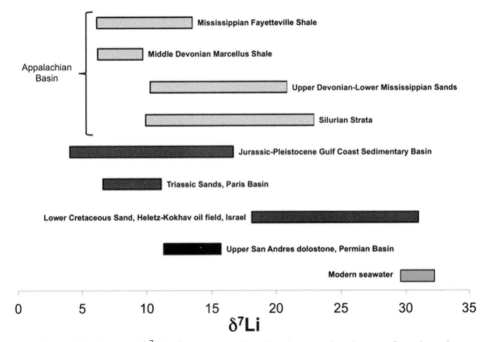

Figure 10.7 Range of δ⁷Li values measured in oil and gas produced waters from throughout the world (modified from Pfister et al. 2017).
Data are from Chan et al. (2002), Millot et al. (2011), Macpherson et al. (2014), Warner et al. (2014), Phan et al. (2016), and Pfister et al. (2017)

Measured δ^7Li values of gas- and oilfield brines from around the world range from +4.2‰ to +30.8‰, with the high value close to that of modern seawater (Figure 10.7). In the Appalachian Basin, there is a reasonable separation of Marcellus Shale from Upper Devonian produced water values, similar to that seen for ^{87}Sr/^{86}Sr (Macpherson et al. 2014; Warner et al. 2014; Phan et al. 2016). Because produced waters tend to be highly enriched in Li relative to stream or near-surface groundwaters, Li can be a sensitive tracer for freshwater contamination from brine. When combined with Sr isotope data, in some cases contamination from Marcellus Shale produced water can be distinguished from Upper Devonian produced water at levels of 0.001% (Figure 10.8). However, rivers in urban regions could be affected by other sources of Li, such as pharmaceuticals and Li battery leakage or waste. For example, Li in the Han River in South Korea is strongly affected as the river passes through the Seoul metropolitan region, with the concentration increasing by a factor of six and the isotopic composition decreasing by more than 10‰ (Choi et al. 2019). Lithium in reagents can be greatly enriched in ^{7}Li as a result of previous extraction of ^{6}Li used in nuclear devices (Qi et al. 1997; Tomascak 2004). Before using Li isotopes to track ground or surface waters near urban areas or waste disposal sites, the long-term (>12 month) baseline for Li concentration and isotope ratio must be established.

Warner et al. (2014) proposed that Li in unconventional shale gas produced water is released by desorption from exchangeable sites on clays caused by the injection of

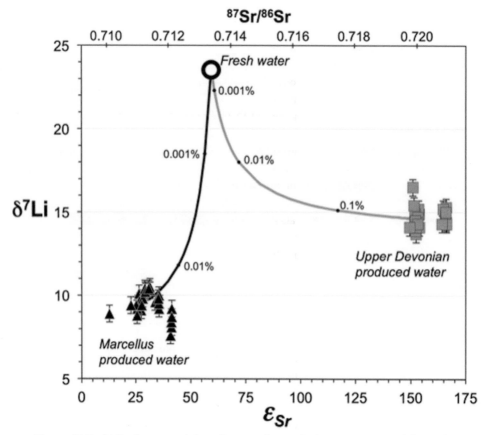

Figure 10.8 Li-Sr isotope mixing diagram for a freshwater source mixing with unconventional Marcellus Shale and conventional Upper Devonian produced water in the Appalachian Basin. The percentage values labeled on the curves indicate the amount of produced water contaminant added to the fresh water. The freshwater Sr endmember is the spring sample from Kolesar Kohl et al. (2014) and the Li endmember is the global average value for rivers. Produced water Sr data are from Chapman et al. (2012), Kolesar Kohl et al. (2014), and Capo et al. (2014), with corresponding Li data from Phan et al. (2016)

relatively fresh water during hydraulic fracturing. Phan et al. (2020) concluded that isotopically light Li observed in samples collected within the first few days after hydraulic fracturing was probably the result of release of Li from clay and organic matter caused by freshwater influx. However, sequential extraction experiments by Phan et al. (2016) suggest that exchangeable Li is probably not a major source of Li in late stage Marcellus Shale produced water, and that although organic matter and sulfides contribute some Li, most Li (>80%) in Marcellus shale rocks is structurally bound in clay minerals. Williams et al. (2015) suggested that in the Wattenberg Shale Gas Field, Colorado, Li and B were released from organic matter into porewaters during clay diagenesis, and that Li was then incorporated in octahedral sites within the illite lattice, while B resided in tetrahedral sites. K-Ar dating of the illite and comparison of the Li isotope compositions of bentonites inside and outside the gas field suggested an influx of isotopically light Li-rich fluids (δ^7Li = -18 to -4) was associated with gas generation 60 million years ago.

As Li isotope fractionation is a temperature-dependent process, δ^7Li in brines could potentially be used as a temperature probe, which can provide information about the depth of origin of brines. Lithium isotope fractionation decreases with increasing temperature as Li is incorporated into smectite clay (Vigier et al. 2008). A temperature-dependent illitization model for Marcellus Shale suggested that Li concentration was correlated with the diagenetic transition of illite to smectite (Phan et al. 2016). Macpherson et al. (2014) found that the Li isotope ratios of produced waters from the US Gulf Coast and Appalachian Basin varied inversely with temperature, although the isotopic signatures also suggested potential fluid migration or fractionation related to burial diagenesis.

The US Permian Basin is the site of conventional and unconventional hydrocarbon extraction. Produced waters are also related to secondary and enhanced oil recovery techniques (EOR), which involve injection of water, steam, CO_2, or a combination to displace and flush residual oil from subsurface reservoir rocks into wellbores. Because CO_2 injection has been used for decades in the region, high permeability rocks in the Permian Basin have also been identified as potential storage formations for geologic carbon sequestration (DOE-NETL 2010). This is also a region of limited water resources, so the effects of oil and gas recovery and CO_2 injection on groundwater systems must be minimized (Romanak et al. 2012). Pfister et al. (2017) demonstrated that groundwaters and oil and gas formation waters at different depths can maintain distinct δ^7Li values that may reflect their long-term geologic history. At one site with groundwater at different depths and TDS increasing with depth, the effects of water injection EOR can be clearly seen in the water chemistry and δ^7Li of the hydrocarbon-bearing formation; however, the isotope ratios rule out any significant transfer of fluids between formations and can serve as a monitoring tool to detect early upward migration or leakage along wellbores.

10.4.3 Ba Isotopes

Barium consists of seven isotopes, two of which (^{130}Ba and ^{132}Ba) are present at the ~0.1% level and the remainder of which (^{134}Ba to ^{138}Ba) increase in abundance from 2.4% to 72%. Mass-dependent fractionation of Ba leads to measurable variations in the ratios of heavy to light isotopes. The isotope composition is commonly reported as δ^{138}Ba, where the measured ^{138}Ba/^{134}Ba ratio is normalized to that of NIST standard 3104a,

$$\delta^{138}Ba(‰) = 10^3 \left[\frac{\left(^{138}Ba/^{134}Ba\right)_{sample}}{\left(^{138}Ba/^{134}Ba\right)_{NIST3104a}} - 1 \right]$$

The ratio is also sometimes reported as δ^{137}Ba, in which the ^{137}Ba/^{134}Ba is normalized to the same standard. Values can be converted by the relationship δ^{138}Ba ≈ 1.33 δ^{137}Ba. The most precise Ba isotope measurements are obtained by adding a double spike (^{135}Ba + ^{137}Ba or other combinations of light and heavy isotopes) to the sample prior to processing to correct for mass fractionation during sample chemistry and mass spectrometry, followed by measurement by MC-ICPMS or TIMS. The δ^{138}Ba is typically measured to a precision of ±0.01–0.05‰.

Mass fractionation of Ba isotopes has been shown to occur during the processes of barite ($BaSO_4$) and carbonate mineral precipitation, pedogenic processes, diffusion, and adsorption (von Allmen et al. 2010; Böttcher et al. 2012; Bullen and Chadwick 2016; Mavromatis et al. 2016, 2020; Pretet et al. 2016; van Zuilen et al. 2016; Hemsing et al. 2018). Natural geological and biological materials, including river water, seawater, silicate rocks, soil, vegetation, and barite nodules of different geologic ages, have a range of δ^{138}Ba values from about –0.75‰ to +0.65‰.

As Ba isotope ratio measurements have only become routine over the past decade, few studies have focused on the Ba isotope systematics of oil- and gas-related fluids. Produced water from oil and gas wells tends to have high concentrations of Ba, which can lead to the formation of barite scale and necessitates the addition of scale inhibitors to hydraulic fracturing fluids. Relative to most produced waters, those from unconventional Marcellus Shale gas wells are particularly high in Ba, with an average of about 2,200 mg/L, and sometimes exceeding 12,000 mg/L (Chapman et al. 2012; Barbot et al. 2013). The high Ba content in shale gas produced water may partially reflect the reducing, low-sulfate environment of the shale (Engle and Rowan 2014; Ouyang et al. 2017). Barium isotope measurements have the potential to fingerprint fluids from different subsurface sources, to contribute to our understanding of scale formation in oil and gas wells, and to provide information about the source of high-TDS fluids co-produced with gas and oil.

Tieman et al. (2020) investigated the Ba isotope systematics of unconventional and conventional produced waters from the Appalachian Basin, focusing on produced waters from the Marcellus shale and overlying conventional Upper Devonian oil and gas wells. They showed that Ba in produced water contained the most isotopically extreme δ^{138}Ba values measured to date, with the conventional (Upper Devonian) produced waters yielding values as low as -0.83‰, and unconventional (Marcellus Shale) produced waters yielding values as high as +1.49‰. The Upper Devonian values were consistently low (–0.83 to –0.52‰, mean –0.81, n = 4) and the Marcellus Shale values consistently high (+0.36 to +1.49, mean +0.87‰, n = 25). This systematic variation suggests that Ba isotopes could be used to differentiate between sources in the event of a leak or spill. In Figure 10.9, a sample of water from the Ohio River, a major river in the Marcellus and Upper Devonian oil- and gas-producing region, is compared to values measured in produced waters. The δ^{138}Ba of the Ohio River falls between the two produced water types, in a field also consistent with values measured from rivers around the world (Tieman et al. 2020). Mixing curves calculated using the weighted averages of the two types of produced water suggest that very small inputs or slow leakage (<0.001% fluid from a produced water source) could be detected by monitoring Ba isotopes of surface (or ground) waters in this region. Because the application of Ba isotopes to these problems is still in its infancy, the Ba isotope ratios of many potential contaminant endmembers have yet to be determined. For example, Ba from brine treatment plants could reflect the brine source, but if barite is precipitated prior

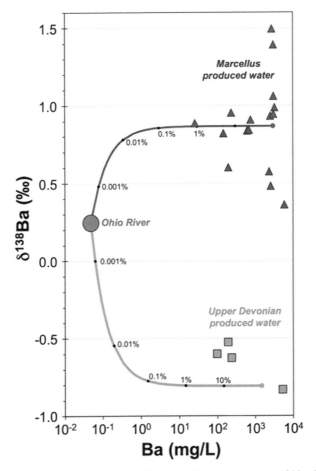

Figure 10.9 Barium isotope mixing diagram for river water (Ohio River, SW Pennsylvania) contaminated with produced water, using average values for Marcellus Shale and Upper Devonian produced waters. Individual data points represent samples collected throughout the Appalachian Basin. The values on the curves indicate the percentage of produced water mixed in to achieve the observed offset. (A black and white version of this figure will appear in some formats. For the colour version, refer to the plate section.) Data are from Tieman et al. (2020)

to discharge (e.g., Veil 2010; Ferrar et al. 2013) then the δ^{138}Ba of the effluent should be greatly elevated owing to mass-dependent fractionation (von Allmen et al. 2010; Hsieh and Henderson 2017), even as the Ba concentration is greatly reduced (Warner et al. 2013). Nonetheless, the Ba isotope system has demonstrated strong potential as a tracer for oil and gas produced water.

The Ba isotope systematics of this region also provided some information about the source of Ba in conventional and unconventional produced water. Analysis of drilling mud and associated produced waters from a single well in West Virginia demonstrated that barite associated with drilling mud is only a minor source of Ba to the produced water, at best. While barite precipitation should increase the δ^{138}Ba values of the residual fluid, the very high Ba concentrations of Marcellus Shale produced waters preclude this as the cause of their high δ^{138}Ba values. Tieman et al. (2020) suggest that the distinct compositions of Marcellus Shale and Upper Devonian produced waters are indicative of formation waters that have experienced geologically dissimilar fluid evolution processes, and have remained separate for potentially tens to hundreds of millions of years. Barium isotope analysis of oil and gas produced waters from other basins around the globe is likely to place important constraints on the origin and evolution of deep formation brines and could represent an important tool for monitoring environmental impacts of hydrocarbon-related fluids.

10.5 Summary

Metal isotope signatures can be used as powerful tools to identify and assess the impact of oil and gas produced water contamination. Strontium, Li, and Ba isotope signatures have been used to distinguish between conventional and unconventional sources of produced water and can detect produced water contamination of freshwater resources at levels of less than 0.01%. This sensitivity is of particular importance when a spill has been substantially diluted and attenuated in a river and can also aid in the rapid detection of subsurface brine intrusion into a shallow aquifer. Discrimination between multiple sources and sensitivity can be enhanced in areas with multiple sources of pollutants, such as the Appalachian region, by the combination of geochemical and isotopic tracers (e.g., Sr/Ca and ^{87}Sr/^{86}Sr, Figure 10.3; Ba and δ^{138}Ba, Figure 10.9) or multi-isotope tracers (e.g., ^{87}Sr/^{86}Sr and δ^{7}Li; Figure 10.8). For a metal isotope tracer to be effective for monitoring hydrocarbon-related fluids interacting with the environment, it should be present at relatively high concentrations in produced water and have measurable isotope variations or be affected by relevant geo/biological processes occurring during contamination. Some metal isotope systems that have potential as environmental tracers but have not yet been extensively applied to oil- and gas-related fluids, include Mg, Fe, K, and Ca. Regional variations in both geochemical and isotopic signatures mean that fluid baseline and end member compositions should be well characterized prior to the application of any type of tracer to optimize sensitivity and signal reliability.

Acknowledgments

We thank our past and present students, postdocs, colleagues, and collaborators for their contributions to the ideas presented here, including E. Chapman, M. Engle, J. Gardiner, J. Graney, J.A. Hakala, R. Hammack, W. Heck, C. Kirby, C. Kolesar Kohl, C. Lopano, G. Macpherson, R. Matecha, S. Pfister, T. Phan, E. Rowan, K. Schroeder, D. Soeder, Z. Tieman, and A. Wall. We appreciate the reviewer comments, which improved the manuscript. Some of the previous studies reported here from our group were supported by the DOE National Energy Technology Laboratory.

References

Andersson PS, Wasserburg GJ, and Ingri J. (1992). The sources and transport of Sr and Nd isotopes in the Baltic Sea. *Earth and Planetary Science Letters*. 113: 459–472.

Bagheri R, Nadri A, Raeisi E, Eggenkamp HGM, Kazemi GA, and Montaseri A. (2014). Hydrochemical and isotopic ($\delta^{18}O$, $\delta^{2}H$, $^{87}Sr/^{86}Sr$, $\delta^{37}Cl$ and $\delta^{81}Br$) evidence for the origin of saline formation water in a gas reservoir. *Chemical Geology*. 384: 62–75.

Balashov VN, Engelder T, Gu X, Fantle MS, and Brantley SL. (2015). A model describing flowback chemistry changes with time after Marcellus Shale hydraulic fracturing. *American Association of Petroleum Geologists Bulletin*. 99: 143–154.

Banner JL. (2004). Radiogenic isotopes: Systematics and applications to earth surface processes and chemical stratigraphy. *Earth Science Reviews*. 65: 141–194.

Barbot E, Vidic NS, Gregory KB, and Vidic RD. (2013). Spatial and temporal correlation of water quality parameters of produced waters from Devonian-age shale following hydraulic fracturing. *Environmental Science and Technology*. 47: 2562–2569.

Bloomberg.com. (2019). MGX Minerals and Eureka Resources Announce Operation of First Commercial Rapid Petrolithium Recovery System in Pennsylvania. www.bloomberg.com/press-releases/2019-10-24/mgx-minerals-and-eureka-resources-announce-operation-of-first-commercial-rapid-petrolithium-recovery-system-in-pennsyl.

Böttcher ME, Geprägs P, Neubert N, von Allmen K, Pretet C, Samankassou E, and Nägler TF. (2012). Barium isotope fractionation during experimental formation of the double carbonate $BaMn[CO_3]_2$ at ambient temperature. *Isotopes in Environmental and Health Studies*. 48: 457–463.

Bottomley DJ, Chan LH, Katz A, Starinsky A, and Clark ID. (2003). Lithium isotope geochemistry and origin of Canadian Shield brines. *Ground Water*. 41, 847–856.

Brandt JE, Lauer NE, Vengosh A, Bernhardt ES, and Di Giulio RT. (2018). Strontium isotope ratios in fish otoliths as biogenic tracers of coal combustion residual inputs to freshwater ecosystems. *Environmental Science & Technology Letters*. 5: 718–723.

Brantley SL, Yoxtheimer D, Arjmand, S., Grieve, P., Vidic, R., Pollack, J., Llewellyn GT, Abad J, and Simon S. (2014). Water resource impacts during unconventional shale gas development: The Pennsylvania experience. *International Journal of Coal Geology*. 126: 140-156.

Brinck EL and Frost CD. (2007). Detecting infiltration and impacts of introduced water using strontium isotopes. *Ground Water*. 45: 554–568.

Bullen TD. (2014). Metal stable isotopes in weathering and hydrology. In Holland HD and Turekian KK (eds.) *Treatise on Geochemistry 7*. Elsevier, 329–359.

Bullen TD and Eisenhauer A. (2009). Metal stable isotopes in low-temperature systems: A primer. *Elements*. 5: 349–352.

Bullen T and Chadwick O. (2016). Ca, Sr and Ba stable isotopes reveal the fate of soil nutrients along a tropical climosequence in Hawaii. *Chemical Geology*. 422: 25–45.

Burgos WD, Castillo-Meza L, Tasker TL, Geeza TJ, Drohan PJ, Liu X, Landis JD, Blotevogel J, McLaughlin M, Borch T, and Warner NR. (2017). Watershed-scale impacts from surface water disposal of oil and gas wastewater in western Pennsylvania. *Environmental Science and Technology*. 51: 8851–8860.

Campbell CE, Pearson BN, and Frost CD. (2008). Strontium isotopes as indicators of aquifer communication in an area of coal bed natural gas production, Powder River Basin, Wyoming and Montana. *Rocky Mountain Geology*. 43: 149–175.

Capo RC, Stewart BW, and Chadwick OA. (1998). Strontium isotopes as tracers of ecosystem processes: Theory and methods. *Geoderma*. 82: 197–225.

Capo RC, Stewart BW, Rowan EL, Kolesar Kohl CA, Wall AJ, Chapman EC, Hammack RW, and Schroeder KT. (2014). The strontium isotopic evolution of Marcellus

Formation produced waters, southwestern Pennsylvania. *International Journal of Coal Geology*. 126: 57–63.

Chan L-H, Starinsky A, and Katz A. (2002). The behavior of lithium and its isotopes in oilfield brines: Evidence from the Heletz–Kokhav field, Israel. *Geochimica et Cosmochimica Acta*. 66: 615–623.

Chan LH, Edmond JM, Thompson G, and Gillis K. (1992). Lithium isotopic composition of submarine basalts: implications for the lithium cycle in the oceans. *Earth and Planetary Science Letters*. 108: 151–160.

Chan L-H, Gieskes JM, You C-F, and Edmond JM. (1994). Lithium isotope geochemistry of sediments and hydrothermal fluids of the Guaymas Basin, Gulf of California. *Geochimica et Cosmochimica Acta*. 58: 4443–4454.

Chapman EC, Capo RC, Stewart BW, Kirby CS, Hammack RW, Schroeder KT, and Edenborn, H.M. (2012). Geochemical and strontium isotope characterization of produced waters from Marcellus Shale natural gas extraction. *Environmental Science and Technology*. 46: 3545–3553.

Chaudhuri S and Clauer N. (1993). Stontium isotopic compositions and potassium and rubidium contents of formation waters in sedimentary basins: Clues to the origin of the solutes. *Geochimica et Cosmochimica Acta*. 57: 429–437.

Choi H-B, Ryu J-S, Shin W-J, Vigier N. (2019). The impact of anthropogenic inputs on lithium content in river and tap water. *Nature Communications*. 105371. https://doi.org/10.1038/s41467-019-13376-y.

Collins AG. (1978). Geochemistry of anomalous lithium in oil-field brines. *Oklahoma Geological Survey Circular*. 79: 95–98.

DePaolo DJ and Wasserburg GJ. (1976). Nd isotopic variations and petrogenetic models. *Geophysical Research Letters*. 3: 249–252.

DOE-NETL. (2010). Carbon dioxide enhance oil recovery: Untapped domestic energy supply and long term carbon storage solution. Department of Energy – National Energy Technology Laboratory, www.netl.doe.gov/file%20library/research/oil-gas/small_CO2_EOR_Primer.pdf.

Eiler JM, Bergquist BA, Bourg IC, Cartigny P, Farquhar J, Gagnon A, Guo W, Halevy I, and Hofmann A et al. (2014). Frontiers of stable isotope geoscience. *Chemical Geology*. 372: 119–143.

Engle MA and Rowan EL. (2013). Interpretation of Na–Cl–Br systematics in sedimentary basin brines: Comparison of concentration, element ratio, and isometric log-ratio approaches. *Mathematical Geosciences*. 45: 87–101.

Engle MA and Rowan EL. (2014). Geochemical evolution of produced waters from hydraulic fracturing of the Marcellus Shale, northern Appalachian Basin: A multivariate compositional data analysis approach. *International Journal of Coal Geology*. 126: 45–56.

Eureka Resources. (2019). MGX Minerals and Eureka Resources announce joint venture to recover lithium from produced water in eastern United States. www.eureka-resources.com/blog1.

Faure G and Mensing TM. (2005). *Isotopes: Principles and Applications*. John Wiley & Sons.

Ferrar KJ, Michanowicz DR, Christen CL, Mulcahy N, Malone SL, and Sharma RK. (2013). Assessment of effluent contaminants from three facilities discharging Marcellus shale wastewater to surface waters in Pennsylvania. *Environmental Science and Technology*. 47: 3472–3481.

Frost CD, Pearson BN, Ogle KM, Heffern EL, and Lyman RM. (2002). Sr isotope tracing of aquifer interactions in an area of accelerating coal-bed methane production, Powder River Basin, Wyoming. *Geology*. 30: 923–926.

Geeza TJ, Gillikin DP, McDevitt B, Van Sice K, and Warner NR. (2018). Accumulation of Marcellus Formation oil and gas wastewater metals in freshwater mussel shells. *Environmental Science & Technology*. 52: 10883–10892.

Haluszczak LO, Rose AW, and Kump LR. (2013). Geochemical evaluation of flowback brine from Marcellus gas wells in Pennsylvania, USA. *Applied Geochemistry*. 28: 55–61.

Hammack R, Zorn E, Harbert W, Capo R, Sharma S, and Siriwardane H. (2013). An evaluation of zonal isolation after hydraulic fracturing; Results from horizontal Marcellus Shale gas wells at NETL's Greene County Test Site in southwestern Pennsylvania. *Society of Petroleum Engineers Conference Paper* DOI: 10.2118/165720-MS, SPE-165720-MS.

Hayes T. (2009). *Sampling and Analysis of Water Streams Associated with the Development of Marcellus Shale Gas*. Report by the Gas Technology Institute, Des Plaines, IL. Marcellus Shale Coalition.

Hemsing F, Hsieh Y-T, Bridgestock L, Spooner PT, Robinson LF, Frank N, and Henderson GM. (2018). Barium isotopes in cold-water corals. *Earth and Planetary Science Letters*. 491: 183–192.

Hsieh Y-T and Henderson GM. (2017). Barium stable isotopes in the global ocean: Tracer of Ba inputs and utilization. *Earth and Planetary Science Letters*. 473: 269–278.

Huang T, Pang Z, Li Z, Li Y, and Hao Y. (2020). A framework to determine sensitive inorganic monitoring indicators for tracing groundwater contamination by produced formation water from shale gas development in the Fuling Gasfield, SW China. *Journal of Hydrology*. 581: 124403.

Huh Y, Chan L-H, Zhang L, and Edmond JM. (1998). Lithium and its isotopes in major world rivers: Implications for weathering and the oceanic budget. *Geochimica et Cosmochimica Acta*. 62: 2039–2051.

Johnson CM, Beard BL, and Albarède F. (2004). Geochemistry of non-traditional stable isotopes. In Rosso JJ (ed.) *Reviews in Mineralogy & Geochemistry*. Mineralogical Society of America, p. 454.

Johnson JD, Graney JR, Capo RC, and Stewart BW. (2015). Identification and quantification of regional brine and road salt sources in watersheds along the New York / Pennsylvania border, USA. *Applied Geochemistry*. 60: 37–50.

Kolesar Kohl CA, Capo RC, Stewart BW, Wall AJ, Schroeder KT, Hammack RW, and Guthrie GD. (2014). Strontium isotopes test long-term zonal isolation of injected and Marcellus Formation water after hydraulic fracturing. *Environmental Science and Technology*. 48: 9867–9873.

Lauer NE, Harkness JS, and Vengosh A. (2016). Brine Spills Associated with Unconventional Oil Development in North Dakota. *Environmental Science & Technology*. 50: 5389–5397.

Lauer NE, Warner NR, and Vengosh A. (2018). Sources of radium accumulation in stream sediments near disposal sites in Pennsylvania: Implications for disposal of conventional oil and gas wastewater. *Environmental Science and Technology 52*, 955-962.

Macpherson GL, Capo RC, Stewart BW, Phan TT, Schroeder KT, and Hammack RW. (2014). Temperature-dependent Li isotope ratios in Appalachian Plateau and Gulf Coast Sedimentary Basin saline water. *Geofluids*. 14: 419–429.

Mavromatis V, van Zuilen K, Blanchard M, van Zuilen M, Dietzel M, and Schott J. (2020). Experimental and theoretical modelling of kinetic and equilibrium Ba isotope fractionation during calcite and aragonite precipitation. *Geochimica et Cosmochimica Acta*. 269: 566–580.

Mavromatis V, van Zuilen K, Purgstaller B, Baldermann A, Nägler TF, and Dietzel M. (2016). Barium isotope fractionation during witherite ($BaCO_3$) dissolution, precipitation and at equilibrium. *Geochimica et Cosmochimica Acta*. 190: 72–84.

McIntosh JC, Hendry MJ, Ballentine C, Haszeldine RS, Mayer B, Etiope G, Elsner M, Darrah TH, and Prinzhofer A et al. (2019). A critical review of state-of-the-art and emerging approaches to identify fracking-derived gases and associated contaminants in aquifers. *Environmental Science and Technology*. 53: 1063–1077.

Millot R, Guerrot C, Innocent C, Négrel P, and Sanjuan B. (2011). Chemical, multi-isotopic (Li–B–Sr–U–H–O) and thermal characterization of Triassic formation waters from the Paris Basin. *Chemical Geology*. 283: 226–241.

Naftz DL, Peterman ZE, and Spangler LE. (1997). Using $\delta^{87}Sr$ values to identify sources of salinity to a freshwater aquifer, Greater Aneth Oil Field, Utah, USA. *Chemical Geology*. 141: 195–209.

Neymark LA, Premo WR, and Emsbo P. (2018). Combined radiogenic ($^{87}Sr/^{86}Sr$, $^{234}U/^{238}U$) and stable ($\delta^{88}Sr$) isotope systematics as tracers of anthropogenic groundwater contamination within the Williston Basin, USA. *Applied Geochemistry*. 96: 11–23.

Ni Y, Zou C, Cui H, Li J, Lauer NE, Harkness JS, Kondash AJ, Coyte RM, and Dwyer GS, et al. (2018). Origin of flowback and produced waters from Sichuan Basin, China. *Environmental Science & Technology*. 52: 14519–14527.

Ono S. (2017). Photochemistry of sulfur dioxide and the origin of mass-independent isotope fractionation in Earth's atmosphere. *Annual Review of Earth and Planetary Sciences*. 45: 301–329.

Osborn SG, McIntosh JC, Hanor JS, and Biddulph D. (2012). Iodine-129, $^{87}Sr/^{86}Sr$, and trace elemental geochemistry of northern Appalachian Basin brines: Evidence for basinal-scale fluid migration and clay mineral diagenesis. *American Journal of Science*. 312: 263–287.

Ouyang B, Akob DM, Dunlap D, and Renock D. (2017). Microbially mediated barite dissolution in anoxic brines. *Applied Geochemistry*. 76: 51–59.

Pfister S, Capo RC, Stewart BW, Macpherson GL, Phan TT, Gardiner JB, Diehl JR, Lopano C, and Hakala JA. (2017). Geochemical and lithium isotope tracking of dissolved solid sources in Permian Basin carbonate reservoir and overlying aquifer waters at an enhanced oil recovery site, northwest Texas, USA. *Applied Geochemistry*. 87: 122–135.

Phan TT, Hakala JA, and Sharma S. (2020). Application of isotopic and geochemical signals in unconventional oil and gas reservoir produced waters toward characterizing in situ geochemical fluid-shale reactions. *Science of the Total Environment*. 714: 136867.

Phan TT, Capo RC, Stewart BW, Macpherson GL, Rowan EL, and Hammack RW. (2016). Factors controlling Li concentration and isotopic composition in formation waters and host rocks of Marcellus Shale, Appalachian Basin. *Chemical Geology*. 420: 162–179.

Phan TT, Capo RC, Stewart BW, Graney JR, Johnson JD, Sharma S, and Toro J. (2015). Trace metal distribution and mobility in drill cuttings and produced waters from Marcellus shale gas extraction: uranium, arsenic, barium. *Applied Geochemistry*. 60: 89–103.

Pollastro RM, Roberts LNR, and Cook TA. (2010). Geologic assessment of technically recoverable oil in the Devonian and Mississippian Bakken Formation. *Assessment of Undiscovered Oil and Gas Resources of the Williston Basin Province of North Dakota, Montana, and South Dakota, 2010*. U.S. Geological Survey Digital Data Series, pp. 1–34.

Porcelli D and Baskaran M. (2012). An overview of isotope geochemistry in environmental studies. In Baskaran M (ed.) *Handbook of Environmental Isotope Geochemistry 1.* Springer, pp. 11–32.

Preston TM, Thamke J., Smith BD, and Peterman ZE. (2014). Chapter B: Brine contamination of Prairie Pothole environments at three study sites in the Williston Basin, United States. In Gleason RA and Tangen BA (eds.) *Brine Contamination to Aquatic Resources from Oil and Gas Development in the Williston Basin, United States.* U.S. Geological Survey Scientific Investigations Report 2014-5017, pp. 21–62.

Pretet C, van Zuilen K, Nägler TF, Reynaud S, Böttcher ME, and Samankassou E. (2016). Constraints on barium isotope fractionation during aragonite precipitation by corals. *The Depositional Record.* 1: 118–129.

Qi HP, Coplen TB, Wang QZ, and Wang YH. (1997). Unnatural isotopic composition of lithium reagents. *Analytical Chemistry.* 69: 4076–4078.

Romanak KD, Smyth RC, Yang C, Hovorka SD, Rearick M, and Lu, J. (2012). Sensitivity of groundwater systems to CO_2: Application of a site-specific analysis of carbonate monitoring parameters at the SACROC CO_2-enhanced oil field. *International Journal of Greenhouse Gas Control.* 6: 142–152.

Rotenberg E, Davis DW, Amelin Y, Ghosh S, and Bergquist BA. (2012). Determination of the decay-constant of ^{87}Rb by laboratory accumulation of ^{87}Sr. *Geochimica et Cosmochimica Acta.* 85: 41–57.

Rowan EL, Engle MA, Kirby CS, and Kraemer TF. (2011). Radium content of oil- and gas-field produced waters in the northern Appalachian Basin (USA): Summary and discussion of data. *United States Geological Survey Scientific Investigations Report.* 2011–5135: 1–31.

Rowan EL, Engle MA, Kraemer TF, Schroeder KT, Hammack RW, and Doughten MW. (2015). Geochemical and isotopic evolution of water produced from Middle Devonian Marcellus Shale gas wells, Appalachian Basin, Pennsylvania. *American Association of Petroleum Geologists Bulletin.* 99: 181–206.

Schauble EA. (2007). Role of nuclear volume in driving equilibrium stable isotope fractionation of mercury, thallium, and other very heavy elements. *Geochimica et Cosmochimica Acta.* 71: 2170–2189.

Shrestha N, Chilkoor G, Wilder J, Gadhamshetty V, and Stone JJ. (2017). Potential water resource impacts of hydraulic fracturing from unconventional oil production in the Bakken shale. *Water Research.* 108: 1–24.

Skalak KJ, Engle MA, Rowan EL, Jolly GD, Conko KM, Benthem AJ, and Kraemer TF. (2014). Surface disposal of produced waters in western and southwestern Pennsylvania: Potential for accumulation of alkali-earth elements in sediments. *International Journal of Coal Geology.* 126: 162–170.

Soeder DJ, Sharma S, Pekney N, Hopkinson L, Dilmore R, Kutchko B, Stewart B, Carter C, Hakala A, and Capo R. (2014). An approach for assessing engineering risk from shale gas wells in the United States. *International Journal of Coal Geology.* 126: 4–19.

Stewart BW, Capo RC, and Chadwick OA. (1998). Quantitative strontium isotope models for weathering, pedogenesis and biogeochemical cycling. *Geoderma.* 82: 173–195.

Stewart BW, Chapman EC, Capo RC, Johnson JD, Graney JR, Kirby CS, and Schroeder KT. (2015). Origin of brines, salts and carbonate from shales of the Marcellus Formation: Evidence from geochemical and Sr isotope study of sequentially extracted fluids. *Applied Geochemistry.* 60: 78–88.

Tasker TL, Warner NR, and Burgos WD. (2020). Geochemical and isotope analysis of produced water from the Utica/Point Pleasant Shale, Appalachian Basin. *Environmental Science: Processes & Impacts.* 22: 1224–1232.

Tasker TL, Burgos WD, Piotrowski P, Castillo-Meza L, Blewett TA, Ganow KB, Stallworth A, Delompré PLM, and Goss GG, et al. (2018). Environmental and human health impacts of spreading oil and gas wastewater on roads. *Environmental Science & Technology*. 52: 7081–7091.

Tieman ZG, Stewart BW, Capo RC, Phan TT, Lopano CL, and Hakala JA. (2020). Barium isotopes track the source of dissolved solids in produced water from the unconventional Marcellus Shale gas play. *Environmental Science and Technology*. 54: 4275–4285.

Tisherman R and Bain DJ. (2019). Alkali earth ratios differentiate conventional and unconventional hydrocarbon brine contamination. *Science of the Total Environment*. 695: 133944.

Tomascak PB. (2004). Developments in the understanding and application of lithium isotopes in the Earth and planetary sciences. In Johnson CM, Beard BL, and Albarède F (eds.) *Geochemistry of Non-traditional Stable Isotopes, Reviews in Mineralogy & Geochemistry 55*. Mineralogical Society of America and Geochemical Society, 153–195.

van Zuilen K, Müller T, Nägler TF, Dietzel M, and Küsters T. (2016). Experimental determination of barium isotope fractionation during diffusion and adsorption processes at low temperatures. *Geochimica et Cosmochimica Acta*. 186: 226–241.

Veil JA. (2010). *Final Report: Water Management Technologies Used by Marcellus Shale Gas Producers*. U. S. Department of Energy-NETL-Argonne National Lab. www.evs.anl.gov/pub/dsp_detail.cfm?PubID=2537.

Vengosh A, Jackson RB, Warner NR, Darrah TH, and Kondash, A. (2014). A critical review of the risks to water resources from unconventional shale gas development and hydraulic fracturing in the United States. *Environmental Science and Technology*. 48: 8334–8348.

Vigier N, Decarreau A, Millot R, Carignan J, Petit S, and France-Lanord C. (2008). Quantifying Li isotope fractionation during smectite formation and implications for the Li cycle. *Geochimica et Cosmochimica Acta*. 72: 780–792.

von Allmen K, Böttcher ME, Samankassou E, and Nägler TF. (2010). Barium isotope fractionation in the global barium cycle: First evidence from barium minerals and precipitation experiments. *Chemical Geology*. 277: 70–77.

Warner NR, Christie CA, Jackson RB, and Vengosh A. (2013). Impacts of shale gas wastewater disposal on water quality in western Pennsylvania. *Environmental Science and Technology*. 47: 11849–11857.

Warner NR, Darrah TH, Jackson RB, Millot R, Kloppmann W, and Vengosh A. (2014). New tracers identify hydraulic fracturing fluids and accidental releases from oil and gas operations. *Environmental Science and Technology*. 48: 12552–12560.

Warner NR, Jackson RB, Darrah TH, Osborn SG, Down A, Zhao K, White A, and Vengosh A. (2012). Geochemical evidence for possible natural migration of Marcellus Formation brine to shallow aquifers in Pennsylvania. *Proceedings of the National Academy of Sciences*. 109: 11961–11966.

Weiss DJ, Rehkämper M, Schoenberg R, McLaughlin M, Kirby J, Campbell PGC, Arnold T, Chapman J, Peel K, and Gioia S. (2008)Application of nontraditional stable-isotope systems to the study of sources and fate of metals in the environment. *Environmental Science and Technology*. 42: 655–664.

Wiederhold JG. (2015). Metal Stable Isotope Signatures as Tracers in Environmental Geochemistry. *Environmental Science & Technology*. 49: 2606–2624.

Williams LB, Elliott WC, and Hervig RL. (2015). Tracing hydrocarbons in gas shale using lithium and boron isotopes: Denver Basin USA, Wattenberg Gas Field. *Chemical Geology*. 417: 404–413.

Zagorski WA, Wrightstone GR, and Bowman DC. (2012). The Appalachian Basin Marcellus gas play: Its history of development, geologic controls on production, and future potential as a world-class reservoir. In Breyer JA (ed.) *Shale Reservoirs: Giant Resources for the 21st Century*. AAPG Memoir, pp. 172–200.

Zhang L, Chan L-H, and Gieskes JM. (1998). Lithium isotope geochemistry of pore waters from Ocean Drilling Program Sites 918 and 919, Irminger Basin. *Geochimica et Cosmochimica Acta*. 62: 2437–2450.

Zheng Z, Zhang H, Chen Z, Li X, Zhu P, and Cui X. (2017). Hydrogeochemical and isotopic indicators of hydraulic fracturing flowback fluids in shallow groundwater and stream water, derived from Dameigou shale gas extraction in the northern Qaidam Basin. *Environmental Science & Technology*. 51: 5889–5898.

11

Isotopes as Tracers of Atmospheric and Groundwater Methane Sources

AMY TOWNSEND-SMALL

11.1 Introduction

11.1.1 How Can Isotopes Be Used to Study Methane Sources?

Methane (CH_4) is the smallest organic carbon molecule. Both carbon and hydrogen have naturally occurring stable and radioactive isotopes that can be used to trace sources of methane. An isotope is an atom of a specific element that differs in the number of neutrons in the nucleus but has the same number of protons and electrons as the other atoms of that element, so isotopes differ in mass but not chemical reactivity.

Because the heavier isotopes are so rare, ratios of the rare to the most abundant stable isotopes are usually expressed in the following "delta" notation, relative to standards assigned by the International Atomic Energy Agency (IAEA):

$$\delta X = [(R_{sample} - R_{standard}/R_{standard}] * 1000$$

Where X equals the ratio of interest ($^{13}C/^{12}C$ or $^{2}H/^{1}H$) and the standards are as defined in Table 11.1. The units of this are per mille (‰).

Both ^{14}C and ^{3}H are produced naturally in the atmosphere as a result of cosmic ray interactions, and they are enhanced by nuclear energy and weapons use. After atmospheric weapons testing took place in the mid-twentieth century, concentrations of both isotopes in the atmosphere were greatly increased.

Radiocarbon ratios are usually expressed either in delta notation, "fraction modern" (FM) or as "percent modern carbon" (pMC). When using delta notation, because a correction factor is applied for the amount of ^{13}C in the sample, this is usually expressed as $\Delta^{14}C$ to account for this correction.

Tritium is an extremely rare isotope of hydrogen (Table 11.1) that is nevertheless a useful isotope in hydrology and hydrogeology applications (Kendall and Doctor 2003). Some studies have utilized tritium as a tracer of dissolved methane from landfills, as landfills have elevated tritium levels resulting from disposal of luminous paint and other tritiated carbon sources (Coleman et al. 1995; Mace et al. 2016). However, the short half-life of tritium makes this a poor tracer of atmospheric methane sources. No studies of methane in oil- and gas-producing regions have utilized tritium as a tracer.

The basis for the utility of stable isotopes as a tracer of atmospheric or groundwater methane sources is that biogenic methane is produced during enzymatic reactions in

Table 11.1. *Properties of C and H isotopes and IAEA standards*

Isotope	Natural abundance	Stable or radioactive	Half life	IAEA standard
^{12}C	98.9%	stable	n/a	n/a
^{13}C	1.1%	stable	n/a	PDB
^{14}C (radiocarbon)	1 part per trillion	radioactive	5730 years	NBS Oxalic Acid
^{1}H (hydrogen)	99.98%	stable	n/a	n/a
^{2}H (deuterium or D)	0.02%	stable	n/a	SMOW
^{3}H (tritium)	10^{-18}%	radioactive	12.32 years	n/a

biological respiration, which discriminate against heavier stable isotopes leading to depleted isotopic signatures relative to the carbon substrate ("isotopic fractionation"). Biogenic methane can be produced from organic matter such as during acetate fermentation in anaerobic environments such as wetlands, landfills, and ruminant digestive systems, or from carbon dioxide, such as in some coal mining and other environments (Whiticar 1999). Methane from fossil fuel sources ("thermogenic" methane), on the other hand, is derived mostly from chemical and physical reactions, so isotopic signatures are closer to those of the original organic matter (Whiticar 1999; Whiticar and Schaefer 2007). Methane produced from incomplete combustion, or "pyrogenic" methane, has a similar isotopic signature to the source material (Reeburgh 2007; Schwietzke et al. 2016). Methane from fossil fuels is devoid of radiocarbon whereas biogenic methane has a radiocarbon content or "age" reflecting the age of the carbon it was produced from: ranging from modern for cattle breath to tens of thousands of years old for permafrost methane.

11.2 Isotopes as Tracers of Atmospheric Methane

For isotopes to be used to determine atmospheric methane sources, either a long-term record or an enhancement of methane over background levels is needed. Several examples of studies are shown in the following sections.

11.2.1 Global Methane Sources

Methane concentrations are monitored throughout the world at "clean air" background sites such as the Mauna Loa Observatory in Hawaii (Figure 11.1). Methane concentrations in the global atmosphere were stable from about 1998 to 2007 but have grown ever since, and carbon stable isotopes are measured at some background monitoring sites (NOAA Global Monitoring Division 2020). The reasons for the change in methane concentration growth rate are of great interest, but this is still hotly debated and there is no consensus on whether the increase in methane emissions is the result of anthropogenic or natural sources.

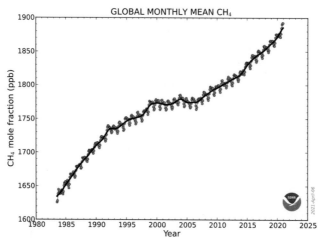

Figure 11.1 Monthly average methane concentrations as measured at the Mauna Loa Observatory in Hawaii since 1983.
Source: www.esrl.noaa.gov/gmd/ccgg/trends_ch4/

Some groups have argued that reduced biogenic sources are the cause of the slowdown in methane emissions at the beginning of this century (Kai et al. 2011), while other groups have concluded that reduced fossil fuel emissions are responsible for the slowdown (Aydin et al. 2011). A database of isotopic composition of fossil fuel, biogenic, and combustion methane sources (Sherwood et al. 2017) was used to construct an atmospheric methane inventory that indicated a larger proportion of fossil fuel emissions than previously thought (Schwietzke et al. 2016). However, others have argued that (Schwietzke et al. 2016) did not consider that biomass burning may have changed over time, and thus may have overestimated biogenic sources (Worden et al. 2017). Also, the same database has also been used to suggest that fossil fuels are not the source of increased methane emissions since 2007 (Milkov et al. 2020), and another analysis used data from both conventional oil and gas wells and shale wells to argue that shale gas has been a major driver of recent increases in methane emissions (Howarth 2019). This is further discussed in Chapter 6. Other data on stable carbon isotopes have also suggested that growth in atmospheric methane since 2007 is predominantly caused by increasing natural and/or anthropogenic biogenic methane emissions from tropical regions (Nisbet et al. 2016). The lack of consensus is the result of limited data on spatial distribution of methane concentrations and other source indicators of methane, such as other isotopes and co-emitted trace gases such as ethane or other alkanes (Nisbet et al. 2019). Carbon-13 is the most abundantly measured tracer in background air, but because many natural and anthropogenic sources have similar $\delta^{13}C$ signatures to each other or to methane in air, it is difficult to use this tracer to determine which sources are increasing (Whiticar and Schaefer 2007). A study that used hydrogen isotopic composition of methane from archived air samples found that increased fossil fuel methane was the likely cause of increasing methane concentrations since 2009 (Rice et al. 2016).

Radiocarbon is an imperfect tracer of global background methane sources. In many ways, this would be an excellent tracer of biogenic versus thermogenic methane sources, but there is an unconstrained "nucleogenic" $^{14}CH_4$ source from nuclear reactors that interferes with the use of this tracer (Lassey et al. 2007a, 2007b). Also, measurements are sparse owing to the larger sample volume requirements and higher cost of radiocarbon measurements versus stable isotope analysis.

11.2.2 City-Level Methane Sources

Stable isotopes have been used to determine methane sources in urban areas. In Los Angeles, isotopic composition of major potential methane sources were characterized using canister sampling, measurement of methane concentration, and measurement of $\delta^{13}C$-CH_4 and δD-CH_4 (Figure 11.2) (Townsend-Small et al. 2012).

Samples were then collected from a high-altitude research station that represented well-mixed urban air: the Mount Wilson observatory. Los Angeles is surrounded by mountain ranges and has predominantly onshore winds. This results in collection of urban, polluted air within the basin and results in the city's notorious afternoon smog events. Air was monitored continuously for methane concentration at the Mount Wilson Observatory during an approximate 48-hour period, while the sampling location was located above the atmospheric boundary layer, and then during the period when polluted air warmed to the point when it rose above the observatory (Figure 11.3). Discrete samples were taken for isotopic analysis throughout these events (Figure 11.3).

When there is a range of concentrations and isotopic compositions, they can be plotted together with isotopic composition versus the inverse of methane concentration in a

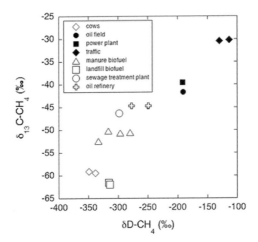

Figure 11.2 Carbon and hydrogen stable isotopic composition of methane collected from various sources in southern California. Solid symbols represent fossil-derived CH_4, and open symbols represent biological sources. Isotope ratios represent source signatures corrected for the presence of background air. (Townsend-Small et al. 2012)

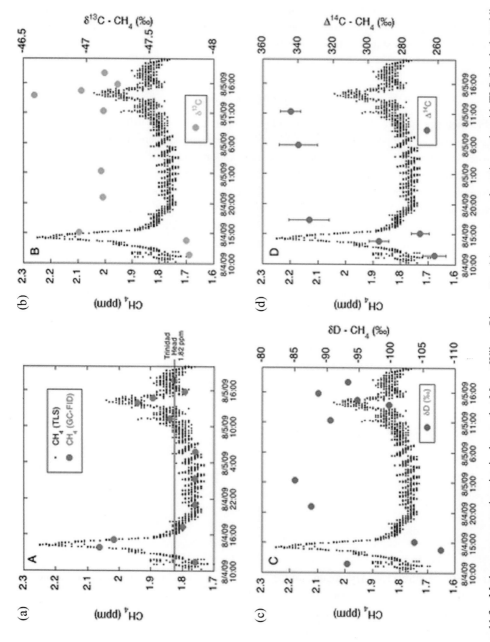

Figure 11.3 Methane concentrations in air observed at Mount Wilson Observatory (a) as measured continuously with TLS (black dotted line) and in discrete samples with GC-FID (large dots). Also shown is the approximate concentration of CH_4 in unpolluted air (1.82 ppm) as measured at Trinidad Head, California, by the NOAA flask sampling network. Shown are (b) $\delta^{13}C$, (c) δD, and (d) $\Delta^{14}C$ in discrete samples, (points) shown with CH_4 concentrations measured by TLS. (Townsend-Small et al. 2012)

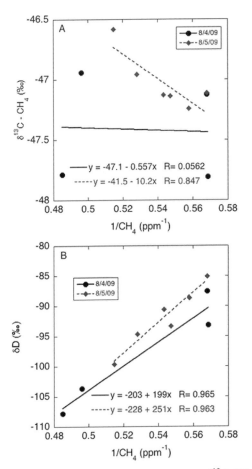

Figure 11.4 Keeling plots of 1/CH$_4$ concentration versus (a) δ^{13}C-CH$_4$ and (b) δD-CH$_4$. (Townsend-Small et al. 2012)

Keeling plot (Keeling 1961, 1958; Pataki et al. 2003). In the case of a significant linear relationship, the value of the *y*-intercept indicates the isotopic composition of the source of excess atmospheric methane above background levels. Figure 11.4 shows Keeling plots for the data in Figure 11.3b and c.

In air sampled at Mount Wilson, the Keeling plot analysis indicated that methane emitted in Los Angeles had a δ^{13}C-CH$_4$ of -41.5‰ and a δD-CH$_4$ of between -203‰ and -228‰ (Figure 11.4). Hydrogen isotopes were a more consistent indicator of methane source than carbon isotopes. Radiocarbon was not a consistent tracer of methane source, with no significant relationship with methane concentration. However, there was a general trend of lower Δ^{14}C-CH$_4$ with higher methane concentration (Figure 11.2d), consistent with a fossil fuel source of methane in Los Angeles.

The stable isotope results clearly show that the dominant source of methane in Los Angeles during the study period have a thermogenic origin, such as from leakage from natural gas pipelines, power plants, oil and gas production wells, or natural seeps (e.g., the La Brea tar pits) (Figure 11.2). Subsequent studies of methane sources in Los Angeles using different tracers of methane sources also indicated that oil and gas sources were probably the major source (Wennberg et al. 2012; Peischl et al. 2013). Of the major anthropogenic methane sources, the Los Angeles area lacks significant dairy or cattle farming or rice agriculture, and it is at the forefront of methane mitigation in landfills. Also, as an arid region, natural sources such as wetlands or lakes are probably small. This study catalyzed a national and international effort to reduce methane emissions across the natural gas supply chain as the United States and other countries transition from coal to gas for electricity (Lamb et al. 2015; Alvarez et al. 2018).

11.2.3 Regional/Basin Level Methane Characterization

Stable isotopes have also been used to distinguish methane sources in areas where oil and gas production is occurring alongside other methane-emitting activities, such as cattle ranching and waste management. In the Denver area, unconventional oil and gas production from shales is occurring alongside older conventional production. There are ~100,000 oil and gas production wells of both types in the region in addition to approximately 1.3 million cattle and approximately 2,000,000 people with attendant landfills and wastewater treatment plants (Townsend-Small et al. 2016).

In the Denver study, many more samples were taken to characterize isotopic composition of sources than in the Los Angeles study, owing to advances in instrumentation that allow for greater sample throughput. Figure 11.5 shows isotopic composition of oil and gas, landfill, and cattle sources from the Denver study. The δ^{13}C-CH$_4$ of oil and gas sources was not consistent, making it a poor tracer of methane sources regionally (Figure 11.4) (Townsend-Small et al. 2016). Methane emitted from landfills and cattle also had a δ^2H-CH$_4$ that was not statistically different (Figure 11.5) (Townsend-Small et al. 2016). But oil and gas wells had a distinct δ^2H-CH$_4$ from biogenic sources, making this a better tracer than δ^{13}C for separating thermogenic and biogenic sources in the atmosphere (Figure 11.5, Townsend-Small et al. 2016).

Methane sources in this region were assessed at several stationary ground sites as well as during aircraft flights around the region (Figure 11.6). One ground site was located in Platteville, at the center of both oil and gas production and cattle feedlot activity. Samples taken at this site indicate about 50% of methane is thermogenic and 50% is biogenic, most likely from cattle (Figure 11.6a). At other ground sites located further from cattle feedlots and oil and gas production sites, δ^2H-CH$_4$ signatures were lighter, indicating a larger proportion of biogenic methane and/or more methane oxidation in the atmosphere (Figure 11.6b and c) (Whiticar and Schaefer 2007). Samples taken in aircraft flights indicate similarly depleted isotopic signatures, representing a large proportion of biogenic methane in the Denver basin. Together, results from ground stations and aircraft flights indicate that methane emissions in the region may be declining, as previous studies (using different source apportionment techniques) indicated that about 27% of methane in the Front Range was biogenic (Pétron et al. 2014). Colorado is at the forefront of regulating

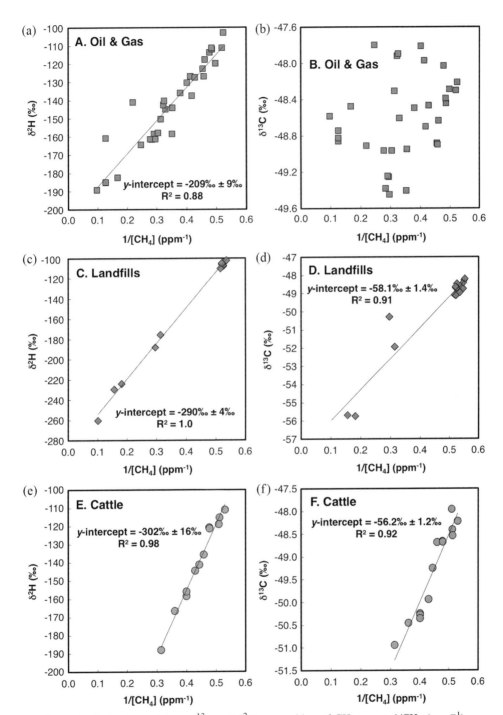

Figure 11.5 Keeling plots of $\delta^{13}C$ and δ^2H composition of CH_4 versus $1/CH_4$ (ppm^{-1}) collected downwind of (a and b) oil and gas, (c and d) landfill, and (e and f) cattle sources in the Denver-Julesburg basin. Each point represents an individual sample. (Townsend-Small et al. 2016)

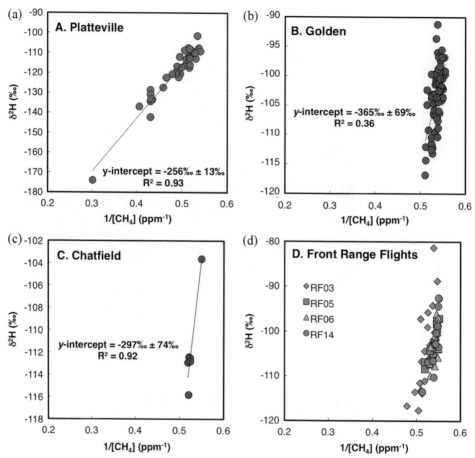

Figure 11.6 Keeling plots of δ^2H composition of CH_4 at three ground sites (a-c) in the Front Range. (d) δ^2H-CH_4 for samples taken during aircraft flights within the Front Range. All flights shown are significant at $p < 0.05$. Flight tracks can be viewed using the following link: www-air.larc.nasa.gov/missions/discover-aq/kmz/FRAPPE_C130_2014_ALL_July26-August18.kmz. (Townsend-Small et al. 2016)

methane emissions from oil and gas production, and these results indicate the regulations may be helping to reduce these emissions: with the caveat that agricultural and urban emissions are still present.

11.3 Isotopes as Tracers of Groundwater Methane

Stable isotopes and radiocarbon can help distinguish among various sources of dissolved methane. Elevated dissolved methane in groundwater can lead to an explosion risk or asphyxiation, although methane itself is nontoxic (Baldassare and Chapman 2018). Natural gas methane in groundwater may also be indicative of the presence of other hydrocarbons from oil and gas sources, or other drilling chemicals (United States Department of the

Interior 2001; Vidic et al. 2013). Natural gas contamination of groundwater owing to shale gas development is a primary concern of homeowners (Vengosh et al. 2014; US EPA National Center for Environmental Assessment 2016).

In geologic or sedimentary settings, biogenic methane can be produced through reduction of carbon dioxide, and this is a common process for production of coalbed methane and groundwater associated with coal seams (Scott et al. 1994; Smith and Pallasser 1996; Clayton 1998; Martini et al. 1998; Schlegel et al. 2011; Golding et al. 2013). Biogenic methane can also be produced by degradation of organic carbon, which has a distinct stable isotopic signature compared to carbonate reduction (Whiticar 1999). Both types of biogenic gas are distinct from thermogenic gas in that they are considered very "dry" gases, in other words, they consist nearly entirely of methane (Baldassare and Chapman 2018). Physical and chemical processes that convert organic matter to thermogenic methane also produce higher molecular weight hydrocarbons such as ethane, propane etc. The presence of these alkanes is also commonly used as a tracer of biogenic versus thermogenic methane in groundwater.

Figure 11.7 shows examples of the carbon and hydrogen isotopic composition of methane that could cause groundwater contamination, plotted against theoretical values from Whiticar (1999). Methane from biogenic organic carbon degradation, also referred to as methyl fermentation or acetate fermentation, is commonly found in landfills and cattle feedlots and has a distinct isotopic signature from other sources (Figure 11.7). Methane from coal bed formations such as the Williston Basin in Montana can be formed from bacterial carbonate reduction and has a lighter $\delta^{13}C$ and δD signature than typical thermogenic natural gas methane such as from the Barnett, Marcellus, and Utica Shales (Figure 11.7). Some producing basins have a mixture of thermogenic and biogenic gas, such as the Antrim Shale, the Denver Basin, and the Bowen Basin (Figure 11.7).

11.3.1 Case Study

A study in the Marcellus Shale drilling region of Pennsylvania near the town of Dimock found that groundwater wells within 1 km of active hydraulic fracking wells had elevated levels of methane and used $\delta^{13}C$-CH_4 and δ^2H-CH_4 to show that this methane was associated with Marcellus shale gas (Osborn et al. 2011; Jackson et al. 2013). These incidents of contamination attracted national and international attention, including features in documentary films such as Gasland (Hammond 2016). A later study of these wells from pre-drilling samples collected by the oil and gas industry indicated that groundwater methane levels in this region were high before the onset of shale gas drilling, but did not measure source indicators such as stable isotopes (Siegel et al. 2015).

Subsequent studies have advocated for the collection of pre-drilling baseline samples that include, among other analytes, the concentration and isotopic composition of methane in order to avoid the controversies that have plagued the Dimock region (Barth-Naftilan et al. 2018; McIntosh et al. 2019). We initiated a time series study in the adjacent state of Ohio to investigate whether hydraulic fracturing led to an increase in the concentration of natural gas methane in groundwater wells located near hydraulic fracturing wells (Botner

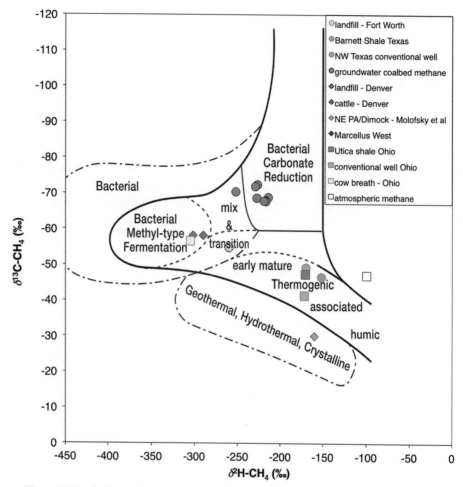

Figure 11.7 Carbon and hydrogen stable isotopic composition of methane from potential groundwater contamination sources, including agricultural and landfill sources as well as oil and gas sources. The isotopic composition of atmospheric methane is also shown. Theoretical range of values of methane sources shown in black and dotted lines from Whiticar (1999). (A black and white version of this figure will appear in some formats. For the colour version, refer to the plate section.)
Data are from Townsend-Small et al. (2016); Sherwood et al. (2017); Botner et al. (2018)

et al. 2018). The data were collected during a free groundwater testing program made available to residents of counties where hydraulic fracturing permits were issued during 2012 to 2015 (Botner et al. 2018). Figure 11.8 shows the number of groundwater samples and active hydraulic fracturing wells drilled over the study period.

A total of 180 groundwater samples were taken. Of those, 118 were from 24 drinking water wells that were sampled from two to eight times over the study period (dark blue triangles in Figure 11.8), depending on homeowner availability. Other samples were from groundwater wells sampled only once. Water samples were collected at the sink or well

Isotopes as Tracers of Methane Sources 283

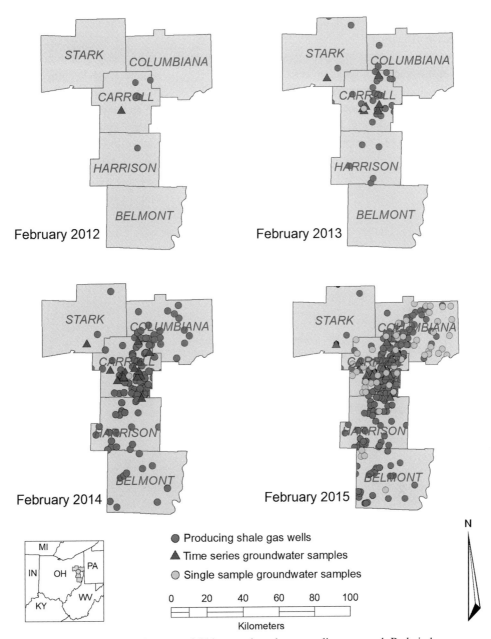

Figure 11.8 Time-series map of Ohio counties where sampling occurred. Red circles are active natural gas wells. Blue diamonds are sites where time series groundwater samples were taken, and light blue circles represent sites where single sample groundwater measurements were made. Groundwater sample locations are noted when samples were taken between the years noted in each map. There was a large increase in active natural gas wells from 2013 to 2014. (Botner et al. 2018) (A black and white version of this figure will appear in some formats. For the colour version, refer to the plate section.)

into serum vials, preserved, and sealed with septa and crimp seals for later analysis via headspace extraction.

There was no consistent pattern of increasing methane concentration in groundwater wells as the number of fracking wells increased in the region (Figure 11.8). In some wells, the concentration of methane was above the 10 mg L^{-1} level that can cause an explosion hazard in enclosed spaces (United States Department of the Interior 2001), and this concentration persisted as fracking increased. The dissolved methane concentration versus the carbon isotopic composition of groundwater methane is shown in Figure 11.9. Also

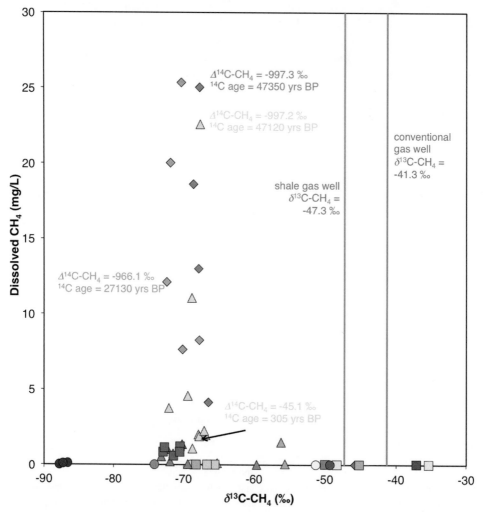

Figure 11.9 Carbon stable isotopic composition of CH$_4$ versus CH$_4$ concentration in samples collected along the time series. Each symbol represents a different groundwater well. Also included are measurements of the δ^{13}C of CH$_4$ from a shale gas well (− 47.3‰) and a conventional gas well (− 41.3‰) in the study area for comparison. The radiocarbon content of CH$_4$ (Δ^{14}C-CH$_4$) and radiocarbon age (in years bp) of four water samples is also shown. (Botner et al. 2018)

shown in Figure 11.9 is the isotopic composition of methane from a shale gas well and a conventional gas well in the same region. Groundwater wells with the highest methane concentrations have δ^{13}C-CH$_4$ signatures clustered around -70‰, clearly in the biogenic methane range (Figure 11.2). Natural gas from samples taken in the region has a δ^{13}C signature between -41.3‰ and -47.3‰ (Figure 11.9). Radiocarbon ages (labeled on the graph) of four samples with similar δ^{13}C signatures but varying methane concentrations indicated that carbon substrates ranged in age from fossil (Δ^{14}C–CH$_4 \sim -997$‰) to close to modern (Δ^{14}C–CH$_4 \sim -45.1$‰) (Figure 11.9).

Groundwater samples from sites measured only once did not show the expected trend (as shown in [Osborn et al. 2011; Jackson et al. 2013]) of higher methane within 1 km of an active gas well (Figure 11.10a), as some sites that were further than 1 km from active gas wells had high methane. Carbon stable isotopic composition of these samples (Figure 11.10b) also indicates that sites with higher methane concentrations have isotopic signatures consistent with a biogenic methane source.

Figure 11.11 shows the carbon and hydrogen stable isotopic composition of all groundwater samples collected during the study, color coded according to their methane concentration (Botner et al. 2018). The shape of the symbol corresponds to the nature of the sample, whether taken as a time series sample or from a groundwater well that was sampled only once. Also shown are several sources from the sampling region, including cow breath and natural gas from conventional and fracking wells, as well as the isotopic composition of methane in air, and literature values of methane sources from (Whiticar 1999). Most samples, including the highest in methane concentration, fall into the category of "bacterial carbonate reduction," which can occur in coal seams when CO_2 is converted into CH$_4$ (Osborn and McIntosh 2010; Golding et al. 2013; Vigneron et al. 2017; Vinson et al. 2017). A previous study of 15 groundwater wells throughout the state of Ohio, including some sites in Appalachia, also found that biogenic methane from carbonate reduction was the predominant source of methane, and that some groundwater wells had

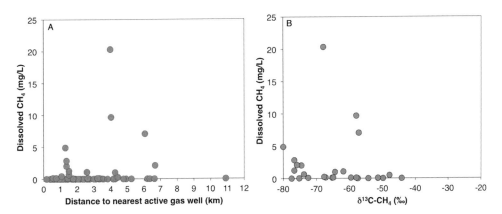

Figure 11.10 (a) Relationship of dissolved CH$_4$ concentration with distance from the nearest active shale gas well in groundwater wells sampled only once. (b) Carbon stable isotopic composition of CH$_4$ in the same samples. (Botner et al. 2018)

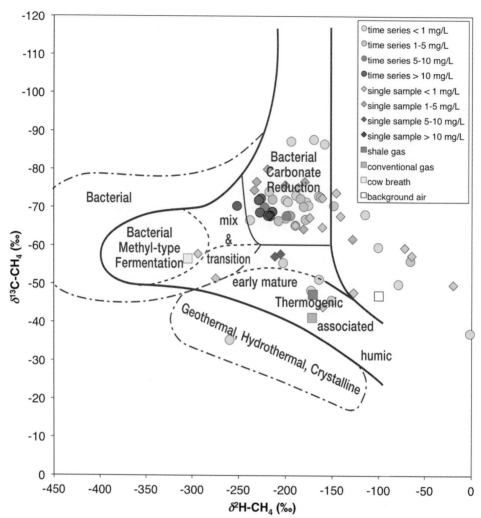

Figure 11.11 Carbon and hydrogen isotopic compositions of groundwater samples taken during this study, along with isotopic composition of several sources (value for background air is from Townsend-Small et al. (2016)). Also shown are approximate literature values for different endmembers of CH_4 sources from Whiticar (1999). Groundwater samples are shown in blue circles, for time series samples, and purple diamonds, for single samples, with shading corresponding to sample CH_4 concentration, with increasing shading corresponding to increasing concentration. (Botner et al. 2018) (A black and white version of this figure will appear in some formats. For the colour version, refer to the plate section.)

methane levels high enough to cause a flammability hazard (Thomas 2018). Figure 11.11 illustrates the importance of measuring both C and H stable isotopes for determining methane sources, as C stable isotopic analysis alone may lead to conflation of biogenic acetate fermentation and biogenic carbonate reduction sources. However, single isotope analysis does help distinguish between natural gas and other sources of methane in groundwater (Botner et al. 2018).

11.4 Conclusions

Isotopes can be powerful tools in elucidating methane sources in groundwater and the atmosphere. However, there are some caveats. Carbon stable isotopes have not provided consistent answers on atmospheric methane sources at the global or regional level, and there is still no scientific consensus on whether anthropogenic sources are the cause of recent increases in global methane levels. This is because there are several methane sources with similar carbon isotopic sources, and because natural gas has a similar carbon isotopic source to methane in air. Hydrogen isotopes of methane can be a more powerful tool, but they are less widely measured at this point. It is important to measure both isotopes in studies of groundwater methane as well, since there are multiple biogenic methane sources with similar carbon isotopic signatures in natural gas extraction areas. Radiocarbon is a potential tool but the high cost and large sample volume as well as unconstrained atmospheric nuclear radiomethane emissions are deterrents; it is a more useful tool for groundwater studies.

References

Alvarez RA, Zavala-Araiza D, Lyon DR, Allen DT, Barkley ZR, Brandt AR, Davis KJ, Herndon SC, Jacob D., Karion A, Kort EA, Lamb BK, Lauvaux T, Maasakkers JD, Marchese AJ, Omara M, Pacala SW, Peischl J, Robinson AL, Shepson PB, Sweeney C, Townsend-Smal, A, Wofsy SC, and Hamburg SP. (2018). Assessment of methane emissions from the U.S. oil and gas supply chain. *Science*. 7204. https://doi.org/10.1126/science.aar7204

Aydin M, Verhulst KR, Saltzman ES, Battle MO, Montzka SA, Blake DR, Tang Q, and Prather MJ. (2011). Recent decreases in fossil-fuel emissions of ethane and methane derived from firn air. *Nature*. 476: 198–201. https://doi.org/10.1038/nature10352

Baldassare F and Chapman E. (2018). Chapter 4 - The application of isotope geochemistry in stray gas investigations: Case studies. In Stout SA and Wang Z (eds.) *Oil Spill Environmental Forensics Case Studies*. Butterworth-Heinemann, pp. 67–86. https://doi.org/10.1016/B978-0-12-804434-6.00004-5

Barth-Naftilan E, Sohng J, and Saiers JE. (2018). Methane in groundwater before, during, and after hydraulic fracturing of the Marcellus Shale. *PNAS*. 115: 6970–6975. https://doi.org/10.1073/pnas.1720898115

Botner EC, Townsend-Small A, Nash DB, Xu X, Schimmelmann A, and Miller JH. (2018). Monitoring concentration and isotopic composition of methane in groundwater in the Utica Shale hydraulic fracturing region of Ohio. *Environmental Monitoring and Assessment*. 190: 322. https://doi.org/10.1007/s10661-018-6696-1

Clayton JL. (1998). Geochemistry of coalbed gas – A review. *International Journal of Coal Geology*. 35: 159–173. https://doi.org/10.1016/S0166-5162(97)00017-7

Coleman DD, Liu C-L, Hackley KC, and Pelphrey SR. (1995). Isotopic Identification of Landfill Methane. *Environmental Geosciences*. 2: 95–103.

Golding SD, Boreham CJ, and Esterle JS. (2013). Stable isotope geochemistry of coal bed and shale gas and related production waters: A review. *International Journal of Coal Geology*. 120: 24–40. https://doi.org/10.1016/j.coal.2013.09.001

Hammond PA. (2016). The relationship between methane migration and shale-gas well operations near Dimock, Pennsylvania, USA. *Hydrogeology Journal*. 24: 503–519. https://doi.org/10.1007/s10040-015-1332-4

Howarth RW. (2019). Ideas and perspectives: is shale gas a major driver of recent increase in global atmospheric methane? *Biogeosciences*. 16: 3033–3046. https://doi.org/10.5194/bg-16-3033-2019

Howarth RW. (2021). Methane and climate change. In Stolz JF, Griffin WM, and Bain DJ (eds.) *Environmental Impacts from the Development of Unconventional Oil and Gas Reserves*. Cambridge University Press.

Jackson RB, Vengosh A, Darrah TH, Warner NR, Down A, Poreda RJ, Osborn SG, Zhao K, and Karr JD. (2013). Increased stray gas abundance in a subset of drinking water wells near Marcellus shale gas extraction. *PNAS*. 110: 11250–11255. https://doi.org/10.1073/pnas.1221635110

Kai FM, Tyler SC, Randerson JT, and Blake DR. (2011). Reduced methane growth rate explained by decreased Northern Hemisphere microbial sources. *Nature*. 476: 194–197. https://doi.org/10.1038/nature10259

Keeling CD. (1958). The concentration and isotopic abundances of atmospheric carbon dioxide in rural areas. *Geochimica et Cosmochimica Acta*. 13: 322–334. https://doi.org/10.1016/0016-7037(58)90033-4

Keeling CD. (1961) The concentration and isotopic abundances of carbon dioxide in rural and marine air. *Geochimica et Cosmochimica Acta*. 24: 277–298. https://doi.org/10.1016/0016-7037(61)90023-0

Kendall C and Doctor DH. (2003). 5.11 - Stable Isotope Applications in Hydrologic Studies. In Holland HD and Turekian KK (eds.), *Treatise on Geochemistry*. Pergamon, pp. 319–364. https://doi.org/10.1016/B0–08-043751-6/05081-7

Lamb BK, Edburg SL, Ferrara TW, Howard T, Harrison MR, Kolb CE, Townsend-Small A, Dyck W, Possolo A, and Whetstone JR. (2015). Direct measurements show decreasing methane emissions from natural gas local distribution systems in the United States. *Environmental Science & Technology*. 49: 5161–5169. https://doi.org/10.1021/es505116p

Lassey KR, Etheridge DM, Lowe DC, Smith AM, and Ferretti DF. (2007a). Centennial evolution of the atmospheric methane budget: What do the carbon isotopes tell us? *Atmospheric Chemistry and Physics*. 7: 2119–2139. https://doi.org/10.5194/acp-7-2119-2007

Lassey KR, Lowe DC, and Smith AM. (2007b). The atmospheric cycling of radiomethane and the "fossil fraction" of the methane source. *Atmospheric Chemistry and Physics*. 7: 2141–2149. https://doi.org/10.5194/acp-7-2141-2007

Mace EK, Aalseth CE, Day AR, Hoppe EW, Keillor ME, Moran JJ, Panisko ME, Seifert A, Tatishvili G, and Williams RM. (2016). First results of a simultaneous measurement of tritium and 14C in an ultra-low-background proportional counter for environmental sources of methane. *Journal of Environmental Radioactivity*. 155–156: 122–129. https://doi.org/10.1016/j.jenvrad.2016.02.001

Martini AM, Walter LM, Budai JM, Ku TCW, Kaiser CJ, and Schoell M. (1998). Genetic and temporal relations between formation waters and biogenic methane: Upper Devonian Antrim Shale, Michigan Basin, USA. *Geochimica et Cosmochimica Acta*. 62: 1699–1720. https://doi.org/10.1016/S0016–7037(98)00090-8

McIntosh JC, Hendry MJ, Ballentine C, Haszeldine RS, Mayer B, Etiope G, Elsner M, Darrah TH, Prinzhofer A, Osborn S, Stalker L, Kuloyo O, Lu Z-T, Martini A, and Lollar BS. (2019). A critical review of state-of-the-art and emerging approaches to identify fracking-derived gases and associated contaminants in aquifers. *Environmental Science & Technology*. 53: 1063–1077. https://doi.org/10.1021/acs.est.8b05807

Milkov AV, Schwietzke S, Allen G, Sherwood OA, and Etiope G. (2020). Using global isotopic data to constrain the role of shale gas production in recent increases in atmospheric methane. *Scientific Reports*. 10: 1–7. https://doi.org/10.1038/s41598–020-61035-w

Nisbet EG et al. (2016). Rising atmospheric methane: 2007–2014 growth and isotopic shift. *Global Biogeochemical Cycles.* 30: 1356–1370. https://doi.org/10.1002/2016GB005406

Nisbet et al. (2019). Very strong atmospheric methane growth in the 4 Years 2014–2017: Implications for the Paris Agreement. *Global Biogeochemical Cycles.* 33: 318–342. https://doi.org/10.1029/2018GB006009

NOAA Global Monitoring Division. (2020). NOAA ESRL Global Monitoring Division - FTP Navigator [WWW Document]. URL www.esrl.noaa.gov/gmd/dv/data/index.php?parameter_name=C13%252FC12%2Bin%2BMethane (accessed March 1, 2020).

Osborn SG and McIntosh JC. (2010). Chemical and isotopic tracers of the contribution of microbial gas in Devonian organic-rich shales and reservoir sandstones, northern Appalachian Basin. *Applied Geochemistry.* 25: 456–471. https://doi.org/10.1016/j.apgeochem.2010.01.001

Osborn SG, Vengosh A, Warner NR, and Jackson RB. (2011). Methane contamination of drinking water accompanying gas-well drilling and hydraulic fracturing. *PNAS.* 108: 8172–8176. https://doi.org/10.1073/pnas.1100682108

Pataki DE, Ehleringer JR, Flanagan LB, Yakir D, Bowling DR, Still CJ, Buchmann N, Kaplan JO, and Berry JA. (2003). The application and interpretation of Keeling plots in terrestrial carbon cycle research. *Global Biogeochemical Cycles.* 17. https://doi.org/10.1029/2001GB001850

Peischl, J. et al. (2013). Quantifying sources of methane using light alkanes in the Los Angeles basin, California. *Journal of Geophysical Research: Atmospheres.* 118: 4974–4990. https://doi.org/10.1002/jgrd.50413

Pétron, G. et al. (2014) A new look at methane and nonmethane hydrocarbon emissions from oil and natural gas operations in the Colorado Denver-Julesburg Basin. *Journal of Geophysical Research: Atmospheres.* 119: 6836–6852. https://doi.org/10.1002/2013JD021272

Reeburgh WS. (2007). Global methane biogeochemistry. In Holland HD and Turekian KK (eds.) *Treatise on Geochemistry.* Pergamon, pp. 1–32. https://doi.org/10.1016/B0-08-043751-6/04036-6

Rice AL, Butenhoff CL, Teama DG, Röger FH, Khalil MAK, and Rasmussen RA. (2016). Atmospheric methane isotopic record favors fossil sources flat in 1980s and 1990s with recent increase. *PNAS.* 113: 10791–10796. https://doi.org/10.1073/pnas.1522923113

Schlegel ME, McIntosh JC, Bates BL, Kirk MF, and Martini AM. (2011). Comparison of fluid geochemistry and microbiology of multiple organic-rich reservoirs in the Illinois Basin, USA: Evidence for controls on methanogenesis and microbial transport. *Geochimica et Cosmochimica Acta.* 75: 1903–1919. https://doi.org/10.1016/j.gca.2011.01.016

Schwietzke S, Sherwood OA, Bruhwiler LMP, Miller JB, Etiope G, Dlugokencky EJ, Michel SE, Arling VA, Vaughn BH, White JWC, and Tans PP. (2016). Upward revision of global fossil fuel methane emissions based on isotope database. *Nature.* 538: 88–91. https://doi.org/10.1038/nature19797

Scott AR, Kaiser WR, and Ayers WB. (1994). Thermogenic and secondary biogenic gases, San Juan Basin, Colorado and New Mexico: Implications for coalbed gas producibility. *AAPG Bulletin.* 78: 1186–1209. https://doi.org/10.1306/A25FEAA9-171B-11D7-8645000102C1865D

Sherwood OA, Schwietzke S, Arling VA, and Etiope G. (2017). Global inventory of gas geochemistry data from fossil fuel, microbial and burning sources, version 2017. *Earth System Science Data.* 9: 639–656. https://doi.org/10.5194/essd-9-639-2017

Siegel DI, Azzolina NA, Smith BJ, Perry AE, and Bothun RL. (2015). Methane Concentrations in Water Wells Unrelated to Proximity to Existing Oil and Gas Wells in Northeastern Pennsylvania. *Environmental Science & Technology*. 49: 4106–4112. https://doi.org/10.1021/es505775c

Smith JW and Pallasser RJ. (1996). Microbial Origin of Australian Coalbed Methane. *AAPG Bulletin*. 80: 891–897. https://doi.org/10.1306/64ED88FE-1724-11D7–8645000102C1865D

Thomas MA. (2018). Chemical and isotopic characteristics of methane in groundwater of Ohio, 2016, U.S. Geological Survey Scientific Investigations Report 2018–5097.

Townsend-Small A, Tyler SC, Pataki DE, Xu X, and Christensen LE. (2012). Isotopic measurements of atmospheric methane in Los Angeles, California, USA: Influence of "fugitive" fossil fuel emissions. *Journal of Geophysical Research: Atmospheres*. 117. https://doi.org/10.1029/2011JD016826

Townsend-Small A, Botner EC, Jimenez KL, Schroeder JR, Blake NJ, Meinardi S, Blake DR, Sive BC, Bon D, Crawford JH, Pfister G, and Flocke FM. (2016). Using stable isotopes of hydrogen to quantify biogenic and thermogenic atmospheric methane sources: A case study from the Colorado Front Range: Hydrogen Isotopes in the Front Range. *Geophysical Research Letters*. 43: 11,462-11,471. https://doi.org/10.1002/2016GL071438

United States Department of the Interior. (2001). *Technical Measures for the Investigation and Mitigation of Fugitive Methane Hazards in Areas of Coal Mining*. Office of Surface Mining Reclamation and Enforcement.

US EPA National Center for Environmental Assessment, I.O. (2016). Hydraulic Fracturing for Oil and Gas: Impacts from the Hydraulic Fracturing Water Cycle on Drinking Water Resources in the United States (Final Report) [WWW Document]. URL https://cfpub.epa.gov/ncea/hfstudy/recordisplay.cfm?deid=332990 (accessed 5.7.20).

Vengosh A, Jackson RB, Warner N, Darrah TH, and Kondash A. (2014). A critical review of the risks to water resources from unconventional shale gas development and hydraulic fracturing in the United States. *Environmental Science & Technology*. 48: 8334–8348. https://doi.org/10.1021/es405118y

Vidic RD, Brantley SL, Vandenbossche JM, Yoxtheimer D, and Abad JD. (2013). Impact of shale gas development on regional water quality. *Science*. 340. https://doi.org/10.1126/science.1235009

Vigneron A, Bishop A, Alsop EB, Hull K, Rhodes I, Hendricks R, Head IM, and Tsesmetzis N. (2017). Microbial and Isotopic Evidence for Methane Cycling in Hydrocarbon-Containing Groundwater from the Pennsylvania Region. *Frontiers in Microbiology*. 8. https://doi.org/10.3389/fmicb.2017.00593

Vinson DS, Blair NE, Martini AM, Larter S, Orem WH, and McIntosh JC. (2017). Microbial methane from in situ biodegradation of coal and shale: A review and reevaluation of hydrogen and carbon isotope signatures. *Chemical Geology*. 453: 128–145. https://doi.org/10.1016/j.chemgeo.2017.01.027

Wennberg PO, Mui W, Wunch D, Kort EA, Blake DR, Atlas EL, Santoni GW, Wofsy SC, Diskin GS, Jeong S, and Fischer ML. (2012). On the Sources of Methane to the Los Angeles Atmosphere. *Environmental Science & Technology*. 46: 9282–9289. https://doi.org/10.1021/es301138y

Whiticar MJ. (1999). Carbon and hydrogen isotope systematics of bacterial formation and oxidation of methane. *Chemical Geology*. 161: 291–314. https://doi.org/10.1016/S0009–2541(99)00092-3

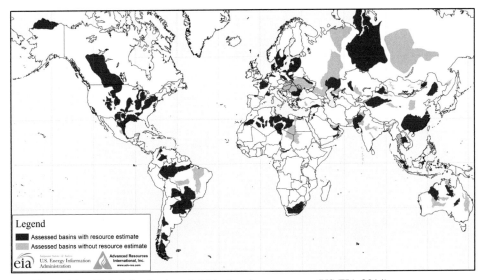

Figure 1.1 Global distribution of tight oil and gas reserves (US EIA 2014).

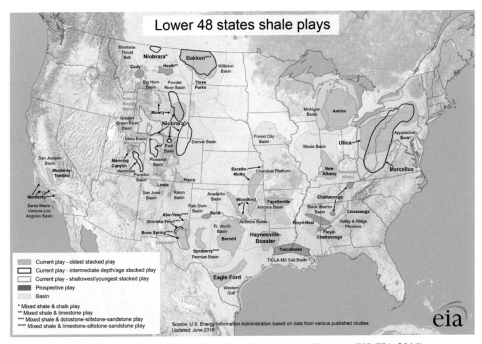

Figure 2.1 Map of United States shale plays in the lower 48 states (US EIA 2016)

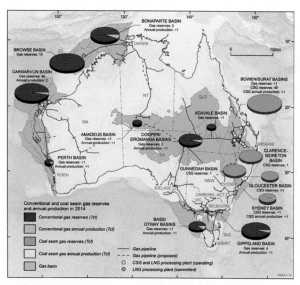

Figure 3.1　Estimates of conventional and unconventional gas reserves, and production in 2014 (tcf)
(Geoscience Australia, 2019)

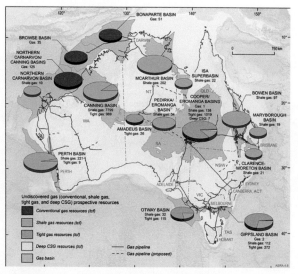

Figure 3.2　Prospective conventional and unconventional gas resources (tcf)
(Geoscience Australia, 2018)

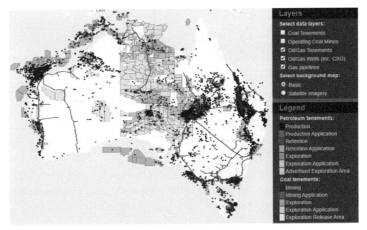

Figure 3.3 Oil and gas tenements (areas with a permit or authority from the government and under the relevant legislation to allow exploration or production of oil and gas) (Energy Resource Insights, 2020)

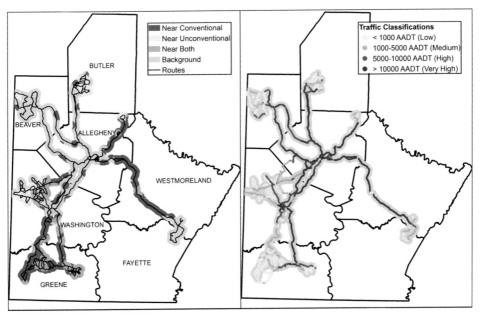

Figure 5.3 The extent of the study area (within 2 km of a sampled point) is shown as shaded. In the left panel, areas within 2 km of a conventional or unconventional well are shown in blue or yellow, respectively. Areas within 2 km of both types of wells are shown in green. Areas within 2 km of any well are classified as "near-well." In the right panel, average traffic density increases as the color gets darker. In general, more heavily traveled roads occur near Pittsburgh (in the center of Allegheny County) and on interstate highways. AADT = Annual Average Daily Traffic reported by Pennsylvania Spatial Data Access (www.pasda.psu.edu/, 2015).
Map generated in ESRI ArcGIS Pro

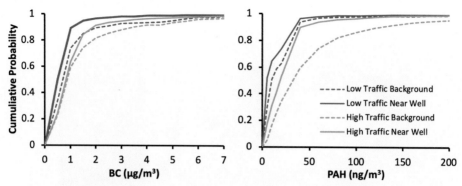

Figure 5.4 Background and near-well measurements for low and high traffic volumes. The amount of traffic on the roads significantly affects the underlying sampled distribution for both black carbon (left) and PAHs (right)

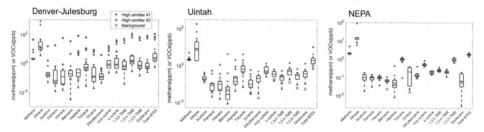

Figure 5.5 VOC (in ppb) and methane concentrations (in ppm) measured in (a) the Denver-Julesburg Basin, (b) the Uintah Basin, and (c) Northeastern PA Marcellus Shale (NEPA). Box-and-whisker plots represent distribution of measurements at gas production facilities. The red line in the middle of the box represents the median concentration; the top and the bottom of the box represent 75th and 25th percentile, respectively. The background measurements of each basin are shown as green crosses

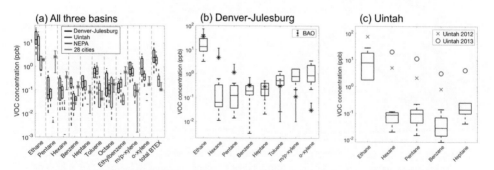

Figure 5.6 Comparison of the VOC concentration measured in this study with (a) VOC measured in 28 US cities (Baker et al. 2008), (b) VOC measured at Boulder Atmospheric Observatory (BAO) (Gilman et al. 2013), and (c) VOC measured in Horse Pool in winter 2012 and winter 2013 (Helmig et al. 2014).

Data from this study are presented using box-whisker plots, and data from previous studies are presented as symbols with standard deviation indicated by the error bars

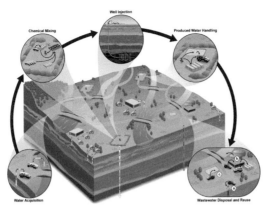

Figure 7.1 Water use and wastewater generation in the development of oil and gas resources. Specific activities in the "Wastewater disposal and reuse" inset are: (a) disposal via injection well, (b) wastewater treatment with reuse or discharge, and (c) evaporation or percolation pit disposal (US EPA 2015)

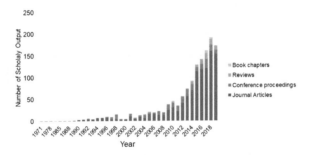

Figure 7.2 Increasing publications related to "oil and gas" and "water" over the past 50 years.
Figure courtesy Xiaoju (Julie) Chen

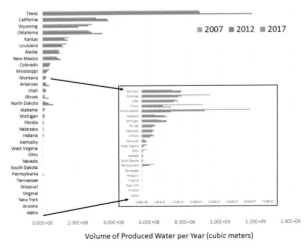

Figure 7.4 Volume of Produced Water by State for 2007, 2012, and 2017. Boxed inset focuses on those states not in the top 10 produced water generators (note x-axis scale).
Data from Clark and Veil (2009); Veil (2015); Veil (2020)

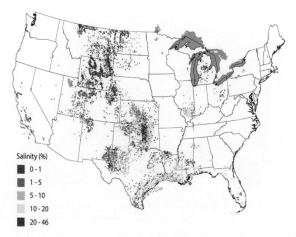

Figure 7.7 Salinity of produced waters in the United States (Allison and Mandler 2018)

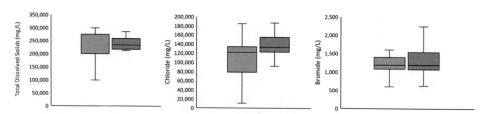

Figure 7.8 Produced water quality (TDS, chloride, and bromide) in Pennsylvania from conventional (blue) and unconventional (orange) wells.
Data from Hayes (2009)

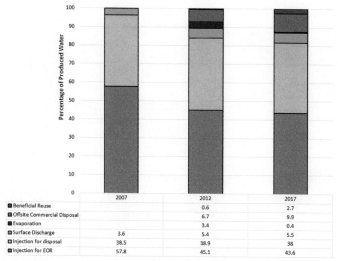

	2007	2012	2017
Beneficial Reuse		0.6	2.7
Offsite Commercial Disposal		6.7	9.9
Evaporation		3.4	0.4
Surface Discharge	3.6	5.4	5.5
Injection for disposal	38.5	38.9	38
Injection for EOR	57.8	45.1	43.6

Figure 7.10 Produced water management practices for 2007, 2012, and 2017.
Data from Clark and Veil (2009); Veil (2015); Veil (2020)

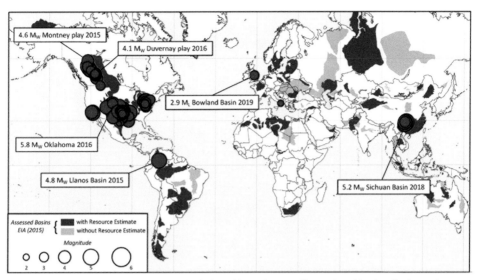

Figure 8.1 Locations of seismicity induced by saltwater disposal and hydraulic fracturing

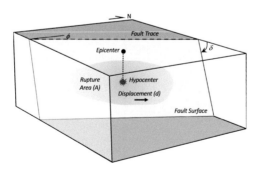

Figure 8.2 Schematic diagram illustrating basic earthquake terminology

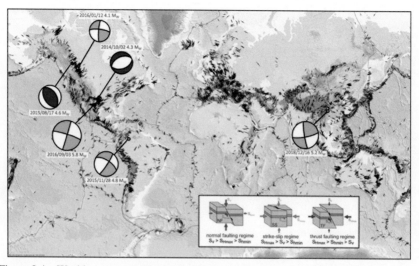

Figure 8.4 World stress map (Heidbach 2016) showing direction of SHmax, stress regime and method of stress determination, with overlay showing selected focal mechanisms for injection-induced events

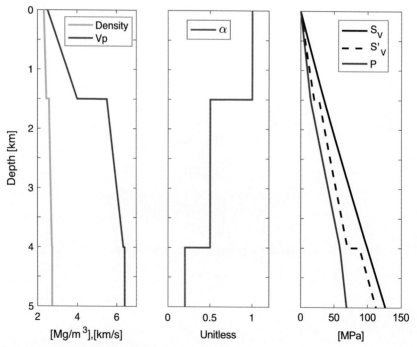

Figure 8.5 Graphs illustrating the calculation of a 1-D effective vertical stress (S'_V) profile for a hypothetical sedimentary basin. Symbols are defined in the text

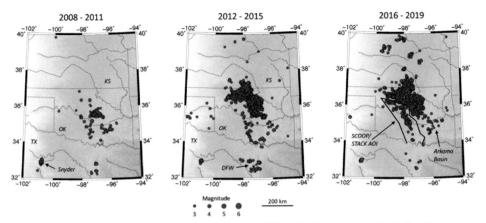

Figure 8.8 $M \geq 2.5$ earthquakes in Oklahoma (OK) and adjacent states including Texas (TX) and Kansas (KS)

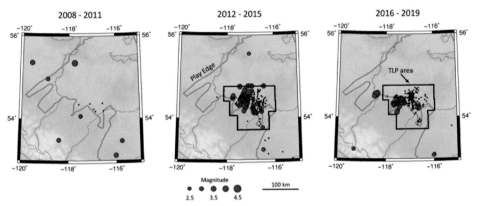

Figure 8.9 $M \geq 2.5$ earthquakes (red circles) in central Alberta, Canada

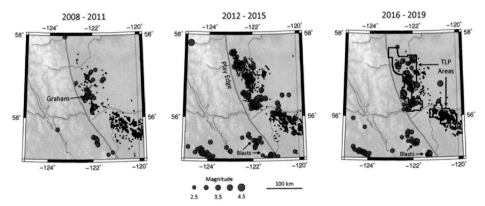

Figure 8.10 $M \geq 2.5$ earthquakes (red circles) in northeastern British Columbia (BC), Canada

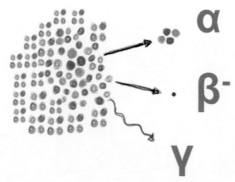

Figure 9.1 Three types of radioactivity that originate in the nucleus of an atom; alpha, a particle with two protons and two neutrons, beta, a negatively charged particle that changes a neutron to a proton, and gamma an electromagnetic energy that originates from the nucleus

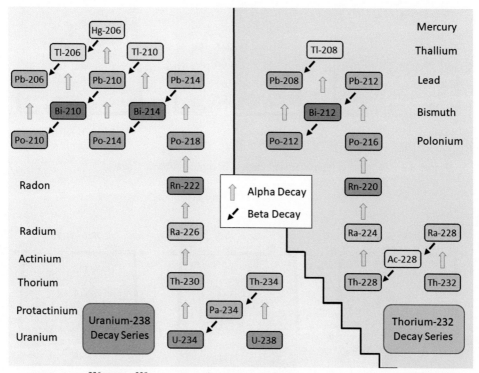

Figure 9.2 ^{226}Ra and ^{228}Ra decay chains showing the primordial parent nuclides, ^{238}U and ^{232}Th.
Adapted from Nelson et al. (2015)

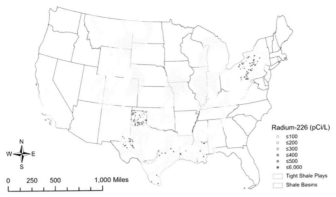

Figure 9.3a Map of the continental United States with published NORM data for produced water. Note the wide range in values reported in (pCi/L) but also the lack of publicly available data for many oil and gas basins. Most data is sourced from the USGS PW database and Fisher 1998

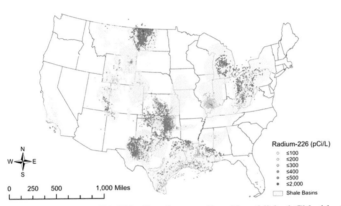

Figure 9.3b There are over 100,000 oil and gas wells with published Chloride (Cl) data available in the USGS PW database. Using the relationship between Cl and radium-226 described by Fisher (1998) for PW from Texas, an estimate of the radium 226 in other basins can be calculated

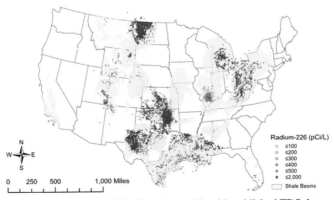

Figure 9.3c There are over 100,000 oil and gas wells with published TDS data available in the USGS PW database. Using the relationship between TDS and radium-226 described by Fisher (1998) for PW from Texas, an estimate of the radium 226 in other basins can be calculated

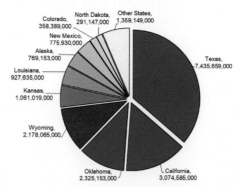

Figure 9.4 Top 10 states of oil and gas produced water volume generation (in barrels/year for 2012)

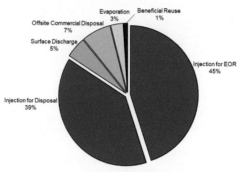

Figure 9.5 Produced water management for 2007 (based on Clark and Veil 2015)

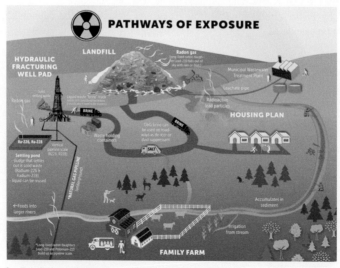

Figure 9.6 Depiction of NORM management and disposal.
Reprinted with permission from Center for Coal Field Justice

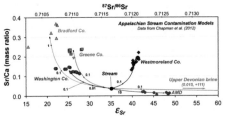

Figure 10.5 Mixing diagram for hypothetical Appalachian Basin stream water contaminated by either Marcellus Shale produced water (values shown here for four different counties in Pennsylvania), Upper Devonian conventional well produced water (endmember values far off scale), or acid mine drainage from SW Pennsylvania. The numbers indicated on the mixing curves are the percent of contaminant added to the stream water. Note that even in this worst-case scenario where the stream isotope composition falls wholly within the Marcellus field, it is still possible to differentiate among contamination sources.
Data are from Chapman et al. (2012)

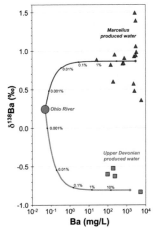

Figure 10.9 Barium isotope mixing diagram for river water (Ohio River, SW Pennsylvania) contaminated with produced water, using average values for Marcellus Shale and Upper Devonian produced waters. Individual data points represent samples collected throughout the Appalachian Basin. The values on the curves indicate the percentage of produced water mixed in to achieve the observed offset.
Data are from Tieman et al. (2020)

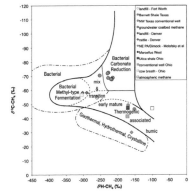

Figure 11.7 Carbon and hydrogen stable isotopic composition of methane from potential groundwater contamination sources, including agricultural and landfill sources as well as oil and gas sources. The isotopic composition of atmospheric methane is also shown. Theoretical range of values of methane sources shown in black and dotted lines from Whiticar (1999).
Data are from Townsend-Small et al. (2016); Sherwood et al. (2017); Botner et al. (2018)

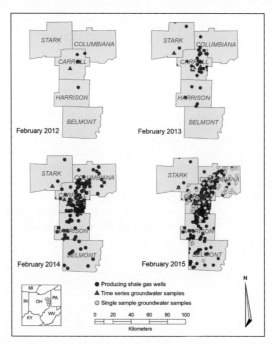

Figure 11.8 Time-series map of Ohio counties where sampling occurred. Red circles are active natural gas wells. Blue diamonds are sites where time series groundwater samples were taken, and light blue circles represent sites where single sample groundwater measurements were made. Groundwater sample locations are noted when samples were taken between the years noted in each map. There was a large increase in active natural gas wells from 2013 to 2014. (Botner et al. 2018)

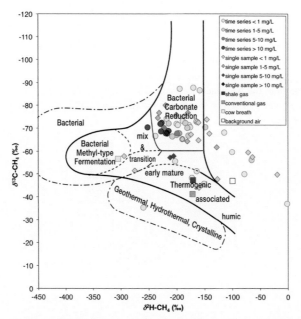

Figure 11.11 Carbon and hydrogen isotopic compositions of groundwater samples taken during this study, along with isotopic composition of several sources (value for background air is from Townsend-Small et al. (2016)). Also shown are approximate literature values for different endmembers of CH_4 sources from Whiticar (1999). Groundwater samples are shown in blue circles, for time series samples, and purple diamonds, for single samples, with shading corresponding to sample CH_4 concentration, with increasing shading corresponding to increasing concentration. (Botner et al. 2018)

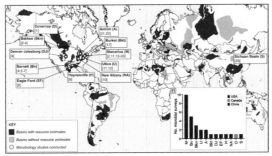

Figure 12.1 (A) Global location of shale formations with assessed resource estimations (dark red) and those with resources yet to be estimated (light brown). Shale formations where microbiology investigations have been assessed are marked with yellow circles (adapted from Mouser et al. 2016), and the numbers refer to the following references (full details in reference list at the end of the chapter): [1] Zhong et al. 2019; [2] Strong et al. 2013; [3] Lipus et al. 2018; [4] Wang et al. 2017; [5] Struchtemeyer et al. 2011; [6] Davis et al. 2012; [7] Struchtemeyer and Elshahed 2012; [8] Santillan et al. 2015; [9] Fichter et al. 2012; [10] Schlegel et al. 2011; [11] Daly et al. 2016; [12] Booker et al. 2017; [13] Murali Mohan et al. 2013a; [14] Murali Mohan et al. 2013b; [15] Cluff et al. 2014; [16] Akob et al. 2015; [17] Tucker et al. 2015; [18] Vikram et al. 2016; [19] Lipus et al. 2017b; [20] Nixon et al. 2019; [21] Wuchter et al. 2013; [22] Kirk et al. 2012; [23] Zhang et al. 2017. Base map modified from the US Energy Information Services Assessed Resources Basin Map, 2013 (public domain). (B) Number of microbiology studies conducted for each formation, where a survey of the population size, composition, or both, has been carried out

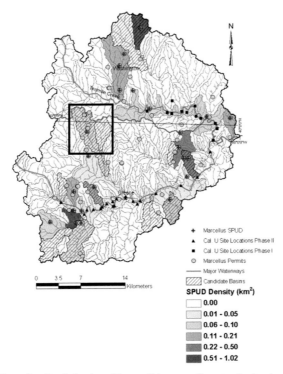

Figure 14.1 Tenmile Creek basin with candidate small watershed-pairs indicated with hatching. Shading indicates the density of Marcellus wells (SPUDs/km^2) in the basin and circles indicate existing permits granted as of 2009. Triangles and squares indicate sites previously sampled by Kimmel and Argent (2010, 2012) for fish and macroinvertebrates. The selected Bates Fork and Fonner Run stream-pair is delineated inside the box.
Figure by Dan Bain with permission

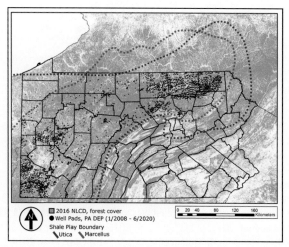

Figure 15.1 Forest cover (green) overlaying Marcellus (blue) and Utica (red) shale plays with wells drilled in PA as of 2019

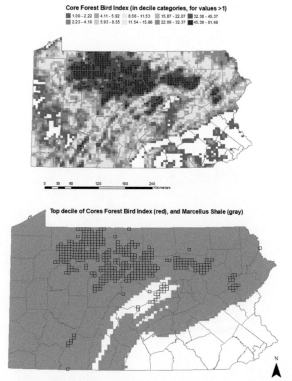

Figure 15.2 Combined abundance of 13 forest interior specialists based on abundance data from the second breeding bird atlas (Shen 2015, top figure). Abundance is greatest in dark red blocks. Bottom shows the blocks which support the greatest abundance of forest interior specialists above the Marcellus shale layer

Whiticar M and Schaefer H. (2007). Constraining past global tropospheric methane budgets with carbon and hydrogen isotope ratios in ice. *Philosophical Transactions of the Royal Society A: Mathematical, Physical and Engineering Sciences.* 365: 1793–1828. https://doi.org/10.1098/rsta.2007.2048

Worden JR, Bloom AA, Pandey, S., Jiang, Z., Worden, H.M., Walker, T.W., Houweling S, and Röckmann T. (2017). Reduced biomass burning emissions reconcile conflicting estimates of the post-2006 atmospheric methane budget. *Nature Communication.* 8: 2227.

12

The Microbiology of Shale Gas Extraction

SOPHIE L. NIXON

12.1 Introduction

Onshore natural gas is recovered from shale formations by means of hydraulic fracturing, a combination of vertical and horizontal drilling and high-pressure fluid injection deep in the subsurface. This technique has transformed the global energy landscape, particularly in the United States, where shale gas is now a major export to nations across the world. Natural gas resides in small, unconnected pores in the rock, the breakdown product of fossilized organic matter in mudstones after hundreds of millions of years of diagenesis. Hydraulic fracturing forces open fractures in the rock, held open by a proppant (typically sand), which allows previously trapped gas to flow freely to the surface in economically recoverable volumes. This process represents a major engineering intervention into the subsurface, involving the high-pressure injection of large volumes of freshwater and chemicals several kilometers deep. One by-product of these operations, and the subject of this chapter, is the creation of unique microbial ecosystems in the deep subsurface. The injected fluids that are recovered at the surface during flowback (flowback waters) and the subsequent fluids produced from the formation (produced fluids) contain microbial signatures that provide a wealth of information about the diversity, function, and adaptation strategies of fractured shale microbial communities.

The purpose of this review is to summarize what is known about the microbiology of the fractured shale environment to date, with a particular focus on the potential for microbial processes that negatively impact on shale gas extraction (collectively termed "biofouling" here). This review builds on others published in recent years (Mouser et al. 2016; Struchtemeyer et al. 2017; Struchtemeyer 2018) by providing an up to date overview of the microbiology of hydraulic fracturing activities, including a number of studies conducted on fractured shale formations outside the United States. The following discussion addresses the habitability of pristine shale formations, the creation of new habitats as a result of hydraulic fracturing, the potential for biofouling during production, including sulfide production and biocide resistance, and some of the major gaps in our knowledge.

12.2 Habitability of Pristine Shales

Onshore natural gas is produced from black shales, fine- to extremely fine-grained sedimentary rocks containing varying amounts of carbonates, pyrite, and high organic matter content. Black shales exhibit extremely low permeabilities in the nanodarcy range, with porosities commonly below 5%. Natural gas is physically trapped in the small pore throats, necessitating artificial fracturing to recover the resource. The main limitation to microbial life in pristine shales is this lack of space, with pore throat sizes often smaller than the average size of a microbial cell and unconnected pores limiting the availability of meteoric water. Pristine shales are therefore considered limited microbial habitats (Mouser et al. 2016).

Despite space limitations, viable microorganisms are known to inhabit pristine shale formations. Several microorganisms have been cultured from shale core material recovered from several kilometers deep, including sulfate-reducing, acetogenic, iron-reducing bacteria, and methanogens (Colwell et al. 1997; Fredrickson et al. 1997; Krumholz et al. 1997; Onstott et al. 1998). Furthermore, methanogens were successfully cultured at high salinity from formation fluids from the Antrim shale, indicating that halotolerant methanogens are native to that formation (Waldron et al. 2007). Indeed, the natural gas in the Antrim shale is partly biogenic in origin (Martini et al. 1998), suggesting that these native methanogenic populations have been long active in the formation. Tucker et al. (2015) also reported evidence of native methanogens in pristine Marcellus shale that appear to become stimulated by hydraulic fracturing. These indigenous microorganisms are thought to use refractory organic matter from the shale formation as carbon and energy sources (Wuchter et al. 2013). Shales typically harbor 2–9% total organic carbon in the form of ancient fossilized organic matter, known as kerogen (Vandenbroucke and Largeau 2007). Evidence for the bioavailability of this material comes from the stimulation of methanogenesis in Antrim communities when water-soluble shale organic matter was supplied as the sole source of carbon (Huang 2008). Further, shale-derived refractory organic matter or its fermentation products may diffuse into more permeable adjacent rocks, such as sandstones, where more abundant microbial activity has been observed (Fredrickson et al. 1997; Krumholz et al. 1997). Together, these studies provide evidence for the use of organic matter as a carbon and energy source for indigenous microbial populations in pristine shale formations.

Not all shale formations are considered habitable. During burial, some formations have been heated to above the known upper temperature limit for life (121°C; Kashefi and Lovely 2003). Unlike the Antrim shale, natural gas in the Barnett shale is entirely thermogenic in origin, generated by high formation temperatures in excess of 150°C, several hundred million years ago (Struchtemeyer et al. 2011). As such, while it may have harbored microbial life prior to these conditions, and present day temperatures are within the habitable range, pristine shale in the Barnett formation is generally considered to be sterile (Struchtmeyer et al. 2011). The Haynesville shale (Texas and Louisiana) has a present-day temperature profile assumed to be inhibitory to life, with some downwell temperature recordings at 3.5 km depth of 127–182°C (Fichter et al. 2012). Similarly, shale gas wells more than 4 km deep in the Eagle Ford shale formation have temperatures above 160°C (Santillan et al. 2015). Hence, while all shale formations are tight rock habitats, in which microbial activity is limited, some formations have been effectively "paleopasteurized" and others remain uninhabitable at depth.

As discussed in this chapter, the high pressure injection of fluids during shale gas extraction acts to open up the tight rock formation and seed it with a diverse array of microorganisms, dwarfing signals of any limited microbial presence in pristine shale. However, it seems feasible that organic matter from the formation itself could sustain nonnative fractured shale communities long after the chemical energy introduced in injected fluids has been exhausted, contributing to their persistence during shale gas production.

12.3 The Fractured Shale Microbial Ecosystem

Over the last decade, an increasing body of literature has shown that hydraulically fractured shales are habitats for microbial life (Figure 12.1; Table 12.1). From these studies, it is clear that hydraulic fracturing creates new microbial ecosystems in the deep terrestrial subsurface. The introduction of water, additives, and microorganisms in injected fluids into newly formed fractures serves to inoculate a previously limited habitat with diverse microbial consortia (Mouser et al. 2016). Over time, an abundance of chemical energy, in concert with subsurface selection pressures, facilitates the establishment of low diversity communities adapted to the extreme conditions that characterize fractured shale environments. Despite the combination of high hydrostatic pressures, moderate to high temperatures, brine-level salinities (Table 12.1), anoxia, and use of biocides, fractured shale formations harbor active microbial communities, dominated by halotolerant taxa, including *Halanaerobium*, *Halomonadaceae*, and *Methanohalophilus* (Davis et al. 2012; Struchtemeyer and Elshahed 2012; Murali Mohan et al. 2013a; Strong et al. 2013; Wuchter et al. 2013; Cluff et al. 2014; Daly et al. 2016; Liang et al. 2016; Evans et al. 2018). Members of these communities have been shown to persist over several hundred days, and can even tolerate surface storage and treatment prior to their re-introduction downhole as recycled produced waters for subsequent fracturing operations (Murali Mohan et al. 2013b; Cliffe et al, 2020). The recovery of strikingly similar microbial communities from distinct shale formations, each with different geological histories and associated physico-geochemical regimes, demonstrates that hydraulic fracturing creates a new environmental niche unique to fractured shales, resulting in active microbial ecosystems kilometers deep in the terrestrial subsurface (Daly et al. 2016; Lipus et al. 2018).

Assessment of microbial community composition in production fluids over time since hydraulic fracturing has highlighted a clear pattern of succession in these ecosystems. Injection fluids are dominated by predominantly aerobic bacteria common to freshwater or surface environments. With time, a less diverse community of facultative and anaerobic microorganisms emerges in flowback and produced waters, dominated by bacteria that can withstand salinities of up to 320,000 mg/L total dissolved solids (TDS) that result from interactions of the injected freshwater and the brine-rich formation (Table 12.1; Davis et al. 2012; Struchtemeyer and Elshahed 2012; Murali Mohan et al. 2013a; Strong et al. 2013; Wuchter et al. 2013; Culff et al. 2014; Akob et al. 2015; Daly et al. 2016). This end-member community of halotolerant organisms highlights salinity as a key selection pressure to this succession. The enrichment of highly oxygen-sensitive methanogens, such as *Methanohalophilus*, indicates that the fractured shale environment also becomes anoxic

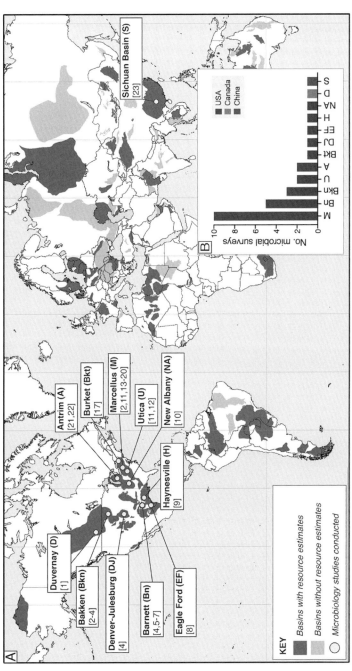

Figure 12.1 (A) Global location of shale formations with assessed resource estimations (dark red) and those with resources yet to be estimated (light brown). Shale formations where microbiology investigations have been assessed are marked with yellow circles (adapted from Mouser et al. 2016), and the numbers refer to the following references (full details in reference list at the end of the chapter): [1] Zhong et al. 2019; [2] Strong et al. 2013; [3] Lipus et al. 2018; [4] Wang et al. 2017; [5] Struchtemeyer et al. 2011; [6] Davis et al. 2012; [7] Struchtemeyer and Elshahed 2012; [8] Santillan et al. 2015; [9] Fichter et al. 2012; [10] Schlegel et al. 2011; [11] Daly et al. 2016; [12] Booker et al. 2017; [13] Murali Mohan et al. 2013a; [14] Murali Mohan et al. 2013b; [15] Cluff et al. 2014; [16] Akob et al. 2015; [17] Tucker et al. 2015; [18] Vikram et al. 2016; [19] Lipus et al. 2017b; [20] Nixon et al. 2019; [21] Wuchter et al. 2013; [22] Kirk et al. 2012; [23] Zhang et al. 2017. Base map modified from the US Energy Information Services Assessed Resources Basin Map, 2013 (public domain). (B) Number of microbiology studies conducted for each formation, where a survey of the population size, composition, or both, has been carried out (A black and white version of this figure will appear in some formats. For the colour version, refer to the plate section.)

295

Table 12.1. *A summary of key microbiology studies carried out on shale gas wells in formations in the USA, Canada, and China, including information on formation age, well depths, sample types and age since hydraulic fracturing, maximum salinity, and dominant microorganisms. Locations of formations are given in Figure 12.1*

Formation	Age	Study	Well depth (m)	Sample types[a]	Oldest PW sample (days)[b]	Max PW (or FW) Cl- (mg/L)	Max PW (or FW) TDS[c] (mg/L)	Dominant PW (or FW) taxa[d]
Marcellus	Middle Devonian	Strong et al. (2013)	-	IF, PW	547	65,452	-	*Thauera*
		Murali Mohan et al. (2013a)	2,450	DM, ISF, IF[$], FW, PW	187	91,800	-	*Halanaerobium**
		Cluff et al. (2014)	2,517	IF[$], FW, PW	-	95,100	-	*Halanaerobium**
		Akob et al. (2015)	1,232-2,555	PW	1,155	184,000	320,000	-
		Daly et al. (2016)	2,500	IF[$], FW, PW	328	95,100	-	*Halanaerobium*, Hatomonadaceae
		Lipus et al. (2017b)	-	PW	1,846	-	223,000	Halanaerobiaceae *Marinilabilia**,
		Nixon et al. (2019)	-	PW	313	-	122,000	*Halanaerobium**, *Methanohalophilus*
Barnett	Mississippi	Davis et al. (2012)	-	PW	-	110,000	-	*Halanaerobium**, *Marinobacter*, *Arcobacter*, *Pseudomonas*
		Struchtemeyer and Elshahed (2012)	-	FW	N/A	-	(3,730)	(Bacillaceae, Planococcaceae)
		Wang et al. (2019)	2,395	FW	N/A	(127,518)	(175,120)	-

Formation	Age	Reference		Sample type	Days			Dominant lineage
Bakken	Late Devonian / Early Mississippian	Strong et al. (2013) Lipus et al. (2017a)	-	FW	N/A	(207,000)	-	(*Halanaerobium, Pseudomonas*)
		Wang et al. (2019)	2,529 - 2,560	FW	N/A	(214,553)	(295,280)	(*Bradyrhizobium, Geobacter, Lactococcus*)
Utica	Middle Ordovician	Booker et al. (2017)	-	IF$, PW	200	-	-	*Halanaerobium**
Antrim	Upper Devonian	Daly et al. (2016)	-	-	302	108,741	-	-
		Wuchter et al. (2013)	-	PW	730	101,000	-	*Halanaerobium, Marinilabilia, Cytophaga*
Burket	Upper Devonian	Akob et al. (2015)	1,330	PW	152	109,000	183,793	-
Denver-Julesburg	Late Cretaceous	Wang et al. (2019)	2,133	PW	140	19,084	30,600	*Rhodococcus*
Eagle Ford	Late Cretaceous	Santillan et al. (2015)	-	PW	-	13,253	-	*Microbacterium*
Duvernay	Upper Devonian	Zhong et al. (2019)	-	ISF, PW	183	119,780	268,000	*Halanaerobium**
Sichuan	-	Zhang et al. (2017)	4,317	PW	304	24,197	-	*Shewanella*

---

[a]DM = drilling mud, ISF = input source fluid, IF = input fluid (§20% produced water, 80% freshwater), FW = flowback water, PW = produced water

[b]approximate time since hydraulic fracturing

[c]Total Dissolved Solids

[d]dominant lineage reported in one or more sample in the study. *denotes >90% of the community

with time (Daly et al. 2016; Nixon et al. 2019). Pressure is likely to contribute to early changes in diversity observed in flowback fluids, as many freshwater microorganisms adapted to life at the surface are unable to withstand the very high pressures of hydraulic fracturing (up to 80 MPa; Mouser et al. 2016) and high hydrostatic pressures in the fractured shale environment (21–48 MPa; Booker et al. 2019).

Given the prevailing high salinities measured in produced waters, osmoprotection is a key requirement for survival in the fractured shale environment. Persistent fractured shale taxa have both the genomic potential (Daly et al. 2016; Nixon et al. 2019) and demonstrated ability (Borton et al. 2018b; Daly et al. 2019) to synthesise and accumulate osmolytes that maintain osmotic balance in high salinity conditions. Some of these osmolytes have been shown to fuel metabolic processes by key lineages. For instance, glycine betaine, an osmolyte produced by *Methanohalophilus* and accumulated widely among osmoadapted fractured shale taxa, can be metabolized by *Halanaerobium*, yielding trimethylamine, which is in turn used by *Methanohalophilus* in methylotrophic methanogenesis (Daly et al. 2016; Borton et al. 2018a; Borton et al. 2018b). Sarcosine is also implicated with interconnected metabolic networks among fractured shale taxa, contributing to the production of organic acids and ammonium (Borton et al. 2018b). The ability to accumulate osmolytes or "compatible solutes" is a widespread trait among fractured shale taxa, including rare low abundance lineages usually found in soils and freshwater or marine environments (Nixon et al. 2019). However, dominant halotolerant lineages such as *Halanaerobium*, *Methanohalophilus*, and *Orenia* have undergone more substantial proteome-wide salinity adaptation. These organisms operate a "salt-in" strategy, whereby inorganic solutes such as KCl are imported into the cytoplasm to balance osmotic stress (Daly et al. 2016). This greater degree of salinity adaptation allows these lineages to outcompete less well-adapted organisms in the fractured shale environment, for which energy demands of the compatible solute strategy can compromise the energetic yields of non-fermentative metabolisms, for instance dissimilatory sulfate reduction (Booker et al. 2017).

While pressure, salinity, and anoxia drive microbial succession in fractured shales, viruses further shape the community. Using genome-resolved metagenomics, Daly et al. (2016) identified a large number of virus–host associations in a late stage fractured shale community, indicating that viruses are an important component of the ecosystem. Furthermore, an increase in the number of CRISPR spacers in key microbial genomes between time points suggests active viral predation in fractured shales (Daly et al. 2016). A more recent study by the same group identified a high diversity of viruses in fractured shale communities, in line with that of peat soil ecosystems, the majority belonging to previously unknown genera (Daly et al. 2019). The abundance of viruses associated with *Halanaerobium* genomes was tightly coupled with host abundance, highlighting the resilience of *Halanaerobium* not only to changes in salinity, pressure, and redox but also viral predation (Daly et al. 2019).

Viruses are likely to shape fractured shale communities in a number of ways. First, strain-level community composition is influenced, as those with CRISPR immunity will resist viral predation and outcompete those without. Indeed, the advent of genome-resolved metagenomics has facilitated the study of CRISPR immunity, and simply by counting the number of CRISPR-Cas arrays, prophage and spacers one can assess resilience to (and history of) viral attack (e.g., Daly et al. 2016; Daly et al. 2019; Nixon et al. 2019). Unsurprisingly, CRISPR immunity appears to be widespread among fractured shale microbes (Daly et al. 2016; Nixon et al. 2019). Second, viral lysis liberates substrates from host cells that can support a network of metabolic interactions in the wider community. Daly et al. (2019) demonstrated that induced prophage within fractured shale isolates of *Halanaerobium* led to the release of amino acids, implicated in a network of Strickland fermentation reactions within fractured shale communities (Daly et al. 2016; Borton et al. 2018b). These fermentation pathways are likely to be important in the long-term persistence of key shale taxa, after more labile substrates present as additives are depleted. As such, active viral populations shape the community by influencing the availability of key substrates for metabolic interactions. Finally, active viral predation may facilitate horizontal gene transfer within fractured shale communities, potentially enabling previously non-adapted strains to better withstand selection pressures.

It is clear from research conducted in the last decade that fractured shales host dynamic and adapted microbial ecosystems, created as a consequence of hydraulic fracturing. Key members of these microbial communities are thought to be introduced in the drilling and fracturing fluids injected into the formation in the process. As such, studying these deep subsurface communities sheds light on microbial survival and adaptation in an extreme environment far from their origins (Daly et al. 2016), and can ultimately further our understanding of microbial resilience in extreme environments and the limits to microbial habitability.

12.4 Microbial Biofouling during Shale Gas Extraction

A number of microbial processes that occur in shale gas wells have the potential to damage well infrastructure, impact gas yields, and increase the risk of environmental damage. These processes include the production of corrosive metabolic by-products, colonization of fractures by microbial biofilms, and the emergence of biocide resistance, which serves to exacerbate microbial issues further. Collectively, these processes are referred to here as "biofouling." Several factors influence these biofouling processes, and these are discussed next.

12.4.1 Use of Biocides

Biocides are a common component of injected fluids, added to prevent microbial growth in shale gas wells. Their general ineffectiveness, however, is evident from the numerous studies reporting the recovery of microbial communities adapted to downwell conditions (Struchtemeyer and Elshahed 2012; Murali Mohan et al. 2013a; Akob et al. 2015; Santillan et al. 2015). While most studies have focused on microbial community composition using DNA-based methods, a number of common fractured shale taxa have been successfully enriched or isolated, including members of the *Halanaerobium* and *Methanohalophilus* genera (e.g., Liang et al. 2016; Booker et al. 2017; Borton et al. 2018a), demonstrating that these halotolerant communities are viable. The reduced efficacy of biocide treatments downwell compared with laboratory tests is attributed to conditions in the fractured shale, namely the high concentrations of salts, metals, and organics (Murali Mohan et al. 2013a). Some have suggested biocide dilution results from their injection into the formation, leading to sublethal levels, and perhaps even their microbial degradation and mineralization (Kahrilas et al. 2015). One study of a biocide-amended shale gas well in the Duvernay formation (Canada) observed a decrease in biocide concentration to below detection after just 10 days of flowback, indicating that biocide doses may be insufficient for the majority of the flowback and production phases (Zhong et al. 2019). Collectively, these findings demonstrate that adapted communities of viable microorganisms are able to persist in shale gas wells in spite of biocide addition.

Several different biocides are used to control microbial growth during shale gas extraction. Readers are referred to Kahrilas et al. (2015) for an in-depth review of their characteristics and modes of action. The most commonly used biocide in disclosed hydraulic fracturing operations is glutaraldehyde (Elsner and Hoelzer 2016). Despite its prevalence, several microbiology-focused studies found glutaraldehyde-based biocides to be particularly ineffective. A study of microbial control in Barnett Shale natural gas wells found glutaraldehyde and glutaraldehyde-based biocides to be ineffective in killing general aerobic and acid-producing bacteria, the dominant microorganisms in fracturing fluid source waters, whereas THPS consistently performed well in killing microorganisms, including fermentative and sulfate-reducing bacteria (Johnson et al. 2008). Murali Mohan and colleagues (2013b) observed that a glutaraldehyde-amended impoundment pond in the Marcellus region, used to store production fluids prior to reuse or disposal, contained higher microbial biomass than either untreated or pretreated and aerated impoundment ponds. The microbial lineages detected in the biocide-treated impoundment pond include members of the *Halanaerobium*, *Marinobacter*, and *Methanohalophilus* genera, all persistent halotolerant members of late-stage fractured shale ecosystems (Daly et al. 2016), in addition to a number of sulfidogenic lineages (Murali Mohan et al. 2013b). In contrast, the impoundment pond that underwent aeration had lower microbial biomass and lacked these anaerobic and sulfidogenic lineages, not only highlighting biocide inefficiencies but also indicating that alternative strategies may be more effective in preventing microbial growth (Murali Mohan et al. 2013b).

Far from preventing colonization of fractured shales, ineffective biocide use may in fact exacerbate microbial activity. Sublethal concentrations of biocides present in shale gas wells can select for biocide-resistant microorganisms. In one study, prolonged exposure of produced water microbial communities to glutaraldehyde led to increased tolerance (Vikram et al. 2014). This selective antimicrobial pressure could stimulate the emergence of antimicrobial resistant genes in shale gas extraction operations, potentially creating antimicrobial hotspots (Campa et al. 2019). The degree to which biocide addition during hydraulic fracturing poses an antimicrobial threat to the environmental and human health is not well understood. Indeed, the mechanisms of biocide resistance are unknown, and hence their potential to be mobilized through lateral gene transfer is also unknown (Campa et al.

2019). Given the prevalence of fractured shale microorganisms in biocide-treated wells, and evidence that sublethal biocides select for resistance, there is clearly a risk that ineffective use not only threatens ongoing biofouling issues but may even contribute to the growing global antimicrobial resistance crisis.

12.4.2 Input Chemistry Degradation

A wide range of additives are used in drilling muds and hydraulic fracturing fluids during shale gas extraction. Individual compounds are added to input fluids to fulfil specific functions. Some of these must be temporally specific, such as crosslinkers and breakers that interact with a base polymer in the fluid, and act to increase and decrease the fracturing fluid viscosity, respectively. Other additives are intended to act for the duration of operations, such as scale and corrosion inhibitors, clay stabilizers, and biocides (Elsner and Hoelzer 2016). Drilling muds, whether oil- or water-based, also contains a myriad of additives, including starch added for viscosity modification and barite (barium sulfate) added as a weighting agent (Struchtemeyer et al. 2011). Large volumes of drilling mud and hydraulic fracturing fluid remain in formation during shale gas production, and hence additives are available to the microbial community as potential substrates.

Several common additives used in drilling and hydraulic fracturing are known to stimulate biofouling. More research has focused on fracturing fluid additives than on drilling mud components, though there is evidence that both can stimulate microbial activity downwell. For instance, barite and sulfonate in Barnett Shale drilling muds stimulated microbial production of sulfide in microcosm experiments, with a concurrent enrichment of multiple sulfidogenic lineages (Struchtemeyer et al. 2011). Organic polymers in these muds are thought to have played a role by supplying electron donors for sulfate reduction, either directly or indirectly, with an enrichment in anaerobic heterotrophs suggesting the generation of fermentation by-products (Struchtemeyer et al. 2011). In other studies, additives added to fracturing fluids have been linked to microbial metabolism. A number of common additives are widely known to be readily degraded by aerobic and anaerobic microbial consortia when supplied as the sole substrate for growth, including ethylene glycol (scale inhibitor), isopropanol (surfactant), and even formaldehyde and glutaraldehyde (biocides) in sublethal concentrations (Kekacs et al. 2015). Laboratory studies conducted at elevated pressures showed that guar gum, a widely used gelling agent, stimulates sulfate-reducing bacteria present in freshwater sources that are often used for fracturing fluids, yielding sulfide and acetic acid, both contributors to corrosion (Nixon et al. 2017). Similarly, *Halanaerobium* can couple guar gum fermentation to the reduction of thiosulfate to the same corrosive products but at elevated salinities (Liang et al. 2016). Another study highlighted the genomic capabilities of fractured shale microorganisms to degrade polyacrylamide, the most commonly used polymer in slickwater fracturing (Nixon et al. 2019). From these studies, and given the large volumes of fluid that remain in formation, fractured shale environments do not appear to be limited in substrates for microbial activity, including biofouling.

Other than stimulating microbial metabolism, the depletion of fracturing fluid additives downwell represents a form of biofouling in itself. Scale inhibitors and clay stabilizers are two classes of additives often included in fracturing fluids to act for the duration of shale gas recovery. The most frequently disclosed scale inhibitors used in US fracturing operations are citric acid, ammonia, erythorbate, and nitriloacetic acid, all of which are biodegradable (Elsner and Hoelzer 2016). Similarly, the most commonly disclosed clay stabilizer, choline chloride, is bioavailable and can be converted to trimethylamine (TMA) by anaerobic microorganisms (Craciun and Balskus 2012). TMA is known to support methanogenesis by halotolerant strains of *Methanohalophilus* in fractured shale communities (Daly et al. 2016, Borton et al. 2018a; Borton et al. 2018b); therefore, choline chloride degradation may contribute substrate for such activity. The microbial degradation of these and other additives intended to prolong shale gas production may therefore shorten the lifespan of the well, negatively impacting on shale gas yields.

12.4.3 Biogenic Sulfide Production

As with conventional oil production, the greatest biofouling threat to shale gas extraction is biogenic sulfide. Not only is sulfide toxic and flammable but it is also highly corrosive and can compromise the steel infrastructure of shale gas production. Further, souring of natural gas is a costly process to remedy (Fichter et al. 2012). As such, the presence, activity, and persistence of sufidogenic microorganisms has been a central theme to research on shale gas microbiology in the past decade. Indeed, the earliest studies of fractured shale microbiology followed observations of biogenic sulfide and microbially induced corrosion in several wells in the Barnett Shale, north central Texas, despite the thermogenic origins of its shale gas (Fichter et al. 2012; Struchtemeyer et al. 2011).

A number of studies have since highlighted the presence of sulfate-reducing lineages in flowback fluids, including members of *Desulfosporosinus* and *Desulfovibrio* genera (Davis et al. 2012; Kirk et al. 2012; Struchtemeyer and Elshahed 2012; Murali Mohan et al. 2013a; Wuchter et al. 2013). These sulfate-reducing bacteria (SRB) are likely to be introduced to the formation during drilling (Struchtmeyer et al. 2011), hydraulic fracturing (Nixon et al. 2017), or may originate from the pre-fractured shale formation itself (Frederickson et al. 1997). Yet, attempts to culture viable SRBs from production fluids have either been unsuccessful (Struchtemeyer and Elshahed 2012) or inconclusive (owing to the presence of cysteine in the growth medium and the enrichment of non-SRB lineages; Akob et al. 2015). Vikram and others (2016) detected genes responsible for dissimilatory sulfate reduction in the metatranscriptome of active communities recovered from Marcellus production fluids, suggesting that sulfidogenesis by SRBs is active in fractured shale communities. However, the low abundance of both these expressed genes (Vikram et al. 2016) and of SRB lineages in 16S rRNA gene sequencing data from numerous studies (Struchtmeyer and Elshahed 2012; Wuchter et al. 2013; Daly et al. 2016) suggests the reduction of sulfate is not a major source of biogenic sulfide in shale gas wells studied to date.

In fact, the dominance of *Halanaerobium* strains in numerous production fluids studied to date poses the greatest souring and corrosion threat during shale gas extraction (Table 12.1; Murali Mohan et al. 2013a; Cluff et al. 2014; Daly et al. 2016). Several studies reported abundances of more than 99% in produced waters (Davis et al. 2012; Murali Mohan et al. 2013a; Cluff et al. 2014; Booker et al. 2017; Nixon et al. 2019). This lineage of strictly anaerobic fermentative bacteria is efficient at accumulating inorganic ions to maintain osmotic balance, giving it a competitive advantage over other sulfidogenic taxa that are less well adapted to the prevailing high salinities (Murali Mohan et al. 2013a; Booker et al. 2017). *Halanaerbium* strains recovered from these produced waters have the capacity to ferment a wide array of organic compounds, including amino acids and sugars, and to reduce thiosulfate to sulfide (Daly et al. 2016; Lipus et al. 2017b). When provided with thiosulfate in culture, one isolate could degrade guar gum and produce acetate and sulfide at a salinity of 10% NaCl (Liang et al. 2016). Sulfide production from thiosulfate reduction appears to increase during the stationary phase of growth, with no growth advantage, indicating thiosulfate is used as an electron sink to remove excess reductant (Booker et al. 2017). Even though sulfide production appears not to be directly linked with growth, the dominance and persistence of *Halanaerobium* in production fluids from multiple shale formations nevertheless highlights the scale of the potential for ongoing production of biogenic sulfide during shale gas extraction. Indeed, the dominance of *Halanaerobium* over a 200-day period in production fluids from the Utica formation coincided with an increase in sulfide and concomitant decrease in thiosulfate, indicative of active thiosulfate reduction (Booker et al. 2017).

The source of thiosulfate in these environments is unclear, but may be generated through the oxidation of sulfur compounds (e.g., pyrite) in the shale or from transformations from additives in the drilling mud and fracturing fluids (Davis et al. 2012). Booker et al. (2017) demonstrated the abiotic leaching of thiosulfate from chips of Utica shale in water, suggesting thiosulfate may directly derive from the formation. The dominance of highly oxygen-sensitive methanogens in late stage production fluids indicates that the in situ environment is anoxic (Nixon et al. 2019), and hence it is feasible that the redox conditions may not be conducive for the complete oxidation of sulfur compounds to sulfate, instead generating thiosulfate. Diagnostic assays to identify souring in conventional hydrocarbon recovery target growth of, or genes used by, sulfate-reducing microorganisms. Given that thiosulfate-reducing microorganisms utilize a different electron acceptor and do so via a different pathway to SRB, these conventional methods of assessment cannot identify a threat from non-SRB sulfidogenic lineages (Murali Mohan et al. 2013a; Booker et al. 2017). There is therefore a critical need for the development of new diagnostic tools that identify thiosulfate-reducing microorganisms and genes in order to control souring during shale gas extraction.

It is important to note that not every shale gas microbiology study to date observed a dominance of *Halanaerobium* (Table 12.1). A notable exception is a well 4.2 km deep in the Sichuan Basin of China, where the dominant genus in produced fluids was found to be *Shewanella* (Zhang et al. 2017). Indeed, no *Halanaerobium* was observed in this well. Several characterized strains of *Shewanella* are known to grow from the reduction of sulfur and thiosulfate, producing sulfide, and can withstand high salinity conditions of up to 10% NaCl (Bowman et al. 1997; Venkateswaran et al. 1999), and hence its dominance and the lack of *Halanaerobium* in the Sichuan basin produced waters may still indicate a corrosion and souring threat, albeit by a different organism. Other shale formations not dominated by *Halanaerobium* include the Denver-Julesburg (Colorado), produced waters from which were dominated by *Rhodococcus* spp. (Wang et al. 2017) and the Eagle Ford (Texas), where *Microbacterium* spp. Dominated (Santillan et al. 2015). One major difference between these and the *Halanaerobium*-dominated Marcellus, Utica, Antrim, Bakken, and Barnett formations is salinity, with approximately ten times lower chloride and TDS concentrations in the former (Table 12.1). This may indicate that formations that give rise to the most saline produced waters also pose the greatest souring threat, given the potential for *Halanaerobium* spp. to outcompete non-halotolerant organisms.

12.4.4 Biofilm Formation

Compared to biocide resistance and sulfidogenesis, much less is known about the potential for biofilm formation in shale gas wells. This is due, in part, to the reliance of using production fluids as proxies for in situ microbial activity, which may in fact reflect these communities less accurately than assumed. However, given the potential for biofilm-related fracture and pipeline clogging to reduce gas yields, this gap in our knowledge is perhaps the most significant to the onshore gas industry.

The first attempts to address biofilm-forming potential assessed the genomic capacity of key taxa. Daly et al. (2016) recovered six near-complete *Halanaerobium* genomes from Marcellus and Utica shale production fluids and identified genes related to biofilm formation in all of them, despite a general lack of this trait in surface-dwelling relatives. These genes encode for flagellar motility and polysaccharide production, associated with cellular aggregation (Daly et al. 2016). Similar genomic capacity was identified in another Marcellus fractured shale-derived *Halanaerobium* genome (Lipus et al. 2017b). These studies suggest *Halanaerobium* can form biofilms, which may represent an important adaptation that allows for their dominance in the fractured shale environment (Daly et al. 2016). Genes implicated with biofilm formation were also identified in a metagenome-assembled *Pseudomonas* genome recovered from Bakken produced waters, suggesting this capacity in other common fractured shale taxa (Lipus et al. 2017a). However, these studies only identify the presence, and not expression, of these genes. The most compelling evidence of active biofilm formation derives from a recent study in which a strain of *Halanaerbium* from the Utica formation was grown at fractured shale pressures (21-48 MPa; Booker et al. 2019). Under these elevated pressures, *Halanaerobium* isolate WG8 produced higher concentrations of exopolysaccharide-related proteins compared with atmospheric pressures, resulting in greater cell aggregation (Booker et al. 2019). These results are consistent with transcriptomic evidence of active microbial communities in Marcellus production fluids expressing genes for aggregation substances, including exopolysaccharides and alginate, indicative of active biofilm formation (Vikram et al. 2016).

Taken together, these studies present clear evidence for the capability of fractured shale microorganisms, including dominant *Halanaerobium* strains, to form biofilms in shale gas wells. Despite this evidence, the implications for shale gas yields remain unclear, and further research is warranted to assess the knock-on impact of biofilm formation on fractured shale permeability. Biofilms will also affect the efficacy of biocides, since biofilms are more resistant to the action of biocides and may facilitate the emergence of resistance (Struchtmeyer et al. 2012; Campa et al. 2019).

12.5 Recycling of Production Fluids

These biofouling issues become more problematic when flowback and produced waters are reused in subsequent hydraulic fracturing operations. Reuse is a common practice, intended to alleviate pressures on freshwater demands by including 10–20% of flowback. However, reuse can accelerate biofouling by reintroducing microbial communities already enriched in lineages adapted to downwell conditions and speeding up their enrichment (Murali Mohan et al. 2013a, b; Zhong et al. 2019). This practice is particularly common in the Marcellus and Utica formations, with several studies reviewed here reporting that injected fluids included 20% produced water from previous operations (Murali Mohan et al. 2013a; Cluff et al. 2014; Daly et al. 2016; Booker et al. 2017). In each case, *Halanaerobium* spp. were dominant in produced waters, and in three of these studies they represented more than 99% of the community. Multiple strains of *Halanaerobium* are viable in production fluids long after collection, highlighting their persistence outside the fractured shale environment as well as within it (Cliffe et al. 2020). Indeed, storage and treatment of production fluids can increase biofouling potential upon reuse, with evidence of further enrichment of sulfidogenic and acid-producing lineages highlighted in previous research (Murali Mohan et al. 2013b; Zhang et al. 2017). The reuse of production fluids can therefore result in a selection-enrichment process (Cluff et al. 2014), increasing the costs associated with microbial control and potentially prolonging the persistence of problematic fractured shale taxa. Given the risks of selecting for persistent biocide resistance through reuse of production fluids, there is an urgent need to better understand the organisms and pathways responsible for the biofouling processes discussed in this chapter in order to develop effective diagnostic tools and control strategies.

12.6 Summary and Knowledge Gap

In summary, hydraulic fracturing of shale formations creates microbial habitats in the deep terrestrial subsurface where pre-fractured shale microbial activity was limited or absent. The high-pressure injection of water, additives, and microorganisms creates the space and introduces the chemical energy required for microbial communities to colonize these environments. Succession in fractured shales is driven by the high prevailing salinities that evolve in situ from the mixing of predominantly freshwater input fluids with salts and brines in the formation. In the most saline formations, the low diversity anaerobic communities that prevail are dominated by halotolerant microorganisms that operate efficient osmoprotectant strategies, metabolize input fluid additives, and have developed adaptive viral immunity. Persistent for hundreds of days post-fracturing and remarkably similar across geographically distinct shale formations (Table 12.1), these anthropogenic deep subsurface ecosystems demonstrate microbial survival, adaptation, and resilience in extreme conditions far from origin. Many microbial processes in fractured shales pose a risk to the efficiency, yields, and environmental impact of shale gas extraction, despite common use of biocides as a means of microbial control. Indeed, ineffective action of biocides in these communities represents one such biofouling process. Other negative processes include sulfidogenesis, which contributes to corrosion and can degrade the quality of natural gas, and biofilm formation, potentially clogging fractures and pipelines and hence reducing gas yields. All of these processes are exacerbated by the reuse of flowback fluid, which acts to reintroduce a pre-selected community, increasing the rate at which these processes occur, and potentially creating antimicrobial resistance hotspots.

Although there is now a substantial body of literature documenting the diversity, function, and activity of fractured shale communities, there are still gaps in our understanding. The study of hydraulically fractured shales has been largely restricted to a study of input and output fluids, although latterly there has been an encouraging trend toward hypothesis-testing in the laboratory (e.g. Borton et al. 2018b; Booker et al. 2019; Daly et al. 2019). Extensive characterization of the geochemistry and microbiology of these input and production fluids has offered a view into fractured shale ecosystems (Mouser et al. 2016), allowing us to summarize their attributes as in this chapter. However, the in situ fractured shale environment remains a black box, and there are several unanswered questions. These mostly impact on the ability to diagnose and control biofouling processes, but they also address persistence in an extreme anthropogenic subsurface environment. These major knowledge gaps are as follows,

1. Biofilm formation: It is not well understood whether fractured shale communities form biofilms in shale gas wells, either in fractures, pipelines, or both. If biofilms are common in these environments, the implications for gas recovery and biocide efficacy may be significant.
2. Biocide resistance: There is clear potential for biocide addition to stimulate resistance; however, it is not understood what the mechanisms of resistance are, or whether they may impact on antimicrobial resistance in the environment more broadly.
3. Sulfidogenesis: Biogenic sulfide production has been observed in a number of shale gas production fluids, yet the chemical drivers of this process, including the contribution of sulfur from shales, remain unclear. Further, given the lack of attention given to the microbiology of shale gas wells beyond the North American continent, the degree to which *Halanaerobium* spp. pose a global souring and corrosion threat is currently unknown.
4. Long-term persistence: Production of shale gas typically continues for years, whereas most studies reviewed here have assessed the microbial ecology of shale gas wells over hundreds of days. The long-term fate of fractured shale communities, including bioavailability of shale-derived organic matter, is not well understood.

This is not an exhaustive list, but rather an indication of what has yet to be resolved using the input-output approach of most studies reviewed here. These gaps in our knowledge will require novel approaches of study to fill, combining field campaigns with laboratory-based experiments, and applying an array of analytical approaches, including geochemistry, meta-omics, fluid mechanics, and modeling to name a few. Addressing these issues will help to ensure shale gas extraction is conducted efficiently, with minimal impact on the environment and human health. Furthermore, issues of biofilm formation, biocide resistance, and long-term persistence are of relevance to other deep subsurface engineering activities that are likely to occur in the coming decades, for instance, carbon capture and storage, geothermal energy, and nuclear waste disposal; all major global challenges that require a deeper level of understanding of the deep biosphere than we currently have. Finally, studying life in extreme environments in the deep subsurface, whether pristine or engineered, helps us to define habitability. This in turn sheds light on the limits of microbial life here on Earth and may even help guide the search for life elsewhere in the solar system, including the Martian subsurface, and in the salt-rich subglacial oceans on the moons of Enceladus and Europa.

Acknowledgments

The author acknowledges support from a Natural Environment Research Council Research Fellowship (NE/R013462/1).

References

Akob DM, Cozzarelli IM, Dunlap DS, Rowan EL, and Lorah MM. (2015). Organic and inorganic composition and microbiology of produced waters from Pennsylvania shale gas wells. *Applied Geochemistry.* 60: 116–125.

Booker AE, Borton MA, Daly RA et al. (2017). Sulfide generation by dominant *Halanaerobium* microorganisms in hydraulically fractured shales. *mSphere.* 2(4): e00257–17.

Booker AE, Hoyt DW, Meulia T et al. (2019). Deep-subsurface pressure stimulates metabolic plasticity in shale-colonizing *Halanaerobium* spp. *Applied and Environmental Microbiology.* 85(12): e00018–19.

Borton MA, Daly RA, O'Banion B et al. (2018a). Comparative genomics and physiology of the genus *Methanohalophilus*, a prevalent methanogen in hydraulically fractured. *Environmental Microbiology.* 20(12): 4596–4611.

Borton MA, Hoyt DW, Roux S et al. (2018b). Coupled laboratory and field investigations resolve microbial interactions that underpin persistence in hydraulically fractured shales. *Proceedings of the National Academy of Sciences.* 115(28): E6585–E6594.

Bowman JP, McCammon SA, Nichols DS et al. (1997). *Shewanella gelidimarina* sp. nov. and *Shewanella frigidimarina* sp. nov., novel Antarctic species with the ability to produce eicosapentaenoic acid (20:5ω3) and grow anaerobically by dissimilatory Fe (III) reduction. *International Journal of Systematic Bacteriology.* 47(4): 1040–1047.

Campa MF, Wolfe AK, Techtmann SM, Harik A, and Hazen TC. (2019). Unconventional oil and gas energy systems: an unidentified hotspot of antimicrobial resistance? *Frontiers in Microbiology.* 10: 2392.

Cliffe L, Nixon SL, Daly RA et al. (2020). Identification of persistent sulfidogenic bacteria in shale gas produced waters. *Frontiers in Microbiology.* 11: 286.

Cluff MA, Hartsock A, MacRae JD, Carter K, and Mouser PJ. (2014). Temporal changes in microbial ecology and geochemistry in produced water from hydraulically fractured Marcellus shale gas wells. *Environmental Science & Technology.* 48(11): 6508–6517.

Colwell FS, Onstott TC, Delwiche ME et al. (1997). Microorganisms from deep, high temperature sandstones: constraints on microbial colonization. *FEMS Microbiology Reviews.* 20: 425–435.

Craciun S and Balskus EP. (2012). Microbial conversion of choline to trimethylamine requires a glycyl radical enzyme. *Proceedings of the National Academy of Sciences.* 109(52): 21307–21312.

Daly RA, Borton MA, Wilkins MJ et al. (2016). Microbial metabolism in a 2.5-km-deep ecosystem created by hydraulic fracturing in shales. *Nature Microbiology.* 1(10): 16146.

Daly RA, Roux S, Borton MA et al. (2019). Viruses control dominant bacteria colonizing the terrestrial deep biosphere after hydraulic fracturing. *Nature Microbiology.* 4: 352–361.

Davis JP, Struchtemeyer CG, and Elshahed MS. (2012). Bacterial communities associated with production facilities of two newly drilled thermogenic natural gas wells in the Barnett Shale (Texas, USA). *Microbial Ecology.* 64(4): 942–954.

Elsner M and Hoelzer K. (2016). Quantitative survey and structural classification of hydraulic fracturing chemicals reported in unconventional gas production. *Environmental Science & Technology.* 50: 3290–3314.

Evans MV, Getzinger G, Luek JL et al. (2019). *In situ* transformation of ethoxylate and glycol surfactants buu shale-colonizing microorganisms during hydraulic fracturing. *ISME Journal.* 13: 2690–2700.

Evans MV, Panescu J, Hanson AJ et al. (2018). Members of *Marinobacter* and *Arcobacter* influence system biogeochemistry during early production of hydraulically fractured natural gas wells in the Appalachian basin. *Frontiers in Microbiology.* 9: 2646.

Fichter J, Wunch K, Moore R et al. (2012). How hot is too hot for bacteria? A technical study assessing bacterial establishment in downhole drilling, fracturing and stimulation operations. In CORROSION 2012, March 11–15, Salt Lake City, Utah. NACE International.

Fredrickson JK, McKinley JP, Bjornstad BN et al. (1997). Pore-size constraints on the activity and survival of subsurface bacteria in a late Cretaceous shale-sandstone sequence, northwestern New Mexico. *Geomicrobiology Journal.* 14: 182–202.

Huang R. (2008). *Shale-Derived Dissolved Organic Matter as a Substrate for Subsurface Methanogenic Communities in the Antrim Shale Michigan Basin, USA.* Masters thesis. Department of Geosciences, University of Massachusetts Amherst.

Johnson K, French K, Fichter JK, and Oden R. (2008). Use of microbiocides in Barnett Shale gas well fracturing fluids to control bacteria related problems. In CORROSION 2008, March 16–20, New Orleans, Louisiana. NACE International.

Kahrilas GA, Blotevogul J, Stewart PS, and Borch T. (2015) Biocides in hydraulic fracturing fluids: A critical review of their usage, mobility, degradation and toxicity. *Environmental Science & Technology.* 49(1): 16–32.

Kashefi K and Lovley DR. (2003) Extending the upper temperature limit for life. *Science.* 301(5635): 934.

Kekacs D, Drollette BD, Brooker M, Plata DL, and Mouser PJ. (2015). Aerobic biodegradation of organic compounds in hydraulic fracturing fluids. *Biodegradation.* 26(4): 271–287.

Kirk MF, Martini AM, Breecker SO et al. (2012). Impact of commercial natural gas production on geochemistry and microbiology in a shale-gas reservoir. *Chemical Geology*. 332–333: 15–25.

Krumholz LR, McKinley JP, Ulrich GA, and Suflita JM. (1997). Confined subsurface microbial communities in cretaceous rock. *Nature*. 386: 64–66.

Liang R, Davidova IA, Marks CR et al. (2016). Metabolic capability of a predominant *Halanaerobium* sp. in hydraulically fractured gas wells and its implication in pipeline corrosion. *Frontiers in Microbiology*. 7: 988.

Lipus D, Ross D, Bibby K, and Gulliver D. (2017a). Draft genome sequence of *Pseudomonas* sp. BDAL1 reconstructed from a Bakken shale hydraulic fracturing-produced water storage tank metagenome. *Genome Announcements*. 5: e00033–17.

Lipus D, Roy D, Khan E et al. (2018). Microbial communities in Bakken region produced water. *FEMS Microbiology Letters*. 365(12): fny107.

Lipus D, Vikram A, Ross D et al. (2017b). Predominance and metabolic potential of Halanaerobium spp. in produced water in hydraulically fractured Marcellus shale wells. *Applied Environmental Microbiology*. 83(8): e02659–16.

Martini AM, Walter LM, Budai JM et al. (1998). Genetic and temporal relations between formation waters and biogenic methane: Upper Devonian Antrim Shale, Michigan Basin, USA. *Geochimica et Cosmochimica Acta*. 62(10): 1699–1720.

Mouser PJ, Borton M, Darrah TH, Hartsock A, and Wrighton KC. (2016). Hydraulic fracturing offers a view of microbial life in the deep terrestrial subsurface. *FEMS Microbiology Ecology*. 92(11).

Murali Mohan A, Hartsock A, Bibby KJ et al. (2013a). Microbial community changes in hydraulic fracturing fluids and produced water from shale gas extraction. *Environmental Science & Technology*. 47(22): 13141–13150.

Murali Mohan A, Hartsock A, Hammack RW, Vidic RD, and Gregory KB. (2013b). Microbial communities in flowback water impoundments from hydraulic fracturing for recovery of shale gas. *FEMS Microbiology Ecology*. 86(3): 567–580.

Nixon SL, Daly RA, Borton MA et al. (2019). Genome-resolved metagenomics extends the environmental distribution of the Verrucomicrobia phylum to the deep terrestrial subsurface. *mSphere*. 4: e00613–19.

Nixon SL, Walker L, Streets MDT et al. (2017). Guar gum stimulates biogenic sulfide production at elevated pressures: Implications for shale gas extraction. *Frontiers in Microbiology*. 8: 679.

Onstott TC, Phelps TJ, Colwell FS et al. (1998). Observations pertaining to the origin and ecology of microorganisms recovered from the deep subsurface of Taylorsville Basin, Virginia. *Geomicrobiology Journal*. 15: 353–385.

Santillan EFU, Choi W, Bennett PC, and Leyris JD. (2015). The effects of biocide use on the microbiology and geochemistry of produced water in the Eagle Ford formation, Texas, U.S.A. *Journal of Petroleum Science and Engineering*. 135: 1–9.

Schlegel ME, McIntosh JC, Bates BL, Kirk MF, and Martini AM. (2011). Comparison of fluid geochemistry and microbiology of multiple organic-rich reservoirs in the Illinois Basin, USA: Evidence for controls on methanogenesis and microbial transport. *Geochimica et Cosmochimica Acta*. 75: 1903–1919.

Strong LC, Gould T, Kasinkas L et al. (2013). Biodegradation in waters from hydraulic fracturing: Chemistry, microbiology, and engineering. *Journal of Environmental Engineering*. 140(5): B4013001.

Struchtemeyer CG. (2018). Microbiology of oil- and natural gas-producing shale formations: An overview. In Steffan R (ed) *Consequences of Microbial Interactions with Hydrocarbons, Oils, and Lipids: Biodegradation and Bioremediation*. Springer Nature Switzerland AG, pp. 215–232.

Struchtemeyer CG and Elshahed MS. (2012). Bacterial communities associated with hydraulic fracturing fluids in thermogenic natural gas wells in North Central Texas, USA. *FEMS Microbiol Ecology*. 81(1): 13–25.

Struchtemeyer CG, Davis JP, and Elshahed MS. (2011). Influence of drilling mud formation process on the bacterial communities in thermogenic natural wells of the Barnett Shale. *Applied and Environmental Microbiology*. 77(14): 4744–4753.

Struchtemeyer CG, Morrison MD, and Elshahed MS. (2012). A critical assessment of the efficacy of biocides used during the hydraulic fracturing process in shale natural gas wells. *International Biodeterioration & Biodegradation*. 71: 15–21.

Struchtemeyer CG, Youssef NH, and Elshahed MS. (2017). Protocols for investigating the microbiology of drilling fluids, hydraulic fracturing fluids, and formations in unconventional natural gas reservoirs. In McGenity TJ, Timmis KN, Fernandez BN (eds.) *Hydrocarbon and Lipid Microbiology Protocols*. Springer-Verlag, pp. 1–25.

Tucker YT, Kotcon J, and Mroz T. (2015). Methanogenic archaea in Marcellus shale: A possible mechanism for enhanced gas recovery in unconventional shale resources. *Environmental Science & Technology*. 49(11): 7048–7055.

Vandenbroucke M and Largeau C. (2007). Kerogen origin, evolution and structure. *Organic Geochemistry*. 38: 719–833.

Venkateswaran K, Moser DP, Dollhopf ME et al. (1999). Polyphasic taxonomy of the genus Shewanella and description of *Shewanella oneidensis* sp. nov. *International Journal of Systematic Bacteriology*. 49: 705–724.

Vikram A, Lipus D, and Bibby K. (2014). Produced water exposure alters bacterial response to biocides. *Environmental Science & Technology*. 48(21): 13001–13009.

Vikram A, Lipus D, and Bibby K. (2016). Metatranscriptome analysis of active microbial communities in produced water samples from the Marcellus Shale. *Microbial Ecology*. 72: 571.

Waldron PJ, Petsch ST, Martini AM, and Nüsslein K. (2007). Salinity constrains on subsurface archaeal diversity and methanogenesis in sedimentary rock rich in organic matter. *Applied and Environmental Microbiology*. 73: 4171–4179.

Wang H, Lu, L, Chen X, Bian Y, and Ren ZJ. (2019). Geochemical and microbial characterizations of flowback and produced water in three shale oil and gas plays in the central and western United States. *Water Res*. 164: 114942.

Wuchter C, Banning E, Mincer T, Drenzek NJ, and Coolen MJ. (2013). Microbial diversity and methanogenic activity of Antrim Shale formation waters from recently fractured wells. *Frontiers in Microbiology*. 4: 367.

Zhang Y, Yu Z, Zhang H, and Thompson IP. (2017). Microbial distribution and variation in produced water from separators to storage tanks of shale gas wells in Sichuan Basin, China. *Environmental Science: Water Research & Technology*. 3(2): 340–351.

Zhong C, Li J, Flynn SL et al. (2019). Temporal changes in microbial community composition and geochemistry in flowback and produced water from the Duvernay formation. *ACS Earth Space Chem*. 3: 1047–1057.

Part III

Case Studies

Part III

13

Evaluation of Potential Water Quality Impacts in Unconventional Oil and Gas Extraction

The Application of Elemental Ratio Approaches to Pennsylvania Pre-Drill Data

DANIEL J. BAIN, TETIANA CANTLAY, REBECCA TISHERMAN, AND JOHN F. STOLZ

13.1 Introduction

Over the last 15 years, unconventional natural gas extraction has grown rapidly and consistently. For example, unconventional wells in Pennsylvania increased from eight in 2005 to more than 12,000 in 2020. (Fracktracker Alliance 2020, chapter 2) Along with the shale gas boom, there has been a substantial increase in water use intensity as unconventional oil and gas operations require significant volumes of water and consequently produce substantial waste waters (See Chapter 8). The composition of unconventional oil and gas waste fluids is generally characterized by high salinity and elevated concentrations of halides, metals, and NORM in particular (Abualfaraj et al. 2014; Akob et al. 2015; Rowan et al. 2015a; Shih et al. 2015; Kim et al. 2016; Vengosh et al. 2017; Cantlay et al. 2020a, 2020b, 2020c). Despite the clear potential for environmental impacts from intentional or unintentional releases of these wastewaters, their composition remains nebulous. This ambiguity is compounded by the wide range of chemistries observed in natural waters. Further, the potential for historical contamination of waters makes detection and quantification a challenge.

Direct and indirect interaction among hydraulic fracturing fluids and ground and surface water could impact the quality of drinking water supplies (Wilson and Van Briesen 2013; Ziemkiewicz et al. 2014; Rester and Warner 2016; Miller 2020) and ecosystem health in general (Colborn et al. 2011; Clark et al. 2012; Lane and Landis 2016; Mrdjen and Lee 2016; Olawoyin et al. 2016; Entrekin et al. 2018). Accidental surface spills, breaches in well-casing integrity and impoundment linings, accidents during transportation, and disposal of produced water have been associated with local surface and groundwater contamination.(Warner et al. 2013; Vengosh et al. 2014; Burgos et al. 2017; Miller 2020). In addition, contamination can occur as a result of intentional and improper release of fracking waste. (Hopey 2011; Milliken 2013). However, the prevalence of these releases is not reported comprehensively (e.g., releases on pads are reported, but releases during transport are not necessarily reported in the same place) and even if these releases were reported comprehensively, many of these release points are hard to monitor (e.g., well casing deterioration). A clear means to detect and evaluate contamination from unconventional wastes is crucial to management of potential impacts from unconventional development.

To date, the search for reliable means to detect and quantify the contributions of unconventional waste waters has relied on a wide variety of approaches. At the state level, most regulatory responses rely on single element criteria defined by use. Domestic well impacts are evaluated by primary and secondary maximum contaminant levels (MCL) defined by the USEPA as required by the Safe Drinking Water Act (SDWA). A dependence on single constituent concentration is not robust when evaluating the entire hydrologic cycle. For example, transfer from solution to soils is not governed by drinking water regulations. Further, while there are MCLs for many contaminants, these MCLs are not necessarily considered when evaluating use/disposal of contaminated waste, as other regulatory frameworks govern these practices (e.g., the use of brines as road treatment, a non-point source, see Chapters 2 and 10). Without consistency in water quality standards and regulatory frameworks, regulation and management can be contradictory. The use of static, single element criteria can miss cases of contamination that do not raise concentrations above health-based MCLs but may suggest consequences in the long run. That is, if concentrations remain below the MCL, health risk is minimized as long as concentrations do not rise further. However, without intensive and expensive characterization of the local hydrogeology, this risk is hard to evaluate over time frames beyond a couple of years for groundwater sources.

At the other end of the continuum are exhaustive, multiparameter evaluations of local waters before and after unconventional activities. In rare cases, this has been done across sites. For example, Harkness et al. (2017) evaluated waters surrounding unconventional drilling sites before, during, and after drilling activity in West Virginia. The waters were evaluated using isotopes of strontium, boron, and lithium, in addition to noble gas composition and tritium, an uncommonly complete set of parameters. However, most comprehensive assessments of water chemistries are limited to smaller sets of wells. Ziemkiewicz and He (2015) built on the Marcellus Shale Energy and Environment Laboratory (MSEEL) site and looked at wells there and across West Virginia over time for total dissolved solids (TDS), organic compounds, metals, and radioactive isotopes. The groundwater data collected from the extensive shallow groundwater monitoring network on the MSEEL site has not apparently been published. Beyond these studies, in most cases, multiparameter evaluations are limited to single experimental wells. For example, Zheng et al. (2017) tested the integration of boron and strontium isotopes with Br/Cl and Ba/Sr at one site in China to determine surface and groundwater contamination from one nearby drilling site. The small number of wells, in addition to access issues precluding randomization and assumptions of independence, challenge statistically rigorous interpretation of these results, limiting our ability to generalize from these approaches. More importantly, the cost burden of this comprehensive set of analyses would be a challenge to implement in modern regulatory environments.

Most approaches to detection and attribution are more limited in access but evaluate potential contamination using multiple lines of parallel evidence. These are the vast majority of the literature, ranging from work early in the boom focusing on halides (Wilson and Van Briesen 2013), strontium isotopes (Chapman et al. 2012), and radium isotopes (Warner et al. 2013). While strontium and radium isotopes have offered unique insight into the geochemistry of brines and the possible environmental impacts, historical and current water data do not always have isotope levels available. Later literature has focused on ratios, such as Br/SO_4 vs Ba/Cl to differentiate AMD from oil and gas brines (Brantley et al. 2014) and Ca/Mg vs. Ca/Sr to distinguish unconventional contamination from conventional contamination (Tisherman and Bain 2019). Ratios clarify chemical dynamics as the influence of dilution on measurements is removed.

Despite all this work, there remains no consensus criteria for the identification of unconventional releases. Part of this arises from the ambiguity introduced by legacy contamination. For example, Pennsylvania is haunted by abandoned mine discharges (Cravotta 2008). Coal wastes can have chemical signatures that overlap with unconventional waste waters (Wilson and Van Briesen 2013). In other cases, modern practices can create similar contamination. Widespread use of halite as a de-icing agent routinely elevates chloride (Cl) concentrations in many near-road environments in colder climates. If water quality evaluations cannot resolve and rule out legacy or modern influences, the origin of contamination remains uncertain.

This ambiguity is heightened by our poor characterization of complex water chemistries. If natural conditions create chemistries similar to those observed in unconventional wastes, chemical criteria will not identify contamination from unconventional wastes. Unfortunately, there is no clear path to clarity in the characterization and regulation of unconventional wastes. This chapter focuses on systematic characterizations of water chemistry prior to local unconventional development, evaluates the waters in the context of proposed water chemistry evaluation criteria, and discusses implications of this context for current and future study and regulation of unconventional activities and water quality.

13.2 Water Quality Evaluation Tools

This chapter focuses on five published methods to evaluate contamination from oil and gas activities: 1) Br/SO_4 vs Ba/Cl (Brantley et al. 2014); 2) Ca/Mg vs Ca/Sr (Tisherman and Bain 2019); 3) Br/SO_4 vs Mg/Li (Cantlay et al. 2020b); and 4) SO_4/Cl vs Mg/Li (Cantlay et al. 2020b). In general, methods that utilize chemical ratios were selected to minimize the confounding influence of dilution. Approaches that compared ratios and concentrations (Wilson and Van Briesen 2013; Tasker et al. 2020) were not used for that reason. In addition, approaches that used sums as terms in the ratios (e.g., (Ba+Sr)/Mg (Johnson et al. 2015)) were not used as they are potentially less parsimonious and precise. Tasker (2018) proposes using the Sr/Ca vs Br/Cl ratio space in their dissertation but this space has not been published in the peer reviewed literature yet, so it is not evaluated here. Stable and radiogenic metal isotopic approaches (Chapman et al. 2012; Capo et al. 2014; Phan et al. 2016; Harkness et al. 2017) can be more precise, but are not widely measured. Further, the costs associated with sample collection appropriate for isotopic analysis and the analysis in general limits the widespread use of (non-gaseous) isotopic approaches to evaluate metal contamination sources in regions with oil and gas extraction activities.

In the application of these tools in this chapter we have preserved differentiations laid out in the original publication of these approaches (e.g., the ratio value separating impacted and AMD values in Brantley et al. 2014). In addition, we utilized both literature chemistry data (Poth 1962; Pennsylvania Department of Environmental Protection, Bureau and of Oil and Gas Management 1991; Cravotta 2008; Blauch et al. 2009; Hayes 2009; Dresel and Rose 2010; Warner et al. 2012; Brantley et al. 2014; Rowan et al. 2015b; Blondes et al. 2018) and data analyzed by Cantlay et al. (2020b), including 73 flowback/produced water samples, nine conventional oil well brines, two impoundment samples, and three mine drainage samples to delineate chemistries observed in potential contamination inputs. These zones of typical chemistry are used throughout this chapter to simplify visualization of pre-drill chemistry gathered as described in the Section 13.3.

13.3 Retrieval of "Public Data" from Existing Data Sources

One of the challenges in assessment of water quality changes during the shale gas boom (2007–present) is the complicated data landscape. Well-known and well-curated sources of water data are routinely used in national scale assessments such as the US Geological Survey's National Water Information System (NWIS) or the US Environmental Protection Agency's STORET database. However, data most appropriate for assessment of potential changes during unconventional gas extraction are best answered at smaller spatial scales and shorter temporal scales than the national networks capture. For example, there are only two wells in the NWIS for Greene and Washington Counties (Pennsylvania) and no water chemistry measurements of water from these wells during the period since unconventional activities began. This is clearly insufficient to assess local changes in water quality.

Water chemistry samples from areas surrounding unconventional oil and gas extraction are much more numerous and much closer to potential impacts. However, many of these data are not easily accessible to citizens or scientists, as data reported to governmental agencies as part of permit requirements are not necessarily public record (particularly owing to privacy concerns) nor are they organized beyond inclusion in permit folders. The Shale Network at Pennsylvania State University has attempted to mediate this disconnect in the Commonwealth of Pennsylvania by making pre-drill water survey data available widely through the Consortium of Universities Allied for Hydrologic Sciences, Inc.'s (CUAHSI) Hydrologic Information System (HIS) (Brantley 2018). This effort presents a potentially powerful means for citizens and scientists to characterize water quality that was previously not feasible.

However, this data system probably presents challenges to a broad examination of water quality characteristics. There are relatively sophisticated portals available for examination of the data (e.g., https://data.cuahsi.org/). These portals are organized around sites and time series data. That is, the map viewer fundamentally relies on selection of individual sampling locations to access the time series of data for that location. When trying to assess changes in water chemistry across space, this interface quickly becomes infeasible for collection of this data. CUAHSI provides packages for modern statistical software to access these data. For this study, the WaterML package (Kadlec et al. 2015) for the statistical language R (R Core Team 2013) was utilized in the scraping of data from the Shale Networks HIS server.

To examine the water quality landscape in pre-unconventional drilling periods we collected water quality data from HIS using the generalized code provided in Box 13.1. This process is not trivial. R is free, but the knowledge necessary to execute the code in Box 13.1 is hard won. While this knowledge can be acquired and the code below should run with minor tweaks, this is a fundamental barrier to use of this knowledge. The speed and fidelity of the data retrieval is a more serious barrier. In particular, this discussion is based on the use of a modern computer in a fast data environment (hard-wired ethernet connection to a university network). All the challenges discussed here would be compounded if a citizen were to use their home data connection to repeat this process.

The first attempt to download these data naively set the code (as shown in Box 13.1) to simply download all the data. This process literally took weeks, and an automatic update of the computer system resulted in loss of these multiple weeks (in hindsight, adding an automatic write in the code would solve this, but would also slow the process down and require additional R competence). Given the potential for wasted time, for this chapter, the code was modified to download PADEP pre-drill data. That is, all SiteCodes with the prefixes listed in Table 13.1 were downloaded using the code in Box 13.1, based on ranges obtained from the table "sites." This process, though shorter, still took the good part of a week (consider the resulting table was a little more than half a million rows. If things are queried, downloaded, and organized at the rate of one row per second, that's greater than five days).

To simplify broad investigation of the chemistry data, we stripped the dataset of any measurements that were not reported because of analytical range issues (e.g., "less than" or "greater than"). After download and removal of less than/greater thans, the data were "pivoted wide" using the tidy package (Wickham et al. 2019) for analysis.

13.4 PADEP Pre-Drill Data and the Robustness of Source Attribution

The robustness and applicability of the ratio spaces that have emerged in the literature were evaluated by placing the PA DEP pre-drill data in these ratio spaces (Figures 13.1–13.5). That is, how do the pre-drill data, assumed to reflect pre-unconventional activity, plot in the ratio spaces proposed to attribute contamination to appropriate sources?

Across these ratio spaces, pre-drill samples are plotted in areas that seem to be influenced by chemistries similar to oil and gas (O&G) brines. For example, 3% of the samples with sufficient data to determine ratios (see discussion later) have Ca:Sr ratios (wt/wt) that are less than 10 (Figure 13.2), values characteristic of unconventional oil and gas brines (Tisherman and Bain 2019). Likewise, 17% of the pre-drill data with reported Br measurements in Bradford County have values that fall into the "Oil and Gas Brines" region of the ratio space proposed by Brantley et al (2014) (Figure 13.1). The Bradford PA DEP pre-drill data plot outside of chemistry ranges observed in the southwestern Pennsylvania for SO_4/Cl vs MgLi and Br/SO_4 vs Mg/Li (Cantlay et al. 2020b). Further, these data plot in regions that are trending toward O&G brine end members (Figure 13.3 and 13.4).

The observation of O&G influence on pre-drill data in multiple ratio spaces is problematic for a number of reasons. If these data reflect the historical contamination of waters by unconventional activities, the pre-drill survey approach is not working on a per-pad basis, as samples that are considered uncontaminated are contaminated by the unconventional activities prior to "pre-drill" sampling. If the waters are contaminated by historical conventional oil and gas activities, the potential measures of contamination will be less sensitive to potential unconventional contamination and therefore may cause contamination events to be harder to detect. If this contamination does not result from historical contamination, these data suggest contamination occurred before local unconventional activities began, potentially from more distant contemporary sources. Available data is not sufficient to evaluate which of these scenarios is most probable.

Table 13.1. *Pennsylvania Department of Environmental Protection Pre-Drill prefixes inferred from Shale Network HIS source description and used in this analysis*[a]

Site CodePrefix	Source description	Count of individual measurements
Various: number only, SPS, apparent pad IDs	Data collected by oil and gas companies and released to USGS then shared to us in excel file	17,779
PADEP_Predrill	Bradford New Dataset; data collected by oil and gas companies as pre-drill data and released to Pennsylvania Department of Environmental Protection to be published online by Shale Network	41,202
	Data collected by oil and gas companies and released to the PA DEP	84,862
	Data collected by oil and gas companies and released to USGS then shared to us in excel file	6,968
	Data collected for groundwater samples by Chesapeake Energy Corp. as pre-drill data and released to Pennsylvania Department of Environmental Protection for publication by Shale Network under supervision of Susan L. Brantley (Penn State)	10,165
	Data collected for groundwater samples by oil and gas companies as pre-drill data and released to Pennsylvania Department of Environmental Protection for publication by Shale Network under supervision of Susan L. Brantley (Penn State)	45,026
	Data collected for groundwater samples from Bradford county by Chesapeake Energy Corp. as pre-drill data and released to Pennsylvania Department of Environmental Protection for publication by Shale Network under supervision of Susan L. Brantley (Penn State)	268,070
Appalachia Consulting 2012	Appalachia Hydrogeologic & Environmental Consulting, Inc.	83
<no prefix, simple number>	Data collected by oil and gas companies and released to the PA DEP	791
NWPA_old	Data collected by oil and gas companies and released to the PA DEP	22,156
WaterWellID	Merged data set by comparing and matching Cabot data set (pre-drill data collected by Cabot and released to PA DEP then to Penn State) and water quality data published by Molofsky et al. 2013	3,145

[a] Note, these counts include data outside of analytical ranges (i.e., less than/greater than) removed for other analyses in this chapter.

Box 13.1
Code for the WaterML package in R to retrieve Shale Network data from HIS

```
library("WaterML") #loads waterML (have to install first)
services <- GetServices() #retrieves a list of web services
     provided by HIS
server <- "http://hydroportal.cuahsi.org/shalenetwork/webapp/
     cuahsi_1_1.asmx?WSDL"
     #selects the Shale Network HIS
variables <- GetVariables(server) #retrieves variables available
     on the Shale Network HIS
sites <- GetSites(server) #retrieves the list of sites available
     on the Shale Network HIS
#setwd("<insert appropriate folder>")
     #Set your working directory here
n = 1 #initialize counter
PennStateSiteData<-GetSiteInfo(server, sites[1,4])
     #initialize dataframe
PennStateValues<-
     cbind(PennStateSiteData[1,],GetValues(server,
     siteCode=PennStateSiteData[1,4],variableCode=
     PennStateSiteData[1,12])) #combines site data and
     chemistry data #for the first record and places in
     the dataframe
zerodata<- c("sites")  #a list of sites that have no data when
     GetValues is executed, with "sites" #as the array name
for (m in 2:nrow(PennStateSiteData)) { #fills data frame for first site
values<-cbind(PennStateSiteData[m,],GetValues(server,siteCode=
     PennStateSiteData[m,4], variableCode=
     PennStateSiteData[m,12])) #combine site data and
     chemistry data
 PennStateValues<-rbind(PennStateValues,values)
     #adds each record to temporary table.
}
for (n in 2:nrow(sites){
     #go through each site, GetValues, reorganize and
     append to datafile
 nsite <-sites[n,4]   #variable for site name
 ndata<-GetSiteInfo(server, nsite) #dataframe of Values
if (nrow(ndata)==0) {   #if no values returned, record site in array
     zerodata<-c(zerodata, nsite)}
 else{
     for (p in 1:nrow(ndata)) { values<-cbind(ndata[p,],
     GetValues(server,siteCode=ndata[p,4],variableCode=
     ndata[p,12]))
     PennStateValues<-rbind(PennStateValues,values)
     }}} #for each Value, append the Site/Value combination to
     final table
```

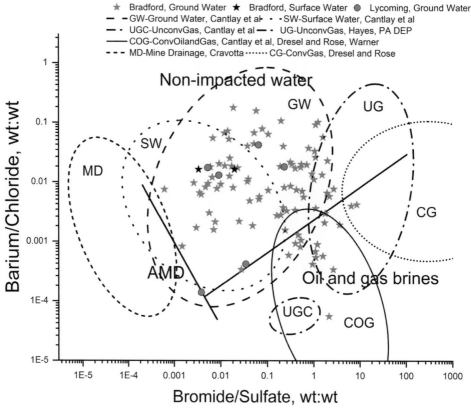

Figure 13.1 Chemistries of "Pre-drill" data reported to the Pennsylvania Department of Environmental Protection by oil and gas companies prior to hydraulic fracturing (likely all wells within 762 meters (2,500 feet) and others at the operator's discretion (58 Pa. C. S. §§ 3218(c)) plotted in Br/SO_4 vs Ba/Cl ratio space (Brantley et al. 2014). Pre-drill groundwater water samples are shown as grey symbols and surface water samples as black symbols. Lines show areas where various end members are expected to plot (see Brantley et al. (2014)). Ellipsis surrounding the clusters (areas) of different sources of contamination and ground and surface water samples were done using the 2D Confidence Ellipse application in OriginPro software (OriginLab, Northampton, Massachusetts)

13.5 Spatial Distribution of Analyses and the Role of Legacy Effects

In the pre-drill data (as defined for this chapter) there appear to be data from six counties and three general areas where counties are not specified (i.e., "SWPA," "NE-non-brad," "NWPA"). In many cases, counties were inferred from the Site Code (e.g., while there were no rows with county "Washington," all SiteCodes with the form "PADEP_Predrill_washington" were assigned the County "Washington"). Table 13.2 shows the distribution of individual pre-drill measurements across the state (these include counts of all analytes, not just sampling events).

Bradford County has an order of magnitude more pre-drill data in the Shale Network HIS than the next biggest county (Table 13.2). With data like the Bradford County data

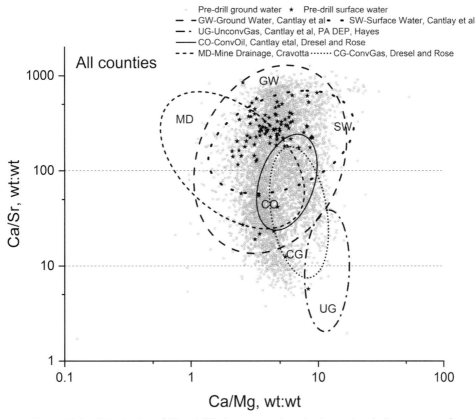

Figure 13.2 Chemistries of "Pre-drill" data reported to the Pennsylvania Department of Environmental Protection by oil and gas companies prior to hydraulic fracturing (likely all wells within 762 meters (2,500 feet) and others at the operator's discretion (58 Pa. C. S. §§ 3218(c)) plotted in Ca/Mg vs Ca/Sr ratio space (Tisherman and Bain 2019). Pre-drill groundwater water samples are shown as grey stars and surface water samples as black stars. Ellipsis surrounding the clusters (areas) of different sources of contamination and ground and surface water samples were done using the 2D Confidence Ellipse application in OriginPro software (OriginLab, Northampton, Massachusetts)

from all areas where unconventional activities have occurred, many of the questions raised through the remainder of this chapter would be clarified. That said, the uneven data density cannot be explained with available data and metadata. Rather than speculate why this occurred, it seems better to strongly encourage continued and increased efforts to provide these data to the public through trusted third parties such as the Shale Network.

Historically, Bradford County had relatively less conventional O&G extraction than other parts of Pennsylvania (Dilmore et al. 2015). In two of the ratio spaces shown in Figures 13.2 and 13.5 there are sufficient data available for multiple counties to allow comparison of water chemistry patterns across space (spatial distributions of parameters measured are discussed in the Section 13.6).

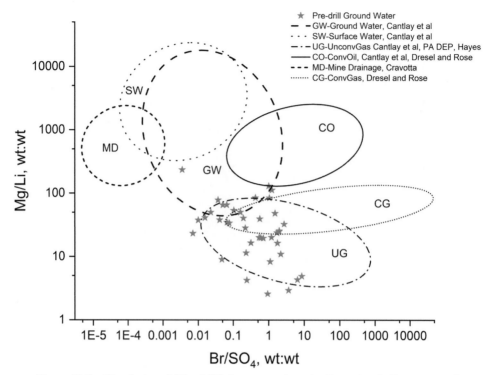

Figure 13.3 Chemistries of "Pre-drill" data reported to the Pennsylvania Department of Environmental Protection by oil and gas companies prior to hydraulic fracturing (likely all wells within 762 meters (2,500 feet) and others at the operator's discretion (58 Pa. C. S. §§ 3218 (c)) plotted in Br/SO$_4$ vs Mg/Li ratio space (Cantlay et al. 2020b). Pre-drill groundwater water samples are shown as grey stars and surface water samples as black stars. Ellipses show areas where various end members plot as summarized and reported in Cantlay et al. (2020b). Only Bradford County data could be plotted here owing to limited Br and Li data

In general, Bradford County has proportionally more samples that plot in portions of these ratio spaces that indicate O&G brine influence relative to Lycoming and Mercer (Figures 13.6 and 13.7). Lycoming County sits south and west of Bradford County, sharing a short border, and Mercer County is located on the other side of the state (Figure 13.8). While Lycoming has some pre-drill samples that plot in contaminant influenced regions, Mercer has very few (Figures 13.6 and 13.7). This challenges the development of consistent, statewide criteria for evaluation of water quality impacts of unconventional gas extraction.

Part of this signal seems to result from typical groundwater chemistries. Historical groundwater chemistries reported in county groundwater reports (Lloyd, Jr. and Carswell 1981; Stoner 1987; Williams et al. 1998) indicate these chemistries were present long before unconventional activities were developed. In general, these waters are attributed to regions where groundwater flow is "restricted" (Williams et al. 1998) and reducing conditions develop, transforming sulfates to sulfides and releasing Ba and Sr. This moves the water down in the Ca/Mg vs Ca/Sr space (Figure 13.9a) and in the Mg/Na vs SO$_4$/Cl space (Figure 13.9b). These waters would also be enriched in constituents such as iron and

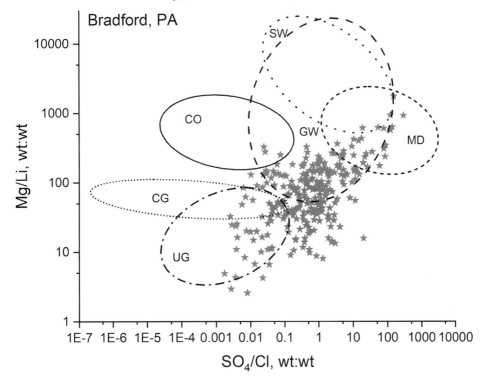

Figure 13.4 Chemistries of "Pre-drill" data reported to the Pennsylvania Department of Environmental Protection by oil and gas companies prior to hydraulic fracturing (likely all wells within 762 meters (2,500 feet) and others at the operator's discretion (58 Pa. C. S. §§ 3218(c)) plotted in SO_4/Cl vs Mg/Li ratio space (Cantlay et al. 2020b). Pre-drill groundwater water samples are shown as grey stars and surface water samples as black stars. Ellipses show areas where various end members plot as summarized and reported in Cantlay et al. (2020a, 2020b, 2020c). Only Bradford County data could be plotted here owing to limited data for Li

manganese, degrading their drinkability or requiring expensive treatment systems. It is not likely some of the groundwaters reported in these studies (i.e., as part of a comprehensive assessment) would ever be developed as a domestic water source. In addition, some of the pre-drill samples were likely collected from livestock watering systems and may never have been used for domestic purposes. This is good policy, as replacement of livestock water sources is expensive, but it is probably necessary to consider the pre-drill samples as a set that includes high TDS, Sr, Ba, Cl, etc. concentrations and this chemistry may confound the tools discussed in this chapter.

While these chemistries preclude the simple application of before/after criteria, they don't preclude the use of these tools in evaluating O&G wastewater impacts. These tools are fundamentally designed to examine the evolution of water chemistry through time. To

Table 13.2. *Counts of all data points (i.e., individual measurement of an individual constituent) per county or region (SWPA, NE-non-brad, NWPA are regions designated in the Shale Network HIS)*[a]

County	No. Data
Bradford	384,979
Greene	33
Lycoming	29,536
Mercer	25,399
NE-non-brad	29,266
NWPA	5,438
Susquehanna	7,965
SWPA	6,719
Washington	10,912

[a] Note, these counts include data outside of analytical ranges (i.e., less than/greater than) removed for other analyses in this chapter.

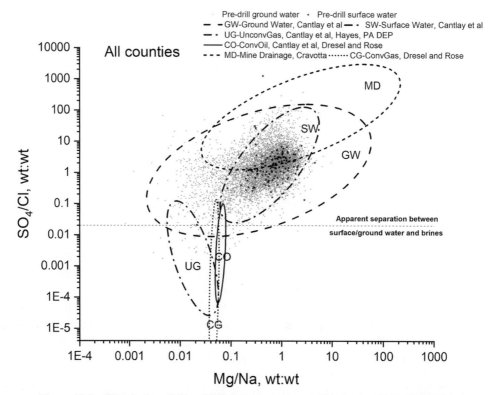

Figure 13.5 Chemistries of "Pre-drill" data reported to the Pennsylvania Department of Environmental Protection by oil and gas companies prior to hydraulic fracturing (likely all wells within 762 meters (2,500 feet) and others at the operator's discretion (58 Pa. C. S. §§ 3218(c)) plotted in Mg/Na vs SO_4/Cl ratio space (Cantlay et al. 2020b). Pre-drill groundwater water samples are shown as grey stars and surface water samples as black stars. Ellipses show areas where various end members plot as summarized and reported in Cantlay et al. (2020a, 2020b, 2020c)

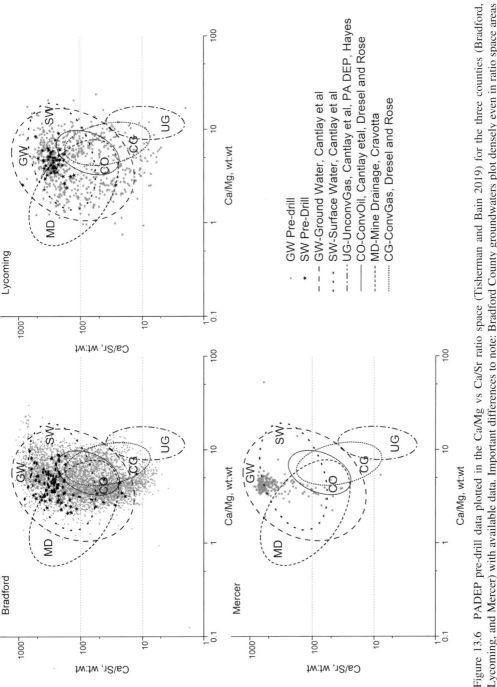

Figure 13.6 PADEP pre-drill data plotted in the Ca/Mg vs Ca/Sr ratio space (Tisherman and Bain 2019) for the three counties (Bradford, Lycoming, and Mercer) with available data. Important differences to note: Bradford County groundwaters plot densely even in ratio space areas characteristic of unconventional brines. Some Lycoming County groundwaters plot in those mixing regions, but the majority of the data are >100 Ca:Sr, or typical of waters not impacted by O&G activities. In Mercer County, almost all these data plot in the "unimpacted portion of the ratio space"

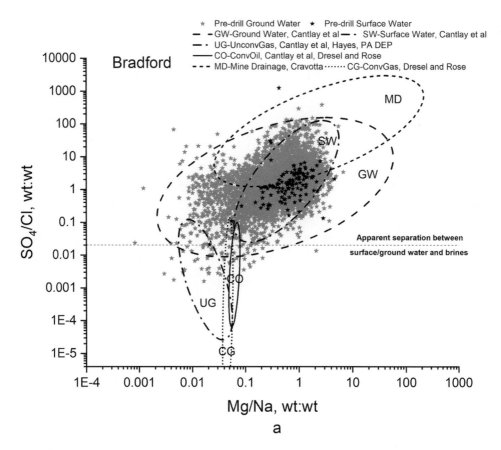

Figure 13.7 PADEP pre-drill data plotted in the Mg/Na vs SO_4/Cl ratio space (Cantlay et al. 2020b) for the two counties (Bradford (panel a) and Lycoming (panel b)) with available data. Note the strong excursion toward lower Mg/Na ratios in the Bradford County data relative to Lycoming County

illustrate, time series data collected during a known release of unconventional brines can be evaluated in these frameworks. Ideally, the pre-drill data discussed through much of this chapter could be compared with post-drill sampling to examine how chemistry evolved during the unconventional gas extraction. However, that is not feasible. There are some determination letter results published on the Shale Network HIS, but the location is obscured for privacy purposes and samples cannot reliably be matched up. Therefore, other data sets were evaluated.

During the period when unconventional wastewater disposal through publicly owned treatment works was ending (late 2010), Tenmile Creek in southwestern PA was sampled (Chapter 14). It is possible to compare water chemistries in this creek during periods when unconventional wastes were being input to the stream with water chemistries following this period. The evolution of Ca/Sr chemistries in this data were evaluated previously (Tisherman and Bain 2019), but here we evaluate this evolution using multiple, parallel lines of evidence.

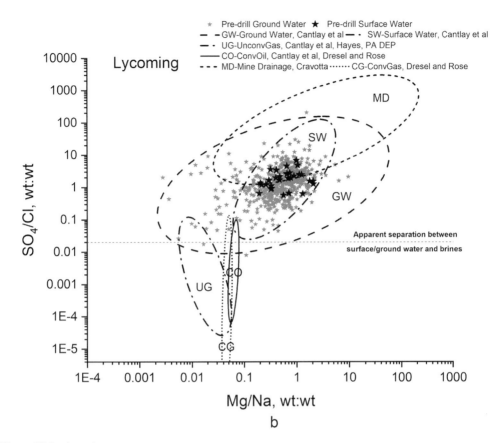

Figure 13.7 (cont.)

When the Tenmile Creek data are plotted in these ratio spaces (Figure 13.10), clear evolution of the water chemistry from more impacted to less impacted is apparent (note, some of these data were previously published by Wilson et al (2014)). The Ca/Sr rapidly increases in late 2010 as releases from the publicly owned treatment works were curtailed (Figure 13.10a). Likewise, Mg/Li ratios grow larger during this period and remain higher (Figure 13.10c). The Mg/Na vs SO_4/Cl trends are more nebulous. Concentrations of Mg increase relative to Na, but the changes in SO_4 are not clear (Figure 13.10b). Part of this probably results from the mine drainage treatment plant situated just upstream of the sampling location, as the SO_4 inputs from that system would obscure changes in the ratio owing to releases from the mine drainage treatment system. Nonetheless, when evaluating time series, substantial changes in water chemistry resulting from a decrease in contamination can be detected in multiple ratio spaces. Additional data are necessary to definitively evaluate these changes, but these frameworks detect the change.

As noted throughout this chapter, both natural and human chemical inputs have the potential to confound these ratio space approaches. Given the limitation in before/after data

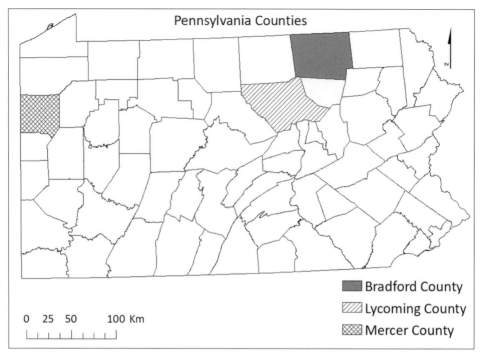

Figure 13.8 Map of Pennsylvania Counties discussed in this chapter

sets with O&G contamination, it is also useful to evaluate the ratio space approaches using data collected from streams with known inputs from natural, saline seeps and road salting activities. Johnson et al. (2015) sampled multiple streams in Pennsylvania and New York over the course of a year to examine deviations in stream chemistry from background streams when (1) road salting was prevalent and (2) in streams with substantial contributions from saline seeps (or salt springs). To examine the impact of these processes on stream chemistry and their influence on stream chemistry evolution in ratio space we plotted four of the streams: Silver Creek (reference), Apalachin-D (road salt), Fall Brook (salt spring), and Wilkes (salt spring) (Figure 13.11).

There are several important aspects to note. The road salt does not seem to strongly influence the Ca/Sr content of the stream water (Apalachin-D, Figure 13.11a), which is not surprising given that halite has a very different chemistry than these brines. In contrast, streams with substantial salt spring inputs do move down toward areas characteristic of O&G brines (e.g., Fall Brook in Figure 13.11a). That said, given this is a seasonal signal (the lowest point is during the maximum moisture deficit), the stream Ca/Sr rebounds as the region wets up in the fall. Therefore, the Ca/Mg vs Ca/Sr space is insensitive to road salting and regional solute inputs may drive the chemistry to appear more like O&G brines, but that evolution is part of the hydrologic cycle. Consider, for example, the evolution observed in Tenmile Creek (Figure 13.10a) during periods when there are no substantial brine inputs. As brine releases are curtailed, the chemistry has a similar upward trajectory, but it does not move down in subsequent years.

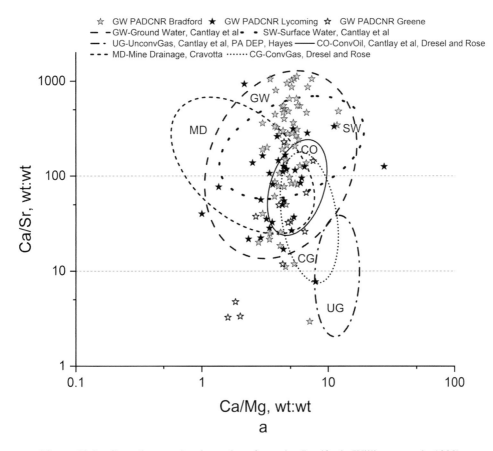

Figure 13.9 Groundwater chemistry data from the Bradford (Williams et al. 1998), Lycoming (Lloyd, Jr. and Carswell 1981), and Greene (Stoner 1987) county groundwater reports plotted in the a) Ca/Mg vs Ca/Sr and b) Mg/Na vs SO_4/Cl ratio spaces. All these chemistries were measured long before unconventional extraction activities began. There were insufficient Br and Li data in these reports to evaluate the other ratio spaces

When examining other ratio spaces, road salting and salt springs do not seem to have strong influence on the Mg/Na vs Cl/SO_4 signature of the stream waters. The chemistries stay well within the ranges observed for surface waters and again cycle over the course of a year (Figure 13.11b). Note that in Tenmile Creek (Figure 13.10c) the chemistry, though probably influenced on the Cl/SO_4 axis by the local acid mine drainage treatment system, moves from a space outside that typical of surface waters into the space typical of surface waters after the brine releases at the sewage treatment outfall ceased. Again, use of multiple, parallel ratio spaces seems to be able to detect O&G brine influences on water chemistry while remaining relatively insensitive to other important regional solute contamination. Therefore, while the pre-drill data does have some samples falling into ratio spaces more typical of brines, tracking the evolution of those waters through time can probably clarify source and impact.

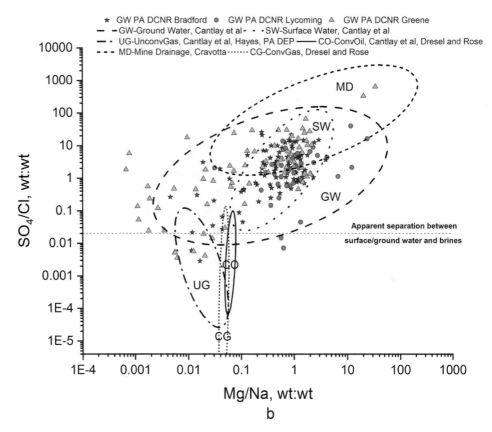

Figure 13.9 (cont.)

13.6 Consistency in Analytes and Barriers to Comprehensive Analysis

One of the most disappointing results of this evaluation was the discovery that only a single county had sufficient data to evaluate all of the potential source attribution tools. The pre-drill data do not, in general, allow comprehensive evaluation of Pennsylvania waters prior to the unconventional boom. For example, in the samples from Washington County only six constituents were measured: "Sodium, Total," "Iron, Total," "Calcium, Total," "Sulfate, Total," "Methane, Dissolved," and "Chloride, Total." Given the understanding of the unusual Marcellus brine chemistry even early in unconventional extraction, requiring only this suite of parameters is hard to justify. It is not sensitive to unusual constituents commonly found in Marcellus brines such as bromide (Wilson and Van Briesen 2013) or barium (Kargbo et al. 2010). These early decisions about sampling suites will probably limit our ability to effectively manage contamination from unconventional activities, particularly at the individual landowner scale.

Moreover, the wide variety of analytical sensitivities observed in these data precludes wide swaths of data from being evaluated in these source attribution frameworks.

Evaluation of Potential Water Quality Impacts in Unconventional Oil and Gas Extraction 331

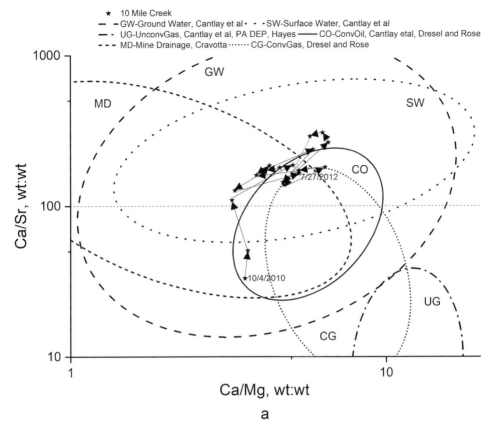

Figure 13.10 Data collected from Tenmile Creek (Greene County, PA) between 2010 and 2012. Individual samples collected through time are shown with start and end dates and arrows indicating direction for the a) Ca/Mg vs Ca/Sr, b) Mg/Na vs SO_4/Cl, and c) SO_4/Cl vs Mg/Li ratio spaces. Note, these plots show a small portion of the spaces shown in the rest of the chapter, blown up to allow visualization of the water chemistry evolution

For example, the detection limits used for bromide generally span up to two orders of magnitude in the pre-drill data (Table 13.3). Given the low environmental levels of bromide and the early recognition of this constituent in unconventional O&G brines from the Marcellus, (Wilson and Van Briesen 2013; Wilson et al. 2014), this analytical insensitivity strongly limits the regulator's and well owner's ability to detect potential contamination: particularly the 5 mg/L detection limit, where contaminant contributions would have to be substantial before detection.

13.7 Implications for Management of Water Quality with Continued Unconventional Extraction: Attribution of Responsibility and the Adequacy of Sampling Regimes

The process of hydraulically fracturing petroleum-rich rocks to enhance recovery creates many opportunities to contaminate water (e.g., Harrison 1983). Some of the potential

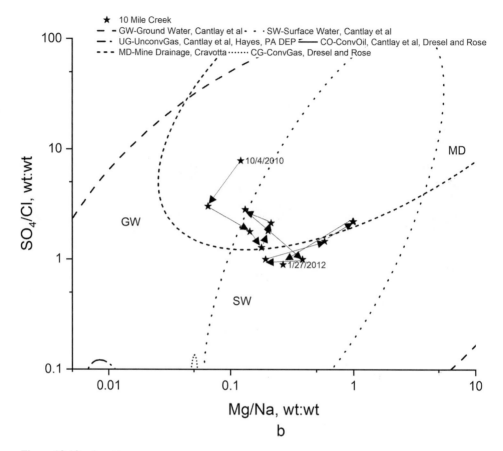

Figure 13.10 (cont.)

scenarios are estimated to be low probability (e.g., up migration of the fracturing fluid, Birdsell et al. 2015). Some of the processes are quite common (e.g., spills, Rahm et al. 2015). However, information asymmetry makes evaluation of comprehensive impacts a fundamental challenge. The activity is widespread, but the information is quite localized.

There are a variety of approaches to solve this data asymmetry. Partnering with O&G operators, regulatory agencies, or governmental organizations collaborating on research and development with both operators and regulators can provide access (Capo et al. 2014; Ziemkiewicz and He 2015; Harkness et al. 2017). However, these collaborations are only able to access a small subset of potential data (e.g., the MSEEL efforts, while useful, concentrate on a single well pad). These data cannot represent the truly randomized sampling we rely on to allow the assumptions of data independence fundamental to rigorous statistical evaluation of these data.

The PADEP pre-drill data allow some of this potential inference. Over 26,000 samples of well and spring waters across Pennsylvania is orders of magnitude larger than other

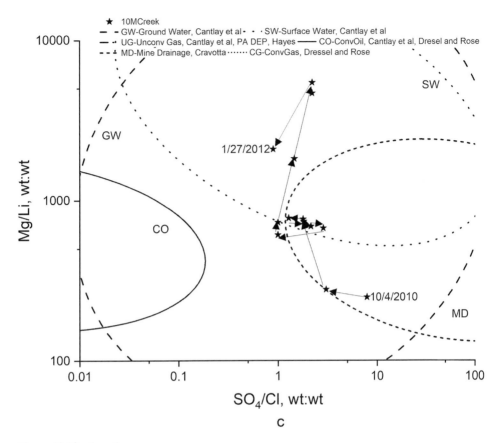

Figure 13.10 (cont.)

potential data sets. And here, using even the simple analyses included in this chapter, we can say fundamental things.

Naturally occurring groundwater chemistry can be similar to the O&G brines generated during unconventional gas extraction. However, the vast majority of pre-drill data has chemistries that are similar to waters unimpacted by O&G contamination. Even if unimpacted waters fall into the more contamination-like ratio space, examination of time series of sample data from these wells can clarify the sources influencing the water chemistry. Further, these approaches rely on a comparatively simpler set of chemical constituents, allowing application to a wider variety of potential data sets (e.g., where isotopic measurements are not possible).

The Shale Network has compiled and organized a substantial data set. However, it only covers a few counties in a meaningful way. Measurement of chemical constituents that allow the application of tools documented in the literature are quite limited, with several tools only possible with Bradford County pre-drill data. It seems that water testing specifications were probably not protective enough. Bromide and barium were clearly

Table 13.3. *Bromide detection limits reported in shale network PADEP pre-drill data*

County	Constituent	Detection limit (mg/L)	Frequency
Bradford	Bromide, total	0.01	1
Bradford	Bromide, total	0.1	229
Mercer	Bromide, total	0.1	60
Bradford	Bromide, total	1	1,596
Lycoming	Bromide, total	1	13
Mercer	Bromide, total	1	43
Bradford	Bromide, total	1.19	1
Bradford	Bromide, total	5	459

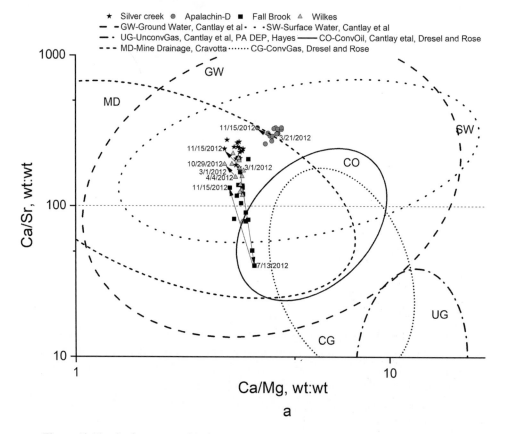

Figure 13.11 Surface water chemistry evolution in New York/Pennsylvania in streams influenced by road salting and salt springs; panel a) Ca/Mg vs Ca/Sr, and panel b) Mg/Na vs SO_4/Cl. Stream water chemistry as reported in Johnson et al. (2015) for Silver Creek (reference conditions), Apalachin-D (road salt), Fall Brook (salt spring), and Wilkes (salt spring). Begin and end dates are shown, and arrows indicate direction of movement

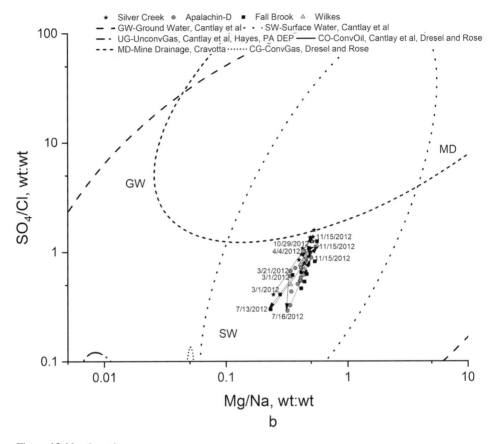

Figure 13.11 (cont.)

recognized as potentially important constituents early on. However, outside of Bradford County, there are no apparent bromide data in the entire pre-drill data set. Likewise, outside of Bradford, Lycoming, and Mercer counties, there is one measurement of barium (in Greene County) in the entire data set. Contamination is often best detected with uncommon constituents. This lack of clarity in pre-drill data will limit our ability to assess future impacts associated with extraction activities.

Finally, the choices of analytical methods seem too insensitive. Five mg/L of bromide is a large concentration (Table 13.3). Even one mg/L is high if we are looking for materials that can provide early warnings of potential contamination. Expanded parameter sets and more sensitive methods would make this pre-drill monitoring program more effective.

Acknowledgments

This work was conducted through generous support from the Heinz Endowments and Colcom Foundations.

References

Abualfaraj N, Gurian PL, and Olson MS. (2014). Characterization of Marcellus Shale flowback water. *Environmental Engineering Science.* **31**(9): 514–524 DOI: 10.1089/ees.2014.0001.

Akob DM, Cozzarelli IM, Dunlap DS, Rowan EL, and Lorah MM. (2015). Organic and inorganic composition and microbiology of produced waters from Pennsylvania shale gas wells. *Applied Geochemistry.* **60**: 116–125 DOI: 10.1016/j.apgeochem.2015.04.011.

Birdsell DT, Rajaram H, Dempsey D, and Viswanathan HS. (2015). Hydraulic fracturing fluid migration in the subsurface: A review and expanded modeling results. *Water Resources Research.* **51**(9): 7159–7188.

Blauch ME, Myers RR, Moore T, Lipinski BA, and Houston NA. (2009). Marcellus shale post-frac flowback waters-Where is all the salt coming from and what are the implications? In *SPE Eastern Regional Meeting*, Society of Petroleum Engineers.

Blondes MS, Gans KD, Engle MA, Kharaka YK, Reidy ME, Saraswathula V, Thordsen JJ, Rowan EL, and Morrissey EA. (2018). U.S. Geological Survey National Produced Waters Geochemical Database (ver. 2.3, January 2018) Available at www.sciencebase.gov/catalog/item/59d25d63e4b05fe04cc235f9

Brantley S. (2018). *Shale Network Data: Consortium for Universities for the Advancement of Hydrologic Sciences, Inc. (CUAHSI).* DOI: 10.4211/his-data-shalenetwork.

Brantley SL, Yoxtheimer D, Arjmand S, Grieve P, Vidic R, Pollak J, Llewellyn GT, Abad J, and Simon C. (2014). Water resource impacts during unconventional shale gas development: The Pennsylvania experience. *International Journal of Coal Geology.* **126**: 140–156 DOI: 10.1016/j.coal.2013.12.017.

Burgos WD, Castillo-Meza L, Tasker TL, Geeza TJ, Drohan PJ, Liu X, Landis JD, Blotevogel J, McLaughlin M, Borch T et al. (2017). Watershed-scale impacts from surface water disposal of oil and gas wastewater in western Pennsylvania. *Environmental Science & Technology.* **51**(15): 8851–8860 DOI: 10.1021/acs.est.7b01696.

Cantlay T, Bain DJ, Curet J, Jack RF, Dickson BC, Basu P, and Stolz JF. (2020a). Determining conventional and unconventional oil and gas well brines in natural sample II: Cation analyses with ICP-MS and ICP-OES. *Journal of Environmental Science and Health, Part A.* **55**(1): 11–23 DOI: 10.1080/10934529.2019.1666561.

Cantlay T, Bain DJ, and Stolz JF. (2020b). Determining conventional and unconventional oil and gas well brines in natural samples III: Mass ratio analyses using both anions and cations. *Journal of Environmental Science and Health, Part A.* **55**(1): 24–32 DOI: 10.1080/10934529.2019.1666562.

Cantlay T, Eastham JL, Rutter J, Bain DJ, Dickson BC, Basu P, and Stolz JF. (2020c). Determining conventional and unconventional oil and gas well brines in natural samples I: Anion analysis with ion chromatography. *Journal of Environmental Science and Health, Part A.* **55**(1): 1–10 DOI: 10.1080/10934529.2019.1666560.

Capo RC, Stewart BW, Rowan EL, Kohl CAK, Wall AJ, Chapman EC, Hammack RW, and Schroeder KT. (2014). The strontium isotopic evolution of Marcellus Formation produced waters, southwestern Pennsylvania. *International Journal of Coal Geology.* **126**: 57–63.

Chapman EC, Capo RC, Stewart BW, Kirby CS, Hammack RW, Schroeder KT, and Edenborn HM. (2012). Geochemical and strontium isotope characterization of produced waters from Marcellus Shale natural gas extraction. *Environmental Science & Technology.* **46**(6): 3545–3553.

Clark CE, Burnham AJ, Harto CB, and Horner RM. (2012). *Introduction: The Technology and Policy of Hydraulic Fracturing and Potential Environmental Impacts of Shale Gas Development*. Taylor & Francis.

Colborn T, Kwiatkowski C, Schultz K, and Bachran M. (2011). Natural gas operations from a public health perspective. *Human and Ecological Risk Assessment: An International Journal.* **17**(5): 1039–1056.

Cravotta CA. (2008). Dissolved metals and associated constituents in abandoned coal-mine discharges, Pennsylvania, USA. Part 1: Constituent quantities and correlations. *Applied Geochemistry.* **23**(2): 166–202 DOI: 10.1016/j.apgeochem.2007.10.011.

Dilmore RM, Sams III JI, Glosser D, Carter KM, and Bain DJ. (2015). Spatial and temporal characteristics of historical oil and gas wells in Pennsylvania: Implications for new shale gas resources. *Environmental Science & Technology.* **49**(20): 12015–12023.

Dresel PE and Rose AW. (2010). Chemistry and origin of oil and gas well brines in western Pennsylvania. *Pennsylvania Geological Survey* (Fourth series): 56.

Entrekin S, Trainor A, Saiers J, Patterson L, Maloney K, Fargione J, Kiesecker J, Baruch-Mordo S, Konschnik K, and Wiseman H. (2018). Water stress from high-volume hydraulic fracturing potentially threatens aquatic biodiversity and ecosystem services in Arkansas, United States. *Environmental Science & Technology.* **52**(4): 2349–2358.

Fracktracker Alliance. (2020). Pennsylvania Shale Viewer. *Pennsylvania Shale Viewer* Available at: www.fractracker.org/map/us/pennsylvania/pa-shale-viewer [Accessed une 10, 2020]

Harkness JS, Darrah TH, Warner NR, Whyte CJ, Moore MT, Millot R, Kloppmann W, Jackson RB, and Vengosh A. (2017). The geochemistry of naturally occurring methane and saline groundwater in an area of unconventional shale gas development. *Geochimica et Cosmochimica Acta.* **208**: 302–334 DOI: 10.1016/j.gca.2017.03.039.

Harrison SS. (1983). Evaluating system for ground-water contamination hazards due to gas-well drilling on the glaciated Appalachian Plateau. *Groundwater.* **21**(6): 689–700.

Hayes T. (2009). *Sampling and Analysis of Water Streams Associated with the Development of Marcellus Shale Gas*. Final Report. Prepared for Marcellus Shale Coalition (Formerly the Marcellus Shale Committee).

Hopey D. (2011). DEP reviewing permit for hauler charged with illegal dumping. *Pittsburgh Post-Gazette.*

Johnson JD, Graney JR, Capo RC, and Stewart BW. (2015). Identification and quantification of regional brine and road salt sources in watersheds along the New York/Pennsylvania border, USA. *Applied Geochemistry.* **60**: 37–50.

Kadlec J, StClair B, Ames DP, and Gill RA. (2015). WaterML R package for managing ecological experiment data on a CUAHSI HydroServer. *Ecological Informatics.* **28**: 19–28.

Kargbo DM, Wilhelm RG, and Campbell DJ. (2010). *Natural gas plays in the Marcellus shale: Challenges and potential opportunities*. ACS Publications.

Kim S, Omur-Ozbek P, Dhanasekar A, Prior A, and Carlson K. (2016). Temporal analysis of flowback and produced water composition from shale oil and gas operations: Impact of frac fluid characteristics. *Journal of Petroleum Science and Engineering.* **147**: 202–210 DOI: 10.1016/j.petrol.2016.06.019.

Lane MK and Landis WG. (2016). An Evaluation of the Hydraulic Fracturing Literature for the Determination of Cause–Effect Relationships and the Analysis of Environmental Risk and Sustainability. In *Environmental and Health Issues in Unconventional Oil and Gas Development* Elsevier; 151–173.

Lloyd, Jr. OB and Carswell LD. (1981). Groundwater resources of the Williamsport region, Lycoming County, Pennsylvania. Water Resource Report 51. Pennsylvania Geological Survey.

Miller BA. (2020). Unconventional oil and gas: Interactions with and implications for groundwater. In *Regulating Water Security in Unconventional Oil and Gas*, Buono RM, López Gunn E, McKay J, and Staddon C (eds.) Springer International Publishing, pp. 267–290. DOI: 10.1007/978-3-030-18342-4_13.

Milliken P. (2013). Brine dumper agrees to cooperate with U.S. Attorney. Youngstown Vindicator.

Mrdjen I and Lee J. (2016). High volume hydraulic fracturing operations: potential impacts on surface water and human health. *International Journal of Environmental Health Research*. **26**(4): 361–380.

Olawoyin R, McGlothlin C, Conserve DF, and Ogutu J. (2016). Environmental health risk perception of hydraulic fracturing in the US. *Cogent Environmental Science*. **2**(1): 1209994.

Pennsylvania Department of Environmental Protection, Bureau, of Oil and Gas Management. (1991). NORM Survey Summary.

Phan TT, Capo RC, Stewart BW, Macpherson GL, Rowan EL, and Hammack RW. (2016). Factors controlling Li concentration and isotopic composition in formation waters and host rocks of Marcellus Shale, Appalachian Basin. *Chemical Geology*. **420**: 162–179.

Poth CW. (1962). The occurrence of brine in western Pennsylvania. *Topographic and Geologic Survey* (Bulletin M 47): 59.

R Core Team. (2013). *R: A Language and Environment for Statistical Computing*. R Foundation for Statistical Computing.

Rahm BG, Vedachalam S, Bertoia LR, Mehta D, Vanka VS, and Riha SJ. 2015. Shale gas operator violations in the Marcellus and what they tell us about water resource risks. *Energy Policy*. **82**: 1–11.

Rester E and Warner SD. (2016). Chapter 4 - A review of drinking water contamination associated with hydraulic fracturing. In, Kaden D and Rose T (eds.) *Environmental and Health Issues in Unconventional Oil and Gas Development*. Elsevier, pp. 49–60. DOI: 10.1016/B978-0-12-804111-6.00004-2.

Rowan EL, Engle MA, Kraemer TF, Schroeder KT, Hammack RW, and Doughten MW. (2015a). Geochemical and isotopic evolution of water produced from Middle Devonian Marcellus shale gas wells, Appalachian basin, Pennsylvania. *AAPG Bulletin*. **99**(2): 181–206 DOI: 10.1306/07071413146.

Rowan EL, Engle MA, Kraemer TF, Schroeder KT, Hammack RW, and Doughten MW. (2015b). Geochemical and isotopic evolution of water produced from Middle Devonian Marcellus shale gas wells, Appalachian basin, Pennsylvania: Geochemistry of produced water from Marcellus Shale water, PA. *Aapg Bulletin*. **99**(2): 181–206.

Shih J-S, Saiers JE, Anisfeld SC, Chu Z, Muehlenbachs LA, and Olmstead SM. (2015). Characterization and analysis of liquid waste from Marcellus Shale gas development. *Environmental Science & Technology*. **49**(16): 9557–9565 DOI: 10.1021/acs.est.5b01780.

Stoner JD. (1987). *Water Resources and the Effects of Coal Mining, Greene County, Pennsylvania*. Pennsylvania Geological Survey.

Tasker TL. (2018). *Tracing the Environmental and Human Health Impacts of Oil and Gas Development*. Pennsylvania State University.

Tasker TL, Warner NR, and Burgos WD. (2020). Geochemical and isotope analysis of produced water from the Utica/Point Pleasant Shale, Appalachian Basin. *Environmental Science: Processes & Impacts*. **22**(5): 1224–1232 DOI: 10.1039/D0EM00066C.

Tisherman R and Bain DJ. (2019). Alkali earth ratios differentiate conventional and unconventional hydrocarbon brine contamination. *Science of The Total Environment*. **695**: 133944 DOI: 10.1016/j.scitotenv.2019.133944.

Vengosh A, Jackson RB, Warner N, Darrah TH, and Kondash A. (2014). A critical review of the risks to water resources from unconventional shale gas development and hydraulic fracturing in the United States. *Environmental Science & Technology*. **48**(15): 8334–8348 DOI: 10.1021/es405118y.

Vengosh A, Kondash A, Harkness J, Lauer N, Warner N, and Darrah TH. (2017). The geochemistry of hydraulic fracturing fluids. *Procedia Earth and Planetary Science*. **17**: 21–24 DOI: 10.1016/j.proeps.2016.12.011.

Warner NR, Christie CA, Jackson RB, and Vengosh A. (2013). Impacts of Shale Gas Wastewater Disposal on Water Quality in Western Pennsylvania. *Environmental Science & Technology*. **47**(20): 11849–11857 DOI: 10.1021/es402165b.

Warner NR, Jackson RB, Darrah TH, Osborn SG, Down A, Zhao K, White A, and Vengosh A. (2012). Geochemical evidence for possible natural migration of Marcellus Formation brine to shallow aquifers in Pennsylvania. *Proceedings of the National Academy of Sciences*. **109**(30): 11961 DOI: 10.1073/pnas.1121181109.

Wickham H, Averick M, Bryan J, Chang W, McGowan L, François R, Grolemund G, Hayes A, Henry L, and Hester J. (2019). Welcome to the Tidyverse. *Journal of Open Source Software*. **4**(43): 1686.

Williams JH, Taylor LE, and Low DJ. (1998). *Hydrogeology and groundwater quality of the glaciated valleys of Bradford, Tioga, and Potter Counties, Pennsylvania*. Pennsylvania Geological Survey.

Wilson JM and Van Briesen JM. (2013). Source Water Changes and Energy Extraction Activities in the Monongahela River, 2009–2012. *Environmental Science & Technology*. **47**(21): 12575–12582 DOI: 10.1021/es402437n.

Wilson JM, Wang Y, and VanBriesen JM. (2014). Sources of high total dissolved solids to drinking water supply in Southwestern Pennsylvania. *Journal of Environmental Engineering*. **140**(5) DOI: 10.1061/(ASCE)EE.1943-7870.0000733.

Zheng Z, Zhang H, Chen Z, Li X, Zhu P, and Cui X. (2017). Hydrochemical and isotopic indicators of hydraulic fracturing flowback fluids in shallow groundwater and stream water, derived from dameigou shale gas extraction in the northern qaidam basin. *Environmental Science & Technology*. **51**(11): 5889–5898.

Ziemkiewicz PF and He YT. (2015). Evolution of water chemistry during Marcellus Shale gas development: A case study in West Virginia. *Chemosphere*. **134**: 224–231.

Ziemkiewicz PF, Quaranta JD, Darnell A, Wise R. (2014). Exposure pathways related to shale gas development and procedures for reducing environmental and public risk. *Journal of Natural Gas Science and Engineering*. **16**: 77–84 DOI: 10.1016/j.jngse.2013.11.003.

14

A Baseline Ecological Study of Tributaries in the Tenmile Creek Watershed, Southwest Pennsylvania

BRADY PORTER, ELIZABETH DAKIN, SARAH WOODLEY, AND JOHN F. STOLZ

14.1 Introduction

In 2009 the extraction of natural gas from the Marcellus Shale formation was gearing up in southwestern Pennsylvania with limited state regulations in place (Act 13 would be enacted three years later in 2012) and few studies on the possible environmental impacts (Lampe and Stolz 2015). No baseline or targeted monitoring studies were in place and no state or federal funding was made available to study the possible environmental impacts. In a commonwealth affected by the legacy of coal mining with over 5,500 miles of streams impacted by abandoned mine drainage, AMD (PADEP 2020), what risks might this new activity pose to the Commonwealth's aquatic ecosystems? The aim of this study was to understand these potential impacts from the rapidly developing Marcellus Shale extraction activities on the water quality and biological communities of streams in southwestern Pennsylvania.

Unconventional gas extraction involves horizontal drilling, hydrological fracturing (fracking) of the shale formation, and the use of proppant (sand or ceramic) to keep the fractures open to allow for the release of the natural gas (see Chapter 2, Stolz and Griffin 2022). The fracking process uses millions of gallons of water containing a mixture of chemical additives (Lampe and Stolz 2015). While the general constituents that went into the mixture were known at the time (i.e., surfactant, biocide, friction reducer, cross-liners, etc.), details on the exact composition were not readily available (GWPC 2009) as the formulation was a carefully guarded industry "trade secret" (Maule et al. 2013). The fracking process creates wastewater in the form of short-term flowback water and long-term produced water, both of which contain the fluids that were pumped into the well, along with formation water and the rock constituents solubilized from the geologic formation (see Chapter 2, Stolz and Griffin 2022). Since local geology is different from one region to the next; the chemical cocktails were adjusted from other plays to optimize local extraction conditions. These factors all contribute to the complexity of understanding the potential impacts to water chemistry, what types of pollution to look for, and how to formulate a comprehensive monitoring plan for stream water quality.

In addition to water quality, there are also water quantity concerns associated with Marcellus Shale extraction (see Chapter 7, Wilson and VanBriesen, 2022). The millions of gallons of water used in the hydraulic fracturing of an unconventional well are typically transported to a centralized impoundment site using tanker trucks. Often, water is withdrawn from local streams, having potential impacts on stream discharge. The Pennsylvania Department of Environmental Protection (PA DEP) regulates surface water withdrawals from lakes, rivers, and streams in western Pennsylvania (Lampe and Stolz 2015). In doing so, they attempt to consider the impact of water withdrawal on overall stream discharge, its impact on living aquatic organisms, and the role that reduced water volume may have on the dilution of downstream pollution sources.

14.2 Study Aims and Objectives

From the onset of this project, we set out to understand the potential impacts Marcellus Shale extraction activities could have on stream health in southwestern Pennsylvania. Against a legacy of coal mining in the region, we attempt to reveal diagnostic parameters to distinguish impacts of the Marcellus Shale play development on the water quality and aquatic biological communities. A multidisciplinary team involving faculty and students from Duquesne University, the University of Pittsburgh, and the Carnegie Museum of Natural History was established, and this collaboration secured funding from the Heinz Foundation for a one-year study. Given limited time and resources, the group decided to focus on a small-scale watershed, where diagnostic signatures of potential Marcellus Shale impacts would not be masked by dilution or confounded by other factors contributing to stream degradation. We adopted a paired-stream basin design, selecting two small watersheds in the Tenmile Creek system that are similar in size, geographic area, and land use. The experimental stream had newly established Marcellus Shale wells and numerous approved unconventional well permits in its watershed, while the control stream did not have any indication of current or future shale gas extraction plans. We then compared this basin-pair for baseline conditions and potential changes in water quality and biotic communities. We also selected a site on the Tenmile Creek main stem near its confluence with the Monongahela, to provide a broader context of the overall system (but data from that site is outside the scope of this chapter). We took samples from fixed stations over the next two years and these data were incorporated into several M.S. thesis projects by students in the Environmental Science and Management Program at Duquesne University (Eastham 2012; Pascuzzi 2012; Rutter 2012; Dugas 2014). The selected parameters covered in this chapter include water chemistry, microbiology, fish surveys, darter genetics, and salamander surveys.

14.3 Site Selection

Tenmile Creek, a tributary to the Monongahela River, drains approximately 2,266 km^2 of Washington and Greene Counties, in southwestern Pennsylvania. From 2007 through 2009, this watershed was the center of intense Marcellus Shale extraction activity, with 124 issued drilling permits. The Tenmile Creek watershed is divided into around 78 sub-basins, and two of these sub-basins were ultimately selected for a paired basin approach.

One of these basins would have a high density of Marcellus Shale permits and the other would act as a control with no activity. Ideally, these paired streams would be geographically adjacent, with similar drainage areas, comparable land use, and previous baseline data available. Additional considerations included property access points and appropriate stream sizes that were large enough to contain stable biological assemblages, but not so large that they might wash out hydrological equipment during flooding.

Geographic Information System (GIS) mapping was initially employed to evaluate site selection of multiple small stream pairs. A map of the sub-basins of Tenmile Creek was created with a layer indicating the Marcellus Shale permits and active drilling (SPUD) sites as of 2009. Twenty-five sub-basins had PA DEP-issued permits ranging from one to 37. The number of SPUD sites in each basin was divided by the corresponding drainage areas to obtain the density of Marcellus Shale wells per square kilometer and reflected on the map by the intensity of shading (Figure 14.1). This spatial analysis narrowed the choices down to six candidate stream-pairs, each of which received an initial site visit for accessibility, suitability of installing hydrological gage stations, and diversity of biological communities.

Ideally, we would have selected multiple stream-pairs, but limited resources and suitability of stream-pairs for multiple sampling techniques made this impractical. Kimmel and Argent (2010, 2012), from California University of Pennsylvania, conducted fish and macroinvertebrate surveys from a number of sites along the main stems of Tenmile Creek "phase I" and the South Fork Tenmile Creek "phase II," and these could potentially be useful for context. Unfortunately, only two of our candidate sites were on streams near the confluence with their sites. Experimental candidate Rush Run was small and would likely be inundated by the main stem South Fork during times of high water producing an ephemeral transient fish population. Experimental candidate Browns Run had a reservoir between the SPUD sites and the Kimmel and Argent site, potentially confounding comparisons to their work (Figure 14.1). Three of the other stream-pairs (Red Run vs. Little Tenmile Creek, Coal Lick Run vs. Lauren Run, and Hargus Creek vs Garner Run) had low baseflow discharge and/or poor habitat for one of the pairs, making them difficult for gage installation and electrofishing surveys.

The only remaining stream-pair, Bates Fork and Fonner Run, had a good access point, a variety of habitats, and were seemingly large enough to support stable vertebrate populations. Fonner Run joins Bates Fork about 200 m downstream of our study sites to make a fourth order tributary to Browns Creek at the community of Sycamore. Browns Creek, designated as a high-quality warm water fishery (PA DEP), has its confluence with the South Fork of Tenmile Creek in West Waynesburg, Franklin Township, Greene County, Pennsylvania.

Fonner Run, the control site, has a drainage area of 8.5 km^2, is 84% covered by forest, and at the time of this study had no active SPUD sites, permits, or leases for Marcellus Shale extraction. The smaller drainage area of Fonner Run corresponded with a smaller stream that was often relegated to a series of partially connected runs and pools during summer baseflow. A large gravel deposit at the mouth of Fonner Run typically interrupted surface flow at its confluence with the Bates Fork, thereby preventing faunal exchange between the stream-pair during normal flow levels. Our sampling site included a sharp

A Baseline Ecological Study of Tributaries in the Tenmile Creek Watershed 343

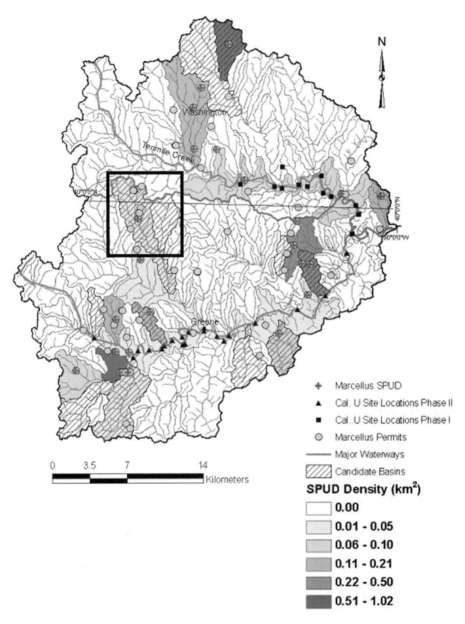

Figure 14.1 Tenmile Creek basin with candidate small watershed-pairs indicated with hatching. Shading indicates the density of Marcellus wells (SPUDs/km^2) in the basin and circles indicate existing permits granted as of 2009. Triangles and squares indicate sites previously sampled by Kimmel and Argent (2010, 2012) for fish and macroinvertebrates. The selected Bates Fork and Fonner Run stream-pair is delineated inside the box. (A black and white version of this figure will appear in some formats. For the colour version, refer to the plate section.)

Figure by Dan Bain with permission

meander that created a deep pool with undercut root wads, offering permanent structure and habitat for larger fish.

The experimental site, Bates Fork, has a drainage area of 15.8 km^2 at its confluence with Fonner Run and was 61% forested (Streamstats.usgs.gov). By July 2010, the upstream drainage had 21 wells distributed over six pads. Over the following year, 16 additional wells were added, including four new pads. No additional wells were drilled in 2012, but there were 20 new permits approved across four new and two existing pads, making this small watershed one of the most rapidly developed plays in the Tenmile Creek system.

There were three violations issued by the PA DEP on Bates Fork well pads that relate to our study dates. The first Environmental Health and Safety (EH&S) violation occurred almost eight months before our first sampling on January 18, 2010 and was cited as "401CAUSEPOLL- Polluting substance(s) allowed to discharge into water of the commonwealth." A second Bates Fork pad received two violations on December 8, 2010 with one EH&S cited as "78.54- Failure to properly control or dispose of industrial or residual waste to prevent pollution of the waters of the commonwealth" and the other one Administrative cited as "78.56PITCNST- Impoundment not structurally sound, impermeable, third party protected, greater than 20" of seasonal high ground water table" (Fractracker.org). This indicates there were several opportunities for polluting substances from Marcellus Shale extraction activities to enter Bates Fork upstream of our study site.

14.4 Water Chemistry

Field measurements of water quality were conducted between July 28, 2010 and May 29, 2012 with a calibrated YSI-Pro Plus handheld multimeter to provide monthly snapshots of water temperature, dissolved oxygen, specific conductivity, and pH. Corresponding grab samples of surface water were collected in one-liter French square glass bottles and kept at 4°C until processed in the lab. A Dionex ICS-1100 with AS-22 anion exchange column, conductivity cell, and UV/Vis detector were used to determine the concentrations of chloride, nitrate, sulfate, and bromide anions following EPA approved methods (Jackson 2006) and previous studies at our facility (Cantlay et al. 2020a, 2020b). Data loggers were deployed on June 5, 2012 to collect water temperature and specific conductivity data at hourly intervals through November 6, 2012, but could not be recovered after a flooding event the following spring.

The Pennsylvania Code (PA Chapter 93) designates the entire Browns Creek drainage as a high-quality warm water fishery. Both tributaries exceeded the maximum water temperature criteria for a warm water fishery (7.78°C) on March 20, 2012, with Bates Fork reading 9.9°C and Fonner Run at 10.5°C. In addition, Bates Fork also exceeded the criteria (22.22°C) for May 29, 2012 with a water temperature of 23.2°C. The continuous data logger did not detect additional water temperature exceedance during the remainder of 2012. Dissolved oxygen is generally inversely correlated with water temperature, but the only violation of the minimum 5 mg/L criterion occurred in Fonner Run on August 3, 2010 with a reading of 4.38 mg/L. Both sites had reduced flows in the summer months, with the smaller Fonner Run occasionally reduced to isolated surface pools that percolated under a cobble bar to reach its confluence with Bates Fork.

Specific conductivity, a measure of the overall ionic load, was slightly higher in Bates Fork, ranging from 113 to 481 µS/cm with an average of 332 ± 37 µS/cm, compared to Fonner Run with a range of 89 to 404 µS/cm and an average of 277 ± 36 µS/cm (Rutter 2012). There were no significant correlations between any of the tested anion concentrations and the overall specific conductivity. All sulfate concentrations were below 51 ppm and the pH remained within an acceptable range of 6.8 to 8.1 standard units, indicating minimal impact of acid mine discharge from legacy coal extraction activities. Elevated chloride and bromide concentrations can indicate pollution from coal bed and/or Marcellus brines. Bromide was generally under the detection limit (0.05 ppm) for all samples, with two exceptions on Bates Fork: August 5, 2011 when it was 0.17 ppm and correlated to a high chloride reading of 69 ppm, and still detectable in the Bates Fork sample at 0.07 ppm on the next sampling date of September 9, 2011 (while chloride returned to baseline at 8 ppm). Otherwise, chloride was under 16 ppm for Bates Fork and 5 ppm for Fonner Run for the duration of the study with no discernable seasonal pattern. It is important to note that the elevated chloride concentration in the Bates Fork was not correlated with a high specific conductivity reading. When chloride is the dominant ion contributing to specific conductance, these two parameters are tightly linked, as is the case for the Pine Creek drainage, an urban tributary to the Allegheny River in Allegheny County, PA (Prettner 2019). However, in Bates Fork other ions may be more dominant, so the continuous specific conductance readings from the data loggers might have missed other "spikes" in chloride concentration in Bates Fork in the latter half of 2012.

14.5 Microbiology

It had been suspected that the Marcellus Shale geological layer might have a unique microbiome, adapted to the high salt environment. This was based on previous studies of other shale formations such as the Barnett Shale in Texas (Fichter et al. 2008; Struchtemeyer and Elshahed 2011; Struchtemeyer et al. 2011) and preliminary investigations of the Marcellus Shale (Mohan et al. 2011). A further discussion of this unique microbiology can be found in Chapter 12 (Nixon, 2022). As part of the baseline study, Bates Fork and Fonner Run microbiology was surveyed. During the study period (2010–2012), two molecular techniques were in common use and used for the survey: ARISA (Automated Ribosomal Intergenic Spacer Analysis, Fisher and Triplett 1999) and clonal libraries constructed from 16S rRNA gene amplification. In addition to the surface water samples from Bates Fork and Fonner Run, the microbiology of flowback and produced water from Marcellus wells in southwestern Pennsylvania were characterized for comparison to the surface waters. A selective medium for the enrichment of Marcellus Shale microbes was designed based on both published data (i.e., Blauch et al. 2009) and lab analysis of the flowback and produced water samples (Eastham 2012). This medium was used to enrich for microorganisms from these samples with the hope of identifying species that might serve as indicator species in the biotic survey (Eastham 2012; Joshi 2013; Dugas 2014).

Enrichment cultures from both the Lone Pine (LP) produced water impoundment and the Southwestern Pennsylvania (SWPA) flowback water sample were obtained (Eastham 2012). Interestingly, both enrichments showed a requirement for NaCl, with at least 30 g/L needed for growth (Eastham 2012). Clonal libraries were constructed and select clones were sequenced. Amplicons from LP were found that were related to several marine and halophilic species including *Halomonas taenensis*, *H. organivorans*, *Cobetia marina*, *Arcobacter marinus*, *A. halophilus*, and *Salinovibrio costicola* (Eastham 2012). A strain of *Salinovibrio costicola*, designated LP-1, was subsequently isolated and characterized (Dugas 2014). Amplicons from SWPA were also found that were related to several marine and halophilic species related to *Halomonas* and *Thalossospira*. Interestingly, one clone was found to be related to *Idiomarina* species (Eastham 2012). Attempts to culture the *Idiomarina* species resulted in the isolation and characterization of a strain of halotolerant *Bacillus firmus* (Joshi 2013). Clonal libraries constructed from the Bates Fork and Fonner Run enrichments yielded only freshwater species including *Arthrobacter*, *Enterobacter*, *Erwinia*, and *Pantoea* species (Eastham 2012).

In a subsequent investigation, DNA was extracted directly from water samples from Bates Fork and Fonner Run (Rutter 2012). Overall, the results indicated that both streams had a microbiome typical of freshwater streams impacted by agriculture. Fonner Run produced the most successfully sequenced clones, possibly owing to the high nutrient inputs from the upstream farmland. The genera that were found included *Escherichia*, *Shigella*, *Bacillus*, *Arthrobacter*, *Massilia*, *Duganella*, and *Curvibacter* (Rutter 2012). The closest species matches for each of the sequenced clones were from goat feces, human feces, manured soil, forest soil, and well water (Ding and Yakota 2004; Li et al. 2004; Zhang et al. 2006; Hariharan et al. 2007; Byrne-Bailey et al. 2010; Pichel et al. 2012; Pelletier and Sygusch 1990). Thus, they were consistent with species found in freshwater samples from forests and farms with livestock. Bates Fork, however, did not yield as many clones. There were three genera that were found in Bates Fork: *Halothermothrix*, *Arthrobacter*, and *Micrococcus* (Rutter 2012). The closest species matches for each of the sequenced clones derived from the Bates Fork samples are related to species found in salt lake sediment, manured soil, and on plant roots (Cayol et al. 1994; Chen et al. 2009; Byrne-Bailey et al. 2010). The identified bacterial species that were found to live in Bates Fork suggest that there is both forested and agricultural land use. Interestingly, *Halothermothrix* is a halophile and would not be expected to be found in a freshwater system. Thus, its detection concurrent with bromide in the water sample could be an indication of a temporary discharge of produced water into the stream.

14.6 Fish Surveys

One of the main environmental concerns raised by Marcellus Shale oil and gas extraction activities is the high levels of total dissolved solids (TDS) and chloride in the wastewater (i.e., produced water), and the potential for this wastewater to contaminate and degrade rivers and stream ecosystems (Chapter 7). Surveys to determine stream fish abundance and community assemblage have become a standard biomonitoring method to detect chronic and episodic degradation of water quality. Kimmel and Argent (2010; 2012) have conducted numerous ichthyological surveys throughout the Tenmile Creek system, mainly concentrated along the South Fork and Tenmile Creek mainstems (Figure 14.1). Darters in the family Percidae include around 200 species of North American freshwater fish that are of particular interest in relation to Marcellus Shale discharge owing to their high sensitivity to TDS and chloride (Kiviat 2013). Furthermore, most darters species lack a swim bladder, do not undergo long-distance spawning migrations, and live their lives on the stream bottom. These characteristics would place darters in direct contact with dense high-salinity water from both point-source discharge and groundwater seepage contamination.

We conducted standard backpack electrofishing surveys on the Bates Fork and Fonner Run sites to obtain baseline fish community data over five survey dates: September 29, 2010, October 20, 2010, June 14, 2011, August 5, 2011, and June 5, 2012. The 100 m sampling reaches were bounded by shallow riffles and sampled with two-pass electrofishing. All fish from each pass were identified to species and enumerated in the field before being returned to the stream.

Fish sampling over the two sites and five sample dates produced a total of 3,766 individuals across 15 species representing four families; Cyprinidae, Catostomidae, Centrarchidae, and Percidae. These are all considered to be primary division freshwater fish, meaning the species in these families are strictly intolerant of saltwater (Myers 1949). Species richness ranged from 10 to 13 for Bates Fork samples, with a cumulative total of 14 species across the five sample dates. Species richness for Fonner Run samples ranged from 11 to 14, with a cumulative total of 15 species. The subtle difference in overall species richness results from a single individual of the greenside darter (*Etheostoma blennioides*) from a single Fonner Run sample date. The greenside darter is considered to be "moderately intolerant" (Ohio EPA 1987) and "intolerant" (Kimmel and Argent 2006) of general stream pollution. The greenside darter, rainbow darter (*E. caeruleum*), and rock bass (*Ambloplites rupestris*) are considered to be intolerant of chloride levels above 24 ppm and therefore receive low chloride Tolerance Indicator Values (TIV) of between two and three (Meador and Carlisle 2007). The additional three darter species captured in these surveys are considered to be intermediate/moderately tolerant of general stream pollution, consistent with their moderate (*E. flabellare* and *Percina maculata*) to tolerant (*E. nigrum*) TIVs for chloride. Shannon's H, a standard measurement of species diversity, was consistently lower for samples from Bates Fork compared to Fonner Run but had no temporal trends (Table 14.1).

The overall health of a stream ecosystem can be inferred from the integrity of its biological community. Metrics assessing species composition, trophic composition, and fish abundance and condition are typically used to establish an Index of Biotic Integrity (IBI) that produces a score that is rated against unimpacted reference sites (Karr et al. 1986). A headwater IBI is appropriate for small tributaries in the Allegheny Plateau ecoregion, using 12 metrics with scorings that scale to drainage area and a maximum possible score of 60 (Ohio EPA 1987). The Ohio headwater IBI scores for Bates Fork fish samples ranged from 30 ("fair") to 46 ("very good") with an average of 39.8 ("marginally good"), while the scores for the Fonner Run samples ranged from 37 ("fair") to 44 ("good")

Table 14.1. Summary of fish survey data from Bates Fork and Fonner Run across five sample dates with IBI scores and designations for both the Ohio Headwater IBI (Ohio EPA 1987) and Monongahela IBI (Kimmel and Argent 2007) and Tolerance Indicator Values for chloride (Meador and Carlisle 2007)

Common name	Scientific name	IBI T OH	IBI T Mon	TIV Cl-	Bates Fork 15.8 sq km					Fonner Run 8.5 sq km				
					29-Sep-10	20-Oct-10	14-Jun-11	5-Aug-11	5-Jun-12	29-Sep-10	20-Oct-10	14-Jun-11	5-Aug-11	5-Jun-12
Minnows	Family Cyprinidae													
Central stoneroller	*Campostoma anomalum*	-	T	7	9	19	14	47	9	72	57	25	55	88
Silverjaw minnow	*Ericymba buccata*	-	M	10	1	1		3		1	5	1		1
Striped shiner	*Luxilus chrysocephalus*	-	M	4	3	21	6	31	76	27	32	5	13	44
Bluntnose minnow	*Pimephales notatus*	T	T	8	14	20	43	24	67	141	72	99	59	212
Western blacknose dace	*Rhinichthys obtusus*	T	T	8	12	48	20	44	10	14	6	13	11	44
Creek chub	*Semotilus atromaculatus*	T	T	9	40	103	94	144	160	69	40	49	93	199
Suckers	Family Catostomidae													
White sucker	*Catostomus commersoni*	T	T	9	2	12	1	8	16	35	39	9	8	42
Northern hog sucker	*Hypentelium nigricans*	MI	M	6			1	1						1
Sunfishes & Basses	Family Centrarchidae													
Rock bass	*Ambloplites rupestris*	-	M	3			6	1	11					1
Green sunfish	*Lepomis cyanellus*	T	T	9			5	1	1	2	2	17	7	29
Perches & Darters	Family Percidae													
Greenside darter	*Etheostoma blennioides*	MI	I	3								1		
Rainbow darter	*Etheostoma caeruleum*	MI	M	2	1	6	24	2	6		3	10	2	2
Fantail darter	*Etheostoma flabellare*	-	M	6	2	25	98	41	212	6	16	60	43	96
Johnny darter	*Etheostoma nigrum*	-	M	10	3	5	16	4	55	10	21	18	8	55
Blackside darter	*Percina maculata*	-	M	6			4		7	2	2	4	2	6

Species Richness	10	10	13	13	12	11	12	13	11	14
Total individuals per sample	87	260	332	351	630	379	295	311	301	820
Total individuals per site	1660					2106				
Total shock time (sec) per sample	1122*	2281	4006	3131	4495	2253	2278	3595	3179	3498
Total shock time(sec) per site	15035					14803				
Shannon's H (Nats)	1.668	1.822	1.917	1.801	1.816	1.756	2.050	2.004	1.874	2.025
Ohio EPA Headwater IBI Score (12-60)	30 Fair	38 Fair	46 Very Good	41 Marg Good	44 Good	37 Fair	44 Good	44 Good	41 Marg Good	42 Marg Good
Kimmel & Argent Mon IBI Score (0-50)	25 Fair/Poor	25 Fair/Poor	27 Fair/Poor	27 Fair/Poor	27 Fair/Poor	21 Poor	25 Fair/Poor	26 Fair/Poor	23 Poor	25 Fair/Poor

Key to IBI Values:
T = Tolerant
- = Intermediate
M = Moderately Tolerant
MI = Moderately Intolerant
I = Intolerant
* single pass electrofishing

with an average of 41.6 ("marginally good") (Table 14.1). Further insight can be gleaned from the individual IBI metrics (not shown). Both sites had an abundance of pollution-tolerant species. The most abundant species in Bates Fork was the creek chub (*Semotilus atromaculatus*), a pollution-tolerant minnow in the generalist feeding guild. The most abundant species in Fonner Run was the bluntnose minnow (*Pimephales notatus*), a pollution-tolerant omnivorous species. These two dominant species, along with the silverjaw minnow (*Ericymba buccata*), green sunfish (*Lepomis cyanellus*), and Johnny darter (*Etheostoma nigrum*) are all pioneering species, being some of the first species to recolonize intermittent streams. Fonner Run generally has a higher percentage of pioneering species, but it has half the drainage area of Bates Fork and was observed as a series of intermittent pools in summer low-flow conditions. These pioneering species are complemented with stable populations of western blacknose dace (*Rhinichthys obtusus*) and fantail darter (*E. flabellare*), two typical headwater species that would be expected in small headwater streams such as Bates Fork and Fonner Run. Simple lithophilic spawners were generally rich in species at both sites, which suggests that siltation is not a serious issue for these streams.

Kimmel and Argent (2006) developed an IBI specifically for tributaries to the Monongahela, which uses a different set of 10 metrics that are uncorrected for drainage area and have a maximum score of 50. The Mon IBI scores for Bates Fork ranged from 25 to 27 ("fair/poor") with an average of 26.2 ("fair/poor"). The Mon IBI scores for Fonner Run ranged from 21 ("poor") to 26 ("fair/poor") with an average of 24 ("poor"). These scores are generally similar between the two sites, do not have a temporal trend, and reveal the same general issues as the Ohio Headwaters IBI. The fish communities are dominated by pollution-tolerant omnivorous species, have a low composition of insectivorous species, and only one top carnivore, the rock bass (*Ambloplites rupestris*). Given these IBI scores, these streams would rank around the middle of the 23 tributaries of the Monongahela River system examined by Kimmel and Argent (2006). One remarkable aspect of both Bates and Fonner compared to these other Monongahela River tributaries is their high fish density. Bates Fork had an average fish density of 1.66 fish /m^2 and Fonner Run had an average density of 4.21 fish /m^2 compared to the reported highest density in Fishpot Run of 0.99 fish/m^2 (Kimmel and Argent 2006).

Despite Bates Fork having nearly double the drainage at the point of confluence with Fonner Run, these are both small headwater streams that have comparable, relatively intact, and stable fish communities. This study has established a solid baseline for fish community structure at the onset of Marcellus Shale extraction activities in the watershed. The slightly lower species richness and diversity in the fish samples of Bates Fork, combined with a lack of temporal decline in any of the IBI metrics, makes it unlikely that these differences result from the Marcellus Shale extraction activities in the small watershed upstream of the Bates Fork site. The fact that Fonner Run did not have any ongoing or planned Marcellus Shale extraction activities at the time of these surveys provides an ideal baseline to detect future potential impacts.

14.7 Darter Genetics

Microsatellites are short (2–4 base pair) tandemly repeated DNA sequences that are frequently used in population genetic studies to quantify and compare levels of genetic diversity and differentiation within and between populations. In this study, we used a suite of microsatellite markers to explore whether fantail and Johnny darters in Bates Fork, with active Marcellus shale extraction activity, showed less genetic diversity than conspecifics in nearby Fonner Run, which at that time did not have any such activity. We also used these markers to measure the amount of genetic differentiation between darter populations in Bates Fork and Fonner Run.

Out of the 15 species of fish collected during fish surveys in Bates Fork and Fonner Run, fantail darters (*Etheostoma flabellare*) and Johnny darters (*E. nigrum*) were among the most abundant species (total fantail darter n = 599, Johnny darter n = 195), and moderate to high numbers of individuals were collected from each stream during multiple surveys conducted in 2010–2012. Fin clips from fantail and Johnny darters were collected during these surveys and preserved in absolute ethanol. Genomic DNA was extracted using a standard phenol:chloroform protocol (Porter et al. 2002), and thirteen microsatellite loci developed for other darter species were tested for amplification in fantail and Johnny darters (Table 14.2). Microsatellites that amplified in at least two thirds of individuals were retained for analysis. Thus, data is presented for fantail darters for four microsatellite loci (Esc153, Eca14, Eca46, and CV24), while Johnny darters have data for seven microsatellite loci (Esc26b, Esc153, Eca11, Eca14, Eca37, Eca48, and CV24). Basic measures of genetic diversity (observed and expected heterozygosity, numbers of alleles/locus, effective number of alleles) were calculated using GENEPOP (Raymond and Rousset 1995). Allelic richness (a measure of genetic diversity corrected for unequal sample sizes) and F_{st} (a measure of population differentiation) were estimated using FSTAT (Goudet 1995).

As shown in Table 14.3, Table 14.4, and Figure 14.2, there was no significant difference ($p>0.05$) in expected heterozygosity, observed heterozygosity, effective number of alleles,

Table 14.2. Microsatellite markers tested in fantail and Johnny darters for inclusion in the study. Markers used in this study for a given species are marked "Y," while those that failed to amplify in more than 33% of individuals were not included in the analysis

Locus name	Fantail darter	Johnny darter	Developed for	Reference
Esc26b	N	Y	*Etheostoma scotti*	Gabel et al. (2008)
Esc153	Y	Y	*Etheostoma scotti*	Gabel et al. (2008)
Eca6	N	N	*Etheostoma caeruleum*	Tonnis, (2006)
Eca11	N	Y	*Etheostoma caeruleum*	Tonnis, (2006)
Eca13	N	N	*Etheostoma caeruleum*	Tonnis, (2006)
Eca14	Y	Y	*Etheostoma caeruleum*	Tonnis, (2006)
Eca24	N	N	*Etheostoma caeruleum*	Tonnis, (2006)
Eca37	N	Y	*Etheostoma caeruleum*	Tonnis, (2006)
Eca44	N	N	*Etheostoma caeruleum*	Tonnis, (2006)
Eca46	Y	N	*Etheostoma caeruleum*	Tonnis, (2006)
Eca48	N	Y	*Etheostoma caeruleum*	Tonnis, (2006)
Eca70	N	N	*Etheostoma caeruleum*	Tonnis, (2006)
CV24	Y	Y	*Ethostoma virgatum*	Porter et al. (2002)

Table 14.3. Locus-specific and overall genetic characteristics of fantail darter populations in Fonner Run and Bates Fork. Observed (H_o) and expected (H_e) heterozygosity, alleles per locus (A), allelic richness (A_r), effective number of alleles per locus (A_e), and genetic differentiation between populations (F_{st})

Locus	H_o/H_e (Bates)	H_o/H_e (Fonner)	A/A_r (Bates)	A/A_r (Fonner)	A_e (Bates)	A_e (Fonner)	F_{st}
Esc153	0.188/0.344	0.750/0.592	4/3.909	6/5.770	1.486	2.369	0.027
Eca14	0.077/0.077	0/0	2/2	1/1	1.079	1.000	0.02
Eca46	0.786/0.885	0.895/0.906	11/10.704	12/10.773	6.639	8.481	0.025
CV24	0.826/0.784	0.727/0.694	7/6.071	5/4.568	4.297	3.115	-0.004
All loci	**0.469/0.523**	**0.593/0.548**	**6/5.671**	**6/5.528**	**2.017**	**2.149**	**0.015**

Table 14.4. Locus-specific and overall genetic characteristics of Johnny darter populations in Fonner Run and Bates Fork. Observed (H_o) and expected (H_e) heterozygosity, alleles per locus (A), allelic richness (A_r), effective number of alleles per locus (A_e), and genetic differentiation between populations (F_{st})

Locus	H_o/H_e (Bates)	H_o/H_e (Fonner)	A/A_r (Bates)	A/A_r (Fonner)	A_e (Bates)	A_e (Fonner)	F_{st}
Esc26b	0.600/0.925	0.571/0.767	6/4.500	10/3.581	5.000	3.9084	0.039
Esc153	1/0.825	1/0.868	6/4.262	11/4.470	4.167	6.666	0.058
Eca11	1/0.917	0.783/0.870	5/5.000	11/4.420	4.495	6.610	-0.004
Eca14	0/0	0/0	1/1	1/1	1.000	1.000	NA
Eca37	0/0	0/0	1/1	1/1	1.000	1.000	NA
Eca48	1/0.917	0.818/0.865	5/5.000	9/4.344	4.495	6.395	0.031
CV24	0.571/0.524	0.640/0.579	2/1.990	3/2.424	1.960	2.319	-0.028
All loci	**0.596/0.587**	**0.545/0.564**	**3.714/3.250**	**6.571/3.034**	**2.062**	**2.226**	**0.023**

or allelic richness between populations for either species of darter. This is due in part to fairly low levels of overall genetic diversity and small sample sizes. In the Johnny darter, two loci (Eca14 and Eca37) were monomorphic, while in the fantail darter Eca14 was monomorphic in the Fonner Run population and had only one heterozygous individual (out of 13 genotyped) in the Bates Fork population. Low genetic diversity and failure to amplify reliably are well-known drawbacks to using microsatellites developed for other species. Because microsatellites tend to be evolving rapidly, changes to the primer binding site will preclude efficient amplification in many cases, and loci that do amplify successfully are more likely to be those that are not mutating as rapidly and thus have less diversity.

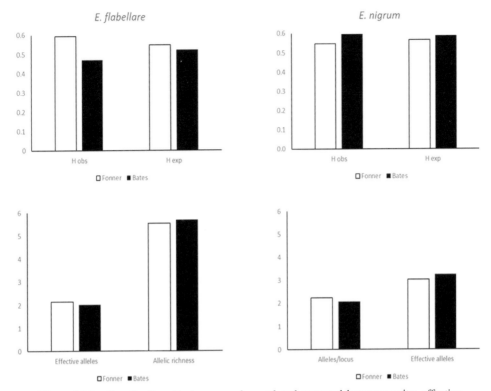

Figure 14.2 Genetic diversity (average observed and expected heterozygosity, effective number of alleles, and allelic richness) of fantail darter (*Etheostoma flabellare*) and Johnny darter (*E. nigrum*) populations in Fonner Run and Bates Fork

We also did not observe significant genetic differentiation between Fonner Run and Bates Fork populations of either fantail darters or Johnny darters (F_{st} = 0.0154 and F_{st} = 0.0229, respectively). However, perhaps these adjacent populations still exchange migrants frequently enough to preclude the formation of genetic differentiation between populations. Although there was a stretch of Fonner Run downstream from our sample site and before the confluence with Bates Fork that was entirely subterranean on the days we were there to sample fish, it is probable that during high flow events the confluence remains above the ground surface and offers a connection between these two stream sections.

14.8 Salamander Surveys

Total dissolved solids (TDS) is a measurement of the concentration of dissolved materials in the water including salts, metals, and minerals. Sources of total dissolved solids in Pennsylvania include natural gas extraction in the Marcellus shale formation, abandoned mine drainage, and road salt run-off. Elevated TDS has been shown to have adverse effects on aquatic biosystems. Amphibians, owing to their permeable skin, are particularly sensitive to water quality and conductivity (Collins and Storfer 2003; Bowles et al. 2006; Karraker et al. 2008; Hilman et al. 2009; Woodley et al. 2014).

In order to determine whether upstream Marcellus Shale activity might impact downstream salamander communities, we collected baseline data on salamander presence in Fonner Run and Bates Fork on seven occasions (3 quadrats per stream) from September 2010 to October 2011. Surveys were done using a method appropriate for patchy habitat such as that found at Bates Fork and Fonner Run (Rocco and Brooks 2000). The method involved placing a 2 × 2 meter quadrat spanning the land–water interface of suitable habitat, with one meter on land and one meter on water. After the quadrat was selected the area was searched thoroughly by hand and dip net. The age (adult, juvenile, or larvae), sex, snout to vent length, and mass of each salamander collected were recorded. Animals were released where they were found. Density of animals was calculated by dividing the abundance of salamanders by the total number of m^2 searched. The TDS concentration, pH, and GPS coordinates were also recorded. Three quadrats were chosen at each stream site (Bates and Fonner). The same quadrats were searched during each survey except the April survey, when rain and high water allowed only two quadrats on Fonner Run and one quadrat on Bates Fork to be searched.

Overall, salamander abundance was low and changed seasonally (Table 14.5). Two-lined Salamanders (*Eurycea bislineata*) were the most prevalent species of salamander found at either site (captured 44 times over the seven survey episodes). One Slimy Salamander (*Plethodon glutinosus*) and one Long-tailed Salamander (*Eurycea longicauda*) were caught as well. There were more *Eurycea bislineata* at Fonner Run than at Bates Fork (average ± sem: 0.14 ± 0.07 salamanders/m^2). The number of males and females at each site was comparable, but Fonner Run had over three times as many juveniles as Bates Fork Creek. There was also an average density of 1.4 ± 0.57 larvae/m^2 found at Fonner Run and zero larvae were found at Bates Fork Creek.

To characterize the habitat at each site, quantification methods from Grossman and Skyfield (2009) and Wolman (1954) were modified and used. Each of the three 4 m^2 quadrats at each stream surveyed was divided into 25 × 25 cm squares. Twenty-five percent of the squares were randomly selected and quantified. Percent sand (< 0.2 cm),

Table 14.5. Number of Two-lined Salamanders (*Eurycea bislineata*) captured at Bates Fork and Fonner Run on seven surveys of three quadrats at each stream. In each cell, the first number indicates Bates Fork and the second number indicates values at Fonner Run. See Pascuzzi 2012 for quadrat GPS coordinates

Date	Adult males	Adult females	Juveniles	Larvae	Total	Density (number per m^2)
9/29/11	2, 0	1, 1	1, 3	0, 3	4, 7	0.33, 0.58
10/20/11	2, 5	3, 3	1, 3	0, 1	6, 12	0.5, 1
4/13/11	0, 0	0, 0	0, 1	0, 0	0, 1	0, 0.08
5/26/11	0, 0	0, 0	0, 0	0, 1	0, 1	0, 0.08
6/30/11	0, 1	0, 0	1, 2	0, 0	1, 3	0.08, 0.25
9/16/11	1, 0	0, 1	0, 2	0, 1	1, 4	0.08, 0.33
10/26/11	0, 0	0, 0	0, 0	0, 4	0, 4	0, 0.33

gravel (0.3-4.5 cm), course gravel (4.6–6.4 cm), cobble (6.5-26 cm), boulder (> 26 cm), silt and clay, and vegetation were determined. Measurements are based on the longest axis (length) of the rock. Each of these categories was then given a code number: 1 = sand, 2 = gravel, 3 = course gravel, 4 = cobble, 5 = boulder, 6 = silt and clay, and 7 = vegetation. The dominant type of these substrates was recorded for each square searched. In addition to the habitat quantification, an analysis of the two streams was done using the United States Geological Survey's (USGS) StreamStats interactive map for Pennsylvania.

The aquatic habitat at Bates Fork Run was largely sand and gravel with little silt and clay (mean = 2.09 ± 0.27, mode = 2). Fonner Run's aquatic habitat was a majority of sand and gravel but also contained some cobble, boulders, and silt and clay (mean = 2.58 ± 0.26, mode = 2). The terrestrial habitat at Bates Fork was nearly homogeneous sand and gravel with little silt and clay and grass (mean = 2.82 ± 0.46, mode = 2), whereas Fonner Run was primarily a heterogeneous mixture of silt and clay and grass with some sand and gravel, and little boulders and cobble (mean = 4.96 ± 0.46, mode = 7). Thus, overall, Bates Fork was largely composed of gravel and sand with little silt and clay, and grass, whereas Fonner Run was an intermediate mixture of all types of substrates. The USGS analysis of Bates Fork and Fonner Run showed that Fonner Run had a smaller drainage area (8.5 km^2) than Bates Fork (15.8 km^2). Also, Fonner Run's river basin was more heavily covered by forest (84%) than Bates Fork (61%).

To conclude, more salamanders were found in Fonner Run than in Bates Fork. Differences in abundance could be caused by a number of factors, such as the differences in habitat that were documented or the differences in the amount of upstream Marcellus Shale-related activity. These baseline data will be useful to assess potential future impacts of TDS and Marcellus Shale drilling. If drilling continues to occur upstream of Bates Fork, baseline data can be used to track any changes to salamander populations. Should drilling begin above Fonner Run, the baseline data can be used to observe any impacts on salamander populations.

14.9 Conclusions

We conducted these studies on a pair of small watersheds in the Tenmile Creek system during the early onset of extraction from the Marcellus Shale Play in southwestern Pennsylvania. Our goals were to understand the potential impacts from this rapidly expanding industry on stream water quality and biological communities and to document existing conditions. We focused on a paired-stream design with the experimental stream, Bates Fork, potentially impacted from extensive Marcellus Shale extraction activities, and the control stream, Fonner Run, without such activity. Although we did not discover any conclusive evidence of stream impacts from this industry, we have established baseline data on water chemistry, fish community structure, darter genetic variation, and salamander density and have gained valuable insight into how to further develop these as effective monitoring tools. We believe all these approaches have the potential to detect diagnostic signals from Marcellus Shale extraction activities in a watershed. Here we present some lessons learned from these studies and suggest ways to improve these monitoring tools.

With limited time and resources, important decisions must be made on the scope and scale of monitoring studies. We decided to conduct multiple, parallel studies at a single site, hedging on an elevated probability of a pollution event, given the large number of wells in the watershed upstream. In theory, water pollution from Marcellus Shale extraction activities can be detected through the examination of the appropriate water chemistry parameters, provided that the frequency of sampling is adequate, the signal is not diluted below the laboratory detection limits, and the signal is not masked by pollution from other sources. Selecting sample sites close to the well pad will limit dilution of the signal from non-impacted tributaries. Increasing the frequency of chemical monitoring of stream samples might increase the chance of detecting an isolated pollution event but also increases the collection effort, analysis time, and cost of chemical analyses. These resource restrictions limit the number of sites that are sampled, and in turn, lower the probability the sampled localities are ones that experience pollution events. Data loggers that record specific conductivity at programmable intervals can be used to supplement water chemistry grab samples, but unless the contaminant is a major component of the ionic load, it will probably not be detected, especially if the overall background ionic load is high from legacy AMD issues, as it is in the Monongahela River system. Bromide, a signature of brine pollution, had concentrations in Bates Fork above our detection limit of 0.05 ppm on only two dates, but it was such a small contribution to the overall ionic load that we would not detect an elevated specific conductance from this anion alone.

At the time of this study in 2010–2012, our analytical technology and knowledge of the most appropriate ions for testing was limited. Currently, our routine water chemistry analyses include the concentrations of anions sampled in this study (chloride, nitrate, sulfate, and bromide) along with fluoride and phosphate (Cantlay et al. 2020a). A second water sample is collected and preserved in purified nitric acid for ICP-MS analysis of cations, including sodium, potassium, calcium, iron, barium, boron, lithium, strontium, magnesium, and manganese (Cantlay et al. 2020b). With this more extensive set of anion and cation concentrations, we are able to perform a series of mass-ratio plots that are useful in identifying the source signature of various types of water pollution (Cantlay et al. 2020b and c). This has been shown to be an effective way to distinguish between contamination caused by conventional versus unconventional drilling or mining (Cantlay et al. 2020c), and could provide a more specific tool to tell whether Bates Fork has been impacted by fracking more than Fonner Run as well as to see if there are temporal trends in either ion concentrations or certain mass ratios that would indicate pollution from unconventional brines.

It also appears possible to detect the bioaccumulation and epigenetic impact of fracking metals in aquatic predators. The Louisiana Waterthrush (*Parkesia motacilla*) is an obligate riparian songbird that consumes aquatic macroinvertebrates (Trevelline et al. 2016). They nest in stream banks and maintain short linear territories along the riparian corridor during breeding season. Latta et al. (2015) discovered that newly grown tail feathers from birds nesting on steams with Marcellus Shale extraction activity have significantly higher levels of barium and strontium compared to birds nesting in streams without shale gas extraction. Further studies indicate that increased barium and strontium levels associated with shale gas extraction may reduce Louisiana waterthrush DNA methylation, which could lead to long-term fitness effects (Frantz et al. 2020). Although the pathways are unclear for how these metals are deposited in the feathers and what epigenetic mechanisms are involved, it suggests they have entered the aquatic food chain and are bioaccumulated to detectable levels in these insectivorous birds (Latta et al. 2015).

We found that it is extremely difficult to identify comparable stream-pairs within a drainage. The Bates Fork and Fonner Run pair was the best out of six of the pairs we examined, but showed differences in drainage area, land use, and aquatic habitats. Bates Fork had almost twice the drainage area of Fonner Run and had a fringe of riparian trees opening to a field along its left descending bank, compared to a fully forested riparian zone on Fonner Run. We attempted to identify a site higher up in the drainage, but still downstream of the Marcellus Shale activity, but getting permission for access was an issue. For fish and salamander surveys, it is important to have similar diversity of aquatic habitats (riffles, runs, and pools). Very small streams such as Fonner Run, probably represent the minimum size to support a stable and diverse fish population. Water withdrawals for use in fracking would devastate such systems when they are in summer baseflow by fragmenting them into intermittent pools with low dissolved oxygen. We saw some evidence of water withdrawals in Bates Fork, about a mile upstream of our sample site, where tankers would pull water from a bridge crossing. We missed an opportunity to use a motion-activated trail camera to correlate these withdrawal events with our hydrological station data downstream. For salamander surveys, the similarity of habitats in both the wetted and dry stream channel are important. Ideally, we would have preferred to include more than one stream-pair in order to make broader generalizations of our results and increase the probability of detecting impacts from Marcellus Shale extraction activities in the Tenmile Creek Watershed.

One of the simplest ways to add to this study would be to resample (including water chemistry, fish community, fish genetics, and salamanders) to see how things have changed over the last 10 years. The data presented here reflects only a short period of time after fracking was initiated upstream of our site in Bates Fork, so subtle or minor effects of a sublethal nature might not have yet had time to have a significant impact on the water and living organisms.

14.10 Summary

Overall, the fish community surveys found minimal differences between the two streams; however, it would be interesting to take additional data to see if there are any distinctions today. Darters can be good indicator species of general pollution and are associated with benthic substrate where dense brines might be concentrated. However, not all darter species should be considered equal when it comes to their tolerance to chloride. Fish survey data from Bates Fork and Fonner Run show that species with low TIVs for chloride (Meador and Carlisle 2007) are relatively rare in the system. This necessitated the use of moderately chloride-tolerant darter species for the genetic analyses in order to provide reasonable sample sizes. This, combined with a relatively short time period between the initiation of fracking in Bates Fork and the beginning of our sampling in 2010, could have led us to underestimate potential effects on chloride-sensitive species.

Although the fish community in our study was relatively diverse, the salamanders were dominated by one species. Systems with limited fish and salamander diversity need to focus on the density of the common species. Keller et al. (2017) attempted to correlate Marcellus Shale well pad density to bioindicators of crayfish, fish, and salamander diversity at 28 heavily forested headwater streams of north-central Pennsylvania. Diversity of these taxonomic groups was naturally low in this region, and the density of the limited species was not significantly correlated to well pad density.

In addition to the fish community data, our study of darter genetics failed to observe any differences in genetic diversity or differentiation between populations. In part, this may be because of small sample sizes, but it may also be the result of using microsatellites developed for other darter species. We chose this tactic because developing novel microsatellite markers for a non-model species was expensive as well as time-consuming. Today, however, high-throughput genomic sequencing has drastically reduced both the cost and time needed (Abdelrim et al. 2018). We expect that microsatellites developed from fantail and Johnny darters would have amplified successfully in more individuals, and that the polymorphism and heterozygosity for those markers would be much greater, thus allowing us to genotype more individuals and to detect more subtle differences between the Fonner Run and Bates Fork. Recent advances in genomics would also allow us to estimate the effective population size and demographic history of darter populations in each stream (Hohenlohe et al. 2020). By sequencing the whole genome of a few individuals, it is possible to estimate the frequency and intensity of population bottlenecks (Li and Durbin 2011), while restriction-site associated DNA sequencing (RADseq) of larger population samples can use single nucleotide polymorphisms (SNPs) to estimate effective population size (Nunziata and Weisrock 2018). These techniques could be used to assess whether sporadic pollution events (that we may not have detected using intermittent sampling) could have caused population die-offs with reduced genetic diversity observed over time.

It is recognized that the methods employed in the microbiology study were not state of the art (high throughput sequencing was prohibitively expensive at the time); nevertheless, they afforded a glimpse at the potential diversity of the two streams as well as fluids associated with shale gas development (i.e., produced water, impoundments). A number of studies over the past decade have since identified a unique shale microbiome, with novel species (Daley et al. 2016, as reviewed in Chapter 12, Nixon, 2022). What was striking in our studies was the robustness (i.e., biocide and antibiotic resistance) and versatility (i.e., tolerance range of pH and salinities) of the microbes associated with shale development.

Despite the lack of significant differences between the two streams selected, this study has successfully provided a wealth of ecological baseline data and helped to clarify what techniques and types of data will allow us to identify fracking pollution events in the future. Identifying paired basins with fixed stations that are suitable for the study of water chemistry as well as microbial and vertebrate diversity is extremely challenging but crucially important for a valid comparison between experimental and control sites.

We hope that the results and techniques presented here can assist in the planning and execution of future investigations of fracking impacts.

References

Abdelkrim J, Robertson BC, Stanton JL, and Gemmell NJ. (2018). Fast, cost-effective development of species-specific microsatellite markers by genomic sequencing. *Biotechniques.* 46(3): 185–191.

Blauch M, Myers R, Moore T, and Houston N. (2009). *Marcellus Shale Post-Frac Flowback Waters - Where is All the Salt Coming From and What are the Implications?* Society of Petroleum Engineers Eastern Regional Meeting, 125740.

Bowles B, Sanders M, and Hansen R. (2006). Ecology of the Jollyville Plateau Salamander (*Eurycea tonkawae*: Plethodontidae) with an assessment of the potential effects of urbanization. *Hydrobiologia.* 553: 111–120.

Byrne-Bailey KG, Gaze WH, Zhang L, Kay P, Boxall A, Hawkey PM, and Wellington EMH. (2010). Integron prevalence and diversity in manured soil. *Applied and Environmental Microbiology.* 77(2): 684–687.

Cantlay T, Eastham JL, Rutter J, Bain DJ, Dickson BC, Basu P, and Stolz JF. (2020a). Determining conventional and unconventional oil and gas well brines in natural samples I: Anion analysis with ion chromatography. *Journal of Environmental Science and Health, Part A: Toxic/Hazardous Substances and Environmental Engineering.* 55(1), 1–10.

Cantlay T, Bain DJ, Curet J, Jack RF, Dickson BC, Basu P, and Stolz JF. (2020b). Determining conventional and unconventional oil and gas well brines in natural sample II: Cation analyses with ICP-MS and ICP-OES. *Journal of Environmental Science and Health, Part A: Toxic/Hazardous Substances and Environmental Engineering.* 55(1): 11–23.

Cantlay T, Bain DJ, and Stolz JF. (2020c). Determining conventional and unconventional oil and gas well brines in natural samples III: Mass ratio analyses using both anions and cations. *Journal of Environmental Science and Health, Part A: Toxic/Hazardous Substances and Environmental Engineering.* 55(1): 24–32.

Cayol JL, Ollivier B, Patel BKC, Prensier G, Guezennec J, and Garcia JL. (1994). Isolation and characterization of *Halothermothrix orenii* gen. nov., sp. nov., a halophilic, thermophilic, fermentative, strictly anaerobic bacterium. *International Journal of Systematic Bacteriology.* 534–540.

Chen HH, Zhao GZ, Park DJ, Zhang YQ, Xu LH, Lee JC, Kim CJ, and Li WJ. (2009). *Micrococcus endophyticus* sp. nov., isolated from surface-sterilized *Aquilaria sinensis* roots. *International Journal of Systematic and Evolutionary Microbiology.* 59: 1070–1075.

Collins J and Storfer A. (2003). Global amphibian declines: Sorting the hypotheses. *Diversity and Distributions.* 9: 89–98.

Daley RA, Borton MA, Wilkins MJ, Hoyt DW, Kountz DJ, Wolfe RA, Welch SA, Marcus DN, Trexler RV, MacRae JD, Krzycki JA, Cole DR, Mouser PJ, and Wrighton KC. (2016). Microbial metabolisms in a 2.5-km-deep ecosystem created by hydraulic fracturing in shales. *Nature Microbiology.* 1: 1–9.

Ding L and Yokota A. (2004). Proposals of *Curvibacter gracilis* gen. nov., sp. nov. and *Herbaspirillum putei* sp. nov. for bacterial strains isolated from well water and reclassification of [*Pseudomonas*] *huttiensis*, [*Pseudomonas*] *lanceolata*, [*Aquaspirillum*] *delicatum* and [*Aquaspirillum*] *autotrophicum* as *Herbaspirillum huttiense* comb. nov., *Curvibacter lanceolatus* comb. nov., *Curvibacter delicates* comb. nov., and *Herbaspirillum* comb. nov. *International Journal of Systematic and Evolutionary Microbiology*. 54: 2223–2230.

Dugas O. (2014). *Isolation and Characterization of Salinivibrio sp. Strain LP-1 from an Impoundment with Marcellus Shale Waste Water*. Master's thesis, Duquesne University.

Eastham JL. (2012). *Enrichment, Characterization, and Identification of Microbial Communities Found in Unconventional Shale Gas Production Water*. Master's thesis, Duquesne University.

Fichter JK, Johnson K, French K, and Oden R. (2008). Use of Microbiocides in Barnett Shale Gas Well Fracturing Fluids to Control Bacteria Related Problems. *NACE International - Corrosion*. 2008: 08658.

Fisher MM and Triplett EW. (1999). Automated approach for ribosomal intergenic spacer analysis of microbial diversity and its application to freshwater bacterial communities. *Applied Environmental Microbiology*. 65: 4630–4636.

Frantz MW, Wood PB, Latta SC, and Welsh AB. (2020). Epigenetic response of Louisiana Waterthrush *Parkesia motacilla* to shale gas development. *Ibis*. 162(4): 1211–1224.

Gabel JM, Dakin EE, Freeman BJ, and Porter BA. (2008). Isolation and identification of eight microsatellite loci in the Cherokee Darter (*Etheostoma scotti*) and their variability in other members of the genera *Etheostoma*, *Ammocrypta*, and *Percina*. *Molecular Ecology Resources*. 8: 149–151.

Goudet J. (1995). FSTAT (Version 1.2): A Computer Program to Calculate F-Statistics, *Journal of Heredity*. 86(6): 485–486.

Grossman GD and Skyfield J. (2009). Quantifying microhabitat availability: Stratified random versus Constrained focal-fish methods. *Hydrobiologia*. 624: 235–240.

Groundwater Protection Council (GWPC). (2009). *Modern Shale Gas Development in the United States: A Primer, prepared for the U.S. Department of Energy, National Energy Technology Laboratory (NETL)*. Oklahoma City, Oklahoma.

Hariharan H, Lopez A, Conboy G, Coles M, and Muirhead T. (2007). Isolation of *Escherichia fergusonii* from the feces and internal organs of a goat with diarrhea. *The Canadian Veterinary Journal*. 48: 630–631.

Hillman S, Withers P, Drewes R, and Hillyard S. (2009). *Ecological and Environmental Physiology of Amphibians*. Oxford University Press.

Hohenlohe PA, Funk WC, and Rajora OP. (2020). Population genomics for wildlife conservation and management. *Molecular Ecology*. 30: 62–82.

Jackson PE. (2006). Ion chromotography in environmental analysis. In Meyers RA (ed.) *Encyclopedia of Analytical Chemistry*. John Wiley & Sons Ltd., pp. 2779–2801.

Joshi SN. (2013). *Isolation and Characterization of a Bacillus Firmus Strain SWPA-1 from Marcellus Shale Flowback Water*. Master's thesis, Duquesne University.

Karr JR, Fausch KD, Andermeier PL, Yant PR, and Schlosser IJ. (1986). Assessing biological integrity in running waters: a method and its rationale. *Illinois Natural History Survey Special Publication*. 5.

Karraker NE, Gibbs JP, and Vonesh JR. (2008). Impacts of road deicing salt on the demography of vernal pool-breeding amphibians. *Ecological Applications*. 18: 724–734.

Keller DH, Horwitz RJ, Mead JV, and Belton TJ. (2017). Natural gas drilling in the Marcellus Shale region: Well pad densities and aquatic communities. *Hydrobiologia*. 795: 49–64.

Kimmel WG and Argent DG. (2006). Development and application of an index of biotic integrity for fish communities of wadeable Monongahela River tributaries. *Journal of Freshwater Ecology*. 21(2): 183–190.

Kimmel WG and Argent DG. (2010). Stream fish community responses to a gradient of specific conductance. *Water, Air, and Soil Pollution*. 206: 49–56.

Kimmel WG and Argent DG. (2012). Status of fish and macroinvertebrate communities in a watershed experiencing high rates of fossil fuel extraction: Tenmile Creek, a major Monongahela River tributary. *Water, Air, and Soil Pollution*. 223: 4647–4657.

Kiviat E. (2013). Risks to biodiversity from hydraulic fracturing for natural gas in the Marcellus and Utica shales. *Annuals of the New York Academy of Sciences*. 1286: 1–14.

Lampe DJ and Stolz JF. (2015). Current perspectives on unconventional shale gas extraction in the Appalachian Basin. *Journal of Environmental Science and Health, Part A: Toxic/Hazardous Substances and Environmental Engineering*. 50: 434–446.

Latta SC, Marshall LC, Frantz MW, and Toms JD. (2015). Evidence from two shale regions that a riparian songbird accumulates metals associated with hydraulic fracturing. *Ecosphere*. 6(9): 144.

Li H and Durbin R. (2011). Inference of human population history from individual whole-genome sequences. *Nature Genetics*. 475: 493–496.

Li WJ, Zhang YQ, Park DJ, Li CT, Xu LH, Kim CJ, and Jiang CL. (2004). *Duganella violaceinigra* sp. nov., a mesophilic bacterium isolated from forest soil. *International Journal of Systematic Evolutionary Microbiology*. 54: 1811–1814.

Maule AL, Makey CM, Benson EB, Burrows IJ, and Scammell MK. (2013). Disclosure of hydraulic fracturing fluid chemical additives: analysis of regulations. *New Solutions*. 23(1): 167–187.

Meador MR and Carlisle DM. (2007). Quantifying Tolerance Indictor Values for Common Stream Fish Species of the United States. *Ecological Indicators*. 7(2): 329–338.

Mohan AM, Gregory KB, Vidic RD, Miller P, and Hammack RW. (2011). Characterization of microbial diversity in treated and untreated flowback water impoundments from gas fracturing operations. *Society of Petroleum Engineers Annual Technical Conference and Exhibition*. 147414.

Myers GS. (1949). Salt-tolerance of fresh-water fish groups in relation to zoogeographical problems. *Bijdragen tot de Dierkunde*. 28: 315–322.

Nixon S. (2022). The microbiology of shale gas extraction. In Stolz JF, Griffin WM, and Bain DJ (eds.) *Environmental Impacts from the Development of Unconventional Oil and Gas Reserves*. Cambridge University Press.

Nunziata SO and Weisrock DW. (2018). Estimation of contemporary effective population size and population declines using RAD sequence data. *Heredity*. 120: 196–207.

Ohio Environmental Protection Agency (Ohio EPA). (1987). *Biological Criteria for the Protection of Aquatic Life: Volumes I–III*. Ohio Environmental Protection Agency.

PADEP (2020). 2020 Pennsylvania Integrated Water Quality Monitoring and Assessment Report. Available online at https://gis.dep.pa.gov/IRStorymap2020/. Accessed March 31, 2021.

Pascuzzi M. (2012). *The Effects of Total Dissolved Solids on Locomotory Behavior and Body Weight of Streamside Salamanders, and a Baseline Survey of Salamander Diversity and Abundance*. Master's thesis, Duquesne University.

Pelletier A and Sygusch J. (1990). Purification and characterization of three chitosanase activities from *Bacillus megaterium* P1. *Applied and Environmental Microbiology*. 56(4): 844–848.

Pichel M, Brengi SP, Cooper KLF, Ribot EM, Al-Busaidy S, Araya P, Fernandez J, Vaz TI, Kam KM, Morcos M, Nielson EM, Nadon C, Pimentel G, Perez-Gutierrez E, Gerner-Smidt P, and Binsztein N. (2012). Standardization and international multi-center validation of a PulseNet Pulsed-Field gel electrophoresis protocol for subtyping *Shigella flexneri* isolates. *Foodborne Pathogens and Disease*. 9(5): 418–424.

Porter BA, Fiumera AC, and Avise JC. (2002). Egg mimicry and allopaternal care: two mate-attracting tactics by which nesting striped darter (*Etheostoma virgatum*) males enhance reproductive success. *Behavioral Ecology and Sociobiology*. 51: 350–359.

Prettner S. (2019). *A Study of Chloride Levels in Pine Creek, Allegheny County, PA*. Master's thesis, Duquesne University.

Raymond M and Rousset F. (1995). GENEPOP (version 1.2): population genetics software for exact tests and ecumenicism. *Journal of Heredity*. 86: 248–249.

Rocco GL and Brooks RP. (2000). *Abundance and Distribution of a Stream Plethodontid Salamander Assemblage in 14 Ecologically Dissimilar Watersheds in the Pennsylvania Central Appalachians*, Final Technical Report No. 2000-4 of the Penn State Cooperative Wetlands Center, Pennsylvania State University.

Rutter J. (2012). *A Baseline Study of Chemical Parameters and Microbial Diversity of Two Streams in the Ten Mile Creek Watershed in Southwestern Pennsylvania*. Master's thesis, Duquesne University.

Stolz JF and Griffin WM. (2022). Unconventional shale gas and oil extraction in the Appalachian Basin. In Stolz JF, Griffin WM, and Bain DJ (eds.) *Environmental Impacts from the Development of Unconventional Oil and Gas Reserves*. Cambridge University Press.

Struchtemeyer CG, Davis JP, and Elshahed MS. (2011). Influence of the drilling mud formulation process on the bacterial communities in thermogenic natural gas wells of the Barnett Shale. *Applied Environmental Microbiology*. 77: 4744–4753.

Struchtemeyer CG and Elshahed MS. (2011). Bacterial communities associated with hydraulic fracturing fluids in thermogenic natural gas wells in North Central Texas, USA. *FEMS Microbiology Ecology*. 81: 13–25.

Tonnis BD. (2006). Microsatellite DNA markers for the rainbow darter, *Etheostoma caeruleum* (Percidae), and their potential utility for other darter species. *Molecular Ecology Notes*. 6(1): 230–232.

Trevelline BK, Latta SC, Marshall LC, Nuttle T, and Porter BA. (2016). Molecular analysis of nestling diet in a long-distance Neotropical migrant, the Louisiana Waterthrush (*Parkesia motacilla*). *The Auk*. 133: 415–428.

Wilson JM and VanBriesen JM. (2022). Water usage and management. In Stolz JF, Griffin WM, and Bain DJ (eds.) *Environmental Impacts from the Development of Unconventional Oil and Gas Reserves*. Cambridge University Press.

Wolman G. (1954). A method of sampling coarse river-bed material. *Transactions, American Geophysical Union*. 35: 951–956.

Woodley SK, Freeman PE, and Ricciardella LF. (2014). Environmental acidification is not associated with altered plasma corticosterone levels in the stream-side salamander, *Desmognathus ochrophaeus*. *General and Comparative Endocrinology*. 201: 8–15.

Zhang YQ, Li WJ, Zhang KY, Tain XP, Jiang Y, Xu LH, Jiang CL, Lai R. (2006). *Massilia dura* sp. nov., *Massilia albidiflaca* sp. nov., *Massilia plicata* sp. nov., and *Massilia lutea* sp. nov., isolated from soils in China. *International Journal of Systematic and Evolutionary Microbiology*. 56: 459–463.

15

The Effects of Shale Gas Development on Forest Landscapes and Ecosystems in the Appalachian Basin

MARGARET C. BRITTINGHAM AND PATRICK J. DROHAN

15.1 Shale Gas Development in the Appalachian Basin

The Marcellus and Utica shale plays underlay extensive areas of forest within the Appalachian basin where overlap between core forest habitat and the Marcellus and Utica plays results in an ecosystem that is at high risk of disturbance (Figure 15.1). These forests have high ecological value with the Central Appalachian ecoregion supporting the greatest diversity of species of all ecoregions in the northeast and mid-Atlantic (Anderson et al. 2012). For example, the large blocks of intact forest support critical populations of Neotropical migrant songbirds (AMJV 2020) as well as the greatest diversity of stream and terrestrial salamanders (Crawford and Semlitsch 2007). In Pennsylvania over 85% of streams in the core gas forest districts (seven state forests where most of the gas development is occurring) are classified as EV (exceptional value) or HQ (high quality, DCNR 2018). The large intact forests provide numerous ecological services and are highly valued by the public for their ecological, recreational, and aesthetic values (Tarrant et al. 2003). In addition, they include some of the nation's most commercially valuable hardwood species (Pearce 2001; DCNR 2018).

Unconventional shale gas development (USG) in the Marcellus shale play began around 2005 as high-volume hydraulic fracturing became technically and financially feasible and increased rapidly across the Appalachian basin beginning in 2008 (Jacquet 2018; MCOR 2020). Between 2008 and 2018, 15,939 unconventional gas wells were drilled, with 11,037 in PA; 2,582 in WV; 2,374 in OH, and 18 in NY and MD (Jacquet 2018). Approximately 57% of Pennsylvania wells are in forest habitat (Drohan et al. 2012), and as of 2019, 18,829 wells have been drilled (MCOR 2020, Figure 15.1). The rapid and extensive development within the core forests of the Appalachian region has raised concerns over the potential for significant landscape disturbance and negative ecological effects (Brittingham et al. 2014b; Farwell et al. 2020).

Extraction of oil and gas from shale requires multiple steps starting with seismic surveys and ending with reclamation of the pads after all wells on the pad are plugged. The process and the associated infrastructure can affect forest ecosystems through habitat loss and fragmentation; disturbance; noise and light pollution; increased levels of traffic; erosion, runoff and sedimentation; changes in water quantity and quality; and potential exposure to contaminants (Northrup and Wittemyer 2013; Brittingham et al. 2014b; Souther et al. 2014;

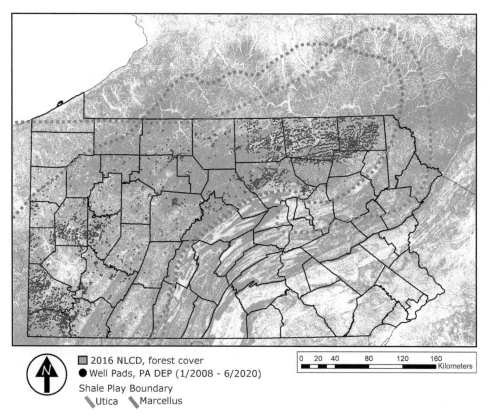

Figure 15.1 Forest cover (green) overlaying Marcellus (blue) and Utica (red) shale plays with wells drilled in PA as of 2019 (A black and white version of this figure will appear in some formats. For the colour version, refer to the plate section.)

DCNR 2018). We provide a description of the infrastructure and brief summaries of some of the general effects of USG that occur within forest ecosystems (e.g., habitat fragmentation, noise disturbance, effects on water), and we then look at effects on specific organisms with an emphasis on at risk species or groups as well as broad community shifts that can affect the health of the forest ecosystem overall. We focus primarily on Pennsylvania because it has the most comprehensive data sets tracking OSG development (e.g., DCNR 2018; MCOR 2020).

15.2 Infrastructure

Construction of natural gas wells in forested areas requires clearing of the land for the well pad, supporting infrastructure, pipelines, and roads needed to access the site. Mean well pad size in Pennsylvania ranges from 1 to 2.2 ha (Drohan and Brittingham 2012; Langlois et al. 2017; DCNR 2018). When associated disturbance is included, mean pad size is 2.7 ha (median 2.2 ha, range 0.1-20.5 ha, Drohan and Brittingham 2012). Pads are also required to support compressor stations, freshwater impoundments, stone pits, and meter valve or tap stations (DCNR,2018). These may be included on the well pad or on a separate pad. The maximum size of well pads generally includes additional infrastructure such as freshwater impoundments.

Unlike most well pads used in conventional oil and gas production, USG pads are covered in stone to enable them to support the heavy equipment and activity required to frack the well. Pads are much larger than what is used for conventional gas development and are generally built to accommodate multiple wells per pad, thus creating a different pattern of disturbance from conventional gas development where there may be fewer smaller pads with only one well per pad. On PA DCNR forest land, typical pads have anywhere from 1–10 wells (DCNR 2018) with pads with more wells tending to be larger. The average number of wells per pad is 3.2 in PA, 2.2 in Ohio, and 2.1 in West Virginia (Jacquet 2018). Interim reclamation can be used to reduce the size of the pad, and this has occurred on 16% of pads across Pennsylvania (Drohan and Brittingham 2012) but on only a few pads on state forest land (DCNR 2018). Consequently, stone pads can be expected to be present for decades.

New, as well as larger and wider, roads are required to provide access and to support heavy truck traffic. Although roads were relatively common throughout the eastern forests of the Appalachian Basin prior to USG development, most were smaller dirt roads with closed canopies, developed primarily for timber sales (Fredericksen 1998; Heilman et al. 2002). Unconventional shale gas development is associated both with an increase in the number of roads and a change in road size and substrate. Because of the need for heavy equipment, and sometimes oversized equipment and its movement, roads built to support USG are wider and have an increased depth of road base than traditional forest roads (DCNR 2018). This results in more open tree canopies over the roadway, which negatively affects habitat connectivity and wild character, and increases habitat fragmentation effects and problems with dust (DCNR 2018). On PA state forest land 425 km of road has been built or modified for USG, resulting in a clearing of 94.4 ha of forest land (DCNR 2018).

Pipelines are needed to transport gas from the well to processing and storage facilities and eventually to the consumer. Smaller diameter gathering pipelines are built to transport the gas from the well pad to the nearest large-diameter, pressurized transfer (or "transmission") pipelines. In Pennsylvania, an estimated 16,000–40,000 km of new pipeline will be necessary to support the projected number of new wells by 2030 (Johnson et al. 2011) with the low estimate being more likely if gas production slows as it has in 2020. In the core forest districts within PA State Forests, 303 km of pipeline were built between 2008 and 2018 (DCNR 2018). Although the pads are one of the most visible features of shale gas development, recent analyses suggest that the pipelines make up the largest component of the footprint (Slonecker et al. 2012; Slonecker et al. 2013; Langlois et al. 2017) and are the largest contributor to loss of core forest, resulting in much more fragmentation than the pads (Langlois et al. 2017).

Loss of core forest from pipelines was twice as high on private land (4.3%) than on public land (2.0%, Langlois et al. 2017). This could reflect a difference in conservation goals between private and public landowners or is more likely the result of the ability of public lands to plan and negotiate better deals with the gas companies because they have larger blocks of land to work with and a support staff. For example, public officials have free access to state experts and legal specialists; private landowners do not have this benefit. Langlois et al. (2017) reported that the length of gathering pipeline per pad was 30% longer for pads on private than public land (2.8 vs 1.7 m). They speculated that this resulted primarily from instances on private lands where two different companies working with adjacent landowners built separate pipelines instead of sharing one. Public land managers have a better chance to avoid such a situation via large lease tract negotiation and incentivizing companies to work together. Pipelines are more likely to be collocated with roads on public land than on private land, which also reduces loss of core forest (Langlois et al. 2017). On public forest land in PA there is an effort to collocate pipelines with roads with 22% co-located with a road or existing corridor (DCNR 2018). Pipeline collocation reduces fragmentation effects on landscapes but is not without costs as pipelines co-located with roads tends to result in wider corridors (Langlois et al. 2017; DCNR 2108), and as road width increases, fragmenting effects of the road also increase (Langlois 2017).

15.3 Landscape-Level Effects on Natural Resource Integrity and Biodiversity

The rapid development of shale gas infrastructure with its associated increase in habitat fragmentation, roads, noise, and disturbance as well as changes in water quantity and quality have been associated with a number of ecosystem-related changes. In this section, we review some of the large-scale changes that may impact numerous resource groups. Most of these factors are intertwined (e.g., habitat fragmentation is associated with the spread of invasive plants and traffic is associated with noise), but for clarity we have separated them into individual sections.

15.3.1 Habitat Fragmentation and Forest Loss

Habitat loss and fragmentation occur at multiple stages of the USG development process including from initial seismic surveys used to locate gas reserves, clearing and construction of well pads and infrastructure, and construction of new and expanded roads and pipelines. Habitat fragmentation and the associated loss of core habitat and increase in edge is a direct result of USG development and is particularly evident in forested ecosystems where it is a primary concern (Drohan et al. 2012; Brittingham et al. 2014b). In Pennsylvania, the Bureau of Forestry reports that between 2008 and 2018, 716 ha of forest were converted to shale gas infrastructure resulting in a loss of 6,124 ha of core forest and an increase of 4,011 ha of edge forest (DCNR 2018). To put this into perspective, Langlois et al. (2017) determined that a new well pad and associated gathering pipeline in Lycoming County, a primarily forested county, would result in a mean loss of 34.04 ha of core forest. If a pipeline was already in place, loss associated with the pad would be 5.19 ha.

Fragmentation is associated with numerous types of development, including urban, agriculture, and other forms of energy development, and most of the research on fragmentation has been associated with these other types of development (e.g., Faaborg et al. 1995,

Robinson et al. 1995; Thomas et al. 2014). Although the causes differ, the results are similar. Habitat fragmentation results in changes in species composition and abundance with habitat generalists and species that do well around edges tending to increase in abundance while habitat specialists tend to decline (Faaborg et al. 1995; Robinson et al. 1995; Walters 1998; Thomas et al. 2014). For example, many generalist nest predators and the brown-headed cowbird, *Molothrus ater*, a brood parasite, tend to increase in abundance with fragmentation resulting in lower nest success, particularly for breeding songbirds as an area becomes fragmented (Faaborg et al. 1995; Robinson et al. 1995).

Studies on the effects of habitat fragmentation specifically associated with USG development in the eastern deciduous forest include studies on the effects on songbirds as well as on the spread of invasive plants (e.g. Barton et al. 2016; Farwell et al. 2016; Barlow et al. 2017; Farwell et al. 2020; also see specific sections in this chapter). Aquatic ecosystems can also be affected by new roads and pipelines, particularly where they cross streams and may impede dispersal of fish and other aquatic organisms (Warren and Pardew 1998; Brittingham et al. 2014b).

15.3.2 Traffic and Disturbance

In addition to creating edges in a landscape, roads also present direct hazards to wildlife. In the United States, 1 million vertebrates are killed per day on roads, and road width, vehicle traffic, and speed all influence mortality rates (Forman and Alexander 1998). Roads are wider and larger than traditional forest roads to support USG needs (see earlier in the chapter). In addition, roads supporting USG are associated with an increase in traffic, particularly during the time when development is occurring (Perry 2012; Abramzon et al. 2014; Cooper et al. 2018). There is little available research on the effects of USG roads on wildlife, but since they tend to be wider, larger, and have faster traffic than traditional forest roads, we speculate that they are also associated with higher rates of mortality. Some species of herpetofauna are particularly susceptible to direct mortality from vehicles, owing to their behavior (i.e., slow-moving), and some species preferentially select road surfaces for thermoregulation (Fahrig and Rytwinski 2009; Brand et al. 2014). For some small mammals and herpetofauna, roads can also be a barrier to movement, potentially creating isolated subpopulations, particularly where roads are wide (Merriam et al. 1989; Clark et al. 2001; Marsh et al. 2005). In addition, road avoidance because of the noise generated by traffic reduces the effective habitat area and may be the biggest ecological determinant for some species. (Forman and Alexander 1998).

15.3.3 Noise

In addition to landscape disturbances from construction and operation of the wells, natural gas production is associated with an increase in noise. Noise can be both short-term, for example, associated with construction, fracking, and roads and traffic, as well as long-term, associated primarily with compressor stations. This long-term source of noise is generally the greatest concern. Natural gas extraction requires large compressor stations to maintain gas pressure and keep gas moving through the pipelines. Compressor stations are associated with both the gathering pipelines that move the gas from the well to the interstate pipelines and also with the larger interstate pipelines that move gas to end users. These stations emit loud, low-frequency sounds between 75 and 90 dB(A) continuously and can reach 105 dB(A), noise levels detectable at distances over 1 km into forested areas (Bayne et al. 2008). Pennsylvania's DCNR has guidelines for compressor noise on forest land, but most compressors exceed the noise levels of these guidelines (DCNR 2018). Compressor stations are generally located every 64–112 km (40 to 70 miles) along a transmission pipeline (Marcellus Shale Coalition 2015). However, the number of compressor stations required for gathering line pipelines may be much higher depending on the location and distribution of the pads. For example, on PA State Forest lands, there are 17 compressor stations associated with 303 km of pipelines, which comes out to one compressor station per 18 km of pipeline (DCNR 2018).

Anthropogenic noise reduces habitat quality for species that rely on acoustic signals for communication and predator avoidance, thereby often negatively affecting their fitness (Bayne et al. 2008; Buxton et al. 2017). Birds are particularly susceptible to increased noise levels. Several studies have looked at the effects of noise from compressor stations on birds. Chronic noise levels may cause birds to avoid these areas so that densities are lower near compressor stations than similar areas without them (Bayne et al. 2008). Changes in species composition often occur as avoidance response varies among species (Bayne et al. 2008). In a study of ovenbirds (*Seiurus aurocapilla*), the area near compressors was not avoided, but older males established territories on quiet sites with younger males relegated to noisier sites, thus creating a shift in age structure (Habib et al. 2007).

Noise may affect the interactions between adults and nestlings. Leonard et al. (2015) found that nestling tree swallows (*Tachycineta bicolor*) modified their begging calls in the presence of high levels of ambient noise (65 dB[A]), presumably to be heard above the noise. The minimum frequency increased, and the range of frequencies decreased, although nestling growth was not affected. Williams et al. (2021) found that eastern bluebirds (*Sialia sialis*) and tree swallows experimentally exposed to noise from compressor stations reduced the time they spent incubating their eggs and consequently had lower hatching and fledging success than individuals nesting in quiet boxes. Although much of the work on noise has been conducted on birds, negative effects occur across taxonomic groups. For a comprehensive review of the effects of noise on wildlife see Francis and Barber (2013).

15.3.4 Spread of Invasive Plants

Invasive plants are nonnative species that spread aggressively, often outcompeting native species, inhibiting tree regeneration, and degrading natural ecosystems and wildlife habitat (Mack et al. 2000; Chornesky et al. 2005). Consequently, the introduction and spread of invasive plants in native ecosystems is one of the largest ecological problems we face today (Mack et al. 2000). Shale gas development, including clearing land for pads and pipelines, new roads, and increased levels of activity, provides multiple opportunities for invasive plant species to become established and spread. Removal of native vegetation and topsoil for pads, pipelines, and roads creates disturbed areas that are readily colonized by invasive plants. In addition, road development may promote invasive species via changes in microenvironment adjacent to the road (Mortensen et al. 2009; Barlow et al. 2017) or cultural factors tied to the road (Rauschert et al. 2017; Rew et al. 2018). Roads and pipelines also serve as corridors for movement and ideal habitat for invasive plants to move into previously uninvaded areas (Mortenson et al. 2009), and seed propagules may be introduced on vehicles, equipment, or animals moving along the corridors (Mortenson et al. 2009; Barlow et al. 2017; DCNR 2018). Invasive species populations often grow exponentially in the early stages of invasion, which can result in relatively rapid changes in species composition and diversity. Once established, invasive species are typically more resistant to roadside conditions than native species, as mowing, herbicide treatments, and soil compaction often prevent the reestablishment of native plants, while having lesser effects on invaders (Forman and Alexander 1998).

In Pennsylvania, 61% of newly created well pads had at least one invasive species present, and 19% had three or more species present (Barlow et al. 2017). The probability of invasion was associated with factors such as local road and well density, as well as the nearness of invasive species source populations (Barlow et al. 2017). Although pre-development surveys were not available, invasive plants were rare or absent away from pads, pipelines, and roads, and there is growing concern over the introduction of invasive species into areas where they were formerly absent (Barlow et al. 2017; DCNR 2018). On state forest lands in Pennsylvania there is an aggressive program that involves early detection of invasive plants and rapid response and treatment to try to keep them from spreading (DCNR 2018). Even with this program, the spread of invasive plants continues to be a challenge. Even more concerning is that on the majority of land where gas development is occurring, including private land, there are no programs for early detection and management, suggesting that the accelerated spread of invasive plants in association with gas development has the potential to degrade the health and functioning of these forests to a much greater extent than reported on PA state forest land.

15.3.5 Water Quality and Quantity

Water quality and quantity can be affected at multiple points within the process of USG development and are specifically addressed in Chapter 7 on water usage and management (Wilson and Van Briesen 2021). We provide a brief summary here of effects on surface water. Well pad construction, as well as excavation and clearing for roads and pipelines, can lead to increased levels of erosion, sedimentation, and runoff as well as increases in the amount of impervious surfaces, all of which may negatively affect water quality (Brittingham et al. 2014b; Burton et al. 2014. DCNR 2018). Water used for fracking can impact water quantity and quality through both the removal of fresh water and how wastewater is managed. The amount of water needed to frack a well in Pennsylvania averages between 15.4–18.9 million L, with the range being 11.3 to 37.8 million L (Schmid and Yoxtheimer 2015). Water for fracking may be obtained from surface sources (including lakes and rivers), groundwater, municipal supplies (Schmid and Yoxtheimer 2015), mine drainage (He et al. 2013), and reused fracturing water (Schmid and Yoxtheimer 2015). Water withdrawal for fracking can negatively impact surface waters depending on when and where the water is withdrawn, and it is considered a consumptive use of water since only a small fraction is returned to the water cycle (Kondash and Vengosh 2015). Water withdrawals are regulated by different state agencies or river basin commissions depending on where the well is located (Abdalla 2009).

The impact of surface water withdrawal from rivers and streams is highly dependent on the source itself (e.g., large river versus stream). Water levels in streams, rivers, and lakes naturally fluctuate throughout the year, and withdrawals for fracking have been identified in the Appalachians as an added threat to water levels, quality (Dillon 2011), and recreation (Kellison et al. 2017). In small streams especially, large water withdrawals may concentrate water downstream, potentially to the point where the water quality is low enough in the withdrawal area to adversely affect the aquatic ecosystem (Entrekin et al. 2011). The socioeconomic impact of fracking on water, and potential contamination to freshwater supplies, is a politically charged topic (Finewood and Stroup 2012; Poole and Hudgins 2014). However, research in the Marcellus Play tied to proactively managing water supplies and associated ecosystems affected by fracking has shown the potential to identify ahead of time where management can help avert aquatic resource degradation (DCNR 2018; Maloney et al. 2018).

Fracturing liquids (water and chemical additives) contain a variety of chemicals and sand, which acts as a proppant to keep fissures in the shale open during extraction (USDOE 2009). The USG industry has expanded its use of recycled wastewater, but the disposal of flowback water from the fracking process continues to be a challenge and a potential source of contamination, particularly to aquatic resources (Maloney and Yoxtheimer 2012; Wilson and Van Briesen 2013; Akob et al. 2016; Burgos et al. 2017; Hill et al. 2019). Surface waters can become contaminated through accidental spills or through treatment of oil and gas wastewater and discharge into streams (Wilson and Van Briesen 2013). In addition, the use of USG wastewater on roads has also been a practice in parts of the Marcellus Play region and has the potential to negatively impact water quality (Tasker et al. 2018). The wastewater and component chemicals often remain on or near the road surface, often at elevated levels, for up to four years after application or accidental spills (Lauer et al. 2016). Over time, rain may wash the chemicals into local surface water sources, which can pose additional risks.

Increasing use of flowback water and better wastewater management can reduce contamination of surface water, but it is essential that monitoring water quality takes place to detect potential problems as soon as they occur (Wilson and Van Briesen 2013). Pennsylvania's Department of Natural Resources did not find impacts to stream water quality in association with shale gas development on state forest lands, primarily as a result of the significant measures they take to reduce erosion, sedimentation, and runoff and manage USG wastewater (DCNR 2018). However, in other areas, changes in water quality have been noted, and there are concerns over the effect of these additional stressors on stream health (e.g., Entrekin et al. 2015).

15.4 Effects on at Risk Species or Groups

In the following section, we focus on groups of species that are of conservation concern within the forests of the Appalachian basin and may be negatively affected by shale gas development.

15.4.1 Forest Birds

Birds are frequently used as ecological indicators to monitor changes that are occurring in response to varying forms of disturbance (O'Connell et al. 2000). Because they are diurnal, visible, vocal, and have a range of habitat requirements, surveys of birds are often used to not only show changes in bird populations but also to reflect expected changes in other groups that may be less visible and more difficult to survey (O'Connell et al. 2000). As is frequently the case, there have been more studies on the effects of shale gas development on birds in the Appalachian basin than on other vertebrate groups. We review some of the key findings in this section.

The extensive area of contiguous forest in the Appalachian basin is considered a key habitat for many Neotropical migrant songbirds that breed in the eastern deciduous forest and winter in Central and South America (AMJV 2020). These insect-eating birds have co-evolved with the forest and make up a majority of the bird species and individuals in the forest within the breeding season, contributing directly to the health of the forest (Marquis and Whelan 1994; Whelan et al. 2015; Nyffeler et al. 2018). Many of these are considered priority species for the Appalachian Mountains Bird Conservation Region, which encompasses the Marcellus and Utica shale regions (AMJV 2020). In addition, many of these species have exhibited steep declines in abundance with eastern forest birds as a group exhibiting an overall decline in abundance of 17% since 1970 (Rosenberg et al. 2019). As a group, forest interior specialists are at risk because of the broad overlap between the shale layers and forest habitat (Figures 15.1 and 15.2). For example, in Pennsylvania researchers used abundance data associated with the PA second breeding bird atlas to map the abundance of 13 forest interior specialists and to identify critical habitat blocks for forest interior specialists (Figure 15.2, Shen 2015). When the Marcellus shale layer is added to the map, 96% of critical forest habitat blocks overlap the Marcellus shale layer, suggesting potential for high risk to forest specialists (Figure 15.2).

A number of studies have looked at the response of forest breeding birds to shale gas development, and these show that forest specialists such as ovenbirds tend to decline in abundance near gas infrastructure while generalist species such as American robins (*Turdus migratorius*) and chipping sparrows (*Spizella passerina*) that tend to do well around people and anthropogenic structures and disturbance increase in abundance (Barton et al. 2016; Farwell et al. 2016, 2019). Early successional forest species have shown mixed responses. Barton et al. (2016) found that early successional species benefitted from the disturbance in northern hardwood forests but less so in oak forest. Farwell et al. (2016, 2019) studied

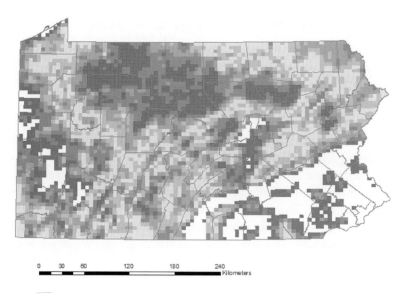
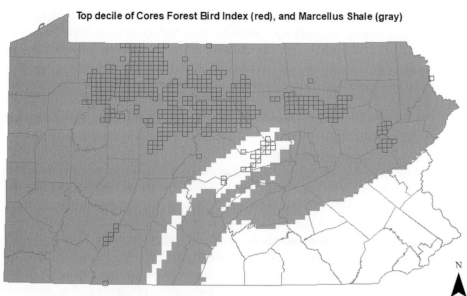

Figure 15.2 Combined abundance of 13 forest interior specialists based on abundance data from the second breeding bird atlas (Shen 2015, top figure). Abundance is greatest in dark red blocks. Bottom shows the blocks which support the greatest abundance of forest interior specialists above the Marcellus shale layer (A black and white version of this figure will appear in some formats. For the colour version, refer to the plate section.)

forest birds in West Virginia and showed that as shale gas development increased, there was a reduction in forest specialists and an increase in synanthropic or human associated species as well as those associated with disturbance and young forest. These shifts occurred with only a loss of 4.5 % in forest cover although the loss to core forest was 12.5%.

The Louisiana Waterthrush (*Parkesia motacilla*) has served as a focal species for studies because it is both a forest interior specialist and a riparian habitat species breeding on high quality streams and feeding on macroinvertebrates (Frantz et al. 2018). As such, it serves as an indicator of changes to both forest and riparian habitats. Recent research has shown that streams in landscapes where shale gas development is occurring show changes in macroinvertebrate populations with species intolerant of pollution declining in number while generalist species that are more tolerant of pollution are increasing (Wood et al. 2016). The waterthrush breeding on streams within the shale landscapes showed reduced productivity associated with shale gas development even though development affected <4% of the landscape (Frantz et al. 2018). Latta et al. (2015) found that waterthrush breeding on streams within watersheds where hydraulic fracturing was occurring in both the Marcellus and Fayetteville shale regions had higher levels of barium and strontium, heavy metals associated with hydraulic fracturing activities, than those on streams in watersheds where drilling was not occurring, suggesting that the surface waters where the birds were feeding was contaminated by the fracking process.

Most of the studies on either focal species of bird communities have been at relatively small localized scales (e.g., Barton et al. 2016; Farwell et al. 2016, 2019). One exception was a basin-wide study conducted by Farwell et al. (2020). In this study, researchers conducted avian point counts across forested areas within the Marcellus/Utica region and analyzed the data in 1 km (314 ha) circles to determine tipping points for forest lost resulting from both shale gas infrastructures but also other forms of disturbance such as exurban development. For the bird community as a whole, 1 km radius circles retaining at least 83% forest cover supported most of the forest interior specialists with a tipping point at 17% loss of forest cover, while early successional and human associated species increased with non-forest habitat and showed a tipping point at approximately 30% non-forest (Farwell et al. 2020). These are mean community values with some species declining at much lower levels of forest loss. Although conducted at different scales, these results are in agreement with Shen (2015), who documented the importance of forest cover at a much larger scale (25 km^2) and found that critical habitat blocks for forest interior specialists had mean forest cover $\geq$95% and mean core forest $\geq$73%, highlighting the importance of contiguous forest for forest interior specialists.

The Appalachian basin is also home to many raptor species that are species of conservation concern and priority species within the Appalachian Mountains Bird Conservation Region, including northern goshawks (*Accipiter gentilis*) and broad-winged hawks (*Buteo platypterus*, AMJV 2020). These are both species that depend on large blocks of forest and are sensitive to disturbance (Goodrich et al. 2020; Squires et al. 2020), and shale gas development is predicted to have negative effects on them consistent with their response to other forms of disturbance.

15.4.2 Forest Salamanders

Forest salamanders (Family *Plethodontidae*) are keystone species within the eastern deciduous forest performing key ecological functions in their role as mid-level predators (Davik et al. 2004). They are one of the most abundant group of vertebrates within the eastern deciduous forest (reviewed in Davic et al. 2004), and the southern Appalachians are recognized as a global hotspot of diversity for this group (Kozak 2017). These forest salamanders are lungless, respire dermally, and must remain cool and moist to avoid desiccation. There has been very little published research on the effects of shale gas development on forest salamanders, but from studies examining their response to other forms of disturbance, they are considered to be a group sensitive to fragmentation and at risk from shale gas development (Kiviat 2013).

Brand et al. (2014) developed occurrence models for five species and then predicted habitat loss dependent on future Marcellus shale development. Brand et al. (2014) predicted forest loss of 4% with 10,000 new wells and up to 20% with 50,000 new wells. Depending on how the play is developed, *P. electromorphus*, *P. wehrlei*, and *P. richmondi* will all lose significant portions of their range. These were based on available data, and there is a lot of uncertainty, emphasizing the need for additional data on abundance and distribution. Studies in Pennsylvania found warmer temperatures and lower abundances of red-backed salamanders (*P. cinereus*) in association with pads and pipelines (Brittingham unpublished data, Brittingham et al. 2014a).

15.4.3 Bats

Northeastern bats are insectivorous and provide vital ecological services. Forest dwelling bats including Myotis species, such as the federally endangered Indiana bat (*Myotis sodalis*) and the federally threatened Northern long-eared bat (*Myotis septentrionalis*), forage and roost within forest habitat overlaying the Marcellus and Utica shale basins. This is also the region where populations of many northeastern bats have been decimated by White nose syndrome (WNS), a fungus that can decimate bat populations, with some populations having declined by 80–90% (Frick et al. 2010). There are few published studies specifically examining the effects of shale gas development on bats, but the overlap between the shale plays and the epicenter of WNS suggests the additional disruption and fragmentation of forest habitat by shale gas development is a concern (Hein 2012). Other potential effects include exposure to contaminants, particularly if bats are foraging over open wastewater pits or contaminated streams (Hein 2012), which have been documented in the region (e.g., Latta et al. 2015).

15.4.4 Aquatic Species

Natural gas development can affect aquatic ecosystems through water withdrawal and contamination, as well as land clearing in close proximity to streams resulting in increases in sedimentation as well as changes in water temperature and oxygen levels (Kiviat 2013;

Weltman-Fahs and Taylor 2013; Souther et al. 2014). In addition, stream crossings for roads and pipelines may reduce stream connectivity (Warren and Pardew 1998). Species or groups of species that are vulnerable to disturbance from USG include species that are sensitive to hydrology, require high water quality and cool well-oxygenated streams as well as those whose range broadly overlaps the Marcellus and Utica shale regions (Kiviat 2013). Example species include brook trout (*Salvelinus fontinalis*) and freshwater mussels (*Unionoidea*). Freshwater mussels are of interest because of the already high numbers of listed species, their sensitivity to toxicants, and the diversity of this group within the Marcellus-Utica region (Lydeard 2004; Kiviat 2013; Brittingham et al. 2014b).

The brook trout (*Salvelinus fontinalis*) is the only native breeding salmonid in the Eastern United States and it is a species associated with high quality forested streams with clear, cold water. Brook trout declines are associated with anthropogenic disturbance and habitat fragmentation, and they have been used as indicators of anthropogenic disturbance (reviewed in Weltman-Fahs and Taylor 2013). Merriam et al. (2018) modeled the occurrence and predicted loss of brook trout in streams in relation to unconventional gas development. Merriam et al. (2018) found that the effects of shale gas development were cumulative with other forms of anthropogenic disturbance, and the probability of loss of occupancy was greatest in streams that were intermediate in terms of other stressors and already close to the tipping point. Modeling brook trout occupancy and current permits in Pennsylvania, they predicted that 4% of streams that currently have brook trout will lose them as a result of gas development, including four Pennsylvania Class A designated streams (Merriam et al. 2018).

15.5 Reclamation

How we refer to a landscape being managed to improve its state from that of support for USG is important to consider. For example, the use of the terminology to refer to the process of developing an interim or post USG use (such as restore versus reclaim) can bring about very different perceptions, confuse participants given the language of different scientific disciplines (Johnson et al. 1997), and infer different values (Vining et al. 2000; Burger 2002). Terminology used among land managers, USG developers, or the public can also play a role in inferring political beliefs, opinions on policy, and values (DiCaglio et al. 2018). We choose to use the term "reclaim" because we feel that the term "restoration" can set unrealistic goals (Ehrenfeld 2000; Hobbs 2007) owing to the evolution of human-affected landscapes and novel ecosystems (Hobbs et al. 2009; Miller and Bestelmeyer 2017).

Reclamation of forested landscapes affected by USG has been limited to date but has great potential. Similar to approaches elsewhere (Rowland et al. 2009), reclamation in the Appalachians follows a government set standard, sometimes modified by a private landowner or a public agency (e.g., DCNR 2018), especially if either holds the mineral or surface rights, or are protected by state government by environmental agency monitoring. The scale of reclamation suitable for USG does not compare to that of minelands (e.g., Rowland et al. 2009; Zipper et al. 2011) given the aerial footprint of USG development (whether pad, pipeline, compressor station etc.) is of a much, much smaller extent than a typical mineland reclamation (Zipper et al. 2011; MacDonald et al. 2015; Miller and Zégre 2016). However, given the extensive development of USG and associated landscape challenges (Drohan and Brittingham 2012) it is critical to devise viable reclamation plans.

In Pennsylvania, researchers from Penn State University have worked closely with the Pennsylvania Department of Conservation and Natural Resources, Bureau of Forestry to assess how well pads might be reclaimed, what plant species are appropriate, and what soil conditions might limit success. Barlow et al. (2020) investigated seed mix success and invasive species occurrence on a pipeline that had been reclaimed to produce soils with an elevated pH compared to that of nearby adjacent undisturbed soils. Barlow et al. (2020) found that restoration seed mixes with native grasses, sedges, and forbs became established and survived and thus should be considered as an alternative to nonnative mixes with a lower potential habitat value. Soil pH was found to strongly influence patterns of plant community composition and should be considered when designing restoration seed mixes to avoid creating soil chemistry different from adjacent undisturbed soils. Barlow et al. (2020) also found that *Microstegium vimineum* spread faster and was more likely to become a dominant species on sites with a soil pH less than 5, supporting the recommendation that future monitoring protocols for *M. vimineum* should include soil pH.

Barlow (2019) evaluated the establishment success of four commonly recommended native perennial grasses on a simulated well pad (decompacted and compacted) as an alternative to commonly used reclamation mixes. Over the study's three years, compaction was found to not reduce above or below ground biomass in native grasses or weedy flora. While native mix species produced very little biomass in the first year, by the third year native mixes significantly outcompeted weedy flora. Barlow (2019) suggests that given the tolerance of her study's native grasses to soil compaction, native grasses could be incorporated in mixtures for highly disturbed soils like that of USG.

Drohan et al. (2020) created a USG reclamation experiment comparing four physical (soil) reclamation treatments (topsoil replacement over pad stone; topsoil replacement over a pad with stone removed (no decompaction); removed stone with 20 cm surface decompaction and topsoil replacement; and removed stone with 50 cm surface decompaction and topsoil replacement) with three vegetation treatments (native and nonnative species; all native species; and native and pollinator-specific species). Results suggest that stone removal provided better protection against drought and invasive species establishment. Across all soil treatments, the all native mix performed as expected, with native grasses comprising three out of the four species with the highest percentage cover for five growing seasons following planting. The increased presence of invasive species in the topsoil over rock treatment indicates how this practice will probably benefit invasive species and weedy natives that can typically take advantage of thin, droughty soils. Across all years, the seed mixes in the treatment in which rock was not removed showed the lowest summed proportion of native species cover. While the species with the highest percentage cover in 2019 didn't differ substantially between soil treatments, in the two decompaction treatments, native warm season grasses showed much higher increases from 2016 to

2019. After five growing seasons, the summed proportion of native cover across all seed mixes is very similar, which may indicate that while the compacted soil and rock left in place is not an ideal growing medium, in five years some deep-rooted native species are able to grow through the rock and find subsoil. After five years, the mix of native and nonnative species has nearly the same percentage cover of natives, and this is probably because of the early competitive advantage that nonnative cool season grasses have over native species. Drohan et al. (2020) conclude that if a reclamation goal is to achieve high density of native species, a conventional mix, as used in their study, will take much longer to achieve this goal and could be at a risk if other environmental stresses limit native species.

15.6 Recommendations and Future Outlook

The forests of the Appalachian basin provide a wealth of benefits that go well beyond the natural gas underlying them. In order to maintain these important, and in many cases irreplaceable, benefits, development must be done in a way that minimizes risk to the health, viability, and sustainability of this ecosystem. Although shale gas development has recently (2019–2020) slowed down, we are still in the early stages of development with estimates of approximately 30–35% of current leases developed (DCNR 2018). The current lull in development gives us a chance to reassess and plan. Based on the recommendations of others (e.g., Souther et al. 2014; Langlois et al. 2017; DCNR 2018; Fawell et al. 2020) and current conditions and concerns (described in this chapter), we provide the following recommendations. These recommendations do not necessarily reflect the opinions of the organizations or individuals cited here.

- Consolidate infrastructure and development into areas where gas development or other forms of development are already present. Avoid blocks of contiguous forest habitat that are currently devoid of infrastructure (Langlois et al. 2017; Farwell et al. 2020). Farwell (2020) provides specific information on tipping points. Langlois et al. (2017) show how development in areas currently devoid of infrastructure can lead to an exponential loss of core forest because of the need to develop associated infrastructure.
- Prioritize development to focus on infill of wells on pads that are currently built over building new pads. In 2018, the average number of wells per pad was just 3.2 in Pennsylvania (Jacquet et al. 2018) and yet the pads were built with the potential to hold an average of six wells with up to 24 wells (DCNR 2018). Increasing the number of wells per pad will extract gas from a larger area while minimizing additional surface disturbance. Pads that do not hold the potential for future development should be reclaimed.
- Continue to monitor development and response to development (e.g., invasive plants, water quality, noise etc.) in a manner similar to that undertaken by the PA DCNR (DCNR 2018) to enable early detection of problems and rapid response. At a regional level expand monitoring to include other public lands as well as private lands. We realize the complexity of this recommendation and that it will require some sort of regional monitoring and management team with a dedicated funding source.

- Monitor water quality and quantity on both public and private lands across the Marcellus and Utica shale plays. PA DCNR (2018) has a water monitoring program on state forest land in conjunction with other state and federal partners that includes monitoring water quality and quantity as well as surveying macroinvertebrates. This monitoring program could serve as a template for other areas of public and private land.
- Consolidate roads and pipelines to minimize fragmentation and use practices to minimize the width of pipelines and roads. Plant pipelines with native seed mixes that are pollinator friendly and minimize mowing pipelines.
- Continue research and monitoring on the effects of well pad construction and pipeline and road construction on soil physical and chemical properties, as well as the effects of best management practices on hydrology and sediment loads.
- Develop invasive plant species management strategies particularly in areas where expanding road networks are going into previously undeveloped areas (Barlow et al. 2017). Monitoring should begin with multi-well pads, as those are the most likely to be invaded and should take into account soil physical and chemical conditions before and after reclamation (Barlow et al. 2017, 2020).
- Avoid placing compressor stations in natural areas that currently lack significant noise disturbance. In addition, compressor stations could include noise suppression measures such as sound-dampening barriers (Francis et al. 2011; Northrup and Wittemyer 2013).
- Continue to conduct research to identify and mitigate effects of development on natural resource conditions and use an adaptive management approach to form guidelines and policies to minimize negative effects of shale gas development on ecological resources. This will be most effective if researchers, managers, industry, and policy makers work together for a common goal.
- It is evident that there are complex tradeoffs when examining ecological impacts. Comprehensive modeling should be done to evaluate cumulative ecological impacts and the tradeoffs between different siting and development scenarios (Evans and Kiesecker 2014; Milt et al. 2016). In addition, ecological costs must be included in cost benefit analyses of where, when, and whether to develop natural gas resources in order to obtain a better estimate of true costs and of the value society places on these forests (Tarrant and Cordell 2002; Tarrant et al. 2003).

References

Abdalla C. (2009). Water Withdrawals for Development of Marcellus Shale Gas in Pennsylvania. Penn State Extension. https://extension.psu.edu/water-withdrawals-for-development-of-marcellus-shale-gas-in-pennsylvania. Accessed August 24, 2020.

Abramzon S, Samaras C, Curtright A, Litovitz A, and Burger N. (2014). Estimating the consumptive use costs of shale natural gas extraction on Pennsylvania roadways. *Journal of Infrastructure Systems*. 20(3).

Akob DM, Mumford AC, Orem W, Engle MA, Klinges JG, Kent DB, and Cozzarelli IM. (2016). Wastewater disposal from unconventional oil and gas development degrades stream quality at a West Virginia injection facility. *Environmental Science & Technology.* 50(11): 5517–5525.

AMJV (Appalachian Mountains Joint Venture. (2020). Priority landbirds. http://amjv.org/wp-content/uploads/2018/09/AMJV-Priority-Species.pdf. Accessed July 1, 2020.

Anderson MG, Clark M, and Sheldon AO. (2012). Resilient sites for terrestrial conservation in the Northeast and Mid-Atlantic region. *The Nature Conservancy, Eastern Conservation Science.* 289.

Barlow KM. (2019). *Restoring Plant Communities for Multiple Ecosystem Functions after Natural Resource Development.* PhD Thesis, The Pennsylvania State University.

Barlow KM, Mortensen DA, and Drohan PJ. (2020). Soil pH influences patterns of plant community composition after restoration with native-based seed mixes. *Restoration Ecology.* https://doi.org/10.1111/rec.13141

Barlow KM, Mortensen DA, Drohan PJ, and Averill KM. (2017). Unconventional gas development facilitates plant invasions. *Journal of Environmental Management.* 202: 208–216.

Barton EP, Pabian SE, and Brittingham MC. (2016). Bird community response to Marcellus shale gas development. *Journal of Wildlife Management.* 80(7): 1301–1313.

Bayne EM, Habib L, and Boutin S. (2008). Impacts of chronic anthropogenic noise from energy-sector activity on abundance of songbirds in the boreal forest. *Conservation Biology.* 22(5): 1186–1193.

Brand AB, Wiewel ANM. and Grant EHC. (2014). Potential reduction in terrestrial salamander ranges associated with Marcellus shale development. *Biological Conservation.* 180: 233–240.

Brittingham MC, Barton E, Fronk N, Bishop J, Sullivam K, and Morreale S. (2014a). Forest birds, reptiles and amphibians – Quantifying Marcellus shale associated effects on habitat and communities. Final report to the Pennsylvania Game Commission State Wildlife grants Program, Agreement # 4000015961 231048.

Brittingham MC, Maloney KO, Farag AM, Harper DD, and Bowen ZH. (2014b). Ecological risks of shale oil and gas development to wildlife, aquatic resources and their habitats. *Environmental Science & Technology.* 48(19): 11034–11047.

Burger J. (2002). Restoration, stewardship, environmental health, and policy: Understanding stakeholders' perceptions. *Environmental Management.* 30(5): 631–640.

Burgos WD, Castillo-Meza L, Tasker TL, Geeza TJ, Drohan PJ, Liu XF, Landis JD, Blotevogel J, McLaughlin M, Borch T, and Warner NR. (2017). Watershed-scale impacts from surface water disposal of oil and gas wastewater in Western Pennsylvania. *Environmental Science & Technology.* 51(15): 8851–8860.

Burton GA, Basu N, Ellis BR, Kapo KE, Entrekin S, and Nadelhoffer K. (2014). Hydraulic "Fracking": Are Surface Water Impacts An Ecological Concern? *Environmental Toxicology and Chemistry.* 33(8): 1679–1689.

Buxton RT, McKenna MF, Mennitt D, Fristrup K, Crooks K, Angeloni L, and Wittemyer G. (2017). Noise pollution is pervasive in US protected areas. *Science.* 356(6337): 531–533.

Chornesky EA et al. (2005). Science priorities for reducing the threat of invasive species to sustainable forestry. *Bioscience.* 55(4): 335–348.

Clark BK, Clark BS, Johnson LA, and Haynie MT. (2001). Influence of roads on movements of small mammals. *Southwestern Naturalist.* 46(3): 338–344.

Cooper J, Stamford L, and Azapagic A. (2018). Social sustainability assessment of shale gas in the UK. *Sustainable Production and Consumption.* 14: 1–20.

Crawford JA and Semlitsch RD. (2007). Estimation of core terrestrial habitat for stream-breeding salamanders and delineation of riparian buffers for protection of biodiversity. *Conservation Biology*. 21(1): 152–158.

Davic RD and Welsh HH. (2004). On the ecological roles of salamanders. *Annual Review of Ecology Evolution and Systematics*. 35: 405–434.

DCNR. (2018). Shale gas monitoring report. PA Department of Conservation and Natural Resources. http://elibrary.dcnr.pa.gov/GetDocument?docId=1743759&DocName= 37999 DCNR Shale Gas Report 2018 Interactive.pdf. Accessed July 13, 2020.

DiCaglio J, Barlow KM, and Johnson JS. (2018). Rhetorical Recommendations Built on Ecological Experience: A Reassessment of the Challenge of Environmental Communication. *Environmental Communication*. 12(4): 438–450.

Dillon M. (2011). Water scarcity and hydraulic fracturing in Pennsylvania: examining Pennsylvania water law and water shortage issues presented by natural gas operations in the Marcellus shale. *Temple Law Review*. 84: 201.

Drohan PJ and Brittingham M. (2012). Topographic and soil constraints to shale-gas development in the Northcentral Appalachians. *Soil Science Society of America Journal*. 76(5): 1696–1706.

Drohan PJ, Brittingham M, Bishop J, and Yoder K. (2012). Early trends in landcover change and forest fragmentation due to shale-gas development in Pennsylvania: A potential outcome for the Northcentral Appalachians. *Environmental Management*. 1–15.

Drohan PJ, Sitch K, Barlow KM, and Gamble B. (2020). Unconventional shale gas site reclamation approaches: soil physical, soil chemical and plant composition outcomes. *Environmental Management*, in review.

Ehrenfeld JG. (2000). Defining the limits of restoration: The need for realistic goals. *Restoration Ecology*. 8(1): 2–9.

Entrekin S, Evans-White M, Johnson B, and Hagenbuch E. (2011). Rapid expansion of natural gas development poses a threat to surface waters. *Frontiers in Ecology and the Environment*. 9(9): 503–511.

Entrekin SA, Maloney KO, Kapo KE, Walters AW, Evans-White MA, and Klemow KM. (2015). Stream vulnerability to widespread and emergent stressors: A focus on unconventional oil and gas. *Plos One*. 10(9).

Evans JS and Kiesecker JM. (2014). Shale gas, wind and water: Assessing the potential cumulative impacts of Energy development on ecosystem services within the Marcellus Play. *Plos One*. 9(2).

Faaborg J, Brittingham M, Donovan T, and Blake J. (1995). Habitat fragmentation in the temperate zone. *Ecology and Management of Neotropical Migratory Birds*. Oxford University Press, pp. 357–380.

Fahrig L and Rytwinski T. (2009). Effects of Roads on Animal Abundance: an Empirical Review and Synthesis. *Ecology and Society*. 14(1).

Farwell LS, Wood PB, Brown DJ, and Sheehan J. (2019). Proximity to unconventional shale gas infrastructure alters breeding bird abundance and distribution. *Condor*. 121(3).

Farwell LS, Wood PB, Dettmers R, and Brittingham MC. (2020). Threshold responses of songbirds to forest loss and fragmentation across the Marcellus-Utica shale gas region of central Appalachia, USA. *Landscape Ecology*. 35(6): 1353–1370.

Farwell LS, Wood PB, Sheehan J, and George GA. (2016). Shale gas development effects on the songbird community in a central Appalachian forest. *Biological Conservation*. 201: 78–91.

Finewood MH and Stroup LJ. (2012). Fracking and the neoliberalization of the hydro-social cycle in Pennsylvania's Marcellus Shale. *Journal of Contemporary Water Research & Education*. 147(1): 72–79.

Forman RTT and Alexander LE. (1998). Roads and their major ecological effects. *Annual Review of Ecology and Systematics*. 29: 207–231.

Francis CD and Barber JR. (2013). A framework for understanding noise impacts on wildlife: An urgent conservation priority. *Frontiers in Ecology and the Environment*. 11(6): 305–313.

Francis C, Paritsis J, Ortega C, and Cruz A. (2011). Landscape patterns of avian habitat use and nest success are affected by chronic gas well compressor noise. *Landscape Ecology*. 26(9): 1269–1280.

Frantz MW, Wood PB, Sheehan J, and George G. (2018). Demographic response of Louisiana Waterthrush, a stream obligate songbird of conservation concern, to shale gas development. *Condor*. 120(2): 265–282.

Fredericksen TS. (1998). Impacts of logging and development on Central Appalachian forests. *Natural Areas Journal*. 18(2): 175–178.

Frick WF et al. (2010). An emerging disease causes regional population collapse of a common North American bat species. *Science*. 329(5992): 679–682.

Goodrich LJ, Crocoll ST, and Senner SE. (2020). Broad-winged Hawk (Buteo platypterus), version 1.0. In Birds of the World (A. F. Poole, Editor). Cornell Lab of Ornithology, Ithaca, NY, USA. https://doi-org.ezaccess.libraries.psu.edu/10.2173/bow.brwhaw.01.

Habib L, Bayne EM, and Boutin S. (2007). Chronic industrial noise affects pairing success and age structure of ovenbirds *Seiurus aurocapilla*. *Journal of Applied Ecology*. 44(1): 176–184.

He C, Zhang T, and Vidic RD. (2013). Use of abandoned mine drainage for the development of unconventional gas resources. *Disruptive Science and Technology*. 1(4): 169–176.

Heilman GE, Strittholt JR, Slosser NC, and Dellasala DA. (2002). Forest fragmentation of the conterminous United States: Assessing forest intactness through road density and spatial characteristics. *Bioscience*. 52(5): 411–422.

Hein CD. (2012). *Potential Impacts of Shale Gas Development on Bat Populations in the Northeastern United States*. An unpublished report submitted to the Delaware Riverkeeper Network, Bristol, Pennsylvania by Bat Conservation International, Austin, Texas. www.delawareriverkeeper.org/sites/default/files/resources/Reports/Impacts_of_Shale_Gas_Development_on_Bats.pdf (Accessed July 9, 2020).

Hill LAL, Czolowski ED, DiGiulio D, and Shonkoff SBC. (2019). Temporal and spatial trends of conventional and unconventional oil and gas waste management in Pennsylvania, 1991–2017. *Science of the Total Environment*. 674: 623–636.

Hobbs RJ. (2007). Setting effective and realistic restoration goals: Key directions for research. *Restoration Ecology*. 15(2): 354–357.

Hobbs RJ, Higgs E, and Harris JA. (2009). Novel ecosystems: Implications for conservation and restoration. *Trends in Ecology & Evolution*. 24(11): 599–605.

Jacquet JB et al. (2018). A decade of Marcellus Shale: Impacts to people, policy, and culture from 2008 to 2018 in the Greater Mid-Atlantic region of the United States. *Extractive Industries and Society: An International Journal*. 5(4): 596–609.

Johnson DL, Ambrose SH, Bassett TJ, Bowen ML, Crummey DE, Isaacson JS, and Johnson DN. (1997). Meanings of environmental terms. *Journal of Environmental Quality*. 26(3): 581–589.

Johnson N, Gagnolet T, Ralls R, and Stevens J. (2011). Natural Gas Pipelines: Excerpt from Report 2 of the Pennsylvania Energy Impacts Assessment. The Nature Conservancy.

Kellison TB, Bunds KS, Casper JM, and Newman JI. (2017). Public parks usage near hydraulic fracturing operations. *Journal of Outdoor Recreation and Tourism-Research Planning and Management.* 18: 75–80.

Kiviat E. (2013). Risks to biodiversity from hydraulic fracturing for natural gas in the Marcellus and Utica shalesI In Schlesinger WH and Ostfeld RS (eds.) *Year in Ecology and Conservation Biology.* Annals of the New York Academy of Sciences, 1–14.

Kondash A and Vengosh A. (2015). Water footprint of hydraulic fracturing. *Environmental Science & Technology Letters.* 2(10): 276–280.

Kozak KH. (2017). What Drives Variation in Plethodontid Salamander Species Richness over Space and Time? *Herpetologica.* 73(3): 220–228.

Langlois LA. (2017). *Effects of Marcellus Shale Gas Infrastructure on Forest Fragmentation and Bird Communities in Northcentral Pennsylvania.* PhD Thesis. The Pennsylvania State University.

Langlois LA, Drohan PJ, and Brittingham MC. (2017). Linear infrastructure drives habitat conversion and forest fragmentation associated with Marcellus shale gas development in a forested landscape. *Journal of Environmental Management.* 197: 167–176.

Latta SC, Marshall LC, Frantz MW, and Toms JD. (2015). Evidence from two shale regions that a riparian songbird accumulates metals associated with hydraulic fracturing. *Ecosphere.* 6(9).

Lauer NE, Harkness JS, and Vengosh A. (2016). Brine spills associated with unconventional oil development in North Dakota. *Environmental Science & Technology.* 50 (10): 5389–5397.

Leonard ML, Horn AG, Oswald KN, and McIntyre E. (2015). Effect of ambient noise on parent-offspring interactions in tree swallows. *Animal Behaviour.* 109: 1–7.

Lydeard C et al. (2004). The global decline of nonmarine mollusks. *Bioscience.* 54(4): 321–330.

Macdonald SE, Landhausser SM, Skousen J, Franklin J, Frouz J, Hall S, Jacobs DF, and Quideau S. (2015). Forest restoration following surface mining disturbance: challenges and solutions. *New Forests.* 46(5–6): 703–732.

Mack RN, Simberloff D, Mark Lonsdale W, Evans H, Clout M, and Bazzaz FA. (2000). Biotic invasions: Causes, epidemiology, global consequences, and control. *Ecological Applications.* 10(3): 689–710.

Maloney KO, Young JA, Faulkner SP, Hailegiorgis A, Slonecker, ET, and Milheim LE. (2018). A detailed risk assessment of shale gas development on headwater streams in the Pennsylvania portion of the Upper Susquehanna River Basin, USA. *Science of the Total Environment.* 610: 154–166.

Maloney KO and Yoxtheimer DA. (2012). Production and disposal of waste materials from gas and oil extraction from the Marcellus shale play in Pennsylvania. *Environmental Practice.* 14(04): 278–287.

Marcellus Shale Coalition. (2015). Pipeline and Midstream Facilities: Getting Natural Gas to Market Safely. https://marcelluscoalition.org/wp-content/uploads/2020/03/Midstream-and-Pipeline-Fact-Sheet_12.16.15.pdf. Accessed August 21, 2020.

Marquis RJ and Whelan CJ (1994). Insectivorous birds increase growth of white oak through consumption of leaf-chewing insects. *Ecology.* 75(7): 2007–2014.

Marsh D, Milam G, Gorham N, and Beckman N. (2005). Forest roads as partial barriers to terrestrial salamander movement. *Conservation Biology.* 19(6): 2004–2008.

MCOR (Marcellus Center for Outreach and Research). (2020). Tri-state unconventional shale wells drilled by year (PA,OH, WV). http://www.marcellus.psu.edu/resources/images/tristate-wells-2019.jpg. Accessed July 20. 2020.

Merriam ER et al. (2018). Brook trout distributional response to unconventional oil and gas development: Landscape context matters. *Science of the Total Environment*. 628–629, 338–349.

Merriam G, Kozakiewicz M, Tsuchiya E, and Hawley K. (1989). Barriers as boundaries for metapopulations and demes of Peromyscus leucopus in farm landscapes. *Landscape Ecology*. 2(4): 227–235.

Miller AJ and Zegre N. (2016). Landscape-scale disturbance: Insights into the complexity of catchment hydrology in the mountaintop removal mining region of the eastern United States. *Land*. 522; doi:10.3390/land5030022.

Miller JR and Bestelmeyer BT. (2017). What the novel ecosystem concept provides: A reply to Kattan et al. *Restoration Ecology*. 25(4): 488–490.

Milt AW, Gagnolet T, and Armsworth PR. (2016). Synergies and tradeoffs among environmental impacts under conservation planning of shale gas surface infrastructure. *Environmental Management*. 57(1): 21–30.

Mortensen D, Rauschert E, Nord A, and Jones B. (2009). Forest roads facilitate the spread of invasive plants. *Invasive Plant Science and Management*. 2(3): 191–199.

Northrup J and Wittemyer G. (2013). Characterising the impacts of emerging energy development on wildlife, with an eye towards mitigation. *Ecology Letters*. 16(1): 112–125.

Nyffeler M, Sekercioglu CH, and Whelan CJ. (2018). Insectivorous birds consume an estimated 400–500 million tons of prey annually. *Science of Nature*. 105(7–8).

O'Connell TJ, Jackson LE, and Brooks RP. (2000). Bird guilds as indicators of ecological condition in the central Appalachians. *Ecological Applications*. 10(6): 1706–1721.

Pearce DW. (2001). The economic value of forest ecosystems. *Ecosystem Health*. 7(4): 284–296.

Perry SL. (2012). Development, land use, and collective trauma: The Marcellus Shale gas boom in rural Pennsylvania. *Culture, Agriculture, Food & Environment*. 34(1): 81–92.

Poole A and Hudgins A. (2014). "I care more about this place, because I fought for it": Exploring the political ecology of fracking in an ethnographic field school. *Journal of Environmental Studies and Sciences*. 4(1): 37–46.

Rauschert ESJ, Mortensen DA, and Bloser SM. (2017). Human-mediated dispersal via rural road maintenance can move invasive propagules. *Biological Invasions*. 19(7) 2047–2058.

Rew LJ, Brummer TJ, Pollnac FW, Larson CD, Taylor KT, Taper ML, Fleming JD, and Balbach HE. (2018). Hitching a ride: Seed accrual rates on different types of vehicles. *Journal of Environmental Management*. 206: 547–555.

Robinson S, Thompson F, Donovan T, Whitehead D, and Faaborg J. (1995). Regional forest fragmentation and the nesting success of migratory birds. *Science*. 267(5206): 1987–1990.

Rosenberg KV et al. (2019). Decline of the North American avifauna. *Science*. 366(6461); 120.

Rowland SM, Prescott CE, Grayston SJ, Quideau SA, and Bradfield GE. (2009). Recreating a Functioning Forest Soil in Reclaimed Oil Sands in Northern Alberta: An Approach for Measuring Success in Ecological Restoration. *Journal of Environmental Quality*. 38(4): 1580–1590.

Schmid K and Yoxtheimer D. (2015). Wastewater recycling and reuse trends in Pennsylvania shale gas wells. *AAPG Environmental Geosciences*. 22(4): pp. 115–125.

Shen KG. (2015). *Defining Critical Forest Habitat for Area-Sensitive Forest Songbirds in Pennsylvania*. MS Hood College.

Slonecker E, Milheim L, Roig-Silva C, Malizia A, Marr D, and Fisher G. (2012). *Landscape Consequences of Natural Gas Extraction in Bradford and Washington Counties, Pennsylvania, 2004–2010*. U.S. Geological survey open file report 2012-1154.

Slonecker ET, Milheim LE, Roig-Silva CM, Malizia AR, and Gillenwater BH. (2013). *Landscape Consequences of Natural Gas Extraction in Fayette and Lycoming Counties, Pennsylvania, 2004–2010*. U.S. Geological survey open file report 2013-1119.

Souther S, Tingley MW, Popescu VD, Hayman DTS, Ryan ME, Graves TA, Hartl B, and Terrell K. (2014). Biotic impacts of energy development from shale: Research priorities and knowledge gaps. *Frontiers in Ecology and the Environment*. 12(6): 330–338.

Squires JR, Reynolds RT, Orta J, and Marks JS. (2020). Northern Goshawk (*Accipiter gentilis*), version 1.0. In Billerman SM (ed.) *Birds of the World*. Cornell Lab of Ornithology. https://doiorg.ezaccess.libraries.psu.edu/10.2173/bow.norgos.01.

Tarrant MA and Cordell HK. (2002). Amenity values of public and private forests: examining the value–attitude relationship. *Environmental Management*. 30(5): 0692–0703.

Tarrant MA, Cordell HK, and Green GT. (2003). PVF: A scale to measure public values of forests. *Journal of Forestry*. 101(6): 24–30.

Tasker TL et al. (2018). Environmental and Human Health Impacts of Spreading Oil and Gas Wastewater on Roads. *Environmental Science & Technology*. 52(12): 7081–7091.

Thomas EH, Brittingham MC, and Stoleson SH. (2014). Conventional oil and gas development alters forest songbird communities. *Journal of Wildlife Management*. 78(2): 293–306.

U.S. Department of Energy. (2009). *Modern Shale Gas Development in the United States: A Primer*. Washington, DC. www.energy.gov/sites/prod/files/2013/03/f0/ShaleGasPrimer_Online_4-2009.pdf.

Vining J, Tyler E, and Kweon BS. (2000). Public values, opinions, and emotions in restoration controversies. Restoring nature: Perspectives from the social sciences and humanities, pp. 143–161.

Walters JR. (1998). The ecological basis of avian sensitivity to habitat fragmentation In Marzlaff JM and Sallabanks R (eds.) *Avian Conservation: Research and Management*. Island Press, pp. 181–192.

Warren ML and Pardew MG. (1998). Road crossings as barriers to small-stream fish movement. *Transactions of the American Fisheries Society*. 127(4): 637–644.

Weltman-Fahs M and Taylor JM. (2013). Hydraulic fracturing and brook trout habitat in the Marcellus Shale region: Potential impacts and research needs. *Fisheries*. 38(1): 4–15.

Whelan CJ, Sekercioglu CH, and Wenny DG. (2015). Why birds matter: From economic ornithology to ecosystem services. *Journal of Ornithology*. 156: S227–S238.

Williams, DP., Avery, JD., Gabrielson, TB. and Brittingham, MC. (2021). Experimental playback of natural gas compressor noise reduces incubation time and hatching success in two secondary cavity-nesting bird species. *Ornithological Applications*. 123(1): 1–11.

Wilson, JM and VanBriesen, JM. (2013). Source Water Changes and Energy Extraction Activities in the Monongahela River, 2009–2012. *Environmental Science & Technology*, 47(21): 12575–12582.

Wilson JM and VanBriesen JM. (2021). Water usage and management. In Stolz JF, Griffin WM, and Bain DJ (eds.) *Environmental Impacts from the Development of Unconventional Oil and Gas Reserves*. Cambridge University Press

Wood PB, Frantz MW, and Becker DA. (2016). Louisiana Waterthrush and Benthic Macroinvertebrate Response to Shale Gas Development. *Journal of Fish and Wildlife Management*. 7(2): 423–433.

Zipper CE, Burger JA, McGrath JM, Rodrigue JA, and Holtzman GI. (2011). Forest Restoration Potentials of Coal-Mined Lands in the Eastern United States. *Journal of Environmental Quality*. 40(5): 1567–1577.

16

Managing TDS and Sulfate in the Monongahela River

Three Rivers QUEST

PAUL ZIEMKIEWICZ, MELISSA O'NEAL, TAMARA VANDIVORT, JOSEPH KINGSBURY, AND RACHEL PELL

16.1 Introduction

In the late summer of 2008, the US Army Corps of Engineers (USACE) in Pittsburgh reported total dissolved solids (TDS) over 800 mg/L in the Monongahela River (USACE 2009). Simultaneously, the Pennsylvania Department of Environmental Protection (PADEP) found sulfate and TDS levels that exceeded the secondary drinking water standards of 250 mg/L and 500 mg/L, respectively (Wilson et al. 2014). The increasing TDS concentrations affected drinking water supplies for both residential and industrial users. Although TDS is not inherently unhealthy for drinking water, it is indicative of other water quality issues, such as high levels of minerals or toxins in the water. These elevated levels can cause issues ranging from bitter and metallic tasting water to more extreme events such as fish kills. The sources, seasonality, and composition of the TDS problem were unknown and required further investigation.

16.2 Investigating High TDS Levels in the Monongahela River Basin

16.2.1 TDS Monitoring on the Monongahela

The West Virginia Water Research Institute (WVWRI) developed a strategic monitoring program for the Monongahela River and began sampling water quality and flow in July 2009, focusing on TDS and its constituents. Funded by the US Geological Survey's (USGS) 104b program, the study included water quality monitoring and sampling on a biweekly basis at 16 locations in the watershed, including four sites on the Monongahela River and the mouths of 12 of its major tributaries. This initial monitoring project was known simply as "Mon WQ." It would later go through multiple expansions to eventually become the Three Rivers QUEST (3RQ) program that continues this work today.

16.2.2 Development and Implementation of TDS Management Plan

Historically, TDS concentrations in the Monongahela River have exceeded 500 mg/L only when flow drops below 2,000 cubic feet per second (cfs). While TDS trends coincided with

Table 16.1. *Example of the Monongahela River Basin TDS model output for dry (left) and wet (right) periods*

Target stream [TDS]	500	mg/L	Target stream [TDS]	500	mg/L
Stream Q (cfs)	200	cfs	Stream Q (cfs)	3600	cfs
Factor of safety	2		Factor of safety	2	
	MODEL OUTPUT			MODEL OUTPUT	
AMD plant	Pumping Rate Q (cfs)	Q (gpm)	AMD plant	Pumping Rate Q (cfs)	Q (gpm)
A	0.7	323	A	5.2	2,322
B	0.1	54	B	0.9	387
C	0.9	387	C	6.2	2,787
etc.			etc.		

increased gas development in the region, the extent to which the coal vs. gas industries contribute to TDS loadings was uncertain (Ziemkiewicz 2011). One of the few ways to segregate the two energy industries' relative contributions is by characterizing the TDS loads that mine drainage treatment plants produce under a range of operating conditions. Preliminary estimates by WVWRI indicated that if all the mine drainage treatment plants along the upper Monongahela River were running at full capacity with the maximum TDS concentrations, they would contribute 500,000 tons per year of TDS (Ziemkiewicz 2011). The initial USGS study, Mon WQ, suggested that the easiest component of the TDS picture to manage was the active deep coal mines. Results indicated that treated mine drainage, rich in Ca, Na, and SO_4 was the controlling factor in the Monongahela River's TDS load (Merriam et al. 2020). In the late fall of 2009, armed with the data from this monitoring program, WVWRI began working with major coal companies in the Upper Monongahela River Basin. They formed a coal industry TDS Working Group to design and implement a "discharge management" system. The discharge management system can be described as a non-mandated total maximum daily load management plan to maintain a level of TDS under 500 mg/L in the Monongahela River (Ziemkiewicz, 2010). Using typical TDS concentrations, the system accounted for the pumping capacities of the 14 major mine pumping and treatment plants in the Upper Monongahela. It then tied the salt output to the flow in the Monongahela on any particular day. The model for the Monongahela is set not to exceed the secondary drinking water standard of 500 mg/L with a safety factor of 2. The system allows the industry plant operators to look at the gauge reading and set their pumps to the indicated rate, thereby coordinating the outflows. Figure 16.1 and Table 16.1 illustrate this model, where Q represents flow.

Operators of mine discharge treatment facilities implemented discharge management in January 2010. The discharge management provided the tools needed to modulate discharge load based on stream flows to ensure the Monongahela River's mainstem would not exceed the secondary drinking water standards for sulfate or TDS (250 and 500 mg/L,

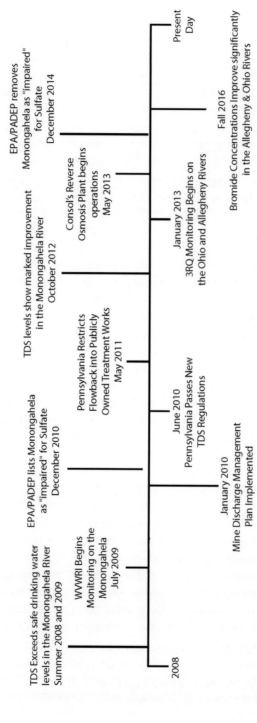

Figure 16.1 Timeline of significant events relating to total dissolved solids in the Upper Ohio River Basin from 2008 to present

respectively). Despite improvements in Monongahela River TDS due to the newly implemented discharge management plan, CONSOL Energy Inc. paid a $5.5 million civil penalty to a Clean Water Act Settlement with USEPA (Consent Decree 2011). In addition, CONSOL was required to construct a centralized wastewater treatment plant to treat water from four mine discharges. Located near Mannington, WV, the reverse osmosis plant came online in 2013 to treat wastewaters from CONSOL's Blacksville No. 2, Loveridge, and Robinson Run mining operations (Coal Age 2013). Flows within the Monongahela River and its tributaries inform management decisions that include the timed discharge of treated mine drainage based on the river's assimilative capacity (Merriam et al. 2020).

Operators of mine discharge treatment facilities implemented discharge management in January 2010. The discharge management provided the tools needed to modulate discharge load based on stream flows to ensure the Monongahela River's mainstem would not exceed the secondary drinking water standards sulfate or TDS (250 and 500 mg/L, respectively).

Despite improvements in Monongahela River TDS due to the newly implemented discharge management plan, CONSOL paid a $5.5 million civil penalty to a Clean Water Act Settlement with USEPA (Consent Decree 2011). In addition, CONSOL was required to construct a centralized wastewater treatment plant to treat water from four mine discharges. The reverse osmosis plant came online in 2013 to treat wastewaters from CONSOL's Blacksville No. 2, Loveridge, and Robinson Run mining operations (Coal Age 2013).

Flows within the Monongahela River and its tributaries (e.g., Dunkard Creek) inform management decisions that include the timed discharge of treated mine drainage based on the river's assimilative capacity (Merriam et al. 2020). Figure 16.3 depicts the reduction in TDS at Dunkard Creek. The dashed vertical lines represent the start of voluntary discharge management in January 2010 (long dash), PA restricting flowback water at POTWs in 2011 (medium dash), and the beginning of operations at CONSOL's reverse osmosis plant (short dash). As noted in Figure 16.3, voluntary discharge management provided an immediate response, improving water quality three years before the reverse osmosis plant began operations.

16.2.3 Watershed Level Management a Success

Since 2010, both sulfate and TDS levels have met EPA standards in the Monongahela River. None of our six monitoring stations (river miles 11, 23, 61, 82, 89, and 102) in the Monongahela River main stem have exceeded 500 mg/L TDS or 250 mg/L sulfate since the initiation of the discharge management program (Figures 16.2 and 16.3). Peak TDS concentrations have been steadily declining at all of our stations.

In December 2014, the USEPA approved the PADEP's report that the river's "instream level of sulfates now meets Pennsylvania's water quality standards." The results meet the Federal Clean Water Act's intent to prevent pollution while restoring polluted waterways without adding new regulations. Noteworthy is that the Monongahela River was listed as contaminated for sulfate in December 2010. Levels were already below 250

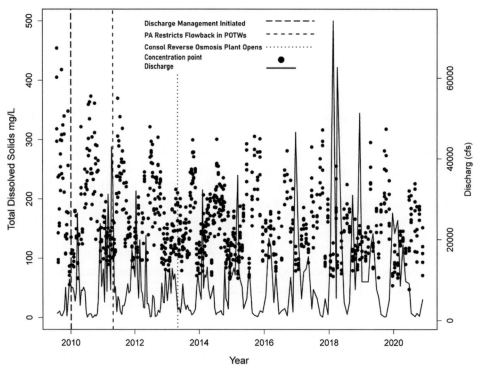

Figure 16.2 TDS concentrations for all Monongahela River mainstem sampling locations (river miles 11, 23, 61, 82, 89, and 102) and discharge (cfs) at USGS gage 03072655 at river mile 82 near Masontown, PA, from 2009 through 2021. The vertical dotted lines represent: long dash = initiation of the discharge management plan in January 2010; medium dash = Pennsylvania restricted flowback water at Public Owned Treatment Works (POTWs) in May 2011; short dash = CONSOL Energy's reverse osmosis plant went online in May 2013

mg/L by implementing the discharge management plan earlier in January 2010 (Figure 16.3). The discharge management plan used a cooperative approach to protect the Monongahela River and provided a timely correction to improve water quality. In resource-rich states such as West Virginia and Pennsylvania, it shows how we can achieve positive results when people come together to resolve problems. In fact, discharge management only works because industry buys into the process, and regular river monitoring validates the outcome.

16.3 Long-Term Water Quality Monitoring in the Ohio River Basin

16.3.1 The Birth of the Three Rivers QUEST

In 2010, Colcom Foundation funded the "Mon River QUEST," which expanded existing water quality monitoring to incorporate important field data collected by volunteer

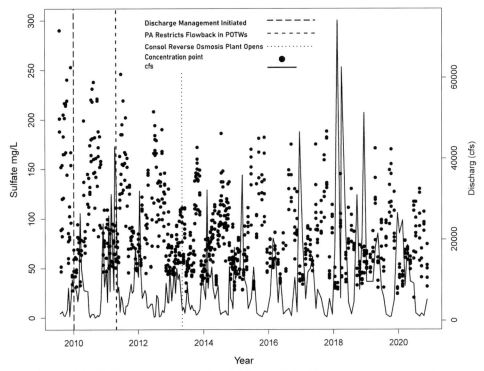

Figure 16.3 Sulfate concentrations for all Monongahela River mainstem sampling locations (river miles 11, 23, 61, 82, 89, and 102) and discharge (cfs) at USGS gage 03072655 at river mile 82 near Masontown, PA, from 2009 through 2021. The vertical dotted lines represent: long dash = initiation of the discharge management plan in January 2010; medium dash = Pennsylvania restricted flowback water at Public Owned Treatment Works (POTWs) in May 2011; short dash = CONSOL Energy's reverse osmosis plant went online in May 2013

watershed groups. Established in 2012, 3RQ broadened to include the Allegheny River and the Upper Ohio River basins by entering into partnerships with Duquesne University, Wheeling Jesuit University, and the Iron Furnace Chapter of Trout Unlimited. In 2019, West Liberty University joined 3RQ as a partner to monitor the Ohio River. These research partners perform routine monitoring and work with volunteer organizations throughout the Upper Ohio River Basin (Figure 16.4).

Each month, 3RQ partners and collaborators collect water samples and/or continuous data from data loggers at the 3RQ sites. Data is analyzed to determine the river's overall health and to identify areas of concern and individual pollution events. The data is displayed in maps on the 3RQ website to provide the general public, federal and state agencies, and industry with timely and accurate information regarding the rivers' overall health. The 3RQ program also provides the data necessary for developing the TDS management plan for the Monongahela River and validating the plan's success.

Figure 16.4 The Three Rivers QUEST coverage of the Upper Ohio River Basin covering parts of Maryland, Ohio, New York, Pennsylvania, and West Virginia.

16.3.2 Site Selection

Forty-six sites within the Ohio River Basin are monitored monthly through this project. Sites were selected in the three river basins to measure the effects of AMD discharges, brine treatment facilities, and powerplants on water quality. Site locations are along the river mainstem and the mouths of major tributaries (Table 16.2). Figure 16.5 illustrates sampling sites and the various input sources related to each site. In the Allegheny headwaters, additional samples are collected during low-flow to evaluate water quality conditions.

The Monongahela River Basin is sampled routinely at 18 sites, including six sites along the Monongahela River and 12 sites distributed along the mouths of major tributaries (Figure 16.6). There are five sites on the mainstem of the Allegheny River and 11 sites distributed among its tributaries. For routine monthly monitoring, the furthest site upstream is A83 near Parker, PA, and the furthest downstream site is A6 near Sharpsburg, PA (Figure 16.7). The Ohio River main stem and tributary sampling occur at 14 sites, including five on the Ohio River's mainstem, beginning in Sewickley, PA, and ending in Ravenswood, WV (Figure 16.8).

The Three Rivers QUEST Project 393

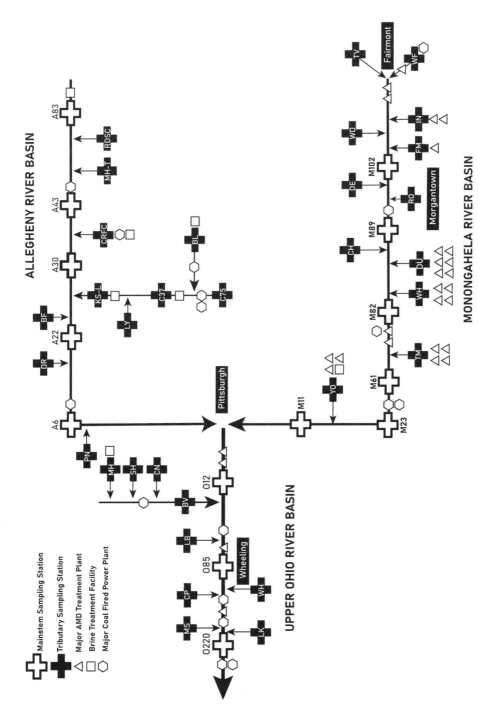

Figure 16.5 A complete diagram of all sites and the relative inputs related to each site. The Allegheny and the Monongahela converge within the city of Pittsburgh, PA, to form the Ohio River

Table 16.2. *Site code and site descriptions by river basin*

Site Code	Site Description	State
Allegheny River Basin		
A83	Allegheny - Parker	PA
RDSC	Redbank Creek at Saint Charles	PA
MHT	Mahoning Creek - Templeton	PA
A45	Allegheny River below L&D 7	PA
CRFC	Crooked Creek - Ford City	PA
A30	Allegheny River above L&D 5	PA
BF	Buffalo Creek	PA
KSL	Kiskiminetas River - Leechburg	PA
LY	Loyalhanna Creek at Kingston	PA
C37	Conemaugh River at Tunnelton	PA
BL	Blacklick Creek	PA
C75	Conemaugh River at Seward	PA
A22	Allegheny - Tarentum	PA
DR	Deer Creek - Harmarville	PA
A6	Allegheny L&D 2	PA
PN	Pine Creek	PA
Monongahela River Basin		
WF	West Fork River	WV
TV	Tygart Valley River	WV
IN	Indian Creek	WV
WD	White Day Creek	WV
FM	Flaggy Meadows Run	WV
M102	Monongahela River at Morgantown	WV
DE	Deckers Creek	WV
RO	Robinson Run	WV
M89	Monongahela River at Point Marion	PA
CH	Cheat River	PA
DU	Dunkard Creek	PA
WH	Whitely Creek	PA
M82	Monongahela River at Masontown	PA
TM	Tenmile Creek	PA
M61	Monongahela River at Brownsville	PA
M23	Monongahela River at Elizabeth	PA
YO	Youghiogheny River	PA
M11	Monongahela River at Homestead	PA
Upper Ohio River Basin		
O12	Ohio River at Sewickley	PA
BV	Beaver River at Beaver Falls	PA
CN	Connoquenessing Creek at Ellwood City	PA
SH	Shenango River at New Castle	PA

Table 16.2. (cont.)

	Allegheny River Basin	
Site Code	Site Description	State
MH	Mahoning River at Lowellsville	OH
LB	Little Beaver Creek at East Liverpool	OH
O85	Ohio River at Pike Island	OH
WH	Wheeling Creek at Wheeling	WV
CP	Captina Creek below Armstrong Mills	OH
MS	Muskingum River at Lowell	OH
LK	Little Kanawha above Parkersburg	WV
O220	Ohio River at Ravenswood	WV

16.3.3 Data Collection

WVWRI and its 3RQ partners perform monthly water sampling for all three river basins at 46 locations, as noted in Figures 16.6–16.8. USGS gauges provide flow measurements for nearby 3RQ sampling sites. Grab samples are analyzed for dissolved alkalinity (mg/L CaCO3 equivalents; EPA method SM-2320B), dissolved Al, Ca, Fe, Mn, Mg, and Na (mg/L; EPA method 3010B), and dissolved Br, Cl, and SO4 (mg/L; EPA method 300.0). TDS is calculated as the sum of the concentrations (referred to as TDSsdc) of all measured dissolved constituents. In situ temperature, electrical conductivity, and pH measures are obtained using a YSI 556 multiprobe (Yellow Springs Instruments, Yellow Springs, Ohio). Additionally, volunteer organizations collect field measurements using handheld meters and data loggers to obtain conductivity readings throughout the Upper Ohio River Basin.

16.3.4 Data Dissemination

Data is uploaded and stored within a 3RQ database called WATERS. Resultant data is made available to the public via web-based maps at https://3riversquest.wvu.edu/data/3rq-maps. Federal and state agencies utilize 3RQ data for providing insight into continuous water quality conditions where such data are unavailable through existing agency programs (Merriam et al. 2020).

16.3.5 Chemical Signatures

Several trends have emerged by studying the ratio of TDS constituents throughout the Monongahela River Basin. The ratio of chloride to sulfate ions can distinguish coal mine water from hydraulic fracturing (frac) water (Wilson et al. 2014). Furthermore, by comparing the mass (mmol/L) of Cl/SO_4, it is possible to distinguish among acid mine drainage, oil & gas produced water, and brine treatment plant water (Wilson et al. 2014). Coal mining, both active and abandoned, brine from gas wells, sewage, and power plants all contribute TDS to the Monongahela River. In some tributaries, such as Tenmile Creek (TM)

3RQ Monongahela River Sites

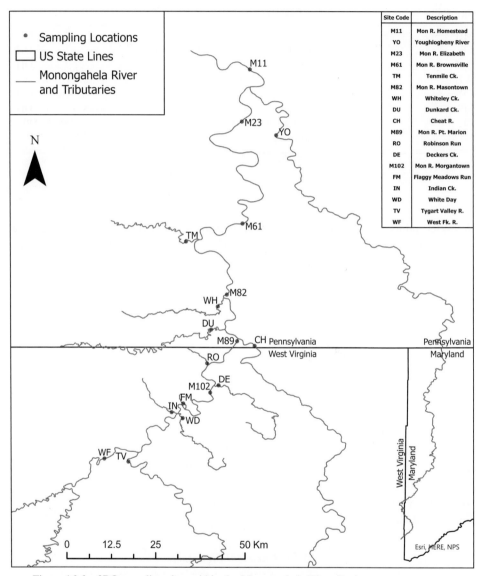

Figure 16.6   3RQ sampling sites within the Monongahela River Basin

and the Youghiogheny River (YO), more than half of the salt is sodium chloride, characteristic of drilling brine. Other tributaries such as Robinson Run (RO), Flaggy Meadows Run (FM), Indian Creek (IN), Dunkard Creek (DU), and West Fork River (WF) have water chemistry signatures more indicative of mine water (Figure 16.9) (Ziemkiewicz 2011).

3RQ Allegheny River Sampling

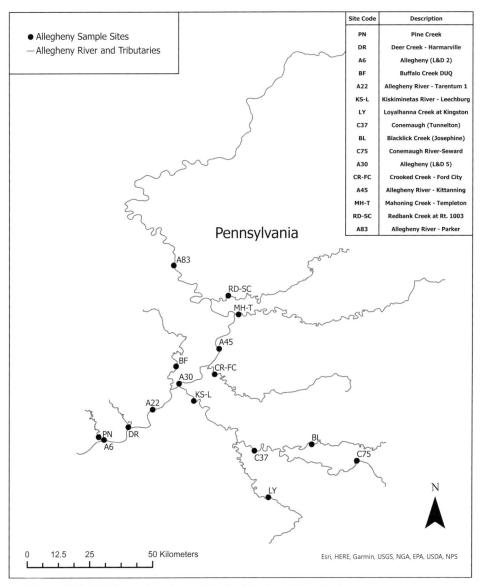

Figure 16.7   3RQ sampling sites within the Allegheny River Basin

16.3.6 Discussion

Throughout the study period of July 2009–March 2021 (Monongahela), January 2013–March 2021 (Allegheny), and January 2013–December 2020 (Ohio), TDS values have not exceeded 500 mg/L at mainstem sampling locations (Figure 16.10).

3RQ Upper Ohio River Sites

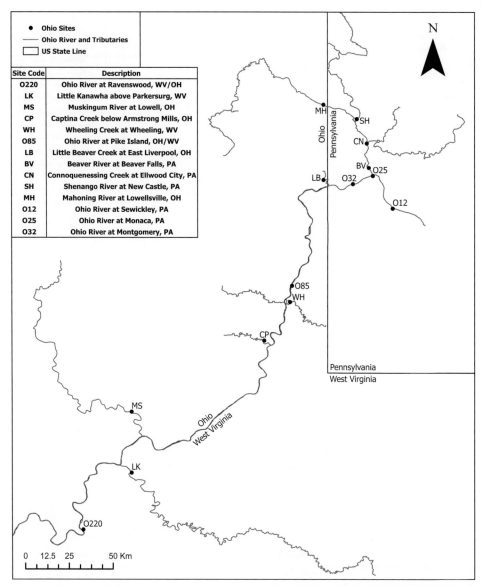

Figure 16.8   3RQ sampling sites within the Ohio River Basin

Long-term data collection provides researchers with the ability to establish trends and monitor for impacts on water quality. Future land-use changes and/or severity of low-flow conditions have the potential to cause dramatic water quality changes (Merriam, et al. 2020).

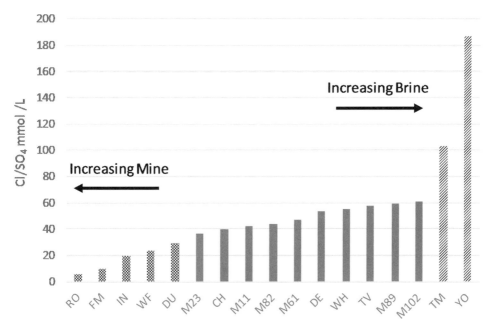

Figure 16.9 The average chloride (Cl) to sulfate (SO_4) ratios in mmol/L between July 2009 and March 2021 in the Monongahela River Basin

TDS data collected throughout the three river systems fluctuate with the season (Figure 16.11). As expected, higher TDS concentrations are seen during late summer/fall when river flow is low. Variability in TDS during low flow conditions is most evident in the Monongahela, where TDS ranges from 56 mg/L to 345 mg/L when flows are below 2,000 cfs (Figure 16.11A). Similarly, Allegheny TDS ranges between 99 mg/L to 342 mg/L when flows are below 6,000 cfs (Figure 16.11B). The more extensive Ohio River system does not show a strong relationship between higher TDS and low flow; TDS ranges between 79 mg/L to 342 mg/L throughout varying flow conditions (Figure 16.11C).

Sulfate levels have dropped in the Monongahela, with levels not exceeding the SWDA limit of 250 mg/L since the start of the discharge management plan. Although sulfate levels approached the SWDA limit during low flows in June 2011 and September 2016 with values of 246 mg/L and 244 mg/L at M61 and M89, respectively, they did not exceed the limit. Concentrations of sulfate increase as flow decreases (Figure 16.12). Values shown above 250 mg/L occurred between July 2009 and December 2010 before implementing the discharge management plan.

The key factor behind the Monongahela River Basin sulfate concentration changes is the implementation of the discharge management plan. The discharge management plan, which began in 2010, is a voluntary, non-regulatory process for controlling TDS from mine discharges discussed earlier in the chapter.

Additional water quality improvements on the Monongahela River were the result of mandatory changes to publicly owned treatment works (POTWs) in Pennsylvania.

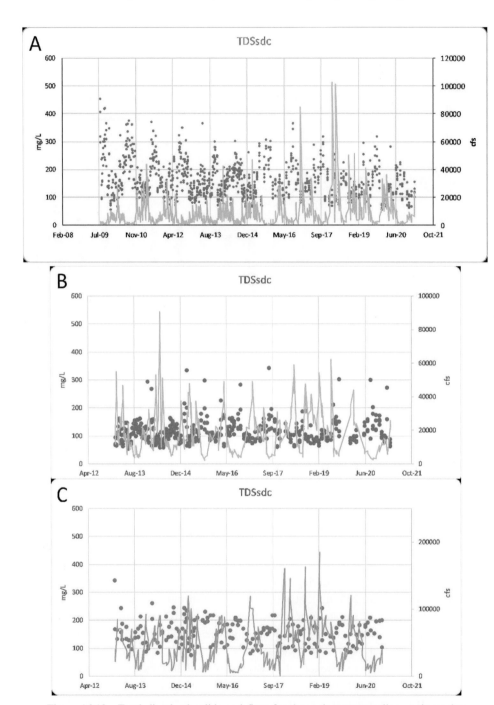

Figure 16.10 Total dissolved solids and flow for the mainstem sampling stations; dots represent TDS (mg/L), solid line represents flow (cfs). (A) Monongahela River from 2009 to 2021 (M11, M23, M61, M82, M89, M102), (B) Allegheny River (A6, A30, A45, A83) from 2013 to 2021, (C) Ohio River from 2013 to 2020 (O12 and O85)

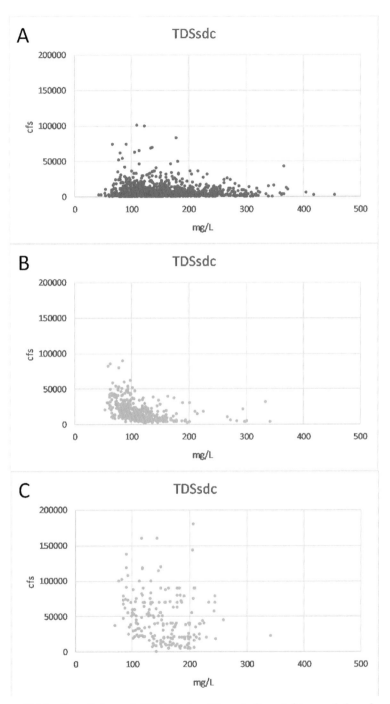

Figure 16.11 Compilation of TDSsdc (mg/L) and flow (cfs) correlation for (A) Monongahela River mainstem sampling locations (M11, M23, M61, M82, M89, M102) from 2009 to 2021, (B) Allegheny River mainstem sampling locations (A6, A30, A45, A83) from 2013 to 2021, (C) Ohio River mainstem sampling locations (O12 and O85) from 2013 to 2020

In April 2011, the Pennsylvania state legislature mandated that POTWs could no longer handle the wastewaters from hydraulic fracturing operations because of previous poor management practices. More specifically, the PADEP instructed 15 POTWs to stop taking flowback fluids from the Marcellus Shale by May 19, 2011 (PADEP 2011). The early wastewater management practices in Pennsylvania exemplify shale gas extraction's potential impacts if not properly regulated. Wastewater treatment facilities could not adequately remove the high levels of TDS found in produced water. Thus, their discharges contributed to elevated TDS levels and bromide in the Monongahela River Basin (US EPA 2016).

Mainstem Allegheny sulfate concentrations (Figure 16.12B) typically remain below 100 mg/L, noticeably lower than the Monongahela, which receives more coal mine drainage. Even during low flows, sulfate levels did not exceed 157 mg/L during the study period (Figure 16.12A). The Ohio River maintains higher flow volumes even during dry periods, resulting in less fluctuation of sulfate (and TDS) concentrations. Seasonal changes are not as dramatic in the larger river systems, with sulfate ranging between 27 mg/L to 110 mg/L (Figure 16.12C).

16.3.7 The Key Role of Volunteer-Based Organizations: The 3RQ REACH Program

The 3RQ REACH program, formally launched in 2015, focuses on outreach activities, engaging volunteer groups, providing technical database assistance, attending local meetings, and facilitating roundtable events. REACH also provided partners with opportunities to disseminate data to audiences such as federal and state agencies and other water-related organizations.

Local volunteer-based organizations enhance the dataset tremendously by providing data on tributaries feeding into the Allegheny, Monongahela, and Ohio rivers. These volunteers collect field data on conductivity, pH, and temperature. Data is stored in a database management system called WATERS and displayed on maps accessible at https://3riversquest.wvu.edu/data/3rq-maps. WATERS provides watershed groups a place to securely store data and a platform for data display via the maps.

Over 25 volunteer groups have contributed data collected from over 1,000 sites to the 3RQ project, including over 50 sites with continuous data loggers. This data provides much-needed information on the health of the headwaters and smaller tributaries in the Ohio River Basin and is valuable in the early detection of a new or worsening water quality issue.

16.3.8 Success through Collaboration

Continued monitoring throughout the Ohio River Basin provides valuable data on the condition of the rivers and an indication of the effectiveness of regulatory and non-regulatory efforts toward healthier waterways. It is the goal of 3RQ to connect our data to meaningful real-world issues, which includes collaborating with other researchers, regulatory agencies, municipal authorities, volunteer-based groups, and others who have

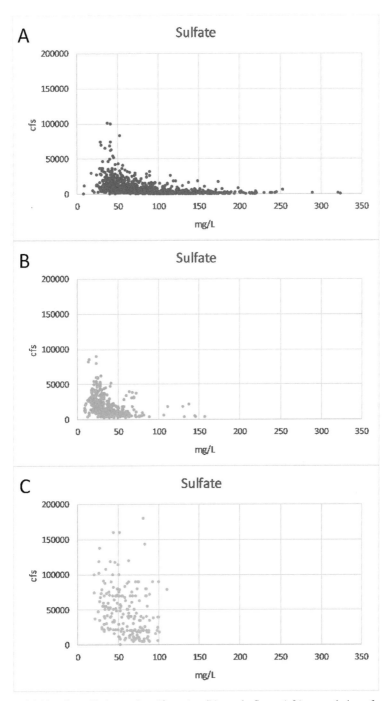

Figure 16.12 Compilation of sulfate (mg/L) and flow (cfs) correlation for (A) Monongahela River mainstem sampling locations (M11, M23, M61, M82, M89, M102) from 2009 to 2021, (B) Allegheny River from mainstem sampling locations (A6, A30, A45, A83) from 2013 to 2021, (C) Ohio River mainstem sampling locations (O12 and O85) from 2013 to 2020

an interest in the health of our rivers. One of the best examples of this is the successful TDS discharge management system on the Monongahela River discussed earlier in this chapter. In addition to the work being completed by 3RQ researchers, some noteworthy applications of 3RQ data include,

- **Data displayed on web-based maps available to the public**: Data is made available to the public via web-based maps on the 3RQ website, www.3riversquest.wvu.edu. Clicking a monitoring site on the map displays a pop-up window with all data recorded for that site. This accessibility has provided detailed data to researchers, state and federal agencies, and the public.
- **Data used to calibrate and validate USACE models**: 3RQ provided over 10 years of TDS data to calibrate and validate models developed by USACE and WVU Forestry. The models characterize long-term spatio-temporal variability in TDS within the Monongahela River Basin and use this information to assess the extent and drivers of vulnerability (Merriam et al. 2020).
- **Delisting of sulfate contamination on the Monongahela River:** In late 2014, the Monongahela River was taken off the PADEP's impaired list for sulfate, in large part thanks to the voluntary discharge management program initiated by WVWRI, which relied on 3RQ data. Additionally, 3RQ data was provided to PADEP to assist in their evaluation of sulfate contamination levels.
- **Development of the *Peters Creek Watershed Assessment and Management Plan***: Peters Creek Watershed Association used data collected through their participation in 3RQ for the creation of the *Peters Creek Watershed Assessment and Management Plan* that was submitted to the Pennsylvania Department of Environmental Protection.
- **Providing data for the Southwestern Pennsylvania Water Quality Network assessments**: Data uploaded to the WATERS database has been central to the Network's aim to assess the condition of the region's surface waters. This will serve as the basis for shaping multi-sectoral water quality improvement efforts in southwestern Pennsylvania.
- **Sharing data with state and federal agencies**: 3RQ shares results with the PADEP, USEPA Region 3, USGS, USEPA HQ, and USACE Pittsburgh district.

3RQ researchers are also involved in several groups aimed toward preserving the Ohio River Basin's health and often give presentations on the 3RQ program at such meetings. These include the Eastern Mine Land Federal Consortium (EMDFC), a group of federal agencies that coordinate federal policy initiatives and research toward basin-wide water quality issues, and the Headwaters Alliance, an informal gathering of organizations set up through Carnegie Mellon University. 3RQ researchers also are involved with the Ohio River Basin Alliance, Ohio River Valley Water Sanitation Commission, the Southwestern PA Water Network, and numerous watershed organizations.

In February 2012, WVWRI was recognized by the National Institutes for Water Resources and awarded a Regional IMPACT award for the 3RQ program. The award recognizes the best research, education, and outreach projects in the nation.

16.4 Targeted Studies

16.4.1 Expanding 3RQ beyond TDS Monitoring: Targeted Studies

In addition to routine water quality monitoring in the Ohio River Basin, 3RQ's "Targeted Studies" have provided the ability for researchers to immediately respond to citizen concerns, develop focused studies, and expand monitoring to parameters other than TDS. Targeted studies have provided reliable information to concerned citizens over potential impacts from shale gas development impacts as well as issues with unknown sources. Additional information on 3RQ targeted studies can be found on the 3RQ website.

16.4.1.1 Kiskiminetas River System

The Kiskiminetas River system, with its major tributaries of Loyalhanna Creek, Blacklick Creek, and the Conemaugh River, has seen major impacts from AMD, making it one of the most impacted watersheds in Pennsylvania. As the system has slowly recovered from this legacy, it experienced new impacts from Marcellus Shale extraction activities, including the disposal of hydraulic fracturing wastewater discharged from publicly owned and industrial water treatment plants (Warner et al. 2013). Water grab samples were collected biweekly from 2013 to 2015 at six sites throughout the system and compared to sites on the Allegheny River mainstem above and below the confluence (Mashuda 2016). Concentration plots of chloride vs. sulfate, chloride vs. bromide, and sulfate/chloride vs. bromide all show site-specific stream clusters that indicate a continuum of pollution impacts from Marcellus and brine treatment water (Blacklick Creek) to AMD signatures of pollution (Conemaugh River). Comparison to historical data shows a reduction in AMD impacts but increases in bromide, chloride, and strontium, indicating the recent impacts from unconventional gas development in the system.

16.4.1.2 Radiological Contaminant Concerns in Tenmile Creek

Targeted studies in Tenmile Creek were initiated from concerns raised by the Izaak Walton League over high radiologicals reported by the PADEP in 2014. 3RQ researchers were able to initiate a two-week study to monitor the Clyde Mine discharge into Tenmile from July 2015 to early August 2015. 3RQ researchers consulted with radiological experts at PACE laboratory to verify appropriate collection and analytical methods. Results from the scientifically sound study showed a maximum concentration of 2.95 pCi/L for radium, while the drinking water limit is 5 pCi/L. The reported values averaged 0.74 pCi/L (WVWRI 2020a). 3RQ reported findings to PADEP, PA Fish and Boat Commission, and US EPA, but researchers could not determine the analytical method utilized for the PADEP samples. During subsequent sampling, PA DEP reported values within a range similar to 3RQ results (PADEP 2011).

16.4.1.3 Elevated Chloride in Pine Creek

Monthly 3RQ sampling showed that Pine Creek, draining 67 square miles of northern Allegheny County, consistently had the highest concentration of chloride of all the 3RQ Allegheny River sites. Therefore, Pine Creek was selected for a targeted study to examine potential sources of contamination (Prettner 2019). Winter road deicing runoff showed acute chloride concentrations reaching 678 mg/L following snowmelt events from December through early April. Chloride continued to be elevated over the remainder of the year and at similar concentrations throughout the system, from the headwaters to the confluence with the Allegheny River, though remaining under the chronic aquatic life criterion of 230 ppm. Samples were taken from residential wells in the watershed (Manley 2017) and compared to adjacent surface waters. Chloride levels in groundwater vary drastically from well to well, but do not appear to be contributing to elevated surface water chlorides. In the surface waters, chloride was the major contributor to the ionic load and its concentration has a direct relationship to specific conductivity ($R^2 = 0.946$) measured with a continuous data logger. Additionally, water samples were analyzed for concentrations of other ions, but no consistent pattern was found that would indicate a common pollution source (Prettner 2019; WVWVRI 2020b). 3RQ continues to work with US EPA and other agencies to monitor Pine Creek and further investigate chloride contamination.

16.4.1.4 Trihalomethanes in Drinking Water

Targeted studies were performed in Greene and Washington counties in Pennsylvania on total trihalomethanes (TTHMs) in response to concerns by the Izaak Walton League's local chapter. Trihalomethanes are a group of organic compounds, including several carcinogenic brominated forms. THMs often occur in drinking water due to chlorine treatment reacting with organic material and, therefore, are also known as "disinfection by-products." Bromide can be an important factor in THM formation, as increased bromide levels have shown increased TTHM formation and a shift toward more brominated THMs (Bird 1979).

In November 2015, 3RQ researchers collected samples from a Southwestern PA Elementary School and the results revealed values that exceeded drinking water limits for TTHMs. Expanded testing included locations upstream of five drinking water systems and at various locations throughout their distribution system. 3RQ performed TTHM studies at several schools and surrounding areas within distribution networks. Water treatment personnel and school superintendents met with the 3RQ team to discuss results from the initial study, which did not show any exceedances of drinking water standards. Low flow conditions in the river system were hypothesized to be a driving factor in the formation of THMs in drinking water systems but required further investigation. An additional study followed, which included sampling at three water authorities and five distribution points from September 2018 to May 2019. The results indicated that the primary constituents in THM formation resulted from the chlorination process (Mirza 2019).

3RQ researchers are currently working to perform more in-depth studies to determine the conditions necessary for THM formation and, most of all, how to reduce THM values below the primary drinking water standard of 80 parts per billion. THM studies have continued to the present day, evaluating treatment systems and looking at ways to provide treatment operators with valuable information to reduce THM formation (WVWRI 2020e).

16.4.1.5 Headwaters of the Allegheny River

In response to shale-gas development in the Allegheny River near Coudersport, Pennsylvania, 3RQ initiated targeted studies in cooperation with the Upper Allegheny Watershed Association (UAWA) and the Potter County Conservation District (PCCD) to develop detailed data on water quality in the headwaters of the Allegheny River around Coudersport in 2016. Shale gas development has generally slowed in the Northern Allegheny River Basin but activity around Coudersport has continued and a groundwater contamination incident occurred in 2015 (PADEP 2011; Stemcosky 2015). The UAWA and the PCCD are monitoring water quality in the Allegheny River and several tributaries around Coudersport with a network of 10 grab sample locations and approximately 20 locations with continuous data loggers. The UAWA and PCCD are actively working with JKLM Energy to monitor water resources as the company moves forward with additional well sites near Coudersport.

16.4.1.6 Pithole Creek

In late 2017, a limited investigation of stream conditions in Pithole Creek, which is located in Venango and Forest counties in the Northern Allegheny River Basin, was completed by 3RQ. Water quality grab samples collected by 3RQ showed elevated levels of parameters associated with oil and gas production (bromide, chloride, sodium) that exceeded almost all other 3RQ sites collected in the Allegheny River Basin since 2013. 3RQ continuous logger data also showed conductivity well above the normal limits for a healthy Allegheny Plateau stream. In 2018, Pithole Creek became a 3RQ targeted study watershed because of the preliminary findings discussed earlier, its history of oil production, current production activity, and brine application on dirt and gravel roads. This provided a research opportunity to shed some light on produced water (brine) impacts on stream integrity. To study these impacts, 3RQ collected grab samples at nine locations and placed continuous data loggers in five locations during 2018. The study found that conductivity in Pithole Creek was higher than that of unimpacted streams in the Appalachian Plateau and the presence of bromide, chloride, and sodium were probably associated with conventional oil and gas production (WVWRI 2020d). 3RQ continues to perform regular sampling at Pine Creek and the other Northern Allegheny sites to monitor water quality.

16.5 Conclusions

Three Rivers Quest is a collaborative effort to collect long-term water quality data in the Monongahela, Allegheny, and Upper Ohio River basins. Resultant data has proven valuable for analyzing long-term trends, identifying pollution sources, and determining remediation options. By working together with industry, academic researchers, and citizen groups, 3RQ has responded to concerns regarding the impacts of the coal and natural gas industries on the Three Rivers Basin. Data shows that TDS and sulfate have decreased in the Monongahela River since the 2010 initiation of a non-regulatory mine discharge management system spearheaded by WVWRI in conjunction with the 3RQ monitoring program. With support from Colcom Foundation and other funding sources, 3RQ intends to maintain a presence in routine water quality monitoring throughout the Upper Ohio River Basin, continue its targeted studies, increase its engagement with federal agencies regarding water quality results and/or concerns, and seek additional collaborations with researchers and watershed groups.

Acknowledgments

The authors would like to thank the other members of the 3RQ consortium, John Stolz, Brady Porter, and Beth Dakin from Duquesne University; James Wood from West Liberty University; Lisa Barreiro, Bruce Dickson, Stanley Kabala, David Saville, and the too many to name undergraduate and graduate students from West Virginia University, Duquesne University, Wheeling Jesuit University, and West Liberty University. We would also like to thank the Colcom Foundation, United States Geological Survey, and West Virginia Water Research Institute for partial support. The contribution is dedicated in memory of Ben Stout (Wheeling Jesuit University) and Carol Zagrocki (Colcom Foundation).

References

Bird James C. (1979). *The Effect of Bromide on Trihalomethane Formation*. Master's thesis, University of Tennessee.
Coal Age. (2013). *CONSOL Energy Installs Advanced Water Treatment Facility in West Virginia*. www.coalage.com/us-news/CONSOL-energy-installs-advanced-water-treatment-facility-in-west-virginia-46950734/
Consent Decree, State of West Virginia through West Virginia Department of Environmental Protection vs CONSOL Energy, Inc.; CONSOLIDATION COAL COMPANY; and WINDSOR COAL COMPANY (2011), at www.epa.gov/sites/production/files/2013-09/documents/consol-cd.pdf
Manley L. (2017). *Analysis of Water Quality in Watersheds of Southwestern Pennsylvania at the Watershed Level*. M.S. Thesis, Duquesne University.
Mashuda E. (2016). *Analysis of the Water Quality of the Kiskiminetas River System and Its Impacts on the Allegheny River, Pennsylvania*. M.S. Thesis, Duquesne University.
Merriam ER, Petty JT, O'Neal M, and Ziemkiewicz PF. (2020). Flow-mediated vulnerability of source waters to elevated TDS in an Appalachian River Basin. *Water 2020*. 12(2): 384.
Mirza Nashid. (2019). *Minimizing Trihalomethane Formation through Source Water Monitoring and Optimizing Treatment Practices*. Graduate Theses, Dissertations, and Problem Reports. 7399. https://researchrepository.wvu.edu/etd/7399
Pennsylvania Department of Environmental Protection (PADEP). (2011). *DEP Calls on Natural Gas Drillers to Stop Giving Treatment Facilities Wastewater*. Press release April 19, 2011.
Prettner Selina. (2019). *A Study of Chloride Levels in Pine Creek, Allegheny County, PA*. Master's thesis, Duquesne University.
Stemcosky K. (2015). *Drilling Operation Leads to Possible Water Contamination in Potter County*. Potter Leader-Enterprise, p. 5. Press Release September 26, 2015.
United States Army Corps of Engineers (USACE). (2009). *Monongahela River Watershed Initial Watershed Assessment*. US Army Corps of Engineers Pittsburgh District.

United States Environmental Protection Agency (USEPA). (2016). *Hydraulic Fracturing for Oil and Gas: Impacts from the Hydraulic Fracturing Water Cycle on Drinking Water Resources in the United States (Executive Summary)*. US Environmental Protection Agency.

Warner NR, Christie CA, Jackson RB, and Vengosh A. (2013). Impacts of shale gas wastewater disposal on water quality in western Pennsylvania. *Environmental Science & Technology*. 47(20): 11849–11857.

West Virginia Water Research Institute (WVWRI). (2020a). *Monongahela: Tenmile Creek*. https://3riversquest.wvu.edu/files/d/31cc31ed-90e3–4289-9742-691116d2bf52/fact sheet_tenmile-pptx.pdf.

West Virginia Water Research Institute (WVWRI). (2020b). *Southern Allegheny: Pine Creek*. https://3riversquest.wvu.edu/files/d/83923a18-de65–4801-b374–6824629bf690/fact sheet_pinecreek-pptx.pdf

West Virginia Water Research Institute (WVWRI). (2020c). *Northern Allegheny: Pithole Creek*. https://3riversquest.wvu.edu/files/d/bbf74ca7–5793-4c38–8b67-cb87ca799cf5/factsheet_pithole-pptx.pdf

West Virginia Water Research Institute (WVWRI). (2020d). *Monongahela: Trihalomethane (THM)*. https://3riversquest.wvu.edu/files/d/bbf74ca7–5793-4c38–8b67-cb87ca799cf5/factsheet_pithole-pptx.pdf

West Virginia Water Research Institute (WVWRI). (2020e). *Ohio: Captina Creek*. https://3riversquest.wvu.edu/files/d/30c1b837–24e3–484a-8d13-e0ba88e19904/captinacreek factsheet-pptx.pdf

Wilson JM, Wang Y, and VanBriesen JM. (2014). Sources of high total dissolved solids to drinking water supply in southwestern Pennsylvania. *Journal of Environmental Engineering*. 140(5). doi: 10.1061/(ASCE)EE.1943-7870.0000733.

Ziemkiewicz P. (2010). *Discharge Management to Control TDS in the Upper Monongahela Basin*. West Virginia Surface Mine Drainage Task Force. https://wvmdtaskforce.files.wordpress.com/2016/01/10-ziemk-tds-water-quality-flow-moni toring-upper.doc

Index

^{12}C, 272–273
^{130}Ba, 261
^{132}Ba, 261
^{134}Ba, 261
^{138}Ba, 261–262, 264
^{13}C, 272–275, 277–278, 281, 285
^{144}Nd, 248
^{147}Sm, 248
^{14}C, 272–273, 277, 285, 288
^{176}Hf, 248
^{176}Lu, 248
^{187}Os, 248
^{187}Re, 248
^{1}H, 272–273
^{223}Ra, 216
^{224}Ra, 216
^{226}Ra, 216–222, 227–229, 233–236, 241, 256
^{228}Ra, 216–221, 228–229, 233, 235, 256
2-butoxyethanol, 160
^{2}H, 272–273, 278, 281
^{3}H, 272–273
40 CFR 192, 222, 227
^{6}Li, 247, 258–259
^{7}Li, 247, 258–259, 261, 264
^{86}Sr, 247–249, 253, 255–256, 258–259, 264–265, 268
^{87}Sr, 247, 249, 253, 255–256, 258–259, 264–265, 268–269

Aarhaus Convention, 82
abandoned mine drainage (AMD), 228, 232, 239, 241, 255, 314–315, 340, 353, 356, 381, 387, 392, 405
Accipiter gentilis, 373, 384
Actinium, 216
Adaptive Environmental Management Regime, (AEMR), 50–51
Ahnet, 5, 10
air quality, 39, 65, 74, 105, 107, 109–110, 114, 123–124, 129, 131
Alabama, 10, 154
Alaska, 10, 16, 154
Alberta, 8, 12, 117, 128, 179, 188–189, 192–194, 197–198, 205–206, 208, 210–213, 383
Algeria, 3, 5, 10, 15, 17
Allegheny County, 106, 108–109, 111, 345, 362, 405
Allegheny River, 345, 391–392, 394, 397, 400–401, 403, 405–407
allelic richness, 351
alpha decay, 180, 214–216, 219–220, 234, 238
Amazonas, 4, 14
Ambloplites rupestris, 347, 350
American Oil and Gas Historical Society, 23
American robins, 371
antimony, 32
Antrim, 20–21, 281, 288, 293, 304, 308–310
Apache, 9, 17
Appalachian Basin, 17, 19, 38, 40–41, 93, 169, 220, 222, 228, 237, 241–243, 246, 251, 255–257, 259–262, 266, 268–269, 271, 289, 338–339, 361–363, 365
Appalachian Fold Belt, 4, 12
Appalachian Mountains Bird Conservation Region, 371, 373
aquatic habitat, 354
aquifer, 29, 58, 60, 154, 168, 239, 245, 255, 258, 264–266, 268
Arab Oil Embargo, 89
Arcobacter halophilus, 346
Arcobacter marinus, 346
Argentina, 3–4, 9, 15, 17
Argonne National Laboratory, 154
Arkansas, 10, 20, 33, 337
Arkoma, 21, 191
arsenic, 228, 230, 268
Arthrobacter, 346
aseismic creep, 186, 188
Austral, 4, 9
Australia, 3, 7, 13, 16–17, 41, 44–45, 47, 52–53, 56–60, 62–65, 67, 69–78, 170
Australian Bureau of Statistics, 53
Australian Energy Market Operator, 52, 71

411

Australian First Peoples, 69
Australian Government Productivity Commission, 49
Australian Petroleum Production and Exploration Association, 16, 50
Automated Ribosomal Intergenic Spacer Analysis (ARISA), 345

Bacillus, 346, 360–361
Bacillus firmus, 346, 360
Bakken, 4, 10, 12, 20–21, 92, 135–136, 147, 152, 169, 257, 268–269, 304, 309
ban, 4, 7, 10, 12–13, 16, 56–57, 69, 83, 86, 88, 91, 94–95, 97
barite, 220–222, 229, 232, 237, 242, 244, 251, 262, 268, 301
barium, 32, 216, 220, 238–240, 242–243, 245–246, 268–270, 301, 330, 333, 356, 373
Barnett, 20–21, 23, 33, 41, 114, 116, 122, 125, 128, 135–136, 146, 154, 160, 281, 293, 300–302, 304, 308, 310, 345, 360, 362
basins, 3, 8–10, 12–14, 20, 44, 114, 116–117, 119, 123, 135, 154, 174, 182, 192, 223–224, 235, 264, 266, 281, 341, 358, 374, 386–392, 394–399, 402, 404, 407
Bates Fork, 342–348, 350–358
bats, 374
Bazhenov, 5, 14
Beaver, 37–38, 41, 96, 109, 111
Becquerel, 214, 218
Beetaloo, 8, 13, 44
beneficial use, 49, 70, 77, 226, 228–229, 231, 239, 241
benzene, 68, 107, 115–117, 119, 122, 127, 129, 160
beta decay, 214–216, 234
bioaccumulation, 229, 356
biocide, 32, 292, 299–300, 304–307, 309, 340, 358
biofouling, 292, 299, 301–302, 305–306
biogenic methane, 272, 274, 278, 285, 287
biogenic sulfide, 302–303, 306, 309
biomass burning, 139
birds, 368, 371, 381
black carbon, 111–112
Blacklick Creek, 228, 394, 397, 405
Blackpool, 85–86
Blacksville No. 2, 389
blenders, 30–31
blow out preventer, 29
bluntnose minnow, 350
boron, 230, 258, 270, 314, 356
Bowen, 44, 58, 60, 71, 281, 379, 381
Bradford County, 320, 322–323, 333
Brazil, 3–4, 14–19
breaker, 32
bridge fuel, 132
Briggs vs Southwestern Energy Production Co, 24
brine trucks, 31
British Columbia, 12, 179, 189, 194–195, 205, 208–209, 213

British Gas, 47–48
British Geological Survey, 85
broad-winged hawks, 373
bromide, 158, 163–169, 239, 330–331, 335, 344–346, 356, 402, 405–407
brook trout, 375
brown-headed cowbird, 367
Browns Creek, 342, 344
Brown, James Gordon, 84
BTEX, 63, 107, 119, 122, 125, 160
Bureau of Forestry, 366, 376
Burgos, 4, 12, 226, 228, 237, 241, 243, 256, 269, 313, 336, 339, 370, 379
Bush, George Herbert Walker, 89
Bush, George W, 89
butane, 24, 33, 35, 38, 114–115
Buteo platypterus, 373, 381
Butler, 111
Byerlee's law, 185

calcium, 32, 158, 166, 216, 220, 229, 246, 256–257, 356
California, 10, 12, 106, 124, 127, 147, 149, 154, 177, 189, 207, 209, 222, 225, 228–229, 240, 266, 275, 289–290, 342
Cambrian, 6, 13, 21, 191, 245
Cambrian-Ordovician Arbuckle Group, 191
Canada, 3–4, 12, 16, 18–19, 88, 102, 128, 134, 160, 168, 170, 174, 189, 191, 193–195, 203–209, 212, 296, 300
Canadian Shield, 258, 265
Canning, 8, 13, 44
carbon dioxide, 23, 38, 53, 63, 78, 98, 132–133, 141–142, 273, 281, 288
carbon monoxide (CO), 65, 105–107
carbon-12, 139
carbon-13, 133, 139–141
carbon-14 radiocarbon dating, 133
Carter, Jimmy, 89
casing, 29–30, 255, 313
cation exchange, 219–220, 222, 243
Catostomidae, 347–348
cellar, 25, 29
centralized waste treatment, 162, 222
Centrarchidae, 347–348
ceramic proppant, 32, 340
Chaco-Parana, 14
Chad, 3, 6
change in fault loading conditions, 186, 188–189
Chapter 32 Title 58 Oil and Gas Act, 25
Cheat River, 394
Cheney, Dick, 19
Chevron, 9, 17, 63
China, 3, 6, 8, 16, 18, 47, 90, 174, 258, 267–268, 296, 303, 310, 314, 362
Chinchilla, 61, 68
chipping sparrows, 371

chloride, 158, 164–166, 170–171, 220, 222, 224, 233, 246, 302, 304, 315, 344–347, 356–357, 362, 395, 399, 405–406
Class II injection well (UIC), 32, 225
Clean Air Act, 89, 105, 125
Clean Water Act, 89, 227, 244, 389
Clinton, Bill, 89
Clinton, Hilary, 90
coal, 14, 25, 29, 32, 44, 47–48, 51, 53, 56, 58, 60–62, 68, 71–75, 77–79, 93, 106, 114–115, 123, 132–133, 138, 141–142, 144–145, 228, 237, 246, 255, 265–266, 273, 278, 281, 285, 287, 290, 336–338, 340–341, 345, 387, 395, 402, 407
coal bed methane, 44
Cobetia marina, 346
coefficient of static friction, 183, 201
Colorado Oil and Gas Conservation Commission (COGCC), 25
Collingham, 6, 13
Colorado, 4, 10, 20, 25, 82, 90, 110, 114–116, 119, 121–122, 125–126, 129, 154, 160, 173, 186, 210, 238, 260, 278, 289–290, 304
completion, 9, 30, 39, 50, 79, 84, 90, 138, 174, 192, 197, 246
compressional waves, 177
compressor stations, 37, 58, 64, 110, 117, 121, 131, 364, 367–368, 378
Compton Effect, 215
compulsory integration, 24
condensate tanks, 29, 31, 117
conditional use permit, 25
conductor hole, 29
Conemaugh River, 256, 394, 397, 405
ConocoPhillips, 47
CONSOL Energy, 389–391
contamination, 61, 68, 242, 249, 269, 335, 338
control center, 30–31
conventional reserves, 10, 19
Cooper-Eromanga, 44
COP21, 132, 142, 144
coprecipitation, 220, 228
Corbett, 95–96
Cordova Embayment, 12
core forest, 363, 365–366, 373, 377
corrosion inhibitor, 32
Council of Australian Governance (GOAG), 48, 56, 72
COVID 19 pandemic, 70
creek chub, 350
Cretaceous, 6, 9–10, 12–13, 21, 308
crosslinker, 32
crude, 19–20, 90
cryogenic plants, 38
coal seam gas (CSG), 44–45, 47, 49–51, 53, 56–58, 60–63, 65, 68, 71–72, 76
Cuadrilla Resources Ltd, 60, 73, 84–86, 101
cube fracking, 30

Curie, Marie, 214, 218
Curvibacter, 346, 359
Centralized Waste Treatment (CWT), 162, 166, 228
Cyprinidae, 347–348

Dallas, 10, 90–91, 101, 106, 189
Darling Downs, 64–65, 75
darters, 347–348, 357
David Cameron, 84
debutanizer, 38
deep well injection, 4, 12, 94–95, 100, 161, 167, 307
deethanizer, 38
deicing, 166, 168, 230–231, 360
deisobutanizer, 38
Delaware River Basin Commission, 12, 24
denitrification, 160
Denton, TX, 10, 91, 101
Denver-Julesburg Basin, 114–115, 117–119, 121, 127, 130, 160, 289
depropanizer, 38
derrick, 29
Desulfosporosinus, 302
Desulfovibrio, 302
deviatoric stress, 183
Devon Energy, 82
Devonian, 9–10, 12, 14, 20–21, 40–42, 191–192, 194, 212, 237, 243, 249–251, 253, 255–256, 259–260, 262, 264–265, 268–269, 288–289, 309, 338
diethanolamine, 38
Diethylene glycol, 38
dip, 177, 191, 354
direct pore-pressure effect, 186
directional drilling, 19–20, 23, 89, 134
dislocation theory, 176
displacement, 40, 176–177
dissolved oxygen, 344, 357
Doctors for the Environment, 49, 65, 72, 74
downstream, 23, 33, 39, 134, 137, 139–141, 144, 163, 166, 227–229, 236, 241, 252, 341–342, 353, 357, 370
drilling log, 30, 32
drilling mud, 62, 221, 246, 263, 301, 303, 310, 362
drilling waste, 25, 30
drip gas, 24, 33, 38
Duganella, 346, 361
Duncan, OK, 23
Dunkard Creek, 389, 394, 396
dust suppression, 62, 166, 226, 230
Duvernay Formation, 192, 209, 211

Etheostoma caeruleum, 347
Etheostoma flabellare, 347, 350
Etheostoma nigrum, 347, 351, 353
E.A.L. Roberts, 20
Eagle Ford, 4, 10, 12, 20–21, 90, 115, 129, 135–136, 152–154, 293, 309

earthquake, 9, 12, 73, 81, 84, 87, 91, 94, 97, 162, 169, 171, 173–174, 176–178, 180, 185–186, 188–189, 191–196, 200, 203–213
East Siberia, 14
eastern bluebirds, 368
Eastern Mine Land Federal Consortium, 404
Eclipse Resources, 29
ecoregion, 347, 363
effective normal stress, 180
effective stress, 173, 182, 204
electrofishing, 342, 347, 349
electrometer, 214
electrons, 214, 272
Ellenburger Formation, 191
Emmanuel Macron, 83
Energy Policy Act of 2005, 19, 39, 43, 89
Enterobacter, 346
Environmental Defense Fund, 137–138
Environmental Protection Act, 48, 52, 77
EOG Resources, 9
epicenter, 176–177, 374
epicentral distance, 177
Epsilon, 7, 13, 16
Ericymba buccata, 348, 350
Ernest Rutherford, 214
erosion, 25, 68, 363, 369–370
Erwinia, 346
Escherichia, 346, 360
estimated ultimate recovery, 3, 33
ethane, 19, 24, 33, 37–39, 42, 96, 116–117, 119, 147, 149, 274, 281, 287
ethane cracker, 37–39, 96
Etheostoma blennioides, 347–348
Etheostoma flabellare, 348, 351, 353
Etheostoma nigrum, 348, 350
ethyl benzene, 107
ethylene, 32, 38, 301
ethylene glycol, 32, 301
Estimated Ultimate Recovery (EUR), 3, 33, 42
European Union, 3
Eurycea bislineata, 354
Eurycea longicauda, 354
evaporation ponds, 61–62
exploding torpedo, 21
ExxonMobil, 9, 38, 63

fantail darter, 350–352
fault plane, 177, 185
fault trace, 177
fault-slip potential, 174, 198
Fayetteville, 20–21, 33, 147, 373
Federal Energy Regulatory Commission, 37, 89
fermentation, 273, 281, 286, 293, 299, 301
fish kill, 386
flare, 92, 133
FLIR camera, 117
flowback, 32, 93, 117, 138, 154, 169–170, 174, 219, 221, 236–239, 242, 246, 255–256, 258, 265, 267–268, 271, 292, 294, 300, 302, 305–306, 309–310, 315, 336–337, 339–340, 345, 360, 370, 389–391, 402
focal mechanism, 177, 180
Fonner Run, 342–348, 350–358
forced pooling, 24
forest, 57, 69, 342, 346, 355, 361, 363, 365–374, 377–383
forest salamanders, 374
formation water, 30, 62, 152, 191, 220, 240, 255–256, 265, 267, 340
FracFocus Chemical Disclosure Registry, 32, 40
fracking, 10, 14–18, 20, 29–30, 33, 39–40, 43, 56–57, 62, 73, 75–76, 78–79, 81–91, 93–97, 117, 125, 134, 152, 168, 170–171, 231, 268, 281, 284–285, 313, 340, 356–358, 367, 369–370, 373, 383
fragmentation, 363, 365–367, 374–375, 378, 380–384
France, 3, 5, 15, 81–83, 87, 96, 98–100, 188, 203, 270, 289
Frasnian shale, 10
freshwater mussels, 256, 375
friction reducer, 32, 340
Friends of the Earth, 82
Front Range, 115, 125, 129, 278, 290
fugitive emissions, 53, 58, 64–65, 72, 119, 147
Fuling gas field, 258

gadolinium, 32
gamma rays, 214–215, 233
gamma spectroscopy, 233–235
Gasfields Commission, 48, 53, 73, 77
gas-in-place, 3
Gasland, 82, 100, 281
gathering pipelines, 23–24, 30
gelling agents, 32
genetic diversity, 351–352, 358
Georgina, 8, 13, 44, 76
Germany, 5, 16, 83, 87, 96, 213
Ghadames/Berkine, 5, 10
global warming, 53, 65, 132, 142, 145–146
glutaraldehyde, 32, 300–301
Goldwyn, 13
Golfo San Jorge, 9
governance, 79, 150, 212
green sunfish, 350
greenhouse gas emissions, 63–64, 71, 98, 132, 140, 144
Greenpeace, 82, 98
greenside darter, 347
groundwater, 29, 42, 50, 59–60, 68, 81, 85, 140, 148, 152, 154, 166, 191, 229, 246–247, 249, 251, 253–255, 258, 261, 267–269, 271–272, 280–282, 284, 287, 290, 313–314, 318, 320, 322, 324, 333, 337, 339, 370
guar gum, 32, 301, 303
Gunnedah, 44
Gutenberg-Richter (G-R) formula, 177

H. organivorans, 346
habitat loss, 363, 374
Halanaerobium, 294, 298–301, 303–307, 309
Halliburton, 19, 23, 41
Halliburton Loophole, 19
Halomonadaceae, 294
Halomonas taenensis, 346
Halothermothrix, 346, 359
Haynesville, 10, 20–21, 33, 90, 135–136, 147, 154, 293
hazardous air pollutants (HAPs), 105–106, 122
helium nuclei, 214
herpetofauna, 367
high energy photons, 214–215
horizontal stress, 180
Horn River, 4, 12
Horton Bluff, 4, 12
Hugoton, KS, 21
hydraulic fracturing (HF), 3, 10, 13, 19–20, 29–31, 33, 39, 42, 56–58, 83–84, 86, 89, 91, 93, 109–110, 114, 123, 131, 134, 136, 140, 150, 154, 157, 160, 163, 167, 169–171, 173–175, 178, 185–186, 188–189, 191–192, 194, 196–198, 201, 204–212, 221, 230–232, 236, 239–240, 242–244, 246, 255–258, 260, 262, 265–267, 269–271, 281, 287, 289, 292–294, 296, 298–302, 305, 308–310, 313, 320–321, 324, 337–340, 359, 361–363, 373, 380, 382, 395, 402, 405
hydrogen, 38, 69, 272, 274–275, 285, 290
hydrogen sulfide, 38
hydrologic cycle, 150, 154, 314, 328
hypocenter, 176, 180
ICP-MS, 336, 356, 359

Idiomarina, 346
Illizi, 5, 10
Impact Fee, 96
impoundments, 24–25, 309, 358, 364
Index of Biotic Integrity, 347
Indiana, 10, 95, 99, 146, 374
Indiana bat, 374
Induced seismicity, 173–174, 206–207, 210
injection-induced seismicity, 173–174, 225
intermediate casing, 29
intermediate string, 29
International Permanent Peoples' Tribunal, 67
Invasive, 368, 383
iridium, 32
iron control, 32
iron roughneck, 29

Jianghan, 6, 8
Johnny darter, 348, 350–352
Johnson, Boris, 86–87
Junggar, 6, 8
Jurassic, 5, 9–10, 12, 14, 21
Jurua Valley, 14

Kangan gas field, 258
Kansas, 10, 154, 186, 189, 191, 210

Karoo, 6, 13, 15
Keeling plot, 277
kick off point, 26, 29
Kiskiminetas, 394, 397, 405
Kyoto Protocol, 142

Lake Eyre Basin Wild Rivers Declaration, 49
Lancashire, 60, 84–86, 101, 206
Land Access Code, 51
Land Access Framework, 48
landfill, 24, 61, 221–222, 231, 278–279, 282
landing point, 29
Lepomis cyanellus, 348, 350
Liard, 4, 12
Libya, 3, 6
liquid nitrogen, 23
liquid scintillation, 234
liquified natural gas, 19, 38
Liquified Natural Gas (LNG), 19, 38, 42, 44, 46–48, 52–53, 57, 63, 70–73, 75–76, 78
lithium, 243, 246, 258, 266, 268–270, 314, 356
Little Kanawha, 395, 398
Lone Pine, 346
Long-tailed Salamander, 354
Louisiana, 10, 20, 33, 90, 154, 308, 356, 360, 362, 373, 381, 385
Louisiana Waterthrush, 356, 360, 362, 373, 381, 385
Loyalhanna Creek, 394, 397, 405
Lycoming County, 322, 338, 366

Manitoba, 12, 92
Mann-Whitney U test, 112–113
Marcellus, 10, 20, 24, 29, 33, 38, 40, 42, 93, 95, 100–101, 106, 109–111, 114, 116–119, 123, 126, 129–131, 135–136, 138, 147, 152–154, 157–158, 160–163, 168–169, 171, 220–221, 234, 236–241, 243, 245, 249–251, 253, 255–258, 260, 262, 264–270, 281, 287–288, 293, 300, 302, 304–305, 308–310, 314, 330–331, 336–342, 344–346, 350–351, 353, 355–361, 363–364, 368, 370–375, 378–384, 402, 405
mass ratio, 336, 359
Massilia, 346, 362
mast, 29
Material Safety Data Sheet, 32
Mauna Loa Observatory, 273
MC-ICPMS, 247, 249, 262
Maximum Contaminant Level (MCL), 314
Melbourne Energy Institute, 64, 74
mental health, 59, 75
Mercer County, 322
methane (CH_4), 23, 33, 38–39, 42, 47, 53, 64, 74, 82, 89–90, 92, 96–98, 115–119, 123–124, 126, 129–130, 132–142, 144–149, 246, 255, 266, 272–273, 275, 277–278, 280–281, 284–285, 287–290, 309, 337
methane emissions, 64, 74, 89–90, 96, 98, 112, 117, 119, 124, 132–134, 136–140, 142, 144–149, 273, 278, 287, 289

methanogen, 307
methanogenic, 293, 310
Methanohalophilus, 294, 298–300, 302, 307
Mexico, 3–4, 10, 12, 15, 17–19, 90, 116, 154, 160, 289, 308
Michigan, 10, 20–21, 154, 230, 288, 308–309
microbiome, 345–346, 358
Micrococcus, 346, 359
Microsatellites, 350–351
Microstegium vimineum, 376
midstream, 23, 33, 38–39
mineral rights, 23, 87–88, 91, 96
Mines Legislation (Streamlining) Amendment Bill 2012, 48
Mississippi, 10, 154, 189
Mississippian, 12, 20–21, 196, 255, 268
Mitchell, George, 23
Mohr diagram, 183
Mohr-Coulomb criterion, 183, 199
Molothrus ater, 367
monoethanolamine, 38
Monongahela River, 27, 94, 101, 164–165, 171, 339, 341, 350, 356, 360, 386–387, 389–392, 394–396, 399, 400–404, 407
Montana, 10, 21, 92, 257, 265, 268, 281
Montney, 189, 194, 196–198, 205–206, 209, 213
Montney Formation, 194, 205
Montney play, 194, 197–198, 205, 209
moratorium, 7, 12–13, 15, 49, 56–57, 84, 86, 95–96
Mount Wilson Observatory, 275
mouse hole, 29
Mouydir, 5, 10
Multiple Land Use Framework, 48, 72
Murteree, 13
Muskingum, 395, 398
Muskwa, 4, 12
Myotis septentrionalis, 374
Myotis sodalis, 374

N,N, Dimethyl formamide, 32
NAAQS, 105, 107, 113–114, 123
Nafe-Drake relationship, 182
naphthalene, 160
Nappamerri, 7, 13
National Ambient Air Quality Standards, 105
National Energy Technology Laboratory, 19, 43, 264, 266, 360
National Pollutant Discharge Elimination System (NPDES), 227–229
National Pollutant Inventory, 65, 76
natural fractures, 30, 186
natural gas, 10, 15, 17, 19–20, 37–41, 43–44, 47, 53, 63–64, 78–79, 82, 84–86, 89–90, 92–93, 95–96, 109, 112, 114–117, 119, 123–126, 129–135, 137–142, 144–148, 156–157, 160, 167, 170–171, 236–237, 240, 246, 265–266, 278, 281, 285, 287, 289, 292–293, 300, 302, 306, 308–310, 313, 336, 340, 353, 361–362, 364, 367, 377–378, 380, 382, 384, 407
Natural Gas Act of 1938, 38
natural gas liquids (NGL), 38, 93, 96, 246
Natural Resources Defense Council, 150
naturally occurring radioactive materials, (NORM), 30, 62, 158, 161, 166, 170, 215–216, 219, 221–223, 225–226, 229–231, 233–235, 238, 241, 246, 313, 338
Nature Conservation Act 1992, 49
Nebraska, 10
NETL, 19, 39, 41, 43, 261, 267, 270, 360
Neuquén, 9
Nevada, 10
New South Wales (NSW), 13, 44, 55–57, 59, 65–67, 71–72, 76
New York, 10, 16–17, 20, 30, 42, 73, 79, 88, 90, 93, 99–102, 115, 134, 140, 142, 145–146, 152, 154, 168, 170, 205, 209, 230, 243–244, 267, 328, 334, 337, 361, 382
Niobrara, 10, 20–21, 219
nitrate, 107, 113, 123, 344, 356
nitroglycerine, 21
NO_2, 105–107, 109–110, 114, 123–124
nondisclosure agreement, 32
North Dakota, 10, 20, 25, 89–90, 92, 95, 98, 101, 116, 147, 152, 154, 170, 228–230, 237, 240, 257, 267–268, 382
North Dakota Industrial Commission, 92
northern goshawks, 373
Northern long-eared bat, 374
Northern Territory, 13, 44, 58, 60, 64, 67, 69–70, 73, 75–76
Northwest Territories, 12
Norway, 19
Nova Scotia, 12

Obama, Barack, 89, 91, 95
Office of Fossil Energy, 89
Office of Groundwater Impact Assessment, 59
Ohio, 10, 20, 25, 29, 35, 37–38, 90, 93–95, 97, 99–100, 152–154, 166, 207, 212, 222, 262–263, 281, 285, 287, 290, 347, 349–350, 361, 365
Ohio Department of Natural Resources (ODNR), 25, 94
Ohio River, 152, 262–263, 388, 390–395, 398–405, 407, 408
oil, 3–4, 8–10, 12–17, 19–20, 23, 30, 32–34, 38–42, 49, 53, 55, 57, 63–65, 68, 72–74, 79, 81–83, 87–97, 105, 107, 109–110, 112–117, 119, 122–132, 135, 138–139, 141–142, 145–147, 149–154, 158, 160–161, 164, 166–171, 173–174, 178, 189, 191–192, 194, 197, 204, 206–207, 210, 216, 219–221, 223, 225, 230, 233–235, 237–239, 241–244, 246–247, 249, 254–259, 261–262, 264, 266–270, 272, 274, 278–282, 287, 289, 301–302,

307, 309, 313–318, 321, 324, 336–337, 346, 359, 362–363, 365, 370, 378–379, 381–383, 395, 407
Oil-in-place, 3
Oklahoma, 10, 20, 90, 152, 154, 169, 171, 174, 179, 186, 188–192, 197–198, 205–212, 225, 266, 360
Oklahoma Corporation Commission (OCC), 191, 197, 210
Olympus Energy, 29
Ontario, 12
ordinance, 10
Ordovician, 6, 10, 12–13, 21, 191, 255
Orenia, 298
Otter Park, 4, 12
ozone, (O_3), 65, 105–107, 110, 114–115, 121, 123–124, 126, 129–130
Osborne, George, 85

P waves, 177, 180
Pacific Institute, 150
Pacific Northwest National Laboratory, 23
pad preparation, 23, 39
Pair Production, 215
Pantoea, 346
Paraná, 9, 14
Parecis, 14
Paris Basin, 5, 82, 268
Parkesia motacilla, 356, 360, 362, 373
Parnaiba, 14
particulate matter, 105, 125, 130, 230
Patchawarra, 7, 13
Pawnee earthquake, 174
Penn State University, 376
Pennsylvania, 10, 20, 23, 25–26, 29, 31–33, 37–42, 79, 82, 88, 90, 93–102, 108–109, 111, 114, 116, 131, 138, 146–147, 152–154, 156–158, 162, 164, 167–169, 171, 221–222, 225, 227–228, 230, 232, 237–240, 242–245, 249–250, 252–253, 255–257, 263, 265–267, 269–270, 281, 287, 290, 307, 313, 315–318, 320–321, 328, 332, 334, 336–342, 344–345, 353, 355, 358, 361–366, 368–371, 374–384, 389–392, 399, 402, 404–407
Pennsylvania Department of Conservation and Natural Resources (PA DCNR), 363–366, 368–370, 375–377, 380
Pennsylvania Department of Environmental Protection (PADEP), 156, 317–318, 320, 325, 332, 334, 340, 361, 386, 389, 402, 404–405, 407
Pennsylvania Supreme Court, 24
Percidae, 347–348, 362
Percina maculata, 347–348
perfing, 30
perforating gun, 30
perforation record, 32
Permian, 5–6, 10, 13, 19–20, 90, 135–136, 160, 194, 205, 261, 268
Permian Basin, 10, 21, 90, 135, 160, 261, 268
permit, x, 13, 25, 38, 83, 93, 95, 197, 316, 337
Perth, 8, 13, 44, 56, 78

Peters Creek Watershed Association, 404
Petroleum and Gas (Production and Safety) Act, 48–49
Petroleum and Gas Inspectorate, 68, 76
Photoelectric Effect, 215
pigging station, 37
Pimephales notatus, 348, 350
Pimienta, 12
Pine Creek, 345, 362, 394, 397, 405–407
pipeline, 25, 28, 30, 34, 37, 39, 81, 92, 229, 304, 309, 365–366, 368, 376, 378
Pipeline and Hazardous Materials Safety Administration, 37
Pithole Creek, 407
Pittsburgh, 10, 15, 100, 108–109, 111–112, 131, 171, 290, 337, 341, 386, 393, 404
play, 3, 13, 20, 40, 80, 97, 106, 152, 154, 160, 180, 189, 191–192, 194, 197, 210–211, 213, 237, 255, 270–271, 341, 363, 374–375, 382
Plethodon cinereus, 374
Plethodon electromorphus, 374
Plethodon glutinosus, 354
Plethodon richmondi, 374
Plethodon wehrlei, 374
Plethodontidae, 359, 374
Pluspetrol SA, 9
PM_{10}, 65, 105
$PM_{2.5}$, 65, 105–107, 110–111, 113–115, 123
Poland, 3, 5, 17, 212, 221, 239
polonium, 214
polycyclic aromatic hydrocarbons (PAHs), 111–112, 160
polyethylene, 38
pore pressure, 94, 173, 182–183, 185–186, 188, 199, 204, 210
poroelasticity, 180, 182–183
porosity, 3, 20, 182, 191, 196
Potiguar, 14
Potter County Conservation District, 407
POTWs, 162–164, 389–391, 399, 402
Powder River Basin, 255, 265–266
Prairie Pothole, 258, 269
precautionary principle, 48, 80
predrill test, 25
Preese Hall, 84
pressure pumps, 30–31
Prince Albert, 6, 13
Prince Edward Island, 12
produced water, 31, 59, 61, 64, 93, 152, 154–155, 157–158, 160–164, 166–167, 170–172, 191, 197, 205, 218–222, 224, 226, 228–243, 245–246, 249–251, 253–260, 262–264, 266, 269, 300, 305, 308–310, 313, 315, 337, 339–340, 345–346, 358, 395, 402, 407
production casing, 29–30
production hole, 29
propane, 24, 33, 35, 38, 114–116, 281
proppant, 23, 27, 30–33, 221, 231, 292, 340, 370
Pseudomonas, 304, 309, 359
PTT Global Petrochemical, 38

public drinking water standards, 166
publicly owned treatment works, 162, 164, 326, 399
pyrogenic, 273

Qaidam Basin, 258
Quebec, 12
Queensland, 13, 44–45, 47–53, 55–56, 58–59, 61–65, 67–70, 72–78
Queensland Gas Company, 47
Queensland Gas Scheme, 48

radiative forcing, 132
radioactive decay, 161, 214, 216–217, 247–248
radiobarite, 220, 228, 232
radiocelestite, 220, 232
radium, 30, 161, 166, 214–222, 224, 227–235, 238–241, 243–245, 267, 314, 405
radium-226, 161, 216–217, 222, 224
radium-228, 161, 166, 216
Radium-228, 216
radon, 30, 62, 78, 215–217, 219
radon-222, 216
Radon-222, 216
Rail Road Commission, 25
rainbow darter, 347, 362
rake, 177
Rangely conventional oil field, 173
rat hole, 29
ratio space, 315, 320–321, 324–325, 327, 333
Reagan, Ronald, 89
reclamation, 363, 365, 375–376, 378, 380
Reconcavo, 14
Red Bank, 394
red-backed salamanders, 374
Reggane, 6, 10
Rendell, Ed, 93, 95
renewable energy, 132
resin-coated sand, 32
Resource Conservation and Recovery Act, 89
reverse osmosis, 61–63, 389
Rhinichthys obtusus, 348, 350
Rhodococcus, 304
rig, 29
rigidity, 177
riparian habitat, 373
riparian trees, 357
Risked OIP/GIP, 3
risk-perception paradigm, 79–81, 83, 87–88, 90, 93, 95, 97
roads, 10, 111, 166, 171, 226, 230, 270, 364–367, 369–370, 375, 378–379, 382
rock bass, 347, 350
Rocky Mountain Arsenal, 173
Röntgen, Wilhelm, 214
Roseneath, 7, 13, 16
royalties, 23, 44, 55, 87–88, 97
rupture, 176–177, 185, 189, 200–201
Russia, 3, 5, 14, 17–18, 79, 99

S waves, 177
Sabinas, 4, 12
Safe Drinking Water Act, 19, 39, 89, 314
salamander, 341, 354–355, 357–358, 361–362, 379, 382
Salinovibrio costicola, 346
Salvelinus fontinalis, 375
Sao Francisco, 14
Nicolas Sarkozy, 81, 83, 99
Saskatchewan, 12, 92, 257
Saudi Arabia, 79
scalar seismic moment, 177
scale inhibitor, 32, 301
Schuepbach Energy, 82
Scotland, 86
secular equilibrium, 217
seismic survey, 25, 84
Selexol, 38
Semotilus atromaculatus, 348, 350
Senate Select Committee on Unconventional Gas Mining, 55, 60, 66, 72, 75, 78
separators, 29, 31, 38, 310
Sergipe-Alagoas, 14
severance tax, 96
shale gas, 3, 8, 10, 13, 15, 17, 19–20, 32, 34, 41–42, 44–45, 58, 64, 81–89, 93, 95–97, 109–110, 112, 116, 124–125, 132–134, 136–139, 141, 144–146, 148, 150, 152–153, 160–161, 164, 169, 171, 206, 239, 243–246, 248, 259, 262, 265, 267–271, 274, 281, 284–285, 287–288, 292–293, 296, 299–300, 302–310, 313, 316, 336–339, 341, 356, 358, 360–363, 365–366, 368, 370–371, 373–375, 377–382, 384, 402, 405, 407
Shale Network, 316, 318–319, 324, 326, 333–334, 336
shale oil, 8–9, 153
shear modulus, 177, 201
shear stress, 183, 188
shear waves, 177
Shell Chemical, 38
Shewanella, 303, 307, 310
Shigella, 346, 361
Sialia sialis, 368
Sichuan, 6, 8, 174, 209, 258, 268, 303, 310
Silurian, 5–6, 10, 257
silverjaw minnow, 350
slick water, 19, 32, 39
Slimy salamander, 354
SO_2, 105–107
Society of Petroleum Engineers, 3, 17, 40–42, 267, 336, 359, 361
sodium, 158, 166, 222, 230, 246, 356
Solimões, 14
Songliao, 6, 8
South Africa, 3, 6, 13, 16–18
South Australia, 13, 44, 58
specific conductivity, 344–345, 356, 406
Spizella passerina, 371
split estate, 23

SPUD, 342
stable isotope, 216, 246, 256, 266, 269, 275, 277
Stanolind, 23
StimuFrac™, 23
storage tanks, 35, 133, 310
STORET database, 316
Strategic Cropping Land Act 2011, 49
strike, 177, 180, 183, 189, 191–192, 194, 196, 200
strontianite, 229
strontium, 32, 166, 216, 220, 243–246, 248, 253, 265–266, 269, 314, 336, 356, 373, 405
Subei, 6, 8
Sulfate, 239, 241–242, 330, 386–387, 389, 391, 395, 399, 402, 404–405, 407
sulfate-reducing bacteria (SRB), 228, 300–303
sulfidogenic, 300–301, 303, 305, 308
Superfund, 89
Surat, 44, 47, 59–60, 65, 70, 74, 76–77
surface casing, 29
surface rights, 23, 375
surfactant, 32, 301, 340
Susquehanna River Basin Commission, 24
Sweden, 5, 19
Sydney, 44, 65, 71, 73–74, 78

Tachycineta bicolor, 368
Tampico, 4, 12
Tannezuft shale, 10
Tarim, 6, 8
Tasmania, 46, 57
Taubate, 14
technically recoverable, 8, 15, 81, 93, 268
technological stigma, 80
technologically enhanced- naturally occurring radioactive material, 221
Tecpetrol SA, 9
Tenappera, 7, 13
Tenmile Creek, 326, 328, 331, 340–342, 344, 347, 355, 357, 361, 394–395, 405
Tennessee, 10, 154
TENORM, 221–222, 242
tensile stress, 188
tensor element, 178
terrestrial habitat, 355, 380
Texas, 10, 20–21, 23, 25, 33, 41, 82–83, 86, 90–94, 99, 114–116, 121, 123–124, 128–129, 152, 154, 160, 170, 189, 191, 207–208, 222, 225, 240, 268, 293, 302, 304, 308–310, 345, 362, 381
Thalossospira, 346
thermogenic, 273, 275, 277–278, 281, 290, 293, 302, 308, 310, 362
thiosulfate, 301, 303–304
thorium, 161, 214–215, 217–218, 220, 222, 227, 233, 240
Three Rivers QUEST (3RQ), 386, 390, 392, 407
thumper truck, 25
tight gas, 20, 23, 45
tight shale, 19
tiltmeters, 23, 40

Timan Pechora, 14
Timimoun, 6, 10
Tindouf, 5–6, 10
Tithonian, 12
Titusville, 93
toluene, 68, 107, 119, 127, 129, 160
Tom Wolf, 96, 98
TOTAL, 82
Total dissolved solids (TDS), 158, 160, 162–163, 219, 222, 224, 226, 228, 230, 233–234, 246, 250–251, 254, 256, 258, 261–262, 294, 304, 314, 323, 346, 354–355, 386, 395, 397, 399–400, 402, 404–405, 407
Total organic carbon (TOC), 185, 188, 293
tracers, 32, 42, 140, 148, 246, 248–249, 251, 254–255, 258, 264–265, 268, 270, 272–273, 278, 280, 289
traffic, 39, 55, 81, 84, 107, 109–113, 123, 129, 131, 176, 196, 204, 208, 363, 365–367
traffic-light protocol, 196
transmission lines, 37
tree swallows, 368, 382
triethylene glycol, 38
trihalomethanes (TTHMs), 169, 406
Tritium, 272
Trump, Donald, 90
Turdus migratorius, 371
Tuxpan, 4, 12
Two-lined Salamanders, 354
Tygart, 394, 396

Uintah, 114, 116–119, 126–127, 135
unconventional, 10, 12–16, 19–20, 27, 34, 38, 40–45, 50–51, 56–57, 59, 62–64, 67–69, 72–74, 76, 78, 81, 89, 105, 107, 109–111, 113–114, 116, 122–124, 138, 154, 157–158, 162–163, 166–167, 169, 171, 173–174, 178, 185, 189, 191–192, 194, 204, 206, 219–222, 230, 239, 244–248, 250, 252, 255, 257–265, 268–270, 278, 308, 310, 313–314, 316–317, 321–322, 326, 330–331, 333, 336–337, 339–341, 346, 356, 359–361, 363, 375, 378, 380–381, 383, 405
underground injection, 19, 94, 161, 164, 167, 171
Unionoidea, 375
unit consolidation, 24
United Arab Emirates, 3, 7
United Kingdom, iv, 8, 19, 84, 96, 101, 146
United States, 3, 9–10, 19–20, 22–23, 37–39, 41–42, 44, 79, 87, 92, 95, 97, 101, 105, 107, 124–126, 128–131, 134, 139, 144, 146–150, 152–153, 156, 158, 164, 168–169, 210, 218–219, 222, 226, 239, 241, 244, 266, 269–270, 278, 280, 284, 288, 290, 292, 337, 339, 355, 360–361, 367, 381, 383–385
United States Army Corps of Engineers, 386, 404
United States Department of Energy (US DOE), 14, 17–19, 38–39, 41, 43, 89, 92, 98, 148, 226, 261, 264, 266, 270, 360, 384
United States Department of the Interior, 281, 284, 290
United States Department of Transportation, 37

United States Environmental Protection Agency (US EPA), 25, 32, 42, 89, 94, 101, 105–106, 113–114, 122–123, 131, 137–138, 140, 146, 148, 150–151, 160, 166–167, 171, 218, 226, 228, 230, 233, 244, 290, 316, 344, 347, 349, 361, 389, 404

United States Energy Information Administration (US EIA) 3–4, 8–10, 12–14, 19–21, 38, 42, 50–51, 79, 99, 134, 136, 148, 207, 216, 244

United States Geological Survey (USGS), 41, 93, 150, 164, 222–223, 226, 239, 242, 269, 318, 355, 386–387, 390–391, 395, 404

Upper Devonian, 12, 249, 251, 253, 255, 259, 262

Upper Green River Basin, 107, 114, 116–117, 125

Upper Monongahela River Basin, 387

upstream, 23, 39, 134–137, 139–141, 144, 166, 256, 327, 344, 346, 350, 354–357

uranium, 30, 161, 214–215, 217–218, 222, 227, 238, 268

Utah, 10, 110, 114, 116, 119, 121, 123, 126–127, 129, 131, 255, 268, 308

Utica, 4, 10, 12, 20, 24, 29, 38, 95, 130, 135, 153, 255, 258, 269, 281, 287, 303–305, 339, 361, 363–364, 371, 373–375, 378, 380, 382

Vaca Muerta, 4, 9, 17
Venezuela, 3–4, 89
Veracruz, 4, 12
vertical stress, 180, 182
Victoria, 13, 46, 56–57, 78
Villard, Paul, 214
Virginia, 10, 35, 100, 309, 314
Volatile organic carbon (VOC), 106–107, 110, 113, 115–119, 121–123, 125–126, 128, 131
Volga-Urals, 14

Water Act 2000, 59–60
water resources, 9, 42, 60–61, 150, 163–164, 166–167, 171, 244, 246, 261, 270
water stress, 152
water tanks, 27, 30–31
WATERS database, 395, 402, 404
well bore, 26, 30

well location plat, 25
West Fork, 394, 396
West Siberia, 14
West Virginia, 10, 20, 25, 30, 34, 37–38, 93–94, 114, 147, 153–154, 236, 239, 263, 314, 339, 365, 373, 379, 390, 392
West Virginia Water Research Institute (WVWRI), 386–387, 395, 404–405, 407
West Virginia, the Department of Environmental Protection, 25
Western Australia, 13, 57, 63
western blacknose dace, 350
wet gas, 35, 37–39
whipstock, 20
White nose syndrome, 374
Whitehill, 6, 13
Williston, 4, 12, 229, 237, 257, 268, 281
witherite, 229, 268
Woodford, 20–21, 191, 194, 206
World Stress Map, 180, 208
Wyoming, 10, 20, 82, 107, 110, 114, 116–117, 119, 121, 123, 125, 129–130, 154, 222, 225, 228, 242, 255, 265–266

x-rays, 214
xylene, 107, 122, 129, 160

Yangtze Platform, 6, 8
Youghiogheny, 394, 396
Youngstown, 94, 100, 338

zipper fracking, 39

$\delta^{13}C$, 276–277, 279, 284
$\delta^{13}C\text{-}CH_4$, 277
δ^2H, 265, 279
$\delta^2H\text{-}CH_4$, 280
$\delta D\text{-}CH_4$, 277
$\delta^{137}Ba$, 261
$\delta^{138}Ba$, 261–262, 264
δD, 275, 277, 281